Pitman Research Notes in Mathematics Series

Submission of proposals for consideration
Suggestions for publication, in the form of outlines and representative samples, are invited by the Editorial Board for assessment. Intending authors should approach one of the main editors or another member of the Editorial Board, citing the relevant AMS subject classifications. Alternatively, outlines may be sent directly to the publisher's offices. Refereeing is by members of the board and other mathematical authorities in the topic concerned, throughout the world.

Preparation of accepted manuscripts
On acceptance of a proposal, the publisher will supply full instructions for the preparation of manuscripts in a form suitable for direct photo-lithographic reproduction. Specially printed grid sheets are provided and a contribution is offered by the publisher towards the cost of typing. Word processor output, subject to the publisher's approval, is also acceptable.

Illustrations should be prepared by the authors, ready for direct reproduction without further improvement. The use of hand-drawn symbols should be avoided wherever possible, in order to maintain maximum clarity of the text.

The publisher will be pleased to give any guidance necessary during the preparation of a typescript, and will be happy to answer any queries.

Important note
In order to avoid later retyping, intending authors are strongly urged not to begin final preparation of a typescript before receiving the publisher's guidelines and special paper. In this way it is hoped to preserve the uniform appearance of the series.

Longman Scientific & Technical
Longman House
Burnt Mill
Harlow, Essex, UK
Tel (0279) 26721)

Titles in this series

1 Improperly posed boundary value problems
A Carasso and A P Stone
2 Lie algebras generated by finite dimensional ideals
I N Stewart
3 Bifurcation problems in nonlinear elasticity
R W Dickey
4 Partial differential equations in the complex domain
D L Colton
5 Quasilinear hyperbolic systems and waves
A Jeffrey
6 Solution of boundary value problems by the method of integral operators
D L Colton
7 Taylor expansions and catastrophes
T Poston and I N Stewart
8 Function theoretic methods in differential equations
R P Gilbert and R J Weinacht
9 Differential topology with a view to applications
D R J Chillingworth
10 Characteristic classes of foliations
H V Pittie
11 Stochastic integration and generalized martingales
A U Kussmaul
12 Zeta-functions: An introduction to algebraic geometry
A D Thomas
13 Explicit *a priori* inequalities with applications to boundary value problems
V G Sigillito
14 Nonlinear diffusion
W E Fitzgibbon III and H F Walker
15 Unsolved problems concerning lattice points
J Hammer
16 Edge-colourings of graphs
S Fiorini and R J Wilson
17 Nonlinear analysis and mechanics: Heriot-Watt Symposium Volume I
R J Knops
18 Actions of fine abelian groups
C Kosniowski
19 Closed graph theorems and webbed spaces
M De Wilde
20 Singular perturbation techniques applied to integro-differential equations
H Grabmüller
21 Retarded functional differential equations: A global point of view
S E A Mohammed
22 Multiparameter spectral theory in Hilbert space
B D Sleeman
24 Mathematical modelling techniques
R Aris
25 Singular points of smooth mappings
C G Gibson
26 Nonlinear evolution equations solvable by the spectral transform
F Calogero
27 Nonlinear analysis and mechanics: Heriot-Watt Symposium Volume II
R J Knops
28 Constructive functional analysis
D S Bridges
29 Elongational flows: Aspects of the behaviour of model elasticoviscous fluids
C J S Petrie
30 Nonlinear analysis and mechanics: Heriot-Watt Symposium Volume III
R J Knops
31 Fractional calculus and integral transforms of generalized functions
A C McBride
32 Complex manifold techniques in theoretical physics
D E Lerner and P D Sommers
33 Hilbert's third problem: scissors congruence
C-H Sah
34 Graph theory and combinatorics
R J Wilson
35 The Tricomi equation with applications to the theory of plane transonic flow
A R Manwell
36 Abstract differential equations
S D Zaidman
37 Advances in twistor theory
L P Hughston and R S Ward
38 Operator theory and functional analysis
I Erdelyi
39 Nonlinear analysis and mechanics: Heriot-Watt Symposium Volume IV
R J Knops
40 Singular systems of differential equations
S L Campbell
41 N-dimensional crystallography
R L E Schwarzenberger
42 Nonlinear partial differential equations in physical problems
D Graffi
43 Shifts and periodicity for right invertible operators
D Przeworska-Rolewicz
44 Rings with chain conditions
A W Chatters and C R Hajarnavis
45 Moduli, deformations and classifications of compact complex manifolds
D Sundararaman
46 Nonlinear problems of analysis in geometry and mechanics
M Atteia, D Bancel and I Gumowski
47 Algorithmic methods in optimal control
W A Gruver and E Sachs
48 Abstract Cauchy problems and functional differential equations
F Kappel and W Schappacher
49 Sequence spaces
W H Ruckle
50 Recent contributions to nonlinear partial differential equations
H Berestycki and H Brezis
51 Subnormal operators
J B Conway

Nest algebras

Kenneth R Davidson
University of Waterloo

Nest algebras

Triangular forms for operator algebras on Hilbert space

Copublished in the United States with
John Wiley & Sons, Inc., New York

Longman Scientific & Technical
Longman Group UK Limited
Longman House, Burnt Mill, Harlow
Essex CM20 2JE, England
and Associated Companies throughout the world.

Copublished in the United States with
John Wiley & Sons, Inc., 605 Third Avenue, New York, NY 10158

First published 1988

AMS Subject Classifications: 47D, 46L, 47C, 47A

ISSN 0269-3674

British Library Cataloguing in Publication Data
Davidson, Kenneth R.
Nest algebras.—(Pitman research notes
in mathematics series ISSN 0269-3674;191).
1. Operator algebras.
I. Title
512'. 55

ISBN 0-582-01993-1

Printed and bound in Great Britain by
Biddles Ltd, Guildford and King's Lynn

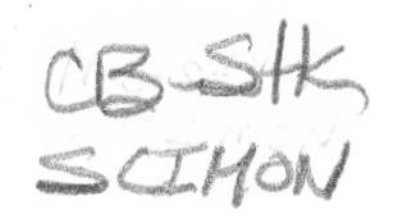

To Virginia

Table of Contents

III Additional Topics

Preface

The study of triangular forms for operators began in the late 50's and early 60's with the work on triangular integrals of the Russian school (Gohberg, Krein, et al) and the work of Ringrose. It was Ringrose who initiated the study of the algebra of all operators with a given triangular form. He coined the term "nest algebra" since the more natural term "triangular algebra" had recently been taken by Kadison and Singer for a related class of algebras. While one cannot claim that the study of these algebras is near completion, one can say that the subject has reached a certain maturity. This is due in part to a complete solution of the similarity problem for nests proposed by Ringrose in his early work. This solution, due to Andersen, Larson, and the author, is the main focus of part II of this monograph.

The intent of these notes is to present this material for the benefit of graduate students with a background in functional analysis. It is hoped that it will also be of use to other experts, but of course one could be more terse for such an audience. Only a basic knowledge of C^*-algebras and von Neumann algebras is assumed. Indeed, many results for C^*-algebras and type I von Neumann algebras are developed carefully here. This subject of non self adjoint operator algebras is attractive, in part, due to the nice blend of single operator theory and self adjoint algebra theory.

There is a larger related class of reflexive algebras (CSL algebras) introduced by Arveson which receives the attention of many researchers interested in nest algebras. The theory of these algebras is much less well understood. So we have chosen only to introduce a small portion of this material at the end of the book. However, wherever possible, the proofs provided in the nest case are chosen so that they extend to other CSL lattices.

The first part of this monograph concentrates on the structure of compact operators. This goes substantially beyond what is normally done in an introductory course in operator theory. But it is elegant and substantial mathematics that should be familiar to more operator theorists. Some of this material is available in the books of Ringrose [5] and Gohberg and Krein [1, 2]. But our viewpoint is more up to date, and many of the proofs are different.

The second part deals with basic structural properties of nest algebras - the radical, unitary invariants, and similarity invariants. The third part deals with further structural properties of nest algebras. It is not always necessary to read what has come before, so a flow chart has been provided for the reader's convenience. The last short part is an introduction to CSL algebras.

An asterisk * on an exercise means that I do not know the answer. It does not necessarily mean that it is an important problem.

These notes in a preliminary form were used for a graduate course taught at the University of Waterloo in winter 1987. There is too much material here for a one semester course. Indeed, it was necessary to pick and choose a narrow path in order to get to the Similarity Theorem in one term. I would like to thank D. Casperson, D. Dicks, L. Marcoux, and F. Zorzitto for numerous helpful criticisms.

A first draft was distributed to many colleagues. I am indebted to those who pointed out various mistakes, oversights and misprints. I mention in particular W. Arveson, J. Erdos, F. Gilfeather, K. Harrison, D. Herrero, A. Hopenwasser, A. Katavolos, J. Kraus, V. Paulsen, D. Pitts and B. Solel.

I have tried hard to get as many out as possible, and take full responsibility for those that escaped my attention. Since I found more than one error per page in the original draft, I can predict with confidence that many more still remain. Still I can hope that most will not severely hamper the reader.

Finally, I must thank Mrs. B. Law for her technical expertise in typing this manuscript. Without the care she took, this book would have been much poorer.

Kenneth R. Davidson
Waterloo, 1988

FLOWCHART OF DEPENDENCE ON EARLIER CHAPTERS

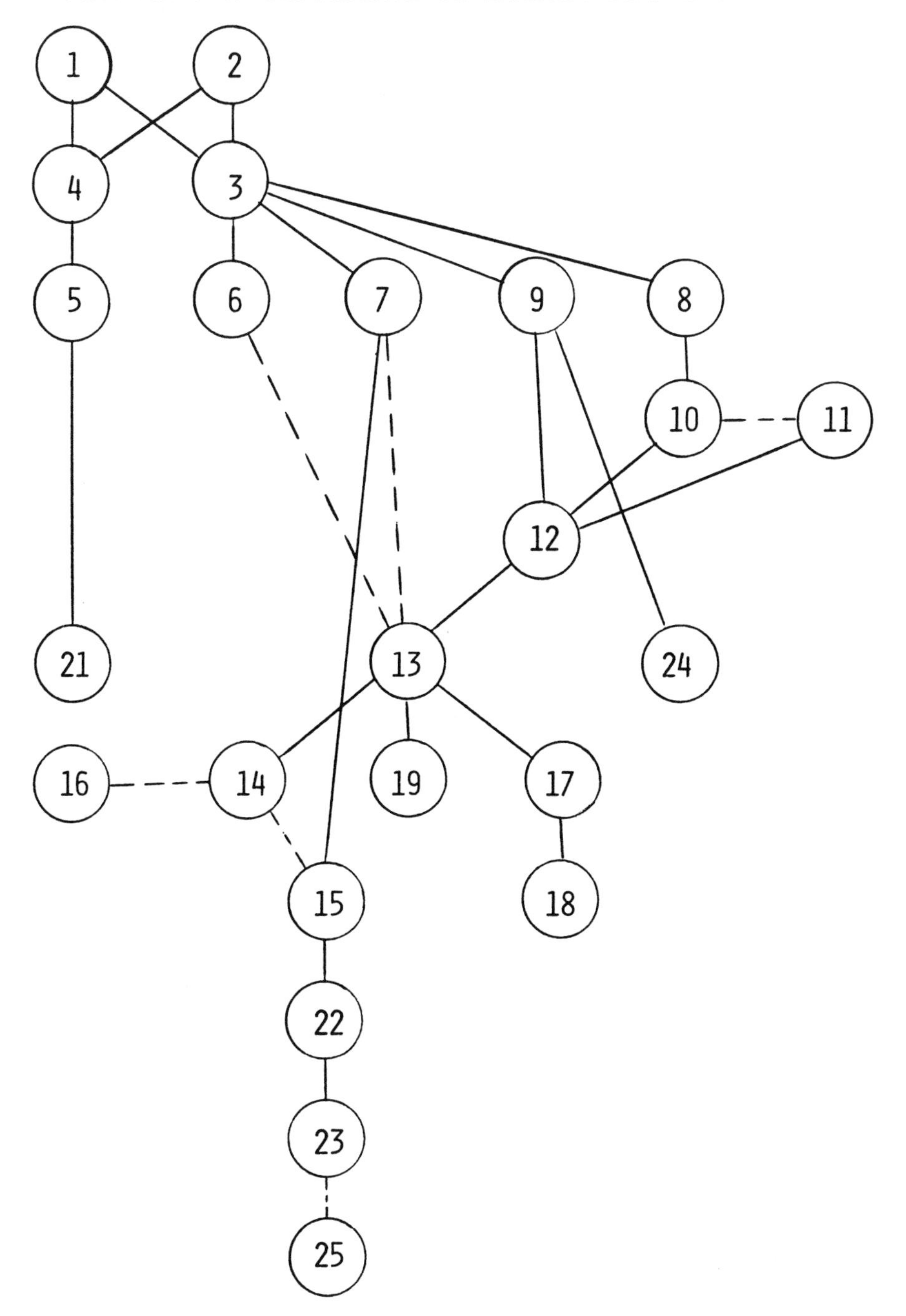

0. Background

In this chapter, we review briefly some of the facts about Hilbert space operators and C^*-algebras which will be needed to read this book. Most of the material can be found in introductory textbooks such as Conway [1], Douglas [2], or Kadison-Ringrose [3].

0.1 Banach Algebras

A *Banach algebra is a Banach space* $\mathcal{A}$ over $\mathbb{C}$ which is also an (associative) algebra over $\mathbb{C}$ such that $\|AB\| \leq \|A\| \, \|B\|$ for all A, B in $\mathcal{A}$. When $\mathcal{A}$ has a unit I, the *spectrum* $\sigma(A)$ (or $\sigma_{\mathcal{A}}(A)$ if $\mathcal{A}$ needs to be clarified) is the set $\{\lambda \in \mathbb{C} : A - \lambda I$ is not invertible in $\mathcal{A}\}$. This is a non-empty compact set. The *spectral radius*

$$spr(A) = \sup\{|\lambda| : \lambda \in \sigma(A)\} = \lim_{n \to \infty} \|A^n\|^{1/n} \quad .$$

Let f be holomorphic in a neighbourhood G of $\sigma(A)$, and let C be a finite union of Jordan curves such that $ind_C(\lambda) = 1$ for every λ in $\sigma(A)$. Define

$$f(A) = \frac{1}{2\pi i} \int_C f(z)(zI - A)^{-1} dz \quad .$$

Let $Hol(\sigma(A))$ denote the set of functions holomorphic in some neighbourhood of $\sigma(A)$.

Riesz Functional Calculus. *Let A be an element of a Banach algebra $\mathcal{A}$ with identity. Then for every f in $Hol(\sigma(A))$, $f(A)$ is well defined independent of the curve C. The map taking f to $f(A)$ is an algebra homomorphism such that each polynomial $p(z) = \sum_{k=0}^{n} c_k z^k$ is taken to $c_0 I + \sum_{k=1}^{n} c_k A^k$.*

Let $\mathcal{A}$ be an abelian Banach algebra with identity. A *multiplicative linear functional* is an algebra homomorphism ϕ into $\mathbb{C}$. Such functionals are always continuous with $\|\phi\| = 1$. They are in a one to one correspondence with the *maximal ideals* of $\mathcal{A}$. The set $M_{\mathcal{A}}$ of maximal ideals is given the weak* topology induced as a subset of the dual space $\mathcal{A}'$. It is easy to verify that $M_{\mathcal{A}}$ is closed, and thus is compact by the Banach-Alaoglu Theorem. There is a natural homomorphism of $\mathcal{A}$ into $C(M_{\mathcal{A}})$, called the *Gelfand transform*, is given by $\hat{A}(\phi) = \phi(A)$.

Gelfand's Theorem. *The Gelfand transform $A \to \hat{A}$ of $\mathcal{A}$ into $C(M_{\mathcal{A}})$ is a continuous algebra homomorphism such that $\|\hat{A}\| \le \|A\|$. Furthermore,*

$$\sigma(A) = \sigma(\hat{A}) = Ran(\hat{A}) = \{\phi(A) : \phi \in M_{\mathcal{A}}\} \ .$$

The kernel of this map is

$$rad(\mathcal{A}) = \bigcap \{ker\, \phi : \phi \in M_{\mathcal{A}}\} = \{A \in \mathcal{A} : \sigma(A) = \{0\}\} \ .$$

An immediate corollary is the:

Spectral Mapping Theorem. *For f in the $Hol(\sigma(A))$, $\sigma(f(A)) = f(\sigma(A))$.*

0.2 C^*-Algebras

A *C^*-algebra* is a Banach algebra with a conjugation operation $*$ such that $(A^*)^* = A$, $(AB)^* = B^*A^*$, $(\alpha A + \beta B)^* = \bar{\alpha}A^* + \bar{\beta}B^*$, and $\|A^*A\| = \|A\|^2$. (That is, $A \to A^*$ is an involution and a conjugate linear anti-isomorphism satisfying the special norm condition $\|A^*A\| = \|A\|^2$.) The main (and in a certain sense only) examples are self-adjoint subalgebras of $\mathcal{B}(\mathcal{H})$.

Gelfand-Naimark Theorem. *If $\mathcal{A}$ is an abelian C^*-algebra with identity, then the Gelfand transform is an isometric $*$-isomorphism of $\mathcal{A}$ onto $C(M_{\mathcal{A}})$.*

An element N of a C^*-algebra $\mathcal{A}$ is *normal* if $N^*N = NN^*$. This occurs precisely when the sub C^*-algebra $C^*(N)$ generated by N and I is abelian. Thus one obtains the immediate corollary:

C^*-Functional Calculus. *If N is a normal element of a C^*-algebra, then $C^*(N)$ is isometrically $*$-isomorphic to $C(\sigma(N))$. The inverse map yields a functional calculus extending the Riesz Functional Calculus such that $\bar{f}(N) = f(N)^*$, and $\sigma(f(N)) = f(\sigma(N))$ for all f in $C(\sigma(N))$.*

Every (two-sided, closed) ideal I of a C^*-algebra $\mathcal{A}$ is self-adjoint, and $\mathcal{A}/I$ is a C^*-algebra. Furthermore,

Proposition. *Let π be a $*$-homomorphism of a C^*-algebra $\mathcal{A}$ into a C^*-algebra $\mathcal{B}$. Then $\|\pi\| = 1$, $Ran(\pi)$ is closed and is isometrically $*$-isomorphic to $\mathcal{A}/ker\,\pi$.*

An element P of a C^*-algebra is *positive* if it is self-adjoint with positive spectrum. By the functional calculus, $P = A^*A$ where $A = P^{1/2}$. Conversely, every operator of the form A^*A is positive (even if A is not normal). A *positive linear functional* is a linear map of $\mathcal{A}$ into $\mathbb{C}$ such that $f(A) \geq 0$ whenever $A \geq 0$. The two variable function $[A,B] = f(B^*A)$ is a sequilinear form on $\mathcal{A} \times \mathcal{A}$. The Cauchy-Schwartz inequality becomes

$$|f(B^*A)|^2 \leq f(A^*A)f(B^*B) \quad .$$

It follows easily from positivity that if $0 \leq A \leq I$, then $0 \leq f(A) \leq f(I)$. Hence if $\|A\| \leq 1$,

$$|f(A)|^2 = |f(I^*A)|^2 \leq f(A^*A)f(I) \leq f(I)^2 \quad .$$

So $\|f\| = \|f(I)\|$ and f is continuous. A *state* is a positive linear functional with $f(I) = 1$.

A representation of a C^*-algebra $\mathcal{A}$ is a $*$-homorphism π of $\mathcal{A}$ into $B(\mathcal{H})$. This representation is *cyclic* if there is a *cyclic vector* x with $\{\pi(A)x : A \in \mathcal{A}\}$ dense in $\mathcal{H}$. A representation is *non-degenerate* if $\pi(\mathcal{A})$ has no kernel. It is easy to check that every non-degenerate representation is unitarily equivalent to a direct sum of cyclic ones.

Gelfand-Naimark-Segal Theorem. *Let $\mathcal{A}$ be a C^*-algebra with identity. Given a state on $\mathcal{A}$, there is a cyclic representation π_f on $\mathcal{H}_f$ with cyclic vector ξ_f such that $f(A) = (\pi_f(A)\xi_f,\xi_f)$. Conversely, given a cyclic representation π with unit cyclic vector ξ, let $f(A) = (\pi(A)\xi,\xi)$. Then there is a unitary U of $\mathcal{H}_f$ onto $\mathcal{H}$ such that $U\xi_f = \xi$, and $\pi(A) = U\pi_f(A)U^*$ for every A in $\mathcal{A}$.*

One can show that every C^*-algebra has enough states $S(\mathcal{A})$ so that $\|A\| = \sup\{|f(A)| : f \in S(\mathcal{A})\}$. Thus by taking a direct sum of the representations from the GNS construction, one obtains:

Corollary. *Every C^*-algebra is isometrically $*$-isomorphic to a subalgebra of $B(\mathcal{H})$.*

0.3 Von Neumann Algebras

The *weak operator topology* (WOT) on $B(\mathcal{H})$ is the weakest topology such that the functionals $\phi(T) = (Tx, y)$ are continuous for every x and y in $\mathcal{H}$. The *strong operator topology* is the weakest topology such that the map $T \to \|Tx\|$ is continuous for all x in $\mathcal{H}$. It will be seen (Theorem 1.15) that $B(\mathcal{H})$ is the dual space of the trace class operators. So it has a *weak* $*$-topology (sometimes called the ultra-weak topology). This is the weakest topology such that the trace $tr(CT)$ is a continuous function of T for every trace class operator C. The weak operator continuous functionals are precisely those where C is finite rank. There are several other related topologies, but these will generally suffice for our purposes.

A von Neumann algebra is a C^*-subalgebra of $B(\mathcal{H})$ containing the identity which is closed in the weak operator topology. If $\mathcal{A}$ is a subset of $B(\mathcal{H})$, the commutant $\mathcal{A}' = \{T \in B(\mathcal{H}): AT = TA \text{ for all } A \text{ in } \mathcal{A}\}$. The von Neumann algebra generated by a single operator T is denoted by $W^*(T)$. There are two basic density theorems that we will require.

The von Neumann Double Commutant Theorem. *Let $\mathcal{A}$ be a unital C^*-subalgebra of $B(\mathcal{H})$. Then the closure of $\mathcal{A}$ in any of the weak operator, strong operator, and weak* topologies is the double commutant $\mathcal{A}''$.*

Kaplansky's Density Theorem. *Let $\mathcal{A}$ be a unital C^*-subalgebra acting on a separable Hilbert space $\mathcal{H}$. Then every operator in the unit ball of $\mathcal{A}''$ is the strong operator limit of a sequence of operators in the unit ball of $\mathcal{A}$. The self-adjoint part of the ball of $\mathcal{A}$ is likewise strongly dense in the self-adjoint part of the ball of $\mathcal{A}''$.*

By the Banach-Alaoglu Theorem, the unit ball of $B(\mathcal{H})$ is compact in the weak* topology. Thus it is also compact in the weak operator topology since it is a weaker topology. It is not, however, compact in the strong operator topology. Multiplication is separately continuous but not jointly continuous in any of these topologies. However, in the strong operator topology, multiplication is jointly continuous on the unit ball. The adjoint operation is continuous in the weak operator and weak* topologies, but not in the strong operator topology. The *strong** topology is the weakest topology containing the strong topology in which adjoint is continuous.

0.4 Compact Operators

An operator K is *compact* if the image of the unit ball of $\mathcal{H}$ under K has compact closure. (In fact, the image of the unit ball under K is closed). Let $\{e_n, n \geq 1\}$ be an orthonormal basis, and let P_n be the orthogonal projection onto $span\{e_k, 1 \leq k \leq n\}$. It is not hard to show that a compact operator K is the norm limit of the finite rank operators $P_n K$. The following results are standard.

Theorem. *For an operator K on Hilbert space, the following are equivalent:*

(i) *K is compact,*

(ii) *K^* is compact,*

(iii) *K is the norm limit of finite rank operators.*

Theorem. *Let K be a compact operator. Then $\sigma(K)$ consists of $\{0\}$ together with a finite or countable set $\{\lambda_n\}$ of eigenvalues with $\{0\}$ as its only limit point. The Riesz projections $E_K\{\lambda_n\}$ are finite rank, and have range equal to $ker(K-\lambda_n I)^{k_n}$ for k_n sufficiently large.*

When K is normal and compact, one obtains:

Spectral Theorem (Compact Case). *Let K be a compact, normal operator on $\mathcal{H}$. Then there is an orthonormal basis $\{e_n\}$ of eigenvectors with eigenvalues $\{\lambda_n\}$ such that $\lim_{n \to \infty} \lambda_n = 0$. So*

$$K\left(\sum_{n=1}^{\infty} a_n e_n\right) = \sum_{n=1}^{\infty} \lambda_n a_n e_n \quad .$$

0.5 Bounded Operators

Let T be a bounded operator on Hilbert space. The *absolute value* of T is the positive operator $|T| = (T^*T)^{1/2}$. It is easy to check that $\| |T| x \| = \|Tx\|$ for every x in $\mathcal{H}$. So one can define a partial isometry U so that $U|T|x = Tx$ and $Uy = 0$ for y in $ker T$. Then UU^* is the projection onto the closure of the *range*, $\overline{Ran(T)}$ and U^*U is the projection onto $(ker T)^{\perp}$. One obtains the *polar decomposition* $T = U|T|$.

For normal operators, one has the Spectral Theorem. More detail will be obtained in Chapter 6.

Spectral Theorem. *Let N be a normal operator on Hilbert space. There is a countably additive, regular Borel measure E_N on $\mathbb{C}$ with values in the projections onto subspaces of $\mathcal{H}$ such that:*

(i) $E_N(\sigma(N)) = I$,

(ii) $N = \int z dE_N(z)$,

(iii) *$E_N(X)$ commutes with every operator commuting with N, for every Borel set X,*

(iv) *if f is bounded Borel function, let $f(N) = \int f(z) dE_N(z)$. This is a homomorphism from the bounded Borel functions on $\mathbb{C}$ onto $W^*(N)$ extending the C^* functional calculus of N.*

Another special property of normal operators is:

Fuglede's Theorem. *Let N be a normal operator, and let A be a bounded operator such that $NA = AN$. Then $N^*A = AN^*$.*

Corollary. *$\{N\}' = \{A : AN = NA\}$ is a von Neumann algebra when N is normal.*

The quotient C^*-algebra $\mathbf{A} = \mathcal{B}(\mathcal{H})/\mathcal{K}$ is called the *Calkin algebra.* The image of an operator T in $\mathbf{A}$ will be denoted by $\widetilde{T}$. An operator T is called *Fredholm* if T has closed range, and both $\ker T$ and $\ker T^* = Ran\,(T)^{\perp}$ are finite dimensional. The *Fredholm index* is defined as

$$ind\, T = dim\, ker\, T - dim\, ker\, T^* \quad .$$

Atkinson's Theorem. *An operator T in $\mathcal{B}(\mathcal{H})$ is Fredholm if and only if $\widetilde{T}$ is invertible in the Calkin algebra.*

Thus the set of Fredholm operators $\mathcal{F}(\mathcal{H})$ is open in $\mathcal{B}(\mathcal{H})$. The Fredholm index is continuous, and hence locally constant, and is invariant under compact perturbations. Furthermore

Theorem. *The Fredholm index is a continuous algebra homomorphism of $\mathcal{F}(\mathcal{H})$ onto $\mathbb{Z}$. The kernel is the connected component of the identity in $\mathcal{F}(\mathcal{H})$.*

The *essential spectrum* of an operator T, denoted $\sigma_e(T)$, is $\sigma(\widetilde{T})$ in the Calkin algebra. The complement $\sigma(T)\backslash\sigma_e(T)$ consists of countably many bounded components of the complement of $\sigma(T)$ on which $T-\lambda I$ is Fredholm of some fixed index, together with countably many isolated points of finite multiplicity clustering at $\sigma_e(T)$.

I. COMPACT OPERATORS

1. Ideals of Compact Operators
2. Triangular Algebras
3. Triangular Forms for Compact Operators
4. Triangular Truncation
5. The Volterra Operator

1. Ideals of Compact Operators

In this chapter, we investigate various ideals of $\mathcal{B}(\mathcal{H})$. The first result justifies the title of this chapter.

1.1 Proposition. *The only proper closed ideal in $\mathcal{B}(\mathcal{H})$ for a separable Hilbert space $\mathcal{H}$ is the set $\mathcal{K}$ of compact operators.*

Proof. Let T be a non-compact operator. By the polar decomposition, $T = U|T|$ and $|T|$ is not compact. So there is an infinite dimensional spectral projection E of $|T|$ on which T is bounded below. The operator $V = E|T| + E^{\perp}$ is invertible, and $E = V^{-1}U^*TE$. Thus E belongs to the two sided ideal generated by T. From this, it easily follows that this ideal is all of $\mathcal{B}(\mathcal{H})$. On the other hand, if K is a non-zero compact operator, a similar argument shows that the ideal generated by K contains a rank one projection. Hence it contains all finite rank operators, which are dense in $\mathcal{K}$. ∎

1.2. Let K be a compact operator. The positive operator $|K| = (K^*K)^{1/2}$ has eigenvalues $s_1 \geq s_2 \geq \dots$ with $\lim_{n\to\infty} s_n = 0$. The *s-numbers* or *singular values* of K are the numbers $s_n = s_n(K)$. Let $\{e_n, n \geq 1\}$ be an orthonormal set of eigenvectors of $|K|$ for the eigenvalues $s_n(K)$. Let $f_n = Ue_n$. Then

$$K = \sum_{n=1}^{\infty} s_n f_n \otimes e_n^*$$

and this sum is norm convergent. Here $x \otimes y^*$ denotes the rank one operator

$$x \otimes y^*(z) = (z,y)x \quad .$$

The identity $A(x \otimes y^*)B = Ax \otimes (B^*y)^*$ is readily verified.

For $1 \leq p \leq \infty$, let $\mathcal{C}_p$ denote the *Schatten p-class* of all compact operators K such that $\{s_n(K)\}$ belongs to ℓ^p. Define a norm on $\mathcal{C}_p$ by

$$\|K\|_p = \Big(\sum_{n\geq 1} s_n(K)^p\Big)^{1/p} \quad .$$

That this is indeed a norm follows from Corollary 1.9. Of particular interest are the *trace class* operators $\mathcal{C}_1$, and the *Hilbert-Schmidt* operators $\mathcal{C}_2$.

1.3 Theorem. *Let* $K = \sum_{n=1}^{\infty} s_n f_n \otimes e_n^*$ *be a trace class operator. Then for every orthonormal bases* $\{\phi_k\}$ *and* $\{\psi_k\}$*, one has*

$$\sum_{k=1}^{\infty} |(K\phi_k,\psi_k)| \leq \|K\|_1$$

and

$$\sum_{k=1}^{\infty}(K\phi_k,\phi_k) = \sum_{k=1}^{\infty} s_n(f_n,e_n) \ .$$

Proof. Using the Cauchy-Schwartz inequality and the Parseval identity,

$$\begin{aligned}
\sum_{k=1}^{\infty} |(K\phi_k,\psi_k)| &\leq \sum_{k=1}^{\infty}\sum_{n=1}^{\infty} s_n \ |(\phi_k,e_n)|\,|(f_n,\psi_k)| \\
&= \sum_{n=1}^{\infty} s_n \sum_{k=1}^{\infty} |(e_n,\phi_k)|\,|(f_n,\psi_k)| \\
&\leq \sum_{n=1}^{\infty} s_n \Big(\sum_{k=1}^{\infty} |(e_n,\phi_k)|^2\Big)^{1/2}\Big(\sum_{k=1}^{\infty} |(f_n,\psi_k)|^2\Big)^{1/2} \\
&= \sum_{n=1}^{\infty} s_n = \|K\|_1 \ .
\end{aligned}$$

Now taking $\psi_k = \phi_k$, one obtains an absolutely convergent series. Thus

$$\begin{aligned}
\sum_{k=1}^{\infty}(K\phi_k,\phi_k) &= \sum_{k=1}^{\infty}\sum_{n=1}^{\infty} s_n(\phi_k,e_n)(f_n,\phi_k) \\
&= \sum_{n=1}^{\infty} s_n \sum_{k=1}^{\infty} (f_n,\phi_k)\overline{(e_n,\phi_k)} = \sum_{n=1}^{\infty} s_n(f_n,e_n) \ . \qquad \blacksquare
\end{aligned}$$

Consequently, one may define a well defined continuous linear functional known as the *trace* on C_1 by

$$tr(K) = \sum_{n=1}^{\infty}(K\phi_n,\phi_n) \ .$$

This is independent of the choice of orthonormal basis.

Now we develop some basic inequalities about s-numbers.

1.4 Lemma. *Let* K *be a compact operator. Then*

$$s_n(K) = \inf\{\|K-F\| : \text{rank} F \leq n-1\} \ .$$

Proof. Let $K = \sum_{n=1}^{\infty} s_n f_n \otimes e_n^*$. Then $F_n = \sum_{k=1}^{n-1} s_k f_k \otimes e_k^*$ has rank $n-1$, and $K-F_n = \sum_{k=n}^{\infty} s_k f_k \otimes e_k^*$. So

$$s_n(K) = \|K-F_n\| \geq \inf\{\|K-F\| : rank\, F \leq n-1\} \ .$$

On the other hand, if $rank\, F \leq n-1$, there is a unit vector x in the intersection $ker\, F \cap span\{e_k, 1 \leq k \leq n\}$. So

$$\|K-F\| \geq \|(K-F)x\| = \|Kx\|$$
$$= \|\sum_{k=1}^{n} s_k(x,e_k)f_k\| \geq s_n \sum_{k=1}^{n} |(x,e_k)|^2 = s_n \ .$$ ∎

1.5 Corollary. *Let K be compact, and let T be a bounded operator. Then*

$$s_n(TK) \leq \|T\| s_n(K) \text{ and } s_n(KT) \leq \|T\| s_n(K) \ .$$

Proof. Let F_n be the operators defined above. Then

$$s_n(TK) \leq \|TK-TF_n\| \leq \|T\|\,\|K-F_n\| = \|T\| s_n(K)$$
$$s_n(KT) \leq \|KT-F_nT\| \leq \|K-F_n\|\,\|T\| = \|T\| s_n(K) \ .$$ ∎

1.6 Corollary. *C_p is a self-adjoint ideal in $B(\mathcal{H})$, and $\|TKS\|_p \leq \|T\|\,\|K\|_p \|S\|$ for all K in C_p and T, S in $B(\mathcal{H})$.*

Proof. Since $s_n(TKS) \leq \|S\|\,\|T\| s_n(K)$, this belongs to ℓ^p and satisfies the desired inequality. That C_p is a linear space follows for if K_1, K_2 belongs to C_p and $F_n^{(i)}$ are the finite rank operators of the previous lemma, then

$$s_{2n}(K_1+K_2) \leq s_{2n-1}(K_1+K_2)$$
$$\leq \|K_1+K_2-F_n^{(1)}-F_n^{(2)}\| \leq s_n(K_1)+s_n(K_2) \ .$$

So $\|K_1+K_2\|_p \leq 2^{1/p}(\|K_1\|_p + \|K_2\|_p)$. If $K = \sum_{n=1}^{\infty} s_n(f_n \otimes e_n^*)$, then $K^* = \sum_{n=1}^{\infty} s_n(e_n \otimes f_n^*)$. So $s_n(K^*) = s_n(K)$ for $n \geq 1$. So $\|K^*\|_p = \|K\|_p$, and therefore C_p is self-adjoint. ∎

To obtain the triangle inequality for the C_p norm, we need some more inequalities.

1.7 Lemma. *Let K be a compact operator. If $\{\phi_k, 1 \le k \le n\}$ and $\{\psi_k, 1 \le k \le n\}$ are orthonormal families, then*

$$\sum_{k=1}^{n} |(K\phi_k, \psi_k)| \le \sum_{k=1}^{n} s_k(K) \ .$$

Proof. Replace ψ_k by a scalar multiple so that $(K\phi_k, \psi_k) \ge 0$ for $1 \le k \le n$. Let U be the *rank* n partial isometry such that $U\psi_k = \phi_k$ for $1 \le k \le n$. Then $s_k(KU) \le s_k(K)$ for $1 \le k \le n$, and $s_k(KU) = 0$ for $k > n$ as *rank* $(KU) \le n$. Hence

$$\sum_{k=1}^{n} (K\phi_k, \psi_k) = \sum_{k=1}^{n} (KU\psi_k, \psi_k) = tr\,(KU)$$

$$\le \|KU\|_1 = \sum_{k=1}^{n} s_k(KU) \le \sum_{k=1}^{n} s_k(K) \ . \qquad \blacksquare$$

1.8 Corollary. *Let K_1 and K_2 be compact operators. Then*

$$\sum_{k=1}^{n} s_k(K_1+K_2) \le \sum_{k=1}^{n} s_k(K_1)+s_k(K_2) \ .$$

Proof. Write $K_1+K_2 = \sum_{k=1}^{\infty} s_k f_k \otimes e_k^*$ where $s_k = s_k(K_1+K_2)$. Then

$$\sum_{k=1}^{n} s_k(K_1+K_2) = \sum_{k=1}^{n} (K_1+K_2 e_k, f_k)$$

$$\le \sum_{k=1}^{n} |(K_1 e_k, f_k)| + |(K_2 e_k, f_k)|$$

$$\le \sum_{k=1}^{n} s_k(K_1)+s_k(K_2) \ . \qquad \blacksquare$$

1.9 Corollary. *Let K_1 and K_2 belong to C_p, $1 \le p \le \infty$. Then*

$$\|K_1+K_2\|_p \le \|K_1\|_p + \|K_2\|_p \ .$$

Proof. The case $p = 1$ is immediate from Corollary 1.8, and $p = \infty$ is the usual operator norm. So suppose $1 < p < \infty$. Let $\lambda_1 \geq \lambda_2 \geq \ldots \geq \lambda_{n+1} = 0$ be a decreasing sequence of positive real numbers. By "summation by parts", one obtains

$$\begin{aligned}\sum_{k=1}^{n} \lambda_k s_k(K_1+K_2) &= \sum_{j=1}^{n}(\lambda_j-\lambda_{j+1})\sum_{k=1}^{j} s_k(K_1+K_2) \\ &\leq \sum_{j=1}^{n}(\lambda_j-\lambda_{j+1})\sum_{k=1}^{j} s_k(K_1)+s_k(K_2) \\ &= \sum_{k=1}^{n}\lambda_k(s_k(K_1)+s_k(K_2)) \ .\end{aligned}$$

Let $1 < q < \infty$ be such that $1/p+1/q = 1$. Let $\lambda_k = s_k(K_1+K_2)^{p/q} = s_k(K_1+K_2)^{p-1}$. By Hölder's inequality,

$$\begin{aligned}\sum_{k=1}^{n} s_k(K_1+K_2)^p &= \sum_{k=1}^{n}\lambda_k s_k(K_1+K_2) \\ &\leq \sum_{k=1}^{n}\lambda_k s_k(K_1)+\sum_{k=1}^{n}\lambda_k s_k(K_2) \\ &\leq (\sum_{k=1}^{n}\lambda_k^q)^{1/q}(\sum_{k=1}^{n} s_k(K_1)^p)^{1/p} + (\sum_{k=1}^{n}\lambda_k^q)^{1/q}(\sum_{k=1}^{n} s_k(K_2)^p)^{1/p} \\ &\leq (\sum_{k=1}^{n} s_k(K_1+K_2)^p)^{1/q}(||K_1||_p+||K_2||_p) \ .\end{aligned}$$

Hence

$$(\sum_{k=1}^{n} s_k(K_1+K_2)^p)^{1/p} \leq ||K_1||_p+||K_2||_p \ .$$

Taking a limit as n increases yields $||K_1+K_2||_p \leq ||K_1||_p+||K_2||_p$. ■

We extract from this proof another technical inequality..

1.10 Lemma. *Let K be a compact operator, and let $\lambda_1 \geq \lambda_2 \geq .. \geq \lambda_{n+1} = 0$ be a decreasing sequence of positive real numbers. Let $\{\phi_k, 1 \leq k \leq n\}$ and $\{\psi_k, 1 \leq k \leq n\}$ be orthonormal sets. Then*

$$\sum_{k=1}^{n}\lambda_k \, |(K\phi_k,\psi_k)\,| \leq \sum_{k=1}^{n}\lambda_k s_k(K) \ .$$

Proof. This is a simple summation by parts argument together with Lemma 1.7.

$$\sum_{k=1}^{n} \lambda_k \, |(K\phi_k,\psi_k)\,| = \sum_{j=1}^{n} (\lambda_j - \lambda_{j-1}) \sum_{k=1}^{j} |(K\phi_k,\psi_k)\,|$$

$$\leq \sum_{j=1}^{n} (\lambda_j - \lambda_{j-1}) \sum_{k=1}^{j} s_k(K) = \sum_{k=1}^{n} \lambda_k s_k(K) \ . \qquad \blacksquare$$

Now we are ready to prove that C_p is complete. Let $\mathcal{F}$ denote the set of finite rank operators.

1.11 Theorem. *C_p is a Banach space for $1 \leq p \leq \infty$, and $\mathcal{F}$ is dense in C_p.*

Proof. It must be shown that C_p is complete. If K is in C_p, then

$$\|K\| = s_1(K) \leq \Big(\sum_{k \geq 1} s_k(K)^p\Big)^{1/p} = \|K\|_p \ .$$

So if K_j is a Cauchy sequence in C_p, then it is Cauchy in $\mathcal{K}$. Thus $K = \lim\limits_{j \to \infty} K_j$ exists in $\mathcal{K}$. A fortiori,

$$(Kx,y) = \lim_{j \to \infty} (K_j x,y)$$

for every pair of vectors x and y in $\mathcal{H}$. Let $s_n = s_n(K)$ and choose orthonormal sets $\{e_n\}$ and $\{f_n\}$ so that $K = \sum\limits_{n=1}^{\infty} s_n(f_n \otimes e_n^*)$. By Lemma 1.10 and Hölder's inequality,

$$\sum_{k=1}^{n} s_k^p = \sum_{k=1}^{n} s_k^{p/q}(Ke_n,f_n) = \lim_{j \to \infty} \Big|\sum_{k=1}^{n} s_k^{p/q}(K_j e_n,f_n)\Big|$$

$$\leq \liminf_{j \to \infty} \sum_{k=1}^{n} s_k^{p/q} s_k(K_j)$$

$$\leq \liminf_{j \to \infty} \Big(\sum_{k=1}^{n} s_k^p\Big)^{1/q} \|K_j\|_p$$

$$= \Big(\sum_{k=1}^{n} s_k^p\Big)^{1/q} \lim_{j \to \infty} \|K_j\|_p \ .$$

Note that $\|K_j\|_p$ converges since $|\,\|K_j\|_p - \|K_k\|_p\,| \leq \|K_j - K_k\|_p$ by the triangle inequality, and so $\|K_j\|_p$ is Cauchy. Thus we obtain

$$\Big(\sum_{k=1}^{n} s_k^p\Big)^{1/p} \leq \lim_{j \to \infty} \|K_j\|_p \ .$$

Taking a limit as n increases yields $\|K\|_p \leq \lim\limits_{j \to \infty} \|K_j\|_p$, so K belongs to C_p. Apply this same inequality to $K - K_n$ to obtain

$$\|K-K_n\|_p \leq \lim_{j\to\infty} \|K_j-K_n\|_p \ .$$

As K_j is Cauchy, the right hand side tends to zero as n increases. So K_n converges to K in the C_p norm.

Given K in C_p represented as above, the finite rank operators $F_n = \sum_{k=1}^{n} s_k(f_k \otimes e_k^*)$ clearly converge to K in C_p. So $\mathcal{F}$ is dense in C_p. ∎

1.12 Theorem. *Let K belong to C_p, $1 \leq p \leq \infty$ and $1/p+1/q = 1$. Then for every T in C_q, KT belongs to C_1 and $\phi_K(T) = tr(KT)$ is a linear functional on C_q with $\|\phi_K\| = \|K\|_p$.*

Proof. Let $s_n = s_n(K)$ and $K = \sum_{n=1}^{\infty} s_n(f_n \otimes e_n^*)$. Let $t_n = s_n(T)$. Also, KT can be written as $KT = \sum_{n=1}^{\infty} s_n(KT)\phi_n \otimes \psi_n^*$. Let U_N be the *rank* N partial isometry such that $U_N\phi_n = \psi_n$ for $1 \leq n \leq N$. Then

$$\begin{aligned}
\sum_{n=1}^{N} s_n(KT) &= \sum_{n=1}^{N} (KT\phi_n, U_N\phi_n) = \sum_{n=1}^{N} (T\phi_n, K^*U_N\phi_n) \\
&= \sum_{n=1}^{N} \sum_{k=1}^{\infty} (T\phi_n, e_k)\overline{(K^*U_n\phi_n, e_k)} \\
&= \sum_{n=1}^{N} \sum_{k=1}^{\infty} (\phi_n, T^*e_k)\overline{(\phi_n, s_k U_N^* f_k)} \\
&= \sum_{k=1}^{\infty} s_k \sum_{n=1}^{N} (U_N^* f_k, \phi_n)\overline{(T^*e_k, \phi_n)} \\
&= \sum_{k=1}^{\infty} s_k(U_N^* f_k, T^*e_k) = \sum_{k=1}^{\infty} s_k(TU_N^* f_k, e_k) \\
&\leq \sum_{k=1}^{\infty} s_k s_k(TU^*) \leq \sum_{k=1}^{N} s_k t_k \\
&\leq (\sum_{k=1}^{\infty} s_k^p)^{1/p}(\sum_{k=1}^{\infty} t_k^q)^{1/q} = \|K\|_p \|T\|_q \ .
\end{aligned}$$

Now let N increase to infinity to obtain $\|KT\|_1 \leq \|K\|_p \|T\|_q$. Hence

$$|\phi_K(T)| = |tr(KT)| \le \|KT\|_1 \le \|K\|_p \|T\|_q$$

so $\|\phi_K\| \le \|K\|_p$.

On the other hand, for $1 < p < \infty$, let $T = \sum_{n=1}^{\infty} s_n^{p/q}(e_n \otimes f_n^*)$, then

$$\phi_K(T) = tr\sum_{n=1}^{\infty} s_n^p (f_n \otimes f_n^*) = \sum_{n=1}^{\infty} s_n^p$$
$$= (\sum_{n=1}^{\infty} s_n^p)^{1/p} (\sum_{n=1}^{\infty} (s_n^{p/q})^q)^{1/q} = \|K\|_p \|T\|_q \quad .$$

So $\|\phi_K\| = \|K\|_p$. If $p = 1$, let $T_N = \sum_{n=1}^{N} e_n \otimes f_n^*$. Then

$$\phi_K(T_N) = tr\sum_{n=1}^{N} s_n (f_n \otimes f_n^*) = \sum_{n=1}^{N} s_n \quad .$$

So $\|\phi_K\| \ge sup\|\phi_K(T_N)\| = \|K\|_1$. And if $p = \infty$, $T = e_1 \otimes f_1^*$ suffices. ■

1.13 Theorem. *If $1 < p \le \infty$ and $1/p+1/q = 1$, then the map of C_q into C'_p given by taking K to ϕ_K is a linear, isometric isomorphism.*

Proof. By the previous theorem, we know that this map is isometric. Clearly it is linear. It must be shown to be surjective. Let ϕ be a continuous linear functional on C_p. Consider the form $<x,y> = \phi(x \otimes y^*)$. It is routine to verify that $<\cdot,\cdot>$ is linear in the first variable and conjugate linear in the second. Furthermore,

$$|<x,y>| \le \|\phi\| \, \|x \otimes y^*\|_p = \|\phi\| \, \|x\| \, \|y\| \quad .$$

Thus there is a bounded linear operator A with $\|A\| \le \|\phi\|$ so that

$$\phi(x \otimes y^*) = <x,y> = (Ax,y) = tr(A(x \otimes y^*)) \quad .$$

By linearity, $\phi(F) = tr(AF)$ for every finite rank operator F.

If A is not compact, then from the proof that the ideal generated by A is all of $B(\mathcal{H})$ one can obtain orthonormal families $\{e_n, n \ge 1\}$ and $\{f_n, n \ge 1\}$ so that $(Ae_n, f_n) \ge \delta > 0$ for all n. Let $F_n = \sum_{k=1}^{n} e_n \otimes f_n^*$. Then

$$n\delta \le \sum_{k=1}^{n} (Ae_n, f_n) = tr(AF_n)$$
$$= \phi(F_n) \le \|\phi\| \, \|F_n\|_p = n^{1/p} \|\phi\| \quad .$$

Hence $\|\phi\| \ge n^{1/q}\delta$ for all $n \ge 1$ which is absurd. So A is compact.

Let $A = \sum_{n=1}^{\infty} s_n(f_n \otimes e_n^*)$, and let $F_n = \sum_{k=1}^{n} s_n^{q/p}(e_n \otimes f_n^*)$ if $1 < p < \infty$ or $F_n = \sum_{k=1}^{n} e_n \otimes f_n^*$ if $p = \infty$. Then

$$\phi(F_n) = tr\sum_{k=1}^{n} s_k^q(f_k \otimes f_k^*) = \sum_{k=1}^{n} s_k^q$$

$$\leq \|\phi\| \, \|F_n\|_p = \|\phi\|(\sum_{k=1}^{n} s_k^q)^{1/p} .$$

Hence $(\sum_{k=1}^{n} s_k^q)^{1/q} \leq \|\phi\|$ for all n. Hence A belongs to C_q and $\| A \|_q \leq \| \phi \|$. By the continuity of ϕ and ϕ_A, the identity $\phi = \phi_A$ follows. ■

1.14 Corollary. *C_2 is a Hilbert space with inner product $(K,L) = tr(L^*K)$.*

1.15 Theorem. *The dual of C_1 is $B(\mathcal{H})$ given by the pairing $\phi_A(T) = tr(AT)$.*

Proof. By Corollary 1.6, AT belongs to C_1 and $\|AT\|_1 \leq \|A\| \, \|T\|_1$. So by Theorem 1.3, $|\phi_A(T)| \leq \|A\| \, \|T\|_1$. Hence $\|\phi_A\| \leq \|A\|$. On the other hand, if ϕ is a linear functional on C_1, then as in the proof of Theorem 1.13, $<x,y> = \phi(x \otimes y^*)$ is a bounded bilinear form given by $\phi(F) = tr(AF)$ for every finite rank operator, and $\|A\| \leq \|\phi\|$. By Theorem 1.11, $\mathcal{F}$ is dense in C_1 so $\phi = \phi_A$ and $\|\phi_A\| = \|A\|$. ■

This shows that $B(\mathcal{H})$ is a dual space, and indeed, it is the double dual of K. The weak* topology in $B(\mathcal{H})$ is the weakest topology in which all the functionals $\phi_T(A) = tr(TA)$ are continuous for all T in C_1.

1.16 Proposition. *Let A belong to C_p, $1 \leq p < \infty$ and let B belong to C_p^*. Then $tr(AB) = tr(BA)$.*

Proof. It follows from Theorems 1.12 and 1.15 that AB and BA belong to C_1. So by Theorem 1.3, the computation of the trace is an absolutely convergent sum. Also, if $\{\phi_k, k \geq 1\}$ is an orthonormal basis,

$$(x,y) = \sum_{k=1}^{\infty} (x,\phi_k)\overline{(y,\phi_k)}$$

converges absolutely for every x and y in $\mathcal{H}$. So

$$tr(BA) = \sum_{n=1}^{\infty}(BA\phi_n,\phi_n) = \sum_{n=1}^{\infty}(A\phi_n,B^*\phi_n)$$

$$= \sum_{n=1}^{\infty}\sum_{k=1}^{\infty}(A\phi_n,\phi_k)\overline{(B^*\phi_n,\phi_k)}$$

$$= \sum_{k=1}^{\infty}\sum_{n=1}^{\infty}(B\phi_k,\phi_n)\overline{(A^*\phi_k,\phi_n)}$$

$$= \sum_{k=1}^{\infty}(B\phi_k,A^*\phi_k) = \sum_{k=1}^{\infty}(AB\phi_k,\phi_k) = tr(AB) \ . \qquad \blacksquare$$

1.17 Corollary. *If T is trace class and S is invertible, then $tr(STS^{-1}) = tr(T)$.*

We finish this chapter with a useful relationship between the strong operator topology on $\mathcal{B}(\mathcal{H})$ and the C_p classes.

1.18 Proposition. *Let K belong to C_p, $1 \le p \le \infty$. If T_α is a bounded net converging to T in the strong operator topology, then $T_\alpha K$ converges to TK in the C_p norm.*

Proof. Let $M = \sup \|T_\alpha\|$ and let $\epsilon > 0$ be given. Choose F finite rank such that $\|K-F\|_p < \epsilon$. Then since the range of F is finite dimensional, $T_\alpha F$ converges to TF in norm. So

$$s_k((T-T_\alpha)F) \le \|(T-T_\alpha)F\|, \quad 1 \le k \le rankF \ .$$

Hence $\|(T-T_\alpha)F\|_p \le (rankF)^{1/p}\|(T-T_\alpha)F\|$ tends to zero. Choose α_0 so that $\|(T-T_\alpha)F\|_p < \epsilon$ for all $\alpha \ge \alpha_0$. Then for $\alpha \ge \alpha_0$,

$$\|TK-T_\alpha K\|_p \le \|T(K-F)\|_p+\|(T-T_\alpha)F\|+\|T_\alpha(F-K)\|_p$$

$$< M\epsilon+\epsilon+M\epsilon = (2M+1)\epsilon \ . \qquad \blacksquare$$

Notes and Remarks.

The C_p classes were introduced by von Neumann and Schatten as completions of the tensor product $\mathcal{H} \otimes \mathcal{H}$ in various norms. These were investigated in a series of papers and this material is contained in Schatten's monograph [1]. The study of ideals of Hilbert spaces was initiated at about the same time by Calkin [1]. He showed that there was a correspondence between ideals and the corresponding "ideals" of s-numbers. The inequalities Lemma 1.7 and Corollary 1.8 are due to Fan [1]. Theorem 1.15 on the predual of $\mathcal{B}(\mathcal{H})$ is due to Dixmier [3]. There are many extensions of these ideas, and a treatment may be found in Gohberg and Krein [1].

Exercises.

1.1 Let T be a Hilbert-Schmidt operator. Let $\{e_n, n \geq 1\}$ be any orthonormal basis. Prove that

$$\|T\|_2 = (\sum_{n=1}^{\infty} \|Te_n\|^2)^{1/2} = (\sum_{n=1}^{\infty} \sum_{m=1}^{\infty} |(Te_n, e_m)|^2)^{1/2} .$$

1.2 Let $\mathcal{H} = L^2(0,1)$ and let $K(x,y)$ be a square integrable function on $(0,1) \times (0,1)$. Let $(Tf)(x) = \int_0^1 f(y)K(x,y)dy$. Prove that T belongs to C_2 and $\|T\|_2 = \|K\|_2$.

1.3 Prove that C_1 coincides with the set of products AB where A and B belong to C_2.

1.4 Let T belong to C_p, $1 \leq p < \infty$, and let $\{e_n, n \geq 1\}$ and $\{f_n, n \geq 1\}$ be arbitrary orthonormal sets. Prove that

$$\sum_{n=1}^{\infty} |(Te_n, f_n)|^p \leq \|T\|_p^p .$$

Show that this inequality is sharp. Conversely, if T is any operator such that

$$\sum_{n=1}^{N} |(Te_n, f_n)|^p \leq M^p$$

for all finite orthonormal sets, then T belongs to C_p and $\|T\|_p \leq M$.

1.5 Let T_α be a bounded net converging to T in the weak operator topology. Show that if K is compact, then KT_α converges to KT in the strong operator topology. Hence if L is also compact, $KT_\alpha L$ converges to KTL in norm.

1.6 Let C be an operator in C_p, $1 \leq p < \infty$ or K. Prove that C can be factored as $C = C_1K$ where C_1 belongs to C_p and K is compact.

2. Triangular Algebras

Any matrix A acting on $\mathbb{C}^n$ has a triangular form. That is, one can find an orthonormal basis $\{e_1,...,e_n\}$ so that the subspaces $N_j = span\{e_1,...,e_j\}$, $1 \leq j \leq n$, are all invariant for A. With respect to this basis, A has a matrix (a_{ij}) in which $a_{ij} = 0$ if $i > j$. Also, the spectrum of A is read off as $\{a_{jj}, 1 \leq j \leq n\}$ including multiplicity. Consider the analogue for an operator T acting on a Hilbert space $\mathcal{H}$. To ask for an orthonormal basis $\{e_n, n \geq 1\}$ so that $N_j = span\{e_1,...,e_j\}$ is invariant is too demanding because this forces T to have lots of eigenvectors. Even fairly nice operators can fail to have eigenvectors. So instead, we ask for a (linearly ordered) chain of (closed) invariant subspaces of T. Even this may not be possible for all operators.

2.1. A *nest* is a chain $\mathcal{N}$ of closed subspaces of a Hilbert space $\mathcal{H}$ containing $\{0\}$ and $\mathcal{H}$ which is closed under intersection and closed span.

In other words, a nest is a chain of subspaces which is complete with respect to the natural lattice operations on the lattice of all subspaces of a Hilbert space. The set $Lat\,T$ of invariant subspaces of an operator T is always a complete lattice. So there is no loss in considering only nests instead of arbitrary chains. Give any chain C of invariant subspaces of T, one can always extend it to a maximal chain of invariant subspaces. This follows from Zorn's Lemma and the fact that the union of an increasing family of chains of invariant subspaces is also a chain of invariant subspaces. This largest chain is necessarily complete, as the completion is also an invariant chain.

Given an orthonormal basis for $\mathbb{C}^n$, one defines the algebra $\mathcal{T}_n$ of all upper triangular matrices. By analogy, one makes the following definition.

The *triangular algebra* or *nest algebra* $\mathcal{T}(\mathcal{N})$ is the set of all operators T such that $TN \subseteq N$ for every element N in $\mathcal{N}$.

2.2 Proposition. *$\mathcal{T}(\mathcal{N})$ is a weak operator closed subalgebra of $B(\mathcal{H})$.*

Proof. It is clear that $\mathcal{T}(\mathcal{N})$ is a linear space containing the scalar operators. If A and B belong to $\mathcal{T}(\mathcal{N})$, and N belongs to $\mathcal{N}$, then

$$AB(N) \subseteq A(N) \subseteq N \quad .$$

So AB belongs to $\mathcal{T}(\mathcal{N})$. Finally, if A_α belong to $\mathcal{T}(\mathcal{N})$ and A is a weak limit point of A_α, then for every N in $\mathcal{N}$, x in N and y in $N^\perp$,

$$0 = \lim(A_\alpha x, y) = (Ax, y) \quad .$$

Thus AN is contained in N, and A belongs to $\mathcal{T}(\mathcal{N})$. ∎

Before proceeding, consider some examples of nests.

2.3 Example. Let P_n be an increasing sequence of finite dimensional subspaces such that their union is dense in $\mathcal{H}$. Then $\mathcal{P} = \{\{0\}, P_n, n \geq 1, \mathcal{H}\}$ is a complete nest. $\mathcal{T}(\mathcal{P})$ consists of all operators which have a block upper triangular matrix with respect to $\mathcal{P}$. The nest $\mathcal{P}$ is maximal exactly when $dim\, P_n = n$ for all n.

2.4 Example. Let $\mathcal{H} = L^2(0,1)$ with Lebesgue measure. For each t in $[0,1]$, let N_t consist of all functions f in $L^2(0,1)$ such that $f(x) = 0$ a.e. on $[t,1]$. Then $\mathcal{N} = \{N_t : 0 \leq t \leq 1\}$ is a "continuous" nest. (See Section 2.7). This nest is known as the *Volterra nest.*

2.5 Example. Let $\mathcal{H} = L^2((0,1)^2)$ be the space of Lebesgue square integrable functions on the unit square. Analogous to the preceding example, let M_t be the set of functions supported on $[0,t] \times [0,1]$. Then $\mathcal{M} = \{M_t, 0 \leq t \leq 1\}$ is a continuous nest. It is maximal (Proposition 2.10) but it is different from Example 2.4 in an essential way (Chapter 7).

2.6 Example. Let μ be counting measure on the set of rational numbers $\mathbb{Q}$. Let $\mathcal{H} = \ell^2(\mathbb{Q}) = L^2(\mu)$. For each t in $\mathbb{R}$, let

$$Q_t^+ = \{f \in \ell^2(\mathbb{Q}) : f(q) = 0 \text{ for } q > t\}$$

and

$$Q_t^- = \{f \in \ell^2(\mathbb{Q}) : f(q) = 0 \text{ for } q \geq t\} \quad .$$

Note that if t is irrational, then $Q_t^- = Q_t^+$. But if t belongs to $\mathbb{Q}$, then $Q_t^+ \ominus Q_t^- = span\{\delta_t\}$ where δ_t is the characteristic function of t. Set $Q_{-\infty} = \{0\}$ and $Q_{+\infty} = \mathcal{H}$. Then $\mathcal{Q} = \{Q_t^\pm : -\infty \leq t \leq +\infty\}$ is a nest. The set $\{\delta_t : t \in \mathbb{Q}\}$ is an orthonormal basis for $\mathcal{H}$. This nest is "atomic" (Section 2.7) because the space is spanned by the gaps between adjacent elements of the nest. This nest is known as the *Cantor nest.*

2.7. Given a collection $\{M_\alpha\}$ of subspaces of a Hilbert space, $\bigvee M_\alpha$ denotes the closed linear span and $\bigwedge M_\alpha$ denotes intersection. This makes the set of subspaces of $\mathcal{H}$ into a lattice. A nest is just a complete, totally ordered sublattice. For N belonging to a nest $\mathcal{N}$, define

$$N_- = \bigvee\{N' \in \mathcal{N} : N' < N\}$$

and

$$N_+ = \bigwedge\{N' \in \mathcal{N}: N' > N\} \ .$$

It is easy to see that N_- is the immediate predecessor to N if there is one. Otherwise $N_- = N$. If $N_- \neq N$, then $N = (N_-)_+$ is the immediate successor of N_-.

The subspaces $N \ominus N_-$ are called the *atoms* of $\mathcal{N}$. If the atoms of $\mathcal{N}$ span $\mathcal{H}$, then $\mathcal{N}$ is *atomic* (e.g. Examples 2.3 and 2.6). If there are no atoms, $\mathcal{N}$ is called *continuous* (e.g. Examples 2.4 and 2.5).

The algebras $\mathcal{T}(\mathcal{N})$ contain an abundance of operators, and this will be used to prove reflexivity. Given an algebra $\mathcal{A}$, the lattice of all invariant subspaces of $\mathcal{A}$ is denoted $Lat\,\mathcal{A}$. Similarly, if $\mathcal{L}$ is a lattice of subspaces, then $Alg\,\mathcal{L}$ denotes the algebra of all operators leaving each element of $\mathcal{L}$ invariant. The lattice $\mathcal{L}$ is *reflexive* if $\mathcal{L} = Lat\,Alg\,\mathcal{L}$; and an algebra $\mathcal{A}$ is *reflexive* if $\mathcal{A} = Alg\,Lat\,\mathcal{A}$. Clearly, $\mathcal{T}(\mathcal{N})$ is reflexive. It turns out that $\mathcal{N}$ is also reflexive.

2.8 Lemma. *Let N belong to $\mathcal{N}$. Then every operator of the form $T = P(N_+)TP(N)^\perp$ belongs to $\mathcal{T}(\mathcal{N})$.*

Proof. Let N' belong to $\mathcal{N}$. Then either $N' \geq N_+$ or $N' \leq N$. In the first case, $P(N')^\perp P(N_+) = 0$ so $P(N')^\perp TP(N') = 0$. In the second case $P(N)^\perp P(N') = 0$ so $P(N')^\perp TP(N') = 0$. Thus N' is invariant for T. ■

2.9 Theorem. *$Lat\,\mathcal{T}(\mathcal{N}) = \mathcal{N}$.*

Proof. Let M be an invariant subspace for $\mathcal{T}(\mathcal{N})$ and let N be the least element of $\mathcal{N}$ containing M. Let N' be any other element of $\mathcal{N}$ with $N' < N$. Since $M \nleq N'$, there is a vector x in M such that $P(N')^\perp x \neq 0$. Let y be any vector in N'_+, and let T be an operator of the form $T = P(N'_+)TP(N')^\perp$ such that $TP(N')^\perp x = y$. By Lemma 2.8, T belongs to $\mathcal{T}(\mathcal{N})$. Thus $y = Tx$ belongs to M. Hence

$$M \geq \bigvee\{N'_+ : N' < N\} = N \ .$$

(To verify this identity, consider two cases, $N = N_-$ and $N = (N_-)_+$.) Thus $M = N$ belongs to $\mathcal{N}$. ■

A nest is *maximal* if it is not contained in any larger nest.

2.10 Proposition. *A nest is maximal if and only if all its atoms are one dimensional.*

Proof. Necessity is clear. Suppose that $\mathcal{N}$ is a nest in which all atoms are one dimensional. Let M be a subspace such that for every N in $\mathcal{N}$, either $N \leq M$ or $N > M$. Let

$$N_0 = \bigvee\{N \in \mathcal{N}: N \leq M\}$$

and

$$N_1 = \bigwedge\{N \in \mathcal{N}: N \geq M\} \ .$$

Clearly, $N_0 \leq M \leq N_1$. Also, if N belongs to $\mathcal{N}$, either $N \leq N_0$ or $N \geq N_1$. If $N_0 = N_1$, then $M = N_0$ belongs to $\mathcal{N}$. If $N_0 \neq N$, then $N_1 \ominus N_0$ is an atom of $\mathcal{N}$ and hence is one dimensional. So M equals N_0 or N_1 and thus belongs to $\mathcal{N}$. (From this, it follows that $N_0 = N_1$ after all.) So $\mathcal{N}$ cannot be expanded to a larger nest. ■

2.11. A subspace N is invariant for T exactly when $N^{\perp}$ is invariant for T^*. Thus it follows that $\mathcal{T}(\mathcal{N})^* = \mathcal{T}(\mathcal{N}^{\perp})$ where $\mathcal{N}^{\perp}$ is the nest $\{N^{\perp}: N \in \mathcal{N}\}$. The intersection $\mathcal{D}(\mathcal{N}) = \mathcal{T}(\mathcal{N}) \cap \mathcal{T}(\mathcal{N})^*$ is the *diagonal* of $\mathcal{T}(\mathcal{N})$. This is the weakly closed, self-adjoint algebra (von Neumann algebra) of all operators T which leave $\mathcal{N}$ and $\mathcal{N}^{\perp}$ invariant. Thus if N belongs to $\mathcal{N}$,

$$TP(N) = P(N)TP(N) = (P(N)T^*P(N))^* = (T^*P(N))^* = P(N)T \ .$$

Conversely, if T commutes with $P(N)$ for all N in $\mathcal{N}$, then

$$TP(N) = P(N)TP(N) \quad \text{and} \quad TP(N)^{\perp} = P(N)^{\perp}TP(N)^{\perp}$$

so T belongs to $\mathcal{D}(\mathcal{N})$. Thus $\mathcal{D}(\mathcal{N}) = \mathcal{N}'$ is the commutant of $\{P(N): N \in \mathcal{N}\}$.

One also defines the *core* of $\mathcal{T}(\mathcal{N})$ to be the von Neumann algebra generated by $\{P(N): N \in \mathcal{N}\}$. By the Double Commutant Theorem, this is just the commutant of $\mathcal{D}(\mathcal{N})$ and will be denoted by $\mathcal{N}''$. Note that $\mathcal{N}''$ is abelian, and hence is the centre of $\mathcal{D}(\mathcal{N})$.

A vector x is a *cyclic* for a von Neumann algebra $\mathcal{A}$ provided $\mathcal{A}x$ is dense in $\mathcal{H}$. A vector x is *separating* for a von Neumann algebra $\mathcal{A}$ provided $Ax = 0$ implies $A = 0$ for A in $\mathcal{A}$.

2.12 Proposition. *Let $\mathcal{M}$ be an abelian von Neumann algebra on a separable Hilbert space. Then $\mathcal{M}'$ has a cyclic vector x. Hence x is a separating vector for $\mathcal{M}$.*

Proof. Let x be any vector in $\mathcal{H}$. Let $\mathcal{H}_x$ denote the closure of $\{Ax : A \in \mathcal{M}'\}$, and let P_x be the orthogonal projection onto $\mathcal{H}_x$. Since $\mathcal{H}_x$ is invariant for $\mathcal{M}'$, P_x commutes with $\mathcal{M}'$ and thus belongs to $\mathcal{M}$. If E is a projection in $\mathcal{M}$, then $A(Ex) = E(Ax)$ for A in $\mathcal{M}$. Thus $\mathcal{H}_{Ex} = E\mathcal{H}_x$ and $P_{Ex} = EP_x$.

Let $\{x_n\}$ be a maximal family of unit vectors (necessarily countable) such that $P_n = P_{x_n}$ are pairwise orthogonal (provided by an application of Zorn's lemma). If $P = \sum P_n$ is not the identity, one could take any unit vector $x = P^{\perp}x$ and obtain a nonzero projection $P_x \leq P^{\perp}$ which extends $\{x_n\}$. Thus $\sum P_n = I$. Set $x = \sum_{n \geq 1} 2^{-n} x_n$. Then $\mathcal{H}_x$ contains $2^n P_n x = x_n$ for all $n \geq 1$. Hence $\mathcal{H}_x$ contains $\sum \mathcal{H}_{x_n} = \mathcal{H}$, and so x is cyclic for $\mathcal{M}'$.

If T belongs to $\mathcal{M}$, then

$$\mathcal{H}_{Tx} = \{ATx : A \in \mathcal{M}'\} = \{TAx : A \in \mathcal{M}'\} = \overline{TH_x} = \overline{Ran\, T} .$$

Thus if $Tx = 0$, one has $Ran\,(T) = \{0\}$ and thus $T = 0$. So x is a separating vector for $\mathcal{M}$. ∎

The nest $\mathcal{N}$ has an order $<$ given by containment. So $(\mathcal{N},<)$ is a topological space when endowed with the order topology. The set $P(\mathcal{N}) = \{P(N) : N \in \mathcal{N}\}$ is ordered as a subset of the positive operators on $\mathcal{H}$, and it is a topological space when endowed with the strong operator topology.

2.13 Theorem. *Let $\mathcal{N}$ be a nest. Then the natural map taking N to $P(N)$ is an order preserving homeomorphism of the compact Hausdorff space $(\mathcal{N},<)$ onto $(P(\mathcal{N}),s)$. If $\mathcal{N}$ has a unit separating vector x, then the map $\Phi_x(N) = (P(N)x,x)$ is an order preserving homeomorphism of $(\mathcal{N},<)$ onto a compact subset of $[0,1]$.*

Proof. First it will be shown that $P(\mathcal{N}) = \{P(N) : N \in \mathcal{N}\}$ is compact in the strong operator topology. Let $\overline{P(\mathcal{N})}$ denote the weak operator closure, which is compact since it is bounded. Let $P_\alpha = P(N_\alpha)$ be a net of projections converging weakly to an operator E. By the Double Commutant Theorem, E is a positive contraction in $\mathcal{N}''$. Let N_0 be the least element of $\mathcal{N}$ such that $P(N_0) \geq E$. Let N be any element of $\mathcal{N}$ with $N < N_0$. By the minimality of N_0, there is a unit vector x in $N_0 \ominus N$ such that $Ex \neq 0$. Thus $(Ex,x) > 0$. Hence there is an α_0 so that $(P_\alpha x,x) > 0$ for all $\alpha \geq \alpha_0$, and consequently $P_\alpha \geq P(N_+)$. Similarly, if N' belongs to $\mathcal{N}$ and $N' > N_0$, let x be any unit vector in $N' \ominus N_0$. So there is an α_0 so that $(P_\alpha x,x) < 1$ for all $\alpha \geq \alpha_0$. Hence $P_\alpha \leq P(N'_-)$. Therefore, for every N, N' in $\mathcal{N}$ with $N < N_0 < N'$ there is an α_0 so that $P_\alpha x = Ex = P(N_0)x$ for every x in $N_+ \oplus (N'_-)^{\perp}$ and $\alpha \geq \alpha_0$. The union of these subspaces is dense in $N_0 \oplus N_0^{\perp} = \mathcal{H}$. It follows that P_α converges to $P(N_0)$ in the strong operator topology. This shows that $P(\mathcal{N})$ is weak operator closed, hence compact, and that the strong operator topology coincides with the weak operator topology on $P(\mathcal{N})$. So it is compact in the strong operator topology as well.

The map taking $P(N)$ to N is an order preserving bijection of a compact Hausdorff space onto another Hausdorff space. To show that it is continuous, it suffices to show that $\{P(N) : N < N_0\}$ and $\{P(N) : N > N_0\}$ are open. If x is a vector, $\|P(N)x\|$ is an increasing function of N which is strong operator continuous, thus $\{P(N):\|P(N)x\| < 1\}$ and $\{P(N):\|P(N)x\| > 1\}$ are open. Such sets separate the points of $\mathcal{N}$. Hence $(P(\mathcal{N}),s)$ and $(\mathcal{N},<)$ are homeomorphic.

If x is a separating vector, the map Φ_x is strongly continuous and strictly increasing on $\mathcal{N}$. Thus $\Phi_x(\mathcal{N})$ is a compact subset of $[0,1]$. The topology on $[0,1]$ is just the order topology, so this is a homeomorphism. ∎

2.14 Example. Now one can talk about the order type of a nest by representing it as a compact subset of $[0,1]$. Consider the examples 2.3 - 2.6. Example 2.3 has order type $\omega = \mathbb{N}_0 \cup \{\infty\}$. Example 2.4 and 2.5 have order type $[0,1]$ and $x(t) = 1$ is a separating vector. In 2.4, it is also a cyclic vector for $\mathcal{N}'' = \mathcal{N}'$. But in 2.5, $\mathcal{M}'$ is much bigger than $\mathcal{M}''$ and $\mathcal{M}''$ has no cyclic vector. Example 2.6 is a bit more complicated. Let a_r be non-zero complex numbers for each r in $\mathbb{Q}$ such that $\sum |a_r|^2 = 1$. Then $x = \sum a_r \delta_r$ is a separating and cyclic vector for $\mathcal{Q}'' = \mathcal{Q}'$. The map Φ_x takes $\mathcal{Q}$ onto a compact subset of $[0,1]$. For each r in $\mathbb{Q}$, $t_{r^-} = \Phi_x(Q_r^-)$ is strictly less than $t_{r^+} = \Phi_x(Q_r^+)$. So there is an interval (t_{r^-},t_{r^+}) in $[0,1]\backslash\Phi_x(\mathcal{Q})$. This set of "gaps" is dense since the atoms are dense in the order of $\mathcal{Q}$. Thus $\Phi_x(\mathcal{Q})$ is a perfect set with no interior and must be order homeomorphic to the Cantor set.

If $N_1 < N_2$ in $\mathcal{N}$, the subspace $E = N_2 \ominus N_1$ is called an *interval* of $\mathcal{N}$. The significance of intervals is due in part to the following two propositions.

2.15 Proposition. *Let E be an interval of $\mathcal{N}$. Then the map Φ_E of $\mathcal{T}(\mathcal{N})$ into $\mathcal{B}(E)$ given by*

$$\Phi_E(T) = P(E)T|E$$

is an algebra homomorphism onto $\mathcal{T}(\mathcal{N}_E)$ where $\mathcal{N}_E$ is the nest $\{N \cap E : N \in \mathcal{N}\}$. This map is continuous in all the natural topologies on $\mathcal{B}(\mathcal{H})$.

Proof. Let $E = N_2 \ominus N_1$ for N_1, N_2 in $\mathcal{N}$ and let $P_i = P(N_i)$. If N is invariant for T, then $TP(N) = P(N)TP(N)$ and $P(N)^\perp T = P(N)^\perp TP(N)^\perp$. Thus for T_1, T_2 in $\mathcal{T}(\mathcal{N})$ one has

$$\begin{aligned} P(E)T_1T_2P(E) &= P_2(P_1^\perp T_1)(T_2P_2)P_1^\perp \\ &= P_2(P_1^\perp T_1 P_1^\perp)(P_2T_2P_2)P_1 \\ &= P(E)T_1P(E)T_2P(E) \ . \end{aligned}$$

Hence $\Phi_E(T_1T_2) = \Phi_E(T_1)\Phi_E(T_2)$. As Φ_E is linear, it is an algebra homomorphism. Clearly, $\|\Phi_E\| \leq 1$ so Φ_E is continuous.

For N in $\mathcal{N}$ and T in $\mathcal{T}(\mathcal{N})$, $\Phi_E(T)$ takes $N \cap E$ into $P(E)N = N \cap E$. So $\Phi_E(T)$ belongs to $\mathcal{T}(\mathcal{N}_E)$. On the other hand, if A belongs to $\mathcal{T}(\mathcal{N}_E)$, let $T = AP(E)$. Clearly, T belongs to $\mathcal{T}(\mathcal{N})$ and $\Phi_E(T) = A$. So Φ_E is surjective. ∎

2.16 Theorem. *Let $\mathcal{A}$ be an algebra of operators, and let E be a subspace such that $\Phi_E(A) = P(E)A|E$ is an algebra homomorphism. Then there are invariant subspaces N_1 and N_2 for $\mathcal{A}$ such that $E = N_2 \ominus N_1$. In particular, if $\mathcal{A} = \mathcal{T}(\mathcal{N})$, then E is an interval of $\mathcal{N}$.*

Proof. Let N_2 be the smallest invariant subspace of A containing E. Thus N_2 is the closure of $\{Ae : A \in \mathcal{A}, e \in E\}$. Set $N_1 = N_2 \ominus E$. It suffices to show that N_1 is invariant for $\mathcal{A}$. Now if A and B belong to $\mathcal{A}$,

$$\begin{aligned} 0 &= \Phi_E(AB) - \Phi_E(A)\Phi_E(B) \\ &= P(E)ABP(E) - P(E)AP(E)BP(E) \\ &= P(E)AP(E)^{\perp}BP(E) \ . \end{aligned}$$

Since the ranges of $BP(E)$ spans N_2 as B runs over $\mathcal{A}$, it follows that the ranges of $P(E)^{\perp}BP(E)$ spans N_1. Thus $P(E)AP(N_1) = 0$ for all A in $\mathcal{A}$. Hence

$$AP(N_1) = P(E)^{\perp}P(N_2)AP(N_1) = P(N_1)AP(N_1) \ .$$

So N_1 is invariant.

In the case of nest algebra, Theorem 2.9 shows that N_1 and N_2 belong to $\mathcal{N}$. So E is an interval of $\mathcal{N}$. ∎

2.17. The atomic part of the diagonal of $\mathcal{T}(\mathcal{N})$ is $\mathcal{D}_a = \sum_{A_\alpha \in \mathbf{A}} \oplus \mathcal{B}(A_\alpha)$ where $\mathbf{A} = \{A_\alpha\}$ is the set of atoms of $\mathcal{N}$. The map $\Delta(T) = \sum P(A_\alpha)TP(A_\alpha)$ is a contractive projection of $\mathcal{B}(\mathcal{H})$ onto $\mathcal{D}_a$. By Proposition 2.15, Δ is multiplicative on $\mathcal{T}(\mathcal{N})$.

Notes and Remarks.

A class of triangular algebras was introduced by Kadison and Singer [1]. This class contained nest algebras with maximal nests and many algebras that are not even norm closed. The order type of a nest, and the diagonal for nest algebra are first used here. The notion of a nest was introduced by Ringrose [1] in connection with triangular forms for compact operators. Proposition 2.10 is proved in that paper. In a sequel, Ringrose [3], Lemma 2.8 and Theorem 2.9 are proven as is Proposition 2.15. Theorem 2.16 is due to Sarason [1]. The notions of *Alg*, *Lat*, and reflexive algebras are due to Halmos.

Exercises.

2.1 Let $\mathcal{L}$ be any collection of subspaces of $\mathcal{H}$. Prove that $Alg\,\mathcal{L}$ is a weakly closed, unital algebra.

2.2 Given a compact subset ω of $[0,1]$, construct a nest $\mathcal{N}$ of order type ω.

2.3 Prove that every von Neumann algebra is the norm closure of the span of its projections. Hence prove that every von Neumann algebra is reflexive, and identify its invariant subspace lattice.

2.4 Let T belong to $\mathcal{T}(\mathcal{N})$, and let $\Delta(T)$ be the projection of section 2.17. Show that the spectrum of $\Delta(T)$ is a subset of $\sigma(T)$.

3. Triangular Forms for Compact Operators

The starting point is the existence of proper invariant subspaces for compact operators. The proof presented here is old fashioned, but illustrates how one can lift the finite dimensional triangular forms to the compact operator case.

3.1 Theorem. *Let K be a compact operator on a Hilbert space $\mathcal{H}$. Then K has a proper invariant subspace.*

Proof. Let P_n be any increasing sequence of finite rank projections increasing to the identity. Then $K_n = P_n K P_n$ converges to K in norm. The restriction of K_n to $P_n \mathcal{H}$ is a finite rank operator and thus can be put in triangular form with respect to some chain $\{P_{n,j}, 0 \leq j \leq rank\, P_n\}$ so that $rank\, P_{n,j} = j$. Fix a unit vector $x = P_1 x$. Let j_n be the least integer such that $(P_{n,j_n} x, x) \geq ½$. Set $Q_n = P_{n,j_n}$ and $Q_n^- = P_{n,j_n - 1}$.

The set $\{Q_n, n \geq 1\}$ is bounded, so there is a subsequence which converges in the weak operator topology to an operator Q. Likewise on a sub-subsequence, one has $Q_{n_j}^-$ converging weakly to an operator Q^-. Relabel this subsequence so that

$$Q = w\text{--}\lim Q_n \quad \text{and} \quad Q^- = w\text{--}\lim Q_n^- \ .$$

Then $0 \leq Q^- \leq Q \leq I$. Furthermore, since $Q_n - Q_n^-$ are rank one, it follows that $Q - Q^-$ has rank at most one. Also,

$$(Qx, x) = \lim (Q_n x, x) \geq ½$$

and

$$(Q^- x, x) = \lim (Q_n^- x, x) \leq ½ \ .$$

Thus $Q \neq 0$ and $Q^- \neq I$. So either Q or Q^- is neither 0 nor I. We shall suppose that $Q \neq I$.

Since K is compact, $\lim\limits_{n \to \infty} K Q_n x = KQx$ exists in norm for every vector x (Exercise 1.5). Thus since $K = \lim\limits_{n \to \infty} K_n$ and $(I - Q_n) K_n Q_n = 0$,

$$\lim_{n\to\infty} Q_n KQx = KQx - \lim_{n\to\infty} (I-Q_n)(KQx - K_n Q_n x) = KQx$$

Consequently,

$$KQx = w-\lim_{n\to\infty} Q_n KQx = QKQx \quad .$$

From this, it follows that K carries the range $Ran(Q)$ of Q into the eigenspace $N = ker(Q-I)$. Hence any subspace M such that $N \subseteq M \subseteq Ran(Q)$ is invariant for K. As $Q \neq I$, N is not all of $\mathcal{H}$; and as $Q \neq 0$, $Ran(Q)$ is not zero. Thus there is a proper subspace M of this form. ∎

3.2 Corollary. *Let K be a compact operator on a Hilbert space $\mathcal{H}$. Then there is a maximal nest $\mathcal{N}$ of invariant subspaces of K.*

Proof. Apply Zorn's lemma to obtain a maximal chain $\mathcal{N}$ of invariant subspaces of K. Since $Lat\,K$ is a complete lattice, $\mathcal{N}$ is complete and thus is a nest. By Proposition 2.10, it is enough to verify that $dim(N \ominus N_-) \leq 1$ for all N in $\mathcal{N}$. If, on the contrary, $dim(N \ominus N_-) \geq 2$, let $K_0 = P(E)K|E$ where $E = N \ominus N_-$. As K_0 is compact, Theorem 3.1 provides a proper closed subspace M of E invariant for K_0. But then $N_- \oplus M$ is invariant for K and is intermediate to N_- and N. This contradicts the maximality of $\mathcal{N}$ in $Lat\,K$. Hence all atoms are one dimensional and $\mathcal{N}$ is a maximal nest. ∎

3.3. As in the matrix case, one can read off information about the spectrum of a compact operator from its triangular form. Let K be compact and let $\mathcal{N}$ be a maximal nest of invariant subspaces of K. Let $\mathbf{A} = \{A_\alpha\}$ be the (one-dimensional) atoms of $\mathcal{N}$. By Proposition 2.15, $\Phi_\alpha(T) = P(A_\alpha)T|A_\alpha$ is an algebra homomorphism of $\mathcal{T}(\mathcal{N})$. In this case, $\Phi_\alpha(T)$ is just a scalar number so Φ_α is a multiplicative linear functional of $\mathcal{T}(\mathcal{N})$. It follows that $\lambda_\alpha = \Phi_\alpha(T)$ belongs to the spectrum of T as an element of $\mathcal{T}(\mathcal{N})$. For otherwise

$$1 = \Phi_\alpha((T-\lambda_\alpha I)(T-\lambda_\alpha I)^{-1}) = \Phi_\alpha(T-\lambda_\alpha I)\Phi((T-\lambda_\alpha I)^{-1}) = 0$$

which is absurd. In general, the spectrum of T in $\mathcal{T}(\mathcal{N})$ is greater than $\sigma(T)$. However, in the case of a compact operator, more can be said.

3.4 Ringrose's Theorem. *Let K be a compact operator. Let $\mathcal{N}$ be a maximal nest of invariant subspaces for K, and let $\mathbf{A} = \{A_\alpha\}$ be the atoms of $\mathcal{N}$. Then $\sigma(K) = \sigma_{\mathcal{T}(\mathcal{N})}(K) = \{0\} \cup \{\Phi_\alpha(K): A_\alpha \in \mathbf{A}\}$ and the non-zero values are repeated according to their algebraic multiplicity.*

First, we need a lemma. A *partition* of a nest $\mathcal{N}$ is a finite set $E_1,\ldots,E_s$ of pairwise orthogonal intervals of $\mathcal{N}$ such that $\sum_{i=1}^{s} E_i = \mathcal{H}$. A partition is always obtained from a finite subnest $\{0\} = M_0 < M_1 < \ldots < M_s = \mathcal{H}$ of $\mathcal{N}$ by $E_i = M_i \ominus M_{i-1}$.

3.5 Lemma. *Let $\mathcal{N}$ be a nest, and let K be a compact operator in $\mathcal{T}(\mathcal{N})$. Given $\epsilon > 0$, there is a partition $E_1,\ldots,E_s$ of $\mathcal{N}$ so that for each $1 \leq i \leq s$, either E_i is an atom or $\|P(E_i)KP(E_i)\| < \epsilon$.*

Proof. Let $N \neq \{0\}$ belong to $\mathcal{N}$. If $N = N_-$, then the net $P(N_\alpha)$ for $\{N_\alpha \in \mathcal{N}: N_\alpha < N\}$ indexed by itself converges to $P(N)$ in the strong* operator topology. Hence by Proposition 1.18, $(P(N)-P(N_\alpha))K$ converges to zero in norm. Thus there is a subspace $L(N) < N$ in $\mathcal{N}$ so that

$$\|P(N \ominus L(N))KP(N \ominus L(N))\| < \epsilon \ .$$

If $N \neq N_-$, set $L(N) = N_-$. Then $N \ominus L(N)$ is an atom. Similarly, if $N \neq \mathcal{H}$, one can find $U(N) > N$ in $\mathcal{N}$ so that either $U(N) \ominus N$ is an atom or

$$\|P(U(N) \ominus N)KP(U(N) \ominus N)\| < \epsilon \ .$$

Let $\mathcal{O}_N = \{N' \in \mathcal{N}: L(N) < N' < U(N)\}$ for N in $\mathcal{N}$ with the exceptions $\mathcal{O}_{\{0\}} = \{N' \in \mathcal{N}: N' < U(\{0\})\}$ and $\mathcal{O}_{\mathcal{H}} = \{N' \in \mathcal{N}: N' > L(\mathcal{H})\}$. These sets are open intervals of $\mathcal{N}$ in the order topology which cover $\mathcal{N}$. By Theorem 2.13, $\mathcal{N}$ is compact. Thus there is a finite subcover $\mathcal{O}_{N_1}, \ldots, \mathcal{O}_{N_k}$. The set $\{N_i, L(N_i), U(N_i), 1 \leq i \leq k\}$ in its natural order is a finite nest $\{0\} = M_0 < M_1 < \ldots < M_s = \mathcal{H}$. Let $E_i = M_i \ominus M_{i-1}$. As each E_i is dominated by one of the intervals $N_j \ominus L(N_j)$ or $U(N_j) \ominus N_j$, it follows that E_i is either an atom or $\|P(E_i)KP(E_i)\| < \epsilon$. ∎

Proof of Theorem 3.4. Let $\Lambda = \{0\} \cup \{\Phi_\alpha(K): A_\alpha \in \mathbf{A}\}$. Fix $\lambda \neq 0$, and let $n \geq 0$ be the number of times λ occurs in Λ. This is finite since K is compact. Let $\epsilon = |\lambda|/2$ and apply Lemma 3.5. One obtains a finite subnest

$$\{0\} = M_0 < M_1 < \ldots < M_s = \mathcal{H}$$

and intervals $E_i = M_i \ominus M_{i-1}$ such that either $\|P(E_i)KP(E_i)\| < \epsilon$ or E_i is an atom. In the latter case $P(E_i)KP(E_i) = \lambda_i P(E_i)$ where $\lambda_i = \Phi_{E_i}(K)$ belongs to Λ.

Let (K_{ij}) be the $s \times s$ upper triangular matrix form of K with respect to this finite nest $\mathcal{M}$. The diagonal entries are $K_{ii} = P(E_i)K|E_i$. By construction, $\|K_{ii}\| < \epsilon$ or $K_{ii} = \lambda_i I_i$ where $I_i = I|E_i$. Note that $\lambda_i = \lambda$ occurs precisely n times, say for i in X. So $K_{ii} - \lambda I_i$ is a one dimensional zero operator in these n instances. Otherwise $K_{ii} - \lambda I_i$ is invertible.

Let C be an oriented circle centered at λ with radius $r = \frac{1}{2} \mathit{dist}\{\lambda, \Lambda \setminus \{\lambda\}\}$. Thus $\sigma(K_{ii})$ is disjoint from $C \cup \operatorname{int} C$ unless i belongs to X. For z in C, $zI-K$ has an $s \times s$ upper triangular form with invertible diagonal entries $K_{ii}-zI_i$. Thus $zI-K$ is invertible, and $(zI-K)^{-1}$ has an $s\times s$ supper triangular form with diagonal entries $(K_{ii}-zI_i)^{-1}$. To see this, let T be the $s\times s$ diagonal matrix with entries $T_{ii} = (K_{ii}-zI_i)^{-1}$. Then $T(K-zI)$ is an upper triangular $s\times s$ matrix with diagonal entries I_i. In other words, $T(K-zI) = I+N$ where N is strictly upper triangular. Since $N^s = 0$, $(I+N)^{-1} = \sum_{k=1}^{S-1} N^k = S$ exists and is upper triangular. Thus $ST = (K-zI)^{-1}$.

Now one can compute the Riesz idempotent for K and $\{\lambda\}$ as

$$P_\lambda = \frac{1}{2\pi i}\int_C (zI-K)^{-1}dz \quad .$$

This has matrix entries given by integration of the matrix entries of $(zI-K)^{-1}$. So it is upper triangular, and its diagonal entries are

$$(P_\lambda)_{ii} = \frac{1}{2\pi i}\int_C (zI_i-K_{ii})^{-1}dz = \begin{cases} 0 & i \notin X \\ I_i & i \in X \end{cases} .$$

Thus $\text{rank}P_\lambda = tr\,P_\lambda = |X| = n$. So the algebraic multiplicity of λ in $\sigma(K)$ is equal to its number of occurences on the diagonal. In particular, $T-\lambda I$ is invertible if λ is not in Λ. Moreover, $(T-\lambda I)^{-1}$ belongs to $\mathcal{T}(\mathcal{M})$. The proof above holds if $\mathcal{M}$ is enlarged to contain any particular element N in $\mathcal{N}$. Thus $(T-\lambda I)^{-1}$ belongs to $\mathcal{T}(\mathcal{N})$. That is, $\sigma_{\mathcal{T}(\mathcal{N})}(K) = \sigma(K) = \Lambda$. ■

3.6. Next, we wish to show that, in a number of ways, $\mathcal{T}(\mathcal{N})$ has an abundance of compact operators. In terms of determining the invariant subspace lattice $\mathcal{N}$, even the rank one operators are sufficient. This is because in the proof of Theorem 2.9, the operator T constructed may be taken to be rank one. The first result is the converse of Lemma 2.8 for rank one operators.

3.7 Lemma. *Let $x \otimes y^*$ be a rank one operator in $\mathcal{T}(\mathcal{N})$. Then there is an element N of $\mathcal{N}$ such that x belongs to N and y belongs to $(N_-)^\perp$.*

Proof. Let N be the least element of $\mathcal{N}$ containing x. Let N' belong to $\mathcal{N}$ with $N' < N$, and set $z = P(N')y$,

$$\begin{aligned} 0 &= P(N')^\perp(x \otimes y^*)P(N')z \\ &= P(N')^\perp x \otimes z^*(z) = \|z\|^2 P(N')^\perp x \quad . \end{aligned}$$

Since x does not belong to N', $P(N')^\perp x \neq 0$. Thus $z = P(N')y = 0$. This is valid for all $N' < N$. Hence $P(N_-)y = 0$ as required. ■

3.8 Proposition. *Let F be a rank n operator in $\mathcal{T}(\mathcal{N})$. Then F is the sum of n rank one operators in $\mathcal{T}(\mathcal{N})$.*

Proof. The proof will proceed by induction on n. The cases $n = 0$ and 1 are immediate. Suppose that F has rank n. For N in $\mathcal{N}$, define $f(N)$ to be the dimension of $N \cap Ran(F)$. Let k be the least non-zero value attained by f. Then $N \cap Ran(F) = X$ is the same k dimensional space for every N in $f^{-1}\{k\}$ since $N \cap Ran(F)$ increases with N. Consequently, the infimum N_0 of $f^{-1}\{k\}$ contains X. So it is the least element of $\mathcal{N}$ with non-zero intersection with the range of F.

Take x to be any unit vector in this intersection, and let $y = F^*x$. For any $N < N_0$ in $\mathcal{N}$, $FP(N)y = P(N)FP(N)y$ belongs to $N \cap Ran\,F = \{0\}$. Hence $P(N)y$ belongs to $ker\,F = (Ran\,F^*)^\perp$. Hence

$$0 = (P(N)y, F^*x) = (P(N)y, y) = \|P(N)y\|^2 \ .$$

Thus y belongs to $\bigcap_{N<N_0} N^\perp = (N_0)_-^\perp$. By Lemma 2.8, $x \otimes y^*$ belongs to $\mathcal{T}(\mathcal{N})$.

The rank of $F - x \otimes y^*$ agrees with the rank of its adjoint $F^* - y \otimes x^*$. Since x belongs to $Ran\,F = (ker\,F^*)^\perp$, this operator annihilates $ker\,F^*$. By design, it also annihilates x. Thus it has rank at most $n-1$. By the induction hypothesis, $F - x \otimes y^*$ is the sum of $n-1$ rank one operators in $\mathcal{T}(\mathcal{N})$, completing the proof. ■

The next lemma sets the stage to prove that $\mathcal{T}(\mathcal{N}) \cap K$ contains a bounded approximate identity for the compact operators. This will yield a number of nice corollaries.

3.9 Lemma. *Let $\mathcal{N}$ be a continuous nest. For any unit vector x in $\mathcal{H}$, let P be the orthogonal projection onto $\overline{\mathcal{N}''x}$. Then P belongs to $\mathcal{N}'$ and there are finite rank operators $R_n = PR_nP$ in $\mathcal{T}(\mathcal{N})$ converging to P in the strong* topology such that $\|R_n\| \leq 1$ for all n.*

Proof. It is clear that the range of P is invariant for $\mathcal{N}''$ and hence P belongs to $\mathcal{N}'$. The map $\phi(N) = \|P(N)x\|^2$ is a continuous non-decreasing function of $\mathcal{N}$ onto $[0,1]$ by Theorem 2.13. For $n \geq 1$ and $0 \leq k \leq 2^n$, let $N_{k,n}$ be the least element of $\mathcal{N}$ such that $\phi(N_{k,n}) = k2^{-n}$. Let $E_{k,n} = N_{k,n} \ominus N_{k-1,n}$ be intervals of $\mathcal{N}$. Then

$$x_{k,n} = 2^{n/2}P(E_{k,n})x$$

is a unit vector in $E_{k,n} \cap P\mathcal{H}$. Set

$$R_n = \sum_{k=2}^{2^n} x_{k-1,n} \otimes x^*_{k,n} \quad .$$

By Lemma 2.8, R_n belongs to $\mathcal{T}(\mathcal{N})$. It is easy to check that $R_n = PR_nP$ and $\|R_n\| = 1$.

Notice that for $m > n$

$$\begin{aligned} R_m x_{k,n} &= R_m(2^{(n-m)/2} \sum_{j=2^{m-n}(k-1)+1}^{2^{m-n}k} x_{j,m}) \\ &= x_{k,n} + 2^{(n-m)/2}(x_{2^{(m-n)}(k-1),m} - x_{2^{(m-n)}k,m}) \quad . \end{aligned}$$

Hence

$$\lim_{m \to \infty} R_m y = y$$

whenever y belongs to $\bigcup_{n \geq 1} span\,\{x_{k,n} : 1 \leq k \leq 2^n\}$. But this is dense in $P\mathcal{H}$ and $\|R_n\| \leq 1$. Thus R_m converges to P in the strong operator topology.

Finally, if y belongs to $P\mathcal{H}$,

$$\|y\|^2 = \lim(y, R_n y) = \lim(R_n^* y, y) \quad .$$

Since $\|R_n^* y\| \leq 1$, it follows that $\lim R_n^* y = y$. So R_n^* converges to P in the strong operator topology as well. ∎

3.10 Theorem. *Let $\mathcal{N}$ be a nest. Then there is a net F_α of finite rank contractions in $\mathcal{T}(\mathcal{N})$ such that F_α converges to the identity in the strong* topology. If $\mathcal{H}$ is a separable space, the net may be taken to be a sequence F_n.*

Proof. Let P_a be the projection into the span of all the atoms of $\mathcal{N}$. It is trivial to construct an increasing net (or sequence in the separable case) of finite rank projections R_α in $\mathcal{N}'$ converging strongly to P_a. The algebra $P_a^\perp \mathcal{T}(\mathcal{N}) | P_a^\perp \mathcal{H}$ is a nest algebra $\mathcal{T}(\mathcal{M})$ where $\mathcal{M}$ is a continuous nest. It suffices now to prove the theorem for a continuous nest.

If x is a unit vector, let P_x be the orthogonal projection onto $\overline{\mathcal{N}''x}$. Let $\{P_{x_\alpha}\}$ be a maximal family of such projections which are pairwise orthogonal (Zorn's lemma). Then $\sum P_{x_\alpha} = I$ for otherwise any vector x orthogonal to each P_{x_α} will extend this set. If $\mathcal{H}$ is separable, then this is a countable family. For each x_α, one has by Lemma 3.9 a sequence $R_{n,\alpha}$ of finite rank contractions converging to P_{x_α} in the strong* topology. Then

$$R_{n,\mathcal{F}} = \sum_{\alpha \in \mathcal{F}} R_{n,\alpha}$$

indexed by $\mathbb{N} \times \Omega$ where Ω is the set of finite subsets of indices is a net of finite rank contractions in $\mathcal{T}(\mathcal{N})$ converging strong* to the identity. In the separable case, $R_{n,n} = \sum_{i=1}^{n} R_{n,\{i\}}$ will suffice. ∎

3.11 Erdos Density Theorem. *The finite rank contractions in* $\mathcal{T}(\mathcal{N})$ *are dense in the unit ball of* $\mathcal{T}(\mathcal{N})$ *in the strong* topology.*

Proof. Let T be a contraction in $\mathcal{T}(\mathcal{N})$, and let R_β be the net constructed in Theorem 3.10. Then TR_β is a finite rank contraction in $\mathcal{T}(\mathcal{N})$, and converges to T in the strong* topology. ∎

3.12 Corollary. *The norm closure of the finite rank contractions in* $\mathcal{T}(\mathcal{N})$ *is the set of compact contractions in* $\mathcal{T}(\mathcal{N})$.

Proof. By Proposition 1.18, KR_β converges to K in norm if K is compact. ∎

3.13 Corollary. *The span of the rank one operators in* $\mathcal{T}(\mathcal{N})$ *is weak* dense in* $\mathcal{T}(\mathcal{N})$.

Proof. Immediate from Proposition 3.8 and Theorem 3.11. ∎

We are now in a position to give a straightforward proof of Lidskii's Theorem about the trace of a trace class operator.

3.14 Corollary. *Let K be a trace class operator in* $\mathcal{T}(\mathcal{N})$. *Then K is the limit in the trace norm of finite rank operators in* $\mathcal{T}(\mathcal{N})$.

Proof. By Proposition 1.18, KR_β converges to K in the trace norm. ∎

3.15 Lidskii's Theorem. *Let K be a trace class operator, and let* $\{\lambda_n, n \geq 1\}$ *be a list of the non zero values in* $\sigma(K)$ *including multiplicity. Then* $tr(K) = \sum_{n \geq 1} \lambda_n$.

Proof. By Corollary 3.2, K is contained in $\mathcal{T}(\mathcal{N})$ for some maximal nest $\mathcal{N}$. By Theorem 3.4, the values $\{\lambda_n\}$ occur as the diagonal entries $\{\Phi_\alpha(K): A_\alpha \in \mathbf{A}\}$ of K with respect to $\mathcal{N}$, including multiplicity. Define a linear map on the trace class operators in $\mathcal{T}(\mathcal{N})$ by

$$\Phi(T) = \sum \Phi_\alpha(T) \ .$$

Let x_α be a unit vector in the atom A_α. Then by Theorem 1.3,

$$|\Phi(T)| = |\sum (Tx_\alpha, x_\alpha)| \leq \sum |(Tx_\alpha, x_\alpha)| \leq \|T\|_1 \ .$$

So Φ is well defined and has norm at most one.

If T is finite rank, then by standard matrix arguments $tr\,T$ equals the sum of its eigenvalues. So $tr\,T = \Phi(T)$. By Corollary 3.14, K is the limit in the trace norm of a sequence T_n of finite rank operators in $\mathcal{T}(\mathcal{N})$. Thus

$$tr\,K = \lim_{n\to\infty} tr\,T_n = \lim_{n\to\infty} \Phi(T_n) = \Phi(K) \ . \qquad \blacksquare$$

Notes and Remarks.

Invariant subspaces for compact operators were obtained independently by von Neumann (unpublished) and Aronszajn and Smith [1]. This was extended to more general operators related to compact operators by Bernstein and Robinson [1], Halmos [1], Arveson-Feldman [1]. Finally, Lomonosov [1] introduced a completely new method to show that if T commutes with a non zero compact operator, then T has a proper invariant subspace. See Radjavi and Rosenthal [2] for a treatment of all these ideas.

Corollary 3.2 and Theorem 3.4 are due to Ringrose [1]. The proof given here is somewhat different than the original, which is done in arbitrary Banach spaces. Theorem 3.8 is in Erdos [3] and is attributed to Ringrose. Theorems 3.10 and 3.11 are due to Erdos. Theorem 3.15 was proved by Lidskii [1] by quite different methods. This proof is due to Erdos [6]. Another proof, due to Power [3], also uses the ideas of nest algebras.

Exercises.

3.1 (Lomonosov, Hilden) Let K be a compact operator, and let $\mathcal{A} = \{K\}'$. Prove that $\mathcal{A}$ has an invariant subspace as follows: Assume that $\mathcal{A}$ has no invariant subspaces.

(i) Show that $\sigma(K) = \{0\}$.

(ii) Let $\|K\| = 1$ and choose x_0 with $\|Kx_0\| = 2$. Let $B = \{x : \|x - x_0\| < 1\}$, and $C = \overline{KB}$. Find open sets $U_1,\dots,U_n$ covering C and operator $A_1,\dots,A_n$ in $\mathcal{A}$ such that $A_j y$ belongs to B for all y in U_j.

(iii) Find a sequence $j_1, j_2, \dots$ from $\{1,\dots,n\}$ such that $A_{j_p} K A_{j_{p-1}} K \dots A_{j_1} K x_0$ belongs to B for every $p \geq 1$. Show that this contradicts (i).

3.2 Let V be the Volterra operator on $L^2(0,1)$ given by

$$V(x) = \int_x^1 f(t)dt \quad .$$

Prove that $\sigma(V) = \{0\}$ by i) computation of $\|V^n\|$, ii) by an application of Ringrose's Theorem, and iii) by explicitly finding $(V-\lambda I)^{-1}$ for $\lambda \neq 0$.

3.3 Prove that every quasinilpotent trace class operator is the limit in C_1 of a sequence of finite rank nilpotent operators.

3.4 Use Lemma 3.5 to give a simple direct proof that the finite rank operators in $\mathcal{T}(\mathcal{N})$ are dense in $\mathcal{T}(\mathcal{N}) \cap \mathcal{K}$.

3.5 (Weyl) Let T be an operator in C_p, $1 < p < \infty$. Let $\{\lambda_n\}$ be the eigenvalues of T including multiplicity. Prove that

$$\left(\sum_{n=1}^{\infty} |\lambda_n|^p\right)^{1/p} \leq \|T\|_p \quad .$$

(Hint: Use Exercise 1.4) Show that equality holds if and only if T is normal.

3.6 (Longstaff) Let $\mathcal{L}$ be a lattice of subspaces. For L in $\mathcal{L}$, define L_- to be the span of $\{M \text{ in } \mathcal{L} : M \not\geq L\}$. Show that the rank one operator $x \otimes y^*$ belongs to $Alg\,\mathcal{L}$ if and only if there is an element L in $\mathcal{L}$ such that x belongs to L and y belongs to $(L_-)^{\perp}$.

3.7 (Lance)

(i) Suppose x and y are vectors in $\mathcal{H}$ and T belongs to $\mathcal{T}(\mathcal{N})$ such that $Tx = y$. Show that $\sup_{N\in\mathcal{N}} \dfrac{\|P(N)^{\perp}y\|}{\|P(N)^{\perp}x\|} \leq \|T\| < \infty$.

(ii) Let $\mathcal{F}$ be a finite nest $0 = F_0 < F_1 < \ldots < F_n = \mathcal{H}$ with interval projection $E_i = P(F_i) - P(F_{i-1})$, $1 \leq i \leq n$. Let x and y be vectors and let $x_j = E_j x$ and $y_j = E_j y$. Suppose that $\sum_{j=k}^{n} \|y_j\|^2 \leq \sum_{j=k}^{n} \|x_j\|^2$ for $1 \leq k \leq n$. Show that there is a finite rank operator T in $\mathcal{T}(\mathcal{F})$ with $\|T\| \leq 1$ and $Tx = y$.

Hint: Use induction on n. Suppose $T_1 = E_1^{\perp}T_1$ in $\mathcal{T}(\mathcal{F})$ has $\|T_1\| \leq 1$ and $T_1 x = E_1^{\perp} y$. Show that $\|(I - T_1^*T_1)^{1/2}x\| \geq \|y_1\|$. Hence form $T = T_1 + E_1 U(1 - T_1^*T)^{1/2}$ where U is a certain rank one partial isometry.

(iii) Let $\mathcal{N}$ be an arbitrary nest. Suppose x and y are vectors such that $\sup_{N\in\mathcal{N}} \dfrac{\|P(N)^{\perp}y\|}{\|P(N)^{\perp}x\|} = M < \infty$. Show that there is an operator T in $\mathcal{T}(\mathcal{N})$ such that $Tx = y$.

4. Triangular Truncation

Let $e_1,\ldots,e_n$ be the standard basis for $\mathbb{C}^n$. The map $\mathcal{T}_n$ which takes a matrix A to its upper triangular part is a bounded linear map. However, as we shall see, $\|\mathcal{T}_n\|$ grows like $\log n$. As a result, triangular truncation is not bounded in infinite dimensions. In this section, we will explore this carefully.

4.1 Example. Consider the $n \times n$ matrix

$$A_n = \begin{bmatrix} 0 & 1 & \frac{1}{2} & \frac{1}{3} & & & \frac{1}{n-1} \\ -1 & 0 & 1 & \frac{1}{2} & & & \frac{1}{n-2} \\ -\frac{1}{2} & -1 & 0 & 1 & & & \\ -\frac{1}{3} & -\frac{1}{2} & -1 & 0 & & & \\ & & & & 0 & 1 & \frac{1}{2} \\ & & & & -1 & 0 & 1 \\ -\frac{1}{n-1} & -\frac{1}{n-2} & & & -\frac{1}{2} & -1 & 0 \end{bmatrix} = [a_{ij}]$$

where $a_{ij} = \dfrac{1}{i-j}$ if $i \neq j$ and $a_{ii} = 0$. Let $T_n = \mathcal{T}_n(A_n)$ be the upper triangular part. It will be shown that $\|A_n\| \leq \pi$ while $\|T_n\| \geq \dfrac{4}{5}\log n$. Let $x = (n-1)^{-1/2}(\sum_{k=2}^{n} e_k)$. Then

$$\|T_n\| \geq \|T_n x\| = (n-1)^{-1/2}\|\sum_{k=1}^{n} b_k e_k\|$$

where $b_k = \sum_{j=1}^{n-k}\dfrac{1}{j} \geq \log(n+1-k) \geq \log(n/3)$ if $k \leq 2n/3$. Hence

$$\|T_n\| \geq n^{-1/2}(\frac{2n}{3})^{1/2}\log(n/3) = (2/3)^{1/2}\log(n/3)$$

whence

$$\lim_{n\to\infty} \frac{\|T_n\|}{\log n} \geq \left(\frac{2}{3}\right)^{1/2} \geq .8 \ .$$

On the other hand, consider the doubly infinite matrix $M = (m_{ij})_{i,j\in\mathbb{Z}}$ where $m_{ij} = (i-j)^{-1}$ if $i \neq j$ and $m_{ii} = 0$ as an operator on $\ell^2(\mathbb{Z})$. Let P_n be the projection of $\ell^2(\mathbb{Z})$ onto the $span\{e_i, 1 \leq i \leq n\}$. Clearly, A_n is unitarily equivalent to $P_n M | P_n \mathcal{H}$ and thus $\|A_n\| \leq \|M\|$ for all n. Next, notice that M is skew adjoint. So $\|M\|^2 = \|M^2\|$. The operator M^2 has a matrix (b_{ij}) where

$$b_{ij} = \sum_{k=-\infty}^{\infty} m_{ik}m_{kj} = \sum_{k\neq i,j} \frac{1}{(i-k)(k-j)} = -\sum_{n\neq 0,j-i} \frac{1}{n(n-j+i)} \ .$$

Thus $b_{ii} = -\sum_{k\neq 0} \frac{1}{k^2} = -2\sum_{k=1}^{\infty} \frac{1}{k^2} = -\pi^2/3$. And, if $j-i = r \neq 0$

$$\begin{aligned} b_{ij} &= -\sum_{n>r} \frac{1}{n(n-r)} + \sum_{n=1}^{r-1} \frac{1}{n(r-n)} - \sum_{n<0} \frac{1}{n(n-r)} \\ &= -2\sum_{n=r+1} \frac{1}{r}\left(\frac{1}{n-r} - \frac{1}{n}\right) + \sum_{n=1}^{r-1} \frac{1}{r}\left(\frac{1}{n} + \frac{1}{n-r}\right) \\ &= -\frac{2}{r}\sum_{k=1}^{r} \frac{1}{k} + \frac{2}{r}\sum_{n=1}^{r-1} \frac{1}{n} = -\frac{2}{r^2} \ . \end{aligned}$$

Hence $M^2 = -\pi^2/3I - \sum_{k\neq 0} 2/k^2 U_k$ where U_k is the unitary operator shifting e_i to e_{i+k}. So

$$\|M^2\| \leq \frac{\pi^2}{3} + \sum_{k\neq 0} \frac{2}{k^2} = 6\sum_{n=1}^{\infty} \frac{1}{n^2} = \pi^2 \ .$$

Hence $\|A_n\| \leq \|M\| \leq \pi$.

4.2. Let $\mathcal{F}$ be a finite nest $\{0\} = N_0 < N_1 < \ldots < N_n = \mathcal{H}$. Set $P_i = P(N_i)$ and $\Delta P_i = P_i - P_{i-1}$ for $1 \leq i \leq n$. For each T in $\mathcal{B}(\mathcal{H})$, define

$$\mathcal{U}_{\mathcal{F}}(T) = \sum_{i=1}^{n} P_i T \Delta P_i$$

$$\mathcal{L}_{\mathcal{F}}(T) = \sum_{i=1}^{n} P_{i-1} T \Delta P_i$$

$$\mathcal{D}_{\mathcal{F}}(T) = \sum_{i=1}^{n} \Delta P_i T \Delta P_i \ .$$

These terms are respectively the upper triangular, strictly upper triangular, and diagonal parts of T with respect to $\mathcal{F}$. The first two terms can be thought of as upper and lower sums in a Riemann sum.

Let $\mathcal{N}$ be a nest, and consider the finite subnests $\mathcal{F}$ of $\mathcal{N}$ as a net ordered by inclusion. An operator T has a *triangular truncation* $\mathcal{U}_{\mathcal{N}}(T)$ if $\lim_{\mathcal{F}} \mathcal{U}_{\mathcal{F}}(T)$ exists in norm. Likewise, define *strictly upper triangular truncation* $\mathcal{L}_{\mathcal{N}}(T)$ and *diagonal* $\mathcal{D}_{\mathcal{N}}(T)$. Similarly define weak versions by replacing norm limits by weak operator limits.

First let us briefly study the diagonal map. Recall from section 2.17 that $\Delta(A)$ is the projection of A onto the atomic part of the diagonal.

4.3 Proposition. *If $T = D+K$ where D belongs to $\mathcal{N}'$ and K is compact, then $\mathcal{D}_{\mathcal{F}}(T)$ converges in norm to $D+\Delta(K)$.*

Proof. Since $\mathcal{D}_{\mathcal{F}}(D) = D$ for every $\mathcal{F}$, it suffices to consider the compact operator K. By Lemma 3.5, given $\epsilon > 0$ there is a finite nest $\mathcal{F}_0$ so that $\|E_iKE_i\| < \epsilon$ for every atom of $\mathcal{F}_0$ which is not an atom of $\mathcal{N}$. It follows that $\|\mathcal{D}_{\mathcal{F}}(K)-\Delta(K)\| < 2\epsilon$ for every $\mathcal{F}$ containing $\mathcal{F}_0$. Thus the proposition is proven. ∎

4.4 Proposition. *If $\mathcal{N}$ is atomic, then $\mathcal{D}_{\mathcal{F}}(T)$ converges in the strong operator topology to $\Delta(T)$ for every T in $\mathcal{B}(\mathcal{H})$.*

Proof. Let $\{e_i\}$ be an orthonormal basis for $\mathcal{H}$ such that for each i, there is an atom A_i containing e_i. Consider a vector $x = \sum_{i=1}^{N} a_i e_i$. If $\mathcal{F}$ is any finite subnest of $\mathcal{N}$ so that each A_i, $1 \le i \le N$, is an atom of $\mathcal{F}$,

$$\mathcal{D}_{\mathcal{F}}(T)x = \sum P(A_i)TP(A_i)(a_ie_i) = \Delta(T)x \quad .$$

Since $\mathcal{D}_{\mathcal{F}}(T)$ is a bounded net (by $\|T\|$) and the algebraic linear span of $\{e_i\}$ is dense in $\mathcal{H}$, $\mathcal{D}_{\mathcal{F}}(T)y$ converges to $\Delta(T)y$ for every y in $\mathcal{H}$. ∎

When $\mathcal{N}$ is not atomic, the net $\mathcal{D}_{\mathcal{F}}(T)$ does not converge weakly for every operator. A more complete analysis will be given in Theorem 8.6 and Theorem 8.11.

Now consider the triangular truncations. The first result is very easy, and is left as an exercise.

4.5 Proposition. *Let $\mathcal{N}$ be a nest and let T be an element of $\mathcal{T}(\mathcal{N})$. Then $\mathcal{U}_{\mathcal{N}}(T) = T = \mathcal{U}_{\mathcal{F}}(T)$ for every finite subnest $\mathcal{F}$ of $\mathcal{N}$. The strictly triangular integral $\mathcal{L}_{\mathcal{N}}(T)$ exists if and only if $\mathcal{D}_{\mathcal{N}}(T)$ exists, and in this case $\mathcal{L}_{\mathcal{N}}(T) = T-\mathcal{D}_{\mathcal{N}}(T)$. Thus, if T is compact, $\mathcal{L}_{\mathcal{N}}(T) = T-\Delta(T)$.*

4.6 Corollary. *If $\mathcal{N}$ is a nest and T is an element of $\mathcal{T}(\mathcal{N})^*$, then $\mathcal{L}_{\mathcal{N}}(T) = \mathcal{L}_{\mathcal{F}}(T) = 0$ for every finite subnest $\mathcal{F}$ of $\mathcal{N}$. The net $\mathcal{U}_{\mathcal{F}}(T) = \mathcal{D}_{\mathcal{F}}(T)$, so $\mathcal{U}_{\mathcal{N}}(T)$ exists exactly when $\mathcal{D}_{\mathcal{N}}(T)$ exists. In particular, if T is compact, $\mathcal{U}_{\mathcal{N}}(T) = \mathcal{D}_{\mathcal{N}}(T) = \Delta(T)$.*

Proof. $\mathcal{T}(\mathcal{N})^* = \mathcal{T}(\mathcal{N}^\perp)$ and $\mathcal{L}_\mathcal{N}(T) = T - \mathcal{U}_{\mathcal{N}^\perp}(T) = 0$. Also,

$$\mathcal{U}_\mathcal{F}(T) = T - \mathcal{L}_{\mathcal{F}^\perp}(T) = \mathcal{D}_{\mathcal{F}^\perp}(T) = \mathcal{D}_\mathcal{F}(T)$$

for every finite subnest of $\mathcal{N}$. Apply Proposition 4.5. ■

4.7. If $\mathcal{U}_\mathcal{N}(T) = \mathcal{L}_\mathcal{N}(T)$, one says that the (upper) *triangular integral* $\int_\mathcal{N} P(N)TdN$ exists. The terms $\mathcal{U}_\mathcal{F}(T)$ and $\mathcal{L}_\mathcal{F}(T)$ are the upper and lower Riemann sums of this integral.

The two preceding rather trivial results have the following more remarkable consequence.

4.8 Theorem. *Let $\mathcal{N}$ be a nest. Suppose that T is a compact operator in $\mathcal{T}(\mathcal{N})$ such that $\Delta(T) = 0$. Then*

$$T = 2\int_\mathcal{N} P(N) ReT dN = 2i \int_\mathcal{N} P(N) ImT dN \quad .$$

Proof. By Proposition 4.5 and 4.6,

$$T = \mathcal{U}_\mathcal{N}(T) = \mathcal{L}_\mathcal{N}(T) + \Delta(T) = \mathcal{L}_\mathcal{N}(T)$$

and

$$0 = \mathcal{L}_\mathcal{N}(T^*) = \mathcal{U}_\mathcal{N}(T^*) - \Delta(T^*) = \mathcal{U}_\mathcal{N}(T^*) \quad .$$

Hence

$$T = \mathcal{U}_\mathcal{N}(T \pm T^*) = \mathcal{L}_\mathcal{N}(T \pm T^*) \quad .$$

So the integrals of both $2ReT = T + T^*$ and $2iImT = T - T^*$ exist and equal T. ■

4.9. It is an easy consequence of Example 4.1 that there exist compact self-adjoint operators whose triangular truncation with respect to some nest of order type $\omega = \mathbb{N} \cup \{\infty\}$ is unbounded. So Theorem 4.8 is not applicable to all compact operators in $\mathcal{B}(\mathcal{H})$. To get a handle on what is going on, it is necessary to consider other ideals of compact operators closed in stronger norms, such as the trace class operators and Hilbert-Schmidt operators. We consider the Hilbert-Schmidt operators first as they are quite easily handled.

4.10 Theorem. *Let $\mathcal{N}$ be any nest. Then the triangular truncation operators $\mathcal{U}_\mathcal{N}$, $\mathcal{L}_\mathcal{N}$ and $\mathcal{D}_\mathcal{N}$ are contractive projections of the Hilbert-Schmidt class C_2 into itself. Thus they are orthogonal projections.*

Proof. Let $\mathcal{F}$ be any finite subnest of $\mathcal{N}$, and let $E_1, \ldots, E_n$ be the orthogonal projections onto the intervals of $\mathcal{F}$. For every operator T in C_2,

$$T = \sum_{i=1}^{n}\sum_{j=1}^{n} E_i T E_j \ .$$

Thus

$$\|T\|_2^2 = tr(TT^*) = tr(\sum_{i,j} E_i T E_j \sum_{k,\ell} E_k T^* E_\ell)$$

$$= tr(\sum_i \sum_j \sum_\ell E_i T E_j T^* E_\ell)$$

$$= tr(\sum_i \sum_j (E_i T E_j)(E_i T E_j)^*) = \sum_{i,j} \|E_i T E_j\|_2^2 \ .$$

Since $\mathcal{U}_{\mathcal{F}}(T) = \sum_{i \le j} E_i T E_j$ and $\mathcal{L}_{\mathcal{F}}(T) = \sum_{i<j} E_i T E_j$, it follows that

$$\|\mathcal{L}_{\mathcal{F}}(T)\|_2^2 \le \|\mathcal{U}_{\mathcal{F}}(T)\|_2^2 \le \|T\|_2^2 \ .$$

Also, if $\mathcal{F}_1 \subset \mathcal{F}_2$ are two finite subnests, then another consideration of the terms $E_i T E_j$ yields

$$\|\mathcal{L}_{\mathcal{F}_1}(T)\|_2^2 \le \|\mathcal{L}_{\mathcal{F}_1}(T)\|_2^2 \le \|\mathcal{U}_{\mathcal{F}_2}(T)\|_2^2 \le \|\mathcal{U}_{\mathcal{F}_1}(T)\|_2^2 \ .$$

and

$$\|\mathcal{L}_{\mathcal{F}_2}(T) - \mathcal{L}_{\mathcal{F}_1}(T)\|_2^2 = \|\mathcal{L}_{\mathcal{F}_2}(T)\|_2^2 - \|\mathcal{L}_{\mathcal{F}_1}(T)\|_2^2$$

and

$$\|\mathcal{U}_{\mathcal{F}_1}(T) - \mathcal{U}_{\mathcal{F}_2}(T)\|_2^2 = \|\mathcal{U}_{\mathcal{F}_1}(T)\|_2^2 - \|\mathcal{U}_{\mathcal{F}_2}(T)\|_2^2 \ .$$

Then $\|\mathcal{L}_{\mathcal{F}}(T)\|_2$ is monotone increasing and bounded above; $\|\mathcal{U}_{\mathcal{F}}(T)\|_2^2$ is monotone decreasing and bounded below. Let $M^2 = \inf_{\mathcal{F}} \|\mathcal{U}_{\mathcal{F}}(T)\|_2^2$. For each $\epsilon > 0$, choose $\mathcal{F}_\epsilon$ so that $\|\mathcal{U}_{\mathcal{F}_\epsilon}(T)\|_2^2 \le M^2 + \epsilon^2$. Then for any finite subnests $\mathcal{F}_1$ containing $\mathcal{F}_\epsilon$,

$$\|\mathcal{U}_{\mathcal{F}_1}(T) - \mathcal{U}_{\mathcal{F}_\epsilon}(T)\|_2^2 \le M^2 + \epsilon^2 - M^2 = \epsilon^2 \ .$$

Hence $\mathcal{U}_{\mathcal{F}}(T)$ converges in the C_2 norm. Likewise, $\mathcal{L}_{\mathcal{F}}(T)$ converges. Thus $\mathcal{U}_{\mathcal{N}}(T)$ and $\mathcal{L}_{\mathcal{N}}(T)$ are well defined, contractive projections.

The space C_2 is a Hilbert space, and every contractive projection on a Hilbert space is self-adjoint. Finally, $\mathcal{D}_{\mathcal{N}} = \mathcal{U}_{\mathcal{N}} - \mathcal{L}_{\mathcal{N}}$ is also a contractive projection. By Proposition 4.3, $\mathcal{D}_{\mathcal{N}}(T) = \Delta(T)$ for all T in C_2. ∎

4.11. In the case of trace class operators, the extreme points of the unit ball are the rank one operators. So it suffices to look at them. Consider, in particular, the rank one operator $1 \otimes 1^*$ in $L^2(0,1)$ and the continuous nest $\mathcal{N} = \{N_t : 0 \le t \le 1\}$ of subspaces of functions supported on $[0,t]$. By Theorem 4.8 and the results of the next section, $\mathcal{U}_{\mathcal{N}}(1 \otimes 1^*) = V$ is the Volterra operator. By Theorem 5.8, the singular values of V are

$\{2/(2n-1)\pi, n \geq 1\}$. Thus V is not trace class, although it belongs to C_p for all $p > 1$. Thus triangular truncation is not bounded in C_1. By the easy duality result below, this yields another proof that truncation is not bounded on the compact operators either. However, this example tells us exactly to which ideal the truncations of trace class operators naturally belong.

Let C_Ω be the normed ideal of compact operators A such that

$$\|A\|_\Omega = \sup_n \left(\sum_{j=1}^{n} 1/j\right)^{-1} \sum_{j=1}^{n} s_j(A)$$

is finite. And define C_ω to be the ideal of all compact operators A such that

$$\|A\|_\omega = \sum_{n=1}^{\infty} \frac{1}{n} s_n(A)$$

is finite. The ideal C_ω is known as the *Macaev ideal*.

The ideal C_Ω is strictly larger than C_1 since it contains all operators whose s-numbers are $O(\frac{1}{n})$. On the other hand, if A belongs to C_Ω,

$$s_n(A) \leq \frac{1}{n} \sum_{j=1}^{n} s_j(A) \leq \frac{1}{n} \|A\|_\Omega \sum_{j=1}^{n} \frac{1}{j} \ .$$

Thus the sequence $s_n(A)$ is $O(\frac{\log n}{n})$. This implies that $s_n(A)$ belongs to ℓ^p for $p > 1$. So C_Ω is contained in the intersection of all C_p classes for $p > 1$. Similarly, note that $(\frac{1}{n}, n \geq 1)$ belong to ℓ^q for all $q > 1$. So if A belongs to C_p, then with $\frac{1}{p} + \frac{1}{q} = 1$,

$$\|A\|_\omega = \sum_{n=1}^{\infty} \frac{1}{n} s_n(A) \leq \|(\frac{1}{n})\|_q \|A\|_p < \infty \ .$$

Thus C_ω contains all the C_p classes for $p < \infty$.

4.12 Lemma. *The map $A \to \phi_A$ given by $\phi_A(X) = tr(AX)$ for X in C_ω is a linear isometry of C_Ω onto $C_\omega{}'$.*

Proof. Let ϕ be a functional on C_ω. Proceed as in Theorem 1.13 by considering the sequilinear form $<x,y> = \phi(x \otimes y^*)$. This satisfies

$$|<x,y>| \leq \|\phi\| \, \|x \otimes y^*\|_\omega = \|\phi\| \, \|x\| \, \|y\| \ .$$

Thus there is a bounded linear operator A with $\|A\| \leq \|\phi\|$ such that

$$\phi(x \otimes y^*) = (Ax,y) = tr(Ax \otimes y^*) \ .$$

Let $\{e_k, 1 \leq k \leq n\}$ and $\{f_k, 1 \leq k \leq n\}$ be two orthonormal sets. Then

$$|\sum_{k=1}^{n}(Af_k,e_k)| = |\phi(\sum_{k=1}^{n} f_k \otimes e_k^*)|$$

$$\leq \|\phi\| \, \|\sum_{k=1}^{n} f_k \otimes e_k^*\|_\omega = \|\phi\|\sum_{k=1}^{n}\frac{1}{k} \; .$$

It follows as in the proof of Theorem 1.13 that A is compact. Let $\{e_k\}$ and $\{f_k\}$ be the orthonormal families such that $A = \sum_{k=1}^{\infty} s_k(e_k \otimes f_k^*)$. Putting this into the inequality above yields

$$\sum_{k=1}^{n} s_k(A) \leq \|\phi\|\sum_{k=1}^{n}\frac{1}{k} \; .$$

Thus $\|A\|_\Omega \leq \|\phi\|$. The functionals ϕ and ϕ_A agree on all finite rank operators. These are easily seen to be dense in C_ω, so $\phi = \phi_A$.

Conversely, let A belong to C_Ω. Let λ_j be positive real numbers such that $\lambda_1 \geq \lambda_2 \geq ... \geq \lambda_n$. Let $\{e_j, 1 \leq j \leq n\}$ and $\{f_j, 1 \leq j \leq n\}$ be orthonormal sets, and let $F = \sum_{j=1}^{n} \lambda_j e_j \otimes f_j^*$. Every rank n operator with s-numbers $\lambda_1,...,\lambda_n$ has this form. Set $\lambda_{n+1} = 0$, and compute using Lemma 1.7 and summation by parts

$$|\phi_A(F)| = |tr(A\sum_{j=1}^{n}\lambda_j e_j \otimes f_j^*)|$$

$$\leq \sum_{j=1}^{n}\lambda_j \, |(Ae_j,f_j)| = \sum_{j=1}^{n}(\lambda_j - \lambda_{j+1})\sum_{k=1}^{j} |(Ae_k,f_k)|$$

$$\leq \sum_{j=1}^{n}(\lambda_j - \lambda_{j+1})\sum_{k=1}^{j} s_k(A) \leq \|A\|_\Omega \sum_{j=1}^{n}(\lambda_j - \lambda_{j+1})\sum_{k=1}^{j}\frac{1}{k}$$

$$= \|A\|_\Omega \sum_{j=1}^{n}\frac{1}{j}\lambda_j = \|A\|_\Omega \|F\|_\omega \; .$$

Since finite rank operators are dense in C_ω, one obtains $\|\phi_A\| \leq \|A\|_\Omega$. Thus equality holds and the theorem is proved. ∎

4.13 Lemma *Let $x \otimes y^*$ be a norm one, rank one operator. For every nest $\mathcal{N}$, $s_k(\mathcal{U}_\mathcal{N}(x \otimes y^*)) \leq \frac{1}{k}$ for all $k \geq 1$.*

Proof. Since $x \otimes y^*$ is in C_2, $\mathcal{U}_\mathcal{N}(x \otimes y^*)$ belongs to C_2 by Theorem 4.10. Consider $k = 1$. Then

$$s_1(\mathcal{U}_\mathcal{N}(x \otimes y^*)) = \|\mathcal{U}_\mathcal{N}(x \otimes y^*)\|$$

$$\leq \|\mathcal{U}_\mathcal{N}(x \otimes y^*)\|_2 \leq \|x \otimes y^*\|_2 = 1 \; .$$

For $k \geq 2$, let $P_0 = 0$, $P_k = I$, and set P_j to be the projection onto the least element N_j of $\mathcal{N}$ such that

$$\|P_jx \otimes y^*P_j\| = \|P_jx\| \, \|P_jy\| \geq \frac{j}{k}$$

for $1 \leq j \leq k-1$. Let $Q_j = P(N_{j-})$. By the minimality of P_j, one has

$$\|(Q_j-P_{j-1})x\| \, \|(Q_j-P_{j-1})y\| \leq \frac{1}{k}$$

for $1 \leq j \leq k-1$. And by the Cauchy-Schwartz inequality,

$$\begin{aligned} 1 = \|x\| \, \|y\| &\geq \|P_{k-1}x\| \, \|P_{k-1}y\| + \|P_{k-1}^{\perp}x\| \, \|P_{k-1}^{\perp}y\| \\ &\geq 1-\frac{1}{k} + \|P_{k-1}^{\perp}x\| \, \|P_{k-1}^{\perp}y\| \ . \end{aligned}$$

Thus $\|P_{k-1}^{\perp}x\| \, \|P_{k-1}^{\perp}y\| \leq \frac{1}{k}$. Write $E_j = Q_j-P_{j-1}$ for $1 \leq j \leq k-1$, and $E_k = P_{k-1}^{\perp}$.

By Lemma 1.4, if K has rank at most $k-1$, then

$$s_k(\mathcal{U}_{\mathcal{N}}(x \otimes y^*)) \leq \|\mathcal{U}_{\mathcal{N}}(x \otimes y^*)-K\| \ .$$

Define $K = \sum_{j=1}^{k-1}(P_j-P_{j-1})x \otimes Q_j^{\perp}y^*$. This has rank at most $k-1$. Let $\mathcal{F}$ be the finite subnest $\{0,N_{1-},N_1,\ldots,N_{k-1-},N_{k-1},\mathcal{H}\}$. One can see that

$$\begin{aligned} \mathcal{U}_{\mathcal{F}}(x \otimes y^*) &= \sum_{j=1}^{n} E_jx \otimes P_{j-1}^{\perp}y^* + (P_j-Q_j)x \otimes Q_j^{\perp}y^* \\ &= K + \sum_{j=1}^{n} E_jx \otimes E_jy^* \ . \end{aligned}$$

Therefore,

$$\mathcal{U}_{\mathcal{N}}(x \otimes y^*)-K = \mathcal{U}_{\mathcal{N}}(\mathcal{U}_{\mathcal{F}}(x \otimes y^*)-K) = \sum_{j=1}^{n} \mathcal{U}_{\mathcal{N}}(E_jx \otimes E_jy^*) \ .$$

This is a sum of orthogonal terms. So by the $k = 1$ case,

$$\begin{aligned} s_k(\mathcal{U}_{\mathcal{N}}(x \otimes y^*)) &\leq \max_{1 \leq j \leq k} \|\mathcal{U}_{\mathcal{N}}(E_jx \otimes E_jy^*)\| \\ &\leq \max_{1 \leq j \leq k} \|E_jx\| \, \|E_jy\| \leq \frac{1}{k} \ . \end{aligned}$$

■

4.14 Theorem. *For every nest $\mathcal{N}$, the operator $\mathcal{U}_{\mathcal{N}}$ carries C_1 into C_Ω and $\|\mathcal{U}_{\mathcal{N}}(T)\|_\Omega \le \|T\|_1$.*

Proof. By Lemma 4.13, this holds for rank one operators. In general, $T = \sum_{k\ge1} s_k e_k \otimes f_k^*$ where $s_k = s_k(T)$ and $\{e_k\}$, $\{f_k\}$ are orthonormal sequences. Thus

$$\|\mathcal{U}_{\mathcal{N}}(T)\|_\Omega \le \sum_{k\ge1} s_k \|\mathcal{U}_{\mathcal{N}}(e_k \otimes f_k^*)\|_\Omega \le \sum_{k\ge1} s_k = \|T\|_1 \ . \qquad \blacksquare$$

Now, we employ duality methods to show that $\mathcal{U}_{\mathcal{N}}$ is bounded from the Macaev ideal C_ω into $\mathcal{K}$.

4.15 Lemma. *Let A belong to $\mathcal{B}(\mathcal{H})$, and let T be trace class. For every finite nest $\mathcal{F}$, $tr(\mathcal{U}_{\mathcal{F}}(A)T^*) = tr(A\mathcal{U}_{\mathcal{F}}(T)^*)$.*

Proof. Let $\mathcal{F}$ be given by projections $0 = P_0 < P_1 < \dots < P_n = I$.

$$\begin{aligned} tr(\mathcal{U}_{\mathcal{F}}(A)T^*) &= \sum_{j=1}^{n} tr(P_i A(P_i - P_{i-1})T^*) \\ &= \sum_{j=1}^{n} tr(A(P_i T(P_i - P_{i-1}))^*) = tr(A\mathcal{U}_{\mathcal{F}}(T)^*) \ . \qquad \blacksquare \end{aligned}$$

4.16 Theorem. *Let $\mathcal{N}$ be a nest. For every A in C_ω, $\mathcal{U}_{\mathcal{N}}(A) = \lim_{\mathcal{F}} \mathcal{U}_{\mathcal{F}}(A)$ exists, and $\|\mathcal{U}_{\mathcal{N}}(A)\| \le \|A\|_\omega$. Thus $\mathcal{U}_{\mathcal{N}}$ is a contractive map from C_ω into $\mathcal{K}$.*

Proof. Let $\mathcal{F}$ be a finite subnest of $\mathcal{N}$. Then $\mathcal{U}_{\mathcal{F}}(A)$ is compact, and there are unit vectors x and y such that

$$\begin{aligned} \|\mathcal{U}_{\mathcal{F}}(A)\| &= (\mathcal{U}_{\mathcal{F}}(A)x, y) = tr(\mathcal{U}_{\mathcal{F}}(A)(y \otimes x^*)^*) \\ &= tr(A(\mathcal{U}_{\mathcal{F}}(y \otimes x^*))^*) \\ &\le \|A\|_\omega \|\mathcal{U}_{\mathcal{F}}(y \otimes x^*)\|_\Omega \le \|A\|_\omega \end{aligned}$$

by Theorem 4.14 and Lemma 4.15.

Let $\epsilon > 0$ be given. Since the finite rank operators are dense in C_ω, choose a finite rank operator X so that $\|A - X\|_\omega < \epsilon$. By Theorem 4.10, there is a finite subnest $\mathcal{F}_\epsilon$ of $\mathcal{N}$ so that if $\mathcal{F}_1$ and $\mathcal{F}_2$ are finite subnests containing $\mathcal{F}_\epsilon$, then $\|\mathcal{U}_{\mathcal{F}_1}(X) - \mathcal{U}_{\mathcal{F}_2}(X)\|_2 < \epsilon$. Thus

$$\|\mathcal{U}_{\mathcal{F}_1}(A)-\mathcal{U}_{\mathcal{F}_2}(A)\| \le \|\mathcal{U}_{\mathcal{F}_1}(A-X)\|+\|\mathcal{U}_{\mathcal{F}_1}(X)-\mathcal{U}_{\mathcal{F}_2}(X)\|+\|\mathcal{U}_{\mathcal{F}_2}(X-A)\|$$
$$\le \|A-X\|_\omega+\|\mathcal{U}_{\mathcal{F}_1}(X)-\mathcal{U}_{\mathcal{F}_2}(X)\|_2+\|A-X\|_\omega \le 3\epsilon \ .$$

Hence $\mathcal{U}_{\mathcal{F}}(A)$ is norm convergent. Thus $\mathcal{U}_{\mathcal{N}}(A)$ is compact, and

$$\|\mathcal{U}_{\mathcal{N}}(A)\| = \lim\|\mathcal{U}_{\mathcal{F}}(A)\| \le \|A\|_\omega \ . \qquad ■$$

4.17 Corollary. *Let $\mathcal{N}$ be a nest. For every A in C_ω, $\mathcal{L}_{\mathcal{N}}(A) = \lim_{\mathcal{F}}\mathcal{L}_{\mathcal{F}}(A)$ exists, and $\|\mathcal{L}_{\mathcal{N}}(A)\| \le \|A\|_\omega$. Thus $\int_{\mathcal{N}}P(N)A\,dN$ exists if and only if $\Delta(A) = 0$.*

Proof. By Proposition 4.3, $\mathcal{D}_{\mathcal{N}}(A) = \Delta(A) = \lim_{\mathcal{F}}\mathcal{D}_{\mathcal{F}}(A)$ in norm. Thus $\mathcal{L}_{\mathcal{N}}(A) = \mathcal{U}_{\mathcal{N}}(A)-\mathcal{D}_{\mathcal{N}}(A)$ exists, and is bounded. To obtain the sharp norm estimate, one must modify Lemma 4.13 for $\mathcal{L}_{\mathcal{F}}$. ■

Theorems 4.14 and 4.16 are evidence that C_ω is the proper ideal for triangular truncation. Theorem 4.19 below is the converse, which shows that no larger ideal will do.

4.18 Lemma. *Let D be a diagonal, self-adjoint operator with eigenvalues $\{\lambda_n, n \in \mathbb{Z}\}$ such that $\lambda_n/(2n-1) \ge 0$ for all n and $\sum_{n=-\infty}^{\infty} \frac{\lambda_n}{2n-1} = \infty$. Then there is a continuous nest $\mathcal{N}$ so that the net $\mathcal{U}_{\mathcal{F}}(D)$ does not converge in the weak operator topology.*

Proof. Let the eigenvector for λ_n be e_n. After a unitary equivalence, we may suppose that $\mathcal{H} = L^2(0,1)$ and $e_n = e_n(t) = e^{(2n-1)\pi i t}$. Let $\mathcal{N}$ be the Volterra nest $\{N_t, 0 \le t \le 1\}$ where N_t is the space of functions supported on $[0,t]$. Let $\mathcal{F}_m$ denote the subnest $\{N_{k/m}, 0 \le k \le m\}$. It will be shown that

$$u_m = (\mathcal{U}_{\mathcal{F}_m}(D)1,1)$$

diverges. Let $P_t = P(N_t)$. Then

$$u_m = (\sum_{k=1}^{m} P_{k/m}D(P_{k/m}-P_{k-1/m})1,1)$$
$$= \sum_{k=1}^{m}\sum_{n=-\infty}^{\infty} \lambda_n \int_{k-1/m}^{k/m} \overline{e_n(t)}dt \int_0^{k/m} e_n(t)dt$$
$$= \sum_{k=1}^{m}\sum_{n=-\infty}^{\infty} \frac{\lambda_n}{(2n-1)^2\pi^2}(e_n(-k/m)-e_n((1-k)/m)))(e_n(k/m)-1)$$
$$= \sum_{n=-\infty}^{\infty} \frac{\lambda_n}{(2n-1)^2\pi^2}\sum_{k=1}^{m} 1-e_n(1/m)+e_n((1-k)/m)-e_n(-k/m)$$
$$= \sum_{n=-\infty}^{\infty} \frac{\lambda_n}{(2n-1)^2\pi^2}(m-me_n(1/m)-2)$$

Thus

$$\text{Im } u_m = -\frac{m}{\pi^2} \sum_{n=-\infty}^{\infty} \frac{\lambda_n}{(2n-1)^2} \sin\frac{(2n-1)\pi}{m} .$$

For convenience, take $m = 4p$. Then since $\sin\theta \geq 2\theta/\pi$ for $0 \leq \theta \leq \pi/2$,

$$\frac{4p}{\pi^2} \sum_{n=-p+1}^{p} \frac{\lambda_n}{(2n-1)^2} \sin\frac{(2n-1)\pi}{4p} \geq \frac{4p}{\pi^2} \sum_{n=-p+1}^{p} \frac{\lambda_n}{(2n-1)^2} \frac{(2n-1)}{2p}$$

$$= \frac{2}{\pi^2} \sum_{n=-p+1}^{p} \frac{\lambda_n}{2n-1} .$$

And

$$\left| \frac{4p}{\pi^2} \Big(\sum_{n \leq -p} + \sum_{n \geq p+1} \Big) \frac{\lambda_n}{(2n-1)^2} \sin\frac{(2n-1)\pi}{4p} \right| \leq \frac{4p}{\pi^2} \|D\| 2 \sum_{n=p+1}^{\infty} \frac{1}{(2n-1)^2}$$

$$\leq \frac{8\|D\|}{\pi^2} < \|D\| .$$

Thus

$$\text{Im } u_{4p} \leq \frac{-2}{\pi^2} \sum_{n=-p+1}^{p} \frac{\lambda_n}{2n-1} + \|D\| .$$

This tends to $-\infty$ by hypothesis. Hence u_m diverges. ■

4.19 Theorem. *Let A be an operator such that for every continuous nest $\mathcal{N}$, $\mathcal{U}_{\mathcal{N}}(A) = w-\lim_{\mathcal{F}} \mathcal{U}_{\mathcal{F}}(A)$ exists in the weak operator topology. Then A belongs to $\mathbb{C}I + C_\omega$.*

Proof. Say that A is universally truncatable if it satisfies the hypothesis of this theorem. If A has this property, so does A^*. For if $\mathcal{F}$ is a finite nest, $\mathcal{U}_{\mathcal{F}}(A^*) = \mathcal{U}_{\mathcal{F}^\perp}(A)^*$ where $\mathcal{F}^\perp$ is the nest of complements of $\mathcal{F}$. Thus if $\mathcal{F}$ are the subnests of $\mathcal{N}$,

$$\omega-\lim_{\mathcal{F}} \mathcal{U}_{\mathcal{F}}(A^*) = \omega-\lim_{\mathcal{F}} \mathcal{U}_{\mathcal{F}^\perp}(A)^* = \mathcal{U}_{\mathcal{N}^\perp}(A)^* = \mathcal{U}_{\mathcal{N}}(A^*)$$

exists for every continuous nest $\mathcal{N}$. So it suffices to consider the self-adjoint case.

If A is universally truncatable, so is $A + \lambda I + K$ for every scalar λ and every K in C_2. So choose λ so that 0 is in the essential spectrum. Then use the Weyl-von Neumann Theorem to perturb A by a C_2 operator K so that it is diagonalizable. A further C_2 perturbation will arrange that infinitely many eigenvalues are positive and infinitely many are negative. If this resulting operator D is not in C_ω, the eigenvalues can be ordered so as to satisfy Lemma 4.18. This is a contradiction. So D is in C_ω, and hence A belongs to $\mathbb{C}I + C_\omega$. ■

Notes and Remarks.

Example 4.1 is based on the fact that the truncation operator sending a Fourier series of a bounded function to its analytic part is unbounded. Indeed, it is a classical theorem of M. Riesz that this map is bounded on L^p for $1 < p < \infty$ but not on L^1 or L^∞. This analogue goes over to the C_p classes (Dunford and Schwartz [1] XI.10). Our example specifically corresponds to the imaginary part of $2\log(1-z)$ which is $2Arg(1-z)$. This is bounded by π which explains $\|M\| = \pi$, since M is unitarily equivalent via the Fourier transform of L^2 onto ℓ^2 to multiplication by $2Arg(1-z)$. The triangular truncation "is" multiplication by the unbounded function $\log(1-z)$, and thus is not a bounded operator. The proof that $\|M\| = \pi$ given here is due to Choi [1].

The book of Gohberg and Krein [2] contains a detailed study of triangular truncation and triangular integrals. The bulk of this chapter - Theorems 4.14, 4.16 and 4.19 are due to Macaev, Gohberg and Krein. The proof of Lemmas 4.13 through 4.16 are based on a much simpler argument due to Erdos [8]. The result of Lemma 4.13 is not sharp. The best bound for s_k is $1/2k-1$, and this is sharp.

Exercises.

4.1 Let T be a quasinilpotent operator such that ImT is trace class. Prove that T belongs to C_Ω.

4.2 Prove that if $\mathcal{F}$ is a finite nest with n atoms, then the norm of $\mathcal{U}_\mathcal{F}$ on $\mathcal{B}(\mathcal{H})$ is at most $\sum_{k=1}^{n} \frac{1}{k}$.

4.3 Modify Lemma 4.13 to prove that $\|\mathcal{L}_\mathcal{N}(A)\| \le \|A\|_\omega$ for every A in C_ω.

4.4* Is there a more elementary way to show that the norm of triangular truncation on M_n increases like $\log n$? For example, can one inductively construct A_n in M_{2^n} with $\|A_n\| \le 1$ but $\|\mathcal{T}(A_n)\| \ge Cn$?

4.5 Let $\mathcal{N}$ be the Volterra nest on $L^2(0,1)$.

(i) Show that to every operator K in C_2, one can associate a unique element $k(x,y)$ in $L^2((0,1)^2)$ such that

$$Kf(x) = \int k(x,y)f(y)dy \ .$$

(ii) Show that the map U taking K to k is unitary.

(iii) Show that $\mathcal{U}_\mathcal{N}(K) = U^*PU(K)$ where P is the projection of $L^2((0,1)^2)$ onto $\{f : supp f \subseteq \{(x,y) : x \le y\}\}$.

5. The Volterra Operator

Consider the classical Volterra operator defined on $L^2(0,1)$ given by

$$Vf(t) = \int_t^1 f(x)dx \quad .$$

The purpose of this section is to characterize V up to unitary equivalence in a rather striking way. It is clear that each element of N_t of the natural continuous nest $\mathcal{N}$ on $L^2(0,1)$ given in Example 2.4 is invariant for V. So V belongs to $\mathcal{T}(\mathcal{N})$. Since $\mathcal{N}$ is a continuous nest, the diagonal map Δ is the zero map. Hence by Ringrose's Theorem 3.4, V is quasinilpotent. Next, compute

$$(V^*f,g) = (f,Vg) = \int_0^1 f(t)\overline{\int_t^1 g(x)dx}dt = \int_0^1(\int_0^x f(t)dt)\overline{g(x)}dx$$

So

$$V^*f(x) = \int_0^x f(t)dt \quad .$$

In particular,

$$V^*x^n = \frac{x^{n+1}}{n+1} \quad .$$

Since the polynomials are dense in $L^2(0,1)$, it follows that 1 is a cyclic vector for V^*.

5.1 Lemma. *V is irreducible.* (i.e. $\{V,V^*\}' = \mathbb{C}I$).

Proof. $2ReV = V+V^*$ is the rank one operator $1 \otimes 1^*$ since

$$(V+V^*)f(t) = \int_0^1 f(t)dt = (f,1)1(t) \quad .$$

If P is a projection commuting with V, then P commutes with $V+V^*$ and hence $\mathbb{C}1$ is invariant. Thus $P1 = 1$ or 0. If $P1 = 1$, consider $P^\perp$ instead. So without loss of generality, $P1 = 0$. Then

$$P\frac{x^n}{n!} = PV^{*^n}1 = V^{*^n}P1 = 0 \quad .$$

So $P = 0$. ∎

An operator T is *completely non-normal* if it has no non-zero reducing subspace M such that $T|M$ is normal. The Volterra operator is completely non-normal by Lemma 5.1.

5.2 Lemma. *Let Q be a completely non-normal quasinilpotent operator such that $Q+Q^*$ is a rank one projection $e \otimes e^*$. Then Q is compact and irreducible and e is a cyclic vector for both Q and Q^*.*

Proof. Since $Q-Q^* = 2Q-(Q+Q^*)$, the essential spectrum $\sigma_e(Q-Q^*) = \sigma_e(2Q) = \{0\}$. But $Q-Q^*$ is normal, thus

$$\|Q-Q^*\|_e = spr_e(Q-Q^*) = 0 \ .$$

That is, $Q-Q^*$ is compact. Hence $Q = ½(Q+Q^*)+½(Q-Q^*)$ is compact. For any vector x,

$$Q^*x = (Q+Q^*)x-Qx = (x,e)e-Qx \ .$$

Thus, the smallest invariant subspace M for Q containing e is also the smallest invariant subspace for Q^*. Hence M reduces Q. Since $Q+Q^*|M^\perp = 0$, $Q\,|M^\perp$ is skew-adjoint. But Q is completely non-normal. Thus $M = \mathcal{H}$, and e is cyclic for Q and Q^*. The remainder of the proof mimics Lemma 5.1. Briefly, any projection P computing with Q must satisfy $Pe = e$ or 0. In the second case, since e is cyclic, one has $PQ^n e = Q^n Pe = 0$ implying $P = 0$. Similarly, in the first case $P = I$. So Q is irreducible. ■

5.3 Lemma. *Let Q be a compact, quasinilpotent, completely non-normal operator such that $Q+Q^*$ is a rank one projection $e \otimes e^*$. Let $\mathcal{M}$ be a maximal nest of invariant subspaces for Q. Then $\mathcal{M}$ is a continuous nest parametrized by the strictly increasing continuous function*

$$t(M) = (P(M)e,e) \ .$$

Proof. Suppose that $M_1 \leq M_2$ are two subspaces in $\mathcal{M}$ such that $t(M_1) = t(M_2)$. Let $A = M_2 \ominus M_1$, and $E = P(A)$. By hypothesis, $(Ee,e) = 0$, whence $Ee = 0$. So,

$$(Q+Q^*)E = 0 = E(Q+Q^*) \ . \tag{5.3.1}$$

Decompose $\mathcal{H}$ as $M_1 \oplus A \oplus M_2^\perp$. Then Q has a 3×3 upper triangular matrix (Q_{ij}). Plugging this into (5.3.1) yields $Q_{12} = 0 = Q_{23}$. So $EQ = QE$. By Lemma 5.2, E is scalar and hence $E = 0$. So $M_1 = M_2$. This shows that t is a strictly increasing function.

Suppose that $\mathcal{M}$ has an atom $A = M^+ \ominus M$. Since $\sigma(Q) = \{0\}$, Ringrose's Theorem 3.4 implies that $P(A)QP(A) = 0$. By adding this to its adjoint, one obtains

$$0 = P(A)(e \otimes e^*)P(A) = P(A)e \otimes (P(A)e)^* \ .$$

Thus $P(A)e = 0$, and

$$t(M^+) = (P(A)e,e)+(P(M)e,e) = t(M) \ .$$

This contradicts the first paragraph. Hence $\mathcal{M}$ is continuous. ∎

5.4 Theorem. *Every quasinilpotent, completely non-normal operator Q such that $Q+Q^*$ is a rank one projection is unitarily equivalent to the Volterra operator V.*

5.5 Theorem. $Lat(V) = \mathcal{N} = \{N_t, 0 \le t \le 1\}$.

Proof. These two theorems will be proved together.

Let $\mathcal{M}$ be a maximal nest of invariant subspaces for Q. Let $e \otimes e^* = Q+Q^*$, and let $t(M)$ be the function of Lemma 5.3. Index $\mathcal{M}$ as $\{M_t : 0 \le t \le 1\}$ such that $t(M_t) = t$. First, we claim that e is a cyclic vector for $\mathcal{M}''$. To see this, let P be the orthogonal projection onto the closed linear span of $\{M_t e : 0 \le t \le 1\}$. This is invariant for $\mathcal{M}$, so P belongs to $\mathcal{M}'$. The operator $T = Q - PQP$ belongs to $\mathcal{T}(\mathcal{M})$. Furthermore, since $Pe = e$,

$$T+T^* = Q+Q^*-P(Q+Q^*)P = 0 \ .$$

Hence T is a compact normal operator in $\mathcal{T}(\mathcal{M})$. By Theorem 3.4, $\sigma(T) = \{0\}$ and hence $T = 0$. It follows that P commutes with Q. By Lemma 5.2, $P = I$ so e is cyclic for $\mathcal{M}''$.

Define a linear map U of $L^2(0,1)$ into $\mathcal{H}$ as follows. Let f be a step function; that is, there are $0 = t_0 < t_1 < \dots < t_n = 1$, x_i are the characteristic functions of $[t_{i-1}, t_i]$, and f has the form $\sum_{i=1}^{n} a_i x_i$. Define $E_i = P(M_{t_i}) - P(M_{t_{i-1}})$, and set

$$Uf = \sum_{i=1}^{n} a_i E_i e \ .$$

It is clear that this is well defined since any refinement of the partition $\{t_i\}$ will yield the same sum. Thus U is a linear operator on the dense subspace of step functions. Let f and g be two step functions, written as $f = \sum_{i=1}^{n} a_i x_i$ and $g = \sum_{i=1}^{n} b_i x_i$ with respect to a common subdivision $\{t_i\}$ of $[0,1]$. Then

$$(Uf,Ug) = (\sum_{i=1}^{n} a_i E_i e, \sum_{j=1}^{n} b_j E_j e) = \sum_{i=1}^{n} a_i \bar{b}_i (E_i e, e)$$
$$= \sum_{i=1}^{n} a_i \bar{b}_i (t_i - t_{i-1}) = \sum_{i=1}^{n} a_i \bar{b}_i (x_i, x_i)$$
$$= (\sum_{i=1}^{n} a_i x_i, \sum_{j=1}^{n} b_j x_j) = (f,g) \ .$$

Thus U is isometric, and therefore extends to an isometry, which we also denote by U, of $L^2(0,1)$ into $\mathcal{H}$. The range of U is spanned by $\{M_t e : 0 \leq t \leq 1\}$ which is dense in $\mathcal{H}$ by the previous paragraph. Thus U is unitary.

By Theorem 4.8, one has

$$V = \int_{\mathcal{N}} P(N)(V+V^*)dN = \lim_{\mathcal{F}} \mathcal{U}_{\mathcal{F}}(1 \otimes 1^*)$$

where $\mathcal{F}$ runs over finite subnests of $\mathcal{N}$, and

$$Q = \int_{\mathcal{M}} P(M)(Q+Q^*)dM = \lim_{\mathcal{G}} \mathcal{U}_{\mathcal{G}}(e \otimes e^*)$$

where $\mathcal{G}$ runs over finite subnests of $\mathcal{M}$. Let $f = N_t 1$, and consider a partition

$$0 = t_0 < \dots < t_{i_0} = t < \dots < t_n = 1 \ .$$

Let $\mathcal{F} = \{N_{t_i}\}$ and $\mathcal{G} = \{M_{t_i}\}$ be the corresponding finite subnests of $\mathcal{N}$ and $\mathcal{M}$. Set $P_i = P(N_{t_i})$ and $R_i = P(M_{t_i})$. Then

$$\mathcal{U}_{\mathcal{F}}(1 \otimes 1^*)f = \sum_{i=1}^{n} (P(N_t)1,(P_i - P_{i-1})1)P_i 1 = \sum_{i=1}^{i_0} (t_i - t_{i-1})P_i 1$$

and

$$\mathcal{U}_{\mathcal{G}}(e \otimes e^*)Uf = \sum_{i=1}^{n} (P(M_t)e,(R_i - R_{i-1})e)R_i e = \sum_{i=1}^{i_0} (t_i - t_{i-1})R_i e \ .$$

Hence $U\mathcal{U}_{\mathcal{F}}(1 \otimes 1^*)f = \mathcal{U}_{\mathcal{G}}(e \otimes e^*)Uf$. Taking this to the limit yields $UVf = QUf$ for each $f = N_t 1$. As these vectors span $L^2(0,1)$, it follows that $UV = QU$, whence $Q = UVU^*$.

It is significant that this unitary U has the further property that $UN_t = M_t$ for all $0 \leq t \leq 1$. Suppose that V has an invariant subspace L. Let $\mathcal{L}$ be a maximal nest of invariant subspaces of V containing L. By the results above, one obtains a unitary U such that $UV = VU$ and $U^*L = N_t$ for some $0 \leq t \leq 1$. By Lemma 5.1, U is scalar and thus L belongs to $\mathcal{N}$. So $\mathcal{N}$ consists of all the invariant subspaces of V. ■

5.6 Remark. The use of Ringrose's Theorem in the first paragraph of the proof is not really necessary. The operator iT is a compact, self-adjoint operator in $\mathcal{T}(\mathcal{M})$, and hence belongs to $\mathcal{M}'$. Thus $\mathcal{M}'$ contains every spectral projection of (iT) which are all finite dimensional unless $T = 0$. But since $\mathcal{M}$ is continuous, $\mathcal{M}'$ contains no finite rank operators.

5.7 Corollary. *V is unitarily equivalent to V^*.*

Proof. This follows as a Corollary or by direct computation using $Uf(x) = f(1-x)$. ■

It is not trivial to compute $\|V\|$. In the next theorem, we compute all the s-numbers of V.

5.8 Theorem. *The eigenvalues of $(V^*V)^{\frac{1}{2}}$ are $\dfrac{2}{(2n-1)\pi}$ for $n \geq 1$. In particular, $\|V\| = 2/\pi$.*

Proof. The spectrum of V^*V consists entirely of eigenvalues since this is a compact self-adjoint operator. So consider this equation for $\lambda \neq 0$:

$$\lambda f(x) = V^*Vf(x) = \int_0^x \left(\int_t^1 f(s)ds\right)dt \quad .$$

The right hand side is differentiable by the Fundamental Theorem of Calculus. Repeated application shows that f is C^∞. Moreover,

$$\lambda f'(x) = \int_x^1 f(s)ds$$

and

$$\lambda f''(x) = -f(x) \quad .$$

The boundary conditions are $f(0) = 0$ and $f'(1) = 0$. This has solutions

$$f_n(x) = \sin\left(\frac{2n-1}{2}\pi x\right) \quad , \quad \lambda_n = \left(\frac{2}{(2n-1)\pi}\right)^2 \quad \text{for } n \geq 1 \quad .$$

It is well known that this forms a complete orthogonal set for $L^2(0,1)$. Consequently, $\sigma((V^*V)^{\frac{1}{2}})$ consists of $\{\frac{2}{(2n-1)\pi}, n \geq 1\}$. ■

5.9. Next, we consider certain algebras related to V. It will be more convenient to work with V^*. It is an easy exercise to show that

$$V^{*n} f(x) = \int_0^x \frac{1}{(n-1)!}(x-t)^{n-1} f(t)dt \quad .$$

Given a polynomial $p(x) = \sum_{j=1}^{n} a_j x^j$, set $\tilde{p}(x) = \sum_{j=1}^{n} \frac{a_j}{(j-1)!} x^{j-1}$. Then

$$p(V^*)f(x) = \int_0^x \tilde{p}(x-t)f(t)dt \quad .$$

Hence

$$\|p(V^*)f\|^2 = \int_0^1 |\int_0^x \tilde{p}(x-t)f(t)dt|^2 dx$$
$$\leq \int_0^1 \|\tilde{p}\|_2^2 \|f\|^2 dt = \|\tilde{p}\|_2^2 \|f\|_2^2 \ .$$

So $\|p(V^*)\| \leq \|\tilde{p}\|_2$.

5.10 Theorem. *Every operator in $\{V^*\}'$ is a limit of a sequence of polynomials in V^* in the strong operator topology. So $\{V^*\}'$ coincides with the weak and weak* closed algebras generated by V^*.*

Proof. Let T be an operator commuting with V^*. First suppose that $T1 = \tilde{p}$ is a polynomial, and let p be the polynomial related to $\tilde{p}$ as above. Then

$$T(\frac{x^n}{n!}) = TV^{*n}1 = V^{*n}T1 = V^{*n}\tilde{p}$$
$$= V^{*n}p(V^*)1 = p(V^*)V^{*n}1 = p(V^*)(\frac{x^n}{n!}) \ .$$

Since polynomials are dense in $L^2(0,1)$, $T = p(V^*)$.

In general, let $T1 = f$ and choose a polynomial $\tilde{p}_n$ so that $\|f-\tilde{p}_n\|_2 \leq 1/n$. Let p_n be such that $p_n(V^*)1 = \tilde{p}_n$. Then

$$\|(T-p_n(V^*))(\frac{x^n}{n!})\| = \|(T-p_n(V^*))V^{*n}1\| = \|V^{*n}(T-p_n(V^*))1\|$$
$$= \|V^{*n}(f-\tilde{p}_n)\| \leq \|V^{*n}\|/n \ .$$

It follows that $p_n(V^*)h$ converges to Th whenever h is a polynomial. Since

$$\|p_n(V^*)\| \leq \|\tilde{p}_n\|_2 \leq \|f\|_2 + 1/n$$

is uniformly bounded, $p_n(V^*)$ converges to T in the strong operator topology.

Thus the unit ball of the algebra generated by V^* is strongly dense in the unit ball of $\{V^*\}'$. A fortiori, it is dense in the weak operator topology. As this topology coincides with the weak* topology on bounded sets, it is also weak* dense. ∎

5.11 Corollary. *$\{V\}'$ coincides with the weak* closed algebra $\mathcal{A}(V)$ generated by V.*

Proof. Immediate from Corollary 5.7 and Theorem 5.10. ∎

The algebra $\mathcal{T}(\mathcal{N})$ is not singly generated as a (weak* closed) algebra since it is not abelian. It is, however, doubly generated. Let M_x be the operator of multiplication by x on $L^2(0,1)$. It follows from the double commutant theorem that

$$\{M_x\}' = \{M_x\}'' = \{M_f : f \in L^\infty(0,1)\} \ .$$

5.12 Theorem. *The unit ball of the algebra generated by $\{V, M_x\}$ is strongly dense in the unit ball of $\mathcal{T}(\mathcal{N})$. In particular, $\mathcal{T}(\mathcal{N})$ is generated by $\{V, M_x\}$ as a weak* closed algebra.*

Proof. It is easy to see that the polynomials $p(M_x) = M_p$ with $\|p\|_\infty \le 1$ are strong* dense in the unit ball of $\{M_x\}' = \mathcal{N}'$. Let N be an element of $\mathcal{N}$, and let f and g be any bounded functions in $L^2(0,1)$ such that $P(N)f = f$ and $P(N)^\perp g = g$. Then

$$\begin{aligned} M_f V M_g &= M_f P(N) V P(N)^\perp M_g \\ &= M_f P(N)(V+V^*)P(N)^\perp M_g \\ &= M_f(1 \otimes 1^*)M_g = f \otimes g^* \ . \end{aligned}$$

Let p_n and q_n be sequences of polynomials so that M_{p_n} and M_{q_n} converge strong* to M_f and M_g. Since V is compact, $M_{p_n}V$ converges in norm to $M_f V$, and $M_f V M_{q_n}$ converges in norm to $M_f V M_g$. Thus $f \otimes g^*$ belongs to the norm closed algebra generated by $\{M_x, V\}$.

By Lemmas 3.7 and 3.8, all the finite rank operators of $\mathcal{T}(\mathcal{N})$ belong to the norm closed algebra of $\{M_x, V\}$. By the Erdos Density Theorem 3.11, the unit ball is strongly dense in the unit ball of $\mathcal{T}(\mathcal{N})$. ∎

Notes and Remarks.

The invariant subspaces of V have been found by Dixmier [1], Donaghue [1], and Brodskii [1]. The usual proof makes use of Titschmarsh's Theorem on the zero divisors in the convolution algebra $L^1(0,1)$. Theorem 5.4 and this approach to the Volterra operator is due to Livsic [1]. Gohberg and Krein [2] show how Titschmarsh's Theorem can be deduced from this (see Exercise 5.4). Another approach to $Lat(V)$ is given by Sarason [4]. Theorem 5.10 is also due to Sarason, but the proof given here is due to Erdos [11]. Theorem 5.12 is a consequence of more general results of Radjavi-Rosenthal [1], and also Arveson [6], on weakly closed algebras with a nest of invariant subspaces (see Chapter 15) See also Sarason [3].

Exercises.

5.1 Verify that $V^{*n} f(x) = \int_0^x f(t)(x-t)^{n-1}/(n-1)!\, dt$.

5.2 Let $\{e_n, n \ge 1\}$ be an orthonormal basis of $\mathcal{H}$. Let D be any diagonal self-adjoint operator with respect to the basis $\{e_n\}$ with distinct eigenvalues. Let $y = \sum_{n=1}^{\infty} a_n e_n$ be a unit vector such that $a_n \ne 0$ for all $n \ge 1$. Set $T = \frac{1}{2} y \otimes y^* + iD$. Then $T + T^* = y \otimes y^*$ is a rank one projection. Prove that T is irreducible and hence completely non-normal. Compare this with Theorem 5.4.

5.3 (Donaghue) Let $\{e_n, n \geq 1\}$ be an orthonormal basis for $\mathcal{H}$. Let A be the weighted shift given by $Ae_0 = 0$ and $Ae_n = 2^{-n} e_{n-1}$ for $n \geq 1$.

(i) Show that $LatA$ is the nest $\mathcal{P} = \{P_n, n \geq 0\}$ where $P_n = span\{e_k, 0 \leq k \leq n\}$. Hint: If $x = \sum_{n=0}^{\infty} a_n e_n$ is not in any P_n and $|a_M| \geq |a_k|$ for $k \geq M$, show that $a_M^{-1} 2^{M(M+1)/2} A^M x = e_0 + \sum_{j=1}^{\infty} \beta_j e_j$ and $|\beta_j| \leq 2^{-Mj}$.

(ii) If $p(x) = \sum_{k=1}^{n} a_k x^k$, show that $\|p(A)\|_2 \leq 3^{-1/2} \|p(A)^* e_0\|$. Hence show that $\{A\}'$ is the norm closed algebra generated by A and I.

(iii) Let D be a diagonal matrix with distinct eigenvalues. Show that the weakly closed algebra generated by A and D is $\mathcal{T}(\mathcal{P})$.

5.4 Recall that $L^1(0,1)$ is a Banach algebra under the convolution product

$$f * g(x) = \int_0^x f(x-t) g(t) dt \quad .$$

Prove the Titchmarsh Convolution Theorem: If f and g belong to $L^1(0,1)$ and $f * g = 0$, then there is an α in $[0,1]$ such that $\operatorname{supp} f \subseteq [\alpha, 1]$ and $\operatorname{supp} g \subseteq [1-\alpha, 1]$. Hint: If f and g are continuous (and hence in L^2), show that $(V^{*n} f) * g = 0$ for all $n \geq 0$.

II. STRUCTURE OF NEST ALGEBRAS

6. The Radical

In the study of any Banach algebra, one is interested in its representation theory. In this section, sufficiently many representations are constructed to fully describe the (Jacobson) radical $\mathcal{R}$ and the quotient algebra $\mathcal{T}(\mathcal{N})/\mathcal{R}$.

The first step is a description of $C^*(\mathcal{N})$, the C^*-algebra generated by $\{P(N):N \in \mathcal{N}\}$. Since the product $\prod_{i=1}^{n} P(N_i) = P(\bigwedge_{i=1}^{n} N_i)$, $C^*(\mathcal{N})$ is the closed linear span of $\{P(N):N \in \mathcal{N}\}$. It is abelian, and thus it is *-isomorphic to $C(M_{\mathcal{N}})$ where $M_{\mathcal{N}}$ is its maximal ideal space. Let $\mathbf{2}$ denote the two element lattice $\{0,1\}$. $Hom(\mathcal{N},\mathbf{2})$ denotes the set of lattice homomorphisms of $\mathcal{N}$ onto $\mathbf{2}$. This set has a natural order given by $\phi \le \psi$ if and only if $\phi(N) \le \psi(N)$ for all N in $\mathcal{N}$. $Hom(\mathcal{N},\mathbf{2})$ is endowed with the *order topology*, which is the smallest topology such that $L_\phi = \{\psi:\psi < \phi\}$ and $U_\phi = \{\psi:\psi > \phi\}$ are open.

6.1 Proposition. *There is a canonical homeomorphism of the maximal ideal space $M_{\mathcal{N}}$ of $C^*(\mathcal{N})$ onto $Hom(\mathcal{N},\mathbf{2})$ given by restriction to $P(\mathcal{N})$.*

Proof. Let ϕ belong to $M_{\mathcal{N}}$. Then $\phi(P(N)) = \phi(P(N))^2$ and thus equals 0 or 1. Also, this is an increasing function of N. Hence $\tilde{\phi}(N) = \phi(P(N))$ is a lattice homomorphism onto $\mathbf{2}$. Conversely, for $\tilde{\phi}$ in $Hom(\mathcal{N},\mathbf{2})$, one defines $\phi(P(N)) = \tilde{\phi}(N)$. Extend ϕ by linearity to the $span\{P(N):N \in \mathcal{N}\}$. Let A be a finite linear combination of $P(N_i)$, $1 \le i \le n$. Order them so that $0 = N_0 < N_1 < N_2 < ...N_n = I$, and set $E_i = P(N_i) - P(N_{i-1})$. Then A can be written as $\sum_{i=1}^{n} a_i E_i$. There will be an integer i_0 so that $\tilde{\phi}(N_{i_0}) = 1 > \tilde{\phi}(N_{i_0 - 1})$, and hence $\phi(E_{i_0}) = 1$ and $\phi(E_i) = 0$ otherwise. So

$$|\phi(A)| = |a_{i_0}| \le \|A\| .$$

Hence ϕ extends to a continuous functional on $C^*(\mathcal{N})$. It is clear that ϕ is uniquely determined by $\tilde{\phi}$, and thus a bijective correspondence is established.

The map Φ taking ϕ to $\tilde{\phi}$ is a bijection of the compact Hausdorff space $M_{\mathcal{N}}$ onto the Hausdorff space $Hom(\mathcal{N},\mathbf{2})$. Thus once it is established that this map is continuous, it will follow that it is a homeomorphism.

Let

$$N_\phi = \sup\{N \in \mathcal{N}: \tilde{\phi}(N) = 0\} \ .$$

If $\tilde{\phi}(N_\phi) = 0$, then $\tilde{\phi}(N) = 0$ if and only if $N \leq N_\phi$. So

$$U_\phi = \{\tilde{\psi} > \tilde{\phi}\} = \{\tilde{\psi}: \tilde{\psi}(N_\phi) = 1\}$$

and

$$L_\phi = \{\tilde{\psi} < \tilde{\phi}\} = \{\tilde{\psi}: N_\psi > N_\phi\} = \bigcup_{N > N_\phi} \{\tilde{\psi}: \tilde{\psi}(N) = 0\} \ .$$

Similarly, if $\tilde{\phi}(N_\phi) = 1$, then $\tilde{\phi}(N) = 0$ if and only if $N < N_\phi$. And

$$U_\phi = \{\tilde{\psi} > \tilde{\phi}\} = \{\tilde{\psi}: N_\psi < N_\phi\} = \bigcup_{N < N_\phi} \{\tilde{\psi}: \tilde{\psi}(N) = 1\}$$

and

$$L_\phi = \{\tilde{\psi} < \tilde{\phi}\} = \{\tilde{\psi}: \tilde{\psi}(N_\phi) = 0\} \ .$$

Now $\Phi^{-1}\{\tilde{\psi}: \tilde{\psi}(N) = 0\} = \{\psi: \psi(P(N)) < 1/2\}$ is open in the weak* topology on $M_{\mathcal{N}}$. Likewise,

$$\Phi^{-1}\{\tilde{\psi}: \tilde{\psi}(N) = 1\} = \{\psi: \psi(P(N)) > 1/2\}$$

is open. So $\Phi^{-1}(U_\phi)$ and $\Phi^{-1}(L_\phi)$ are open for all $\tilde{\phi}$ in $Hom(\mathcal{N},\mathbf{2})$. Thus Φ is continuous, and hence a homeomorphism. ■

6.2. There is a natural parametrization of $M_{\mathcal{N}}$ implicit in the proof above. For ϕ in $M_{\mathcal{N}}$, set $\epsilon_\phi = 1 - \tilde{\phi}(N_\phi)$. The set $X_{\mathcal{N}} = \{(N_\phi, \epsilon_\phi): \phi \in M_{\mathcal{N}}\}$ is a subset of $\mathcal{N} \times \{0,1\}$. Let $<$ be the lexicographic order on $X_{\mathcal{N}}$:

$$(N_1, \epsilon_1) < (N_2, \epsilon_2) \quad \text{if} \quad (N_1 < N_2) \text{ or } (N_1 = N_2, \text{ and } \epsilon_1 < \epsilon_2) \ .$$

The pair (N_ϕ, ϵ_ϕ) determines $\tilde{\phi}$ (and thus ϕ) by the rule

$$\tilde{\phi}(N) = 0 \quad \text{if and only if} \quad (N,0) < (N_\phi, \epsilon_\phi) \ . \tag{6.2.1}$$

The sets L_ϕ and U_ϕ become

$$L_\phi = \{\tilde{\psi}: (N_\psi, \epsilon_\psi) > (N_\phi, \epsilon_\phi)\}$$

and

$$U_\phi = \{\tilde{\psi}: (N_\psi, \epsilon_\psi) < (N_\phi, \epsilon_\phi)\} \ .$$

So the map taking ϕ to (N_ϕ, ϵ_ϕ) is order reversing.

Every pair (N, ϵ) in $\mathcal{N} \times \{0,1\}$ gives rise to an increasing function of $\mathcal{N}$ into $\mathbf{2}$. If one deletes the pairs $(\{0\},0)$ and $(\mathcal{H},1)$ because they imply $\tilde{\phi}(\{0\}) = 1$ and $\tilde{\phi}(\mathcal{H}) = 0$ respectively, the remaining pairs determine an element $\tilde{\phi}$ of $Hom(\mathcal{N},\mathbf{2})$. Some duplication is possible. Consider first a pair $(N,1)$ determining $\tilde{\phi}$ by (6.2.1). Then $\tilde{\phi}(N) = 0$ and $\tilde{\phi}(N') = 1$ for $N' > N$. Thus $(N_\phi, \epsilon_\phi) = (N,1)$. Now, consider $(N,0)$. Then $\tilde{\phi}(N) = 1$

and $\tilde{\phi}(N') = 0$ for $N' < N$. Thus $N_\phi = N_-$. When $N_- = N$, $(N_\phi,\epsilon_\phi) = (N,0)$. But when $N_- < N$, $(N_\phi,\epsilon_\phi) = (N_-,1)$. In this case, $A = N \ominus N_-$ is an atom of $\mathcal{N}$ and ϕ is evaluation at this atom. Thus, one can think of $X_{\mathcal{N}}$ as a quotient of $\mathcal{N} \times \{0,1\}\backslash(\{0\},0) \cup (\mathcal{H},1)$ by identifying $(N_-,1)$ and $(N,0)$ when $N_- \neq N$. When it is convenient to use this parametrization, we will write $\phi \equiv (N_\phi,\epsilon_\phi)$.

6.3. Let us associate to each ϕ in $M_{\mathcal{N}}$, the set $\mathcal{E}_\phi$ of projections $E = P(N_1)-P(N_2)$ with N_1, N_2 in $\mathcal{N}$ such that $\phi(E) = 1$. These projections will be called *test intervals* for ϕ. By Theorem 2.15, the map $\Phi_E(T) = ET|EH$ is an algebra homomorphism. For ϕ in $M_{\mathcal{N}}$, define a seminorm

$$||T||_\phi = \inf\{||ETE||:E \in \mathcal{E}_\phi\} \ .$$

Let I_ϕ denote the set of operators T in $\mathcal{T}(\mathcal{N})$ such that $||T||_\phi = 0$. It is easy to verify that

$$I_\phi = \overline{\bigcup_{E\in\mathcal{E}_\phi} ker\Phi_E} \ .$$

In particular, I_ϕ is a closed ideal. Let $\mathcal{D}_\phi = \mathcal{T}(\mathcal{N})/I_\phi$, and let Φ_ϕ denote the natural quotient map onto $\mathcal{D}_\phi$. I_ϕ will be called a *diagonal ideal*, and $\mathcal{D}_\phi$ its associated *diagonal algebra*.

Let us consider $\mathcal{E}_\phi$ more carefully. If ϕ is evaluation at an atom A, then $E_A = P(A)$ is the least element of $\mathcal{E}_\phi$. Thus $\Phi_\phi = \Phi_A$ is the compression map onto A, and $\mathcal{D}_\phi = \mathcal{B}(A)$. If $\phi \equiv (N,0)$ and $N = N_-$, then every test interval contains a smaller one of the form $P(N)-P(N')$ for $N' < N$. Thus

$$\begin{aligned} ||T||_\phi &= \inf_{N'<N} ||(P(N)-P(N'))T(P(N)-P(N'))|| \\ &= \lim_{N'\uparrow N} ||\Phi_{N\ominus N'}(T)|| \ . \end{aligned}$$

Similarly, if $\phi \equiv (N,1)$ and $N = N_+$, then every test interval contains $P(N')-P(N)$ for some $N' > N$; and

$$\begin{aligned} ||T||_\phi &= \inf_{N>N} ||(P(N')-P(N))T(P(N')-P(N))|| \\ &= \lim_{N'\downarrow N} ||\Phi_{N'\ominus N}(T)|| \ . \end{aligned}$$

There is also a connection with the topology of $M_{\mathcal{N}}$. For each interval $E = P(N_1)-P(N_2)$, let

$$\begin{aligned} \mathcal{O}_E = \{\phi \in M_{\mathcal{N}}:E \in \mathcal{E}_\phi\} &= \{\phi:\tilde{\phi}(N_1) = 1,\tilde{\phi}(N_2) = 0\} \\ &= \{\phi:\tilde{\phi}(N_1) > 1/2 > \tilde{\phi}(N_2)\} \ . \end{aligned}$$

It is clear from the last two formulations that $\mathcal{O}_E$ is both open and closed in $M_{\mathcal{N}}$. Note that $\mathcal{O}_{E_1}, \ldots, \mathcal{O}_{E_n}$ is a finite open cover of $M_{\mathcal{N}}$ if and only if $\sum_{i=1}^{n} E_i \geq I$.

6.4. The (Jacobson) *radical* of a Banach algebra has a number of definitions. One convenient to our purposes is

$$rad(\mathcal{A}) = \{A \in \mathcal{A} : AB \text{ is quasinilpotent for all } B \text{ in } \mathcal{A}\}$$
$$= \{A \in \mathcal{A} : BA \text{ is quasinilpotent for all } B \text{ in } \mathcal{A}\} \ .$$

It is also described as

$$\bigcap \{ker\,\pi : \pi \text{ is an algebraically irreducible representation of } \mathcal{A}\} \ .$$

The equivalence of these notions may be found in any standard text on Banach algebras (cf. Bonsall-Duncan [1], III.17 or Rickart [1] 2.3.2). An algebra is called *semi-simple* if its radical is $\{0\}$.

If $\mathcal{F}$ is a finite nest, $\mathcal{T}(\mathcal{F})$ consists of block upper triangular $n \times n$ matrices. If T in $\mathcal{T}(\mathcal{F})$ has zero diagonal $\Delta(T)$, then AT is strictly upper triangular for every A in $\mathcal{T}(\mathcal{F})$. Thus $(AT)^n = 0$ and T belongs to the radical. Conversely, if $\Delta(T) \neq 0$, then $\Delta(T)^*$ belongs to $\mathcal{T}(\mathcal{F})$. Then since $\Delta(T)^*\Delta T$ is positive,

$$\|(\Delta(T^*)T)^n\|^{1/n} \geq \|\ \Delta((\Delta(T^*)T)^n)\|^{1/n}$$
$$= \|(\Delta(T)^*\Delta(T))^n\|^{1/n} = \|\Delta(T)^*\Delta(T)\| > 0 \ .$$

So $rad\,\mathcal{T}(\mathcal{F}) = \{T \in \mathcal{T}(\mathcal{F}) : \Delta(T) = 0\}$.

If $\mathcal{N}$ is an arbitrary nest and $\mathcal{F}$ is a finite subnest of $\mathcal{N}$, then

$$rad\,\mathcal{T}(\mathcal{F}) \subseteq \mathcal{T}(\mathcal{N}) \subseteq \mathcal{T}(\mathcal{F}) \ .$$

Thus the previous argument implies that $rad\,\mathcal{T}(\mathcal{F}) \subseteq rad\,\mathcal{T}(\mathcal{N})$.

6.5 Lemma. *Let T belong to $\mathcal{T}(\mathcal{N})$ and let ϕ belong to $M_{\mathcal{N}}$. Then there is an operator A in the unit ball of $\mathcal{T}(\mathcal{N})$ such that*

$$\|(AT)^n\|_\phi = \|T\|_\phi^n \quad \text{for } n \geq 1 \ .$$

Proof. If ϕ is evaluation at an atom A, then let UP be the polar decomposition of $\Phi_A(T)$. Then U^* belongs to $\mathcal{T}(\mathcal{N})$ and $\Phi_A(U^*T) = P$ is positive. So

$$\|(U^*T)^n\|_\phi = \|P^n\| = \|P\|^n = \|T\|_\phi^n \ .$$

If $\phi \equiv (N_0, 0)$, $\|T\|_\phi = \lim_{N \uparrow N_0} \|\Phi_{N_0 \ominus N}(T)\|$. We may suppose that $\|T\|_\phi > 0$, and normalize so that $\|T\|_\phi = 1$. Now $P(N_0)$ is the strong operator limit of $P(N_\alpha)$ for $N_\alpha < N$. Thus if $N < N_0$,

$$1 \leq \|(P(N_0) - P(N))T(P(N_0) - P(N))\|$$
$$= \lim_{N_\alpha \uparrow N_0} \|(P(N_\alpha) - P(N))T(P(N_\alpha) - P(N))\| \ .$$

So one may extract a sequence of elements of $\mathcal{N}$, $N_1 < N_2 < \ldots$ so that

$N_0 = \bigvee_{k\geq 1} N_k$ and

$$\|(P(N_{k+1})-P(N_k))T(P(N_{k+1})-P(N_k))\| > 1-1/k$$

for $k \geq 1$. Choose unit vectors x_k and y_k in $N_{k+1} \ominus N_k$ so that

$$(Tx_k, y_k) > 1-1/k \quad .$$

By Lemma 2.8, the rank one operator $x_k \otimes y_{k+1}^*$ belongs to $\mathcal{T}(\mathcal{N})$ for each $k \geq 1$. Thus $A = \sum_{k\geq 1}^{\infty} x_k \otimes y_{k+1}^*$ belongs to $\mathcal{T}(\mathcal{N})$ since it is the strong operator limit of the finite partial sums. Also $\|A\| = 1$.

Notice that AT maps $N_{k+1} \ominus N_k$ into N_k for each $k \geq 1$, and

$$(P(N_k)-P(N_{k-1}))ATx_{k+1} = (Tx_{k+1}, y_{k+1})x_k \quad .$$

Thus

$$\begin{aligned}((AT)^2 x_{k+2}, x_k) &= (P(N_k)^{\perp} ATATP(N_{k+3})x_{k+2}, x_k) \\ &= (ATP(N_{k+1})^{\perp} P(N_{k+2})ATx_{k+2}, x_k) \\ &= (Tx_{k+2}, y_{k+2})(ATx_{k+1}, x_k) = (Tx_{k+2}, y_{k+2})(Tx_{k+1}, y_{k+1}) \\ &> (1-\frac{1}{k+2})(1-\frac{1}{k+1}) \quad .\end{aligned}$$

By induction, one obtains

$$((AT)^n x_{k+n}, x_k) \geq \prod_{j=1}^{n} 1-\frac{1}{k+j} \quad .$$

If $N < N_0$, x_k belongs to $N_0 \ominus N$ for all k sufficiently large. Hence

$$\|\Phi_{N_0 \ominus N}((AT)^n)\| \geq \lim_{k\to\infty} \prod_{j=1}^{n} 1-\frac{1}{k+j} = 1 \quad .$$

So

$$1 \leq \|(AT)^n\|_{\phi} \leq \|A\|^n \|T\|_{\phi}^n \leq \|T\|_{\phi}^n \leq 1$$

as desired. When $\phi \equiv (N_0, 1)$, the proof is analogous. ■

6.6 Corollary *The algebras $\mathcal{D}_{\phi}$ are semi-simple.*

Proof. If $T+I_{\phi} \neq 0$, then with A as constructed in Lemma 6.5, $AT+I_{\phi}$ is not quasinilpotent. ■

6.7 Theorem. *The following are equivalent for T in $\mathcal{T}(\mathcal{N})$.*

1) *T belongs to $rad\,\mathcal{T}(\mathcal{N})$.*
2) *T belongs to $\bigcap\{I_\phi : \phi \in M_\mathcal{N}\}$.*
3) *T belongs to the closure of $\bigcup\{rad\,\mathcal{T}(\mathcal{F}) : \mathcal{F}$ finite subnest of $\mathcal{N}\}$.*
4) *(Ringrose condition) For every $\epsilon > 0$, there is a finite subnest $\mathcal{F}$ of $\mathcal{N}$ such that $\|\Delta_\mathcal{F}(T)\| < \epsilon$.*

Proof. If $\|T\|_\phi > 0$ for some ϕ, Lemma 6.5 provides an A in $\mathcal{T}(\mathcal{N})$ such that AT is not quasinilpotent. Thus 1) implies 2). Suppose 2) holds. For each ϕ in $M_\mathcal{N}$, $\|T\|_\phi = 0$. Hence given $\epsilon > 0$, there is a test interval E_ϕ in $\mathcal{E}_\phi$ such that $\|E_\phi T E_\phi\| < \epsilon$. The collection $\{O_{E_\phi} : \phi \in M_\mathcal{N}\}$ is an open cover of $M_\mathcal{N}$. By compactness, there is a finite subcover $O_{E_1}, \ldots, O_{E_n}$. Thus $\sum_{i=1}^{n} E_i \geq I$. Let $\mathcal{F}$ be the finite nest determined by the endpoints of these E_i. Then

$$\|\Delta_\mathcal{F}(T)\| \leq \max\|E_i T E_i\| < \epsilon \ .$$

This proves 4). If T satisfies 4), then $T - \Delta_\mathcal{F}(T)$ belongs to $rad\,\mathcal{T}(\mathcal{F})$. By hypothesis, there is a sequence $\mathcal{F}_n$ such that $\lim\|\Delta_{\mathcal{F}_n}(T)\| = 0$, so $T = \lim T - \Delta_{\mathcal{F}_n}(T)$. This establishes 3). But $rad\,\mathcal{T}(\mathcal{N})$ is a closed ideal containing each $rad\,\mathcal{T}(\mathcal{F})$, so 3) implies 1). ∎

6.8 Example. Let $\mathcal{P} = \{P_n, n \geq 0, \mathcal{H}\}$ be given by an increasing sequence of subspaces whose union is dense in $\mathcal{H}$. Then $M_\mathcal{P}$ is isomorphic to $\omega = \mathbb{N} \cup \{\infty\}$, and Φ_n is compression to the atom $P_n \ominus P_{n-1}$ for $n \geq 1$. The other ideal is given by the seminorm

$$\|T\|_\infty = \inf\|P(P_n)^\perp T\| \ .$$

An operator T in $\mathcal{T}(\mathcal{P})$ belongs to $rad\,\mathcal{T}(\mathcal{P})$ if and only if $\Delta(T) = 0$ and $\|T\|_\infty = 0$. If all the P_n are finite dimensional, it is easy to see that $\|T\|_\infty = \|T\|_e$ is the essential norm. In this case, $rad\,\mathcal{T}(\mathcal{P})$ consists of the compact operators in $\mathcal{T}(\mathcal{P})$ with zero diagonal.

6.9 Corollary. *A compact operator K in $\mathcal{T}(\mathcal{N})$ belongs to the radical if and only if $\Delta(K) = 0$.*

Proof. Lemma 3.5 and the hypothesis $\Delta(K) = 0$ implies the Ringrose condition. The other implication is trivial. ∎

Let $\mathcal{D}$ be the algebra of all bounded functions (D_ϕ) on $M_\mathcal{N}$ such that D_ϕ belongs to $\mathcal{D}_\phi$, with norm

$$\|(D_\phi)\| = \sup_{\phi \in M_{\mathcal{N}}} \|D_\phi\|_\phi \ .$$

Let $\Phi(T) = (\Phi_\phi(T))$ be the natural homomorphism of $\mathcal{T}(\mathcal{N})$ into $\mathcal{D}$, and let $\mathcal{D}_{\mathcal{N}}$ be the image of Φ.

6.10 Theorem. *For T in $\mathcal{T}(\mathcal{N})$, the map $\phi \to \|T\|_\phi$ is upper semicontinuous. In particular, it is continuous at ϕ_0 if $\|T\|_{\phi_0} = 0$. Furthermore,*

$$\max_{\phi \in M_{\mathcal{N}}} \|T\|_\phi = \|T + rad\,\mathcal{T}(\mathcal{N})\| \ .$$

Thus $\mathcal{T}(\mathcal{N})/rad\,\mathcal{T}(\mathcal{N})$ is isometrically isomorphic to $\mathcal{D}_{\mathcal{N}}$.

Proof. For each ϕ in $M_{\mathcal{N}}$, and $\epsilon > 0$, there is a test interval E_ϕ in $\mathcal{E}_\phi$ such that $\|ETE\| < \|T\|_\phi + \epsilon$. Thus $\|T\|_\psi < \|T\|_\phi + \epsilon$ for all ψ in O_{E_ϕ}. Hence $\phi \to \|T\|_\phi$ is upper semicontinuous. Consequently, if $\|T\|_{\phi_0} = 0$, this function is continuous at ϕ_0. It also follows that $\max\|T\|_\phi$ is attained at some ϕ_0.

By Theorem 6.7, $\|R\|_\phi = 0$ if R belongs to the radical. Thus

$$\|T\|_\phi = \inf\{\|T+R\|_\phi : R \in rad\,\mathcal{T}(\mathcal{N})\} \leq \|T + rad\,\mathcal{T}(\mathcal{N})\| \ .$$

So the map of $\mathcal{T}(\mathcal{N})/rad\,\mathcal{T}(\mathcal{N})$ into $\mathcal{D}$ is contractive. On the other hand, the collection $\{O_{E_\phi} : \phi \in M_{\mathcal{N}}\}$ constructed in the first paragraph is an open cover of $M_{\mathcal{N}}$. As in the proof of Theorem 6.7, a finite subcover yields a finite nest $\mathcal{F}$ such that

$$\|\Delta_{\mathcal{F}}(T)\| < \max\|T\|_\phi + \epsilon \ .$$

Since $T - \Delta_{\mathcal{F}}(T)$ belongs to $rad\,\mathcal{T}(\mathcal{N})$,

$$\|T + rad\,\mathcal{T}(\mathcal{N})\| \leq \|\Delta_{\mathcal{F}}(T)\| < \max\|T\|_\phi + \epsilon \ .$$

Hence the map of $\mathcal{T}(\mathcal{N})/rad\,\mathcal{T}(\mathcal{N})$ into $\mathcal{D}$ is isometric. ∎

As mentioned in section 6.4, the kernel of every algebraically irreducible representation contains the radical, and the intersection of these kernels equals the radical. A *representation* of a Banach algebra $\mathcal{A}$ on $\mathcal{X}$ is a continuous homomorphism π of $\mathcal{A}$ into $\mathcal{B}(\mathcal{X})$. The representation is *topologically irreducible* if the only closed invariant subspaces of $\pi(\mathcal{A})$ are $\{0\}$ and $\mathcal{X}$. Every algebraically irreducible representation of $\mathcal{A}$ is equivalent to a continuous representation of $\mathcal{A}$. It is not known, however, whether the kernel of every topologically irreducible representation is equal to the kernel of an algebraically irreducible one, or even that they contain the radical (cf. Bonsall-Duncan [1], III.25 or Rickart [1], 2.2.3). In the case of nest algebras, the situation is better.

6.11 Theorem. *Let π be a non-zero topologically irreducible representation of $\mathcal{T}(\mathcal{N})$ on a Banach space $\mathcal{X}$. Then $ker\pi$ contains exactly one diagonal ideal I_ϕ.*

Proof. For N in $\mathcal{N}$, let $P = \pi(P(N))$. Then $P = P^2$, and for all T in $\mathcal{T}(\mathcal{N})$, one has

$$\pi(T)P = \pi(TP(N)) = \pi(P(N)TP(N)) = P\pi(T)P \ .$$

Thus P is an idempotent such that $P\mathcal{X}$ is invariant for π. Thus P equals 0 or I. Hence the restriction of π to $\{P(N): N \in \mathcal{N}\}$ is an increasing function of $\mathcal{N}$ into $\{0,I\}$; namely, an element ϕ of $Hom(\mathcal{N},2)$. Hence $\pi(E) = I$ for every E in $\mathcal{E}_\phi$. So

$$\|\pi(T)\| = \|\pi(ETE)\| \leq \|\pi\|\|ETE\|$$

for E in $\mathcal{E}_\phi$. Whence $\|\pi(T)\| \leq \|\pi\|\|T\|_\phi$. In particular, $ker\pi$ contains I_ϕ and π factors through $\mathcal{D}_\phi$.

If ψ is another element of $M_\mathcal{N}$, there is a test interval E in $\mathcal{E}_\phi$ which is not in $\mathcal{E}_\psi$. Thus $\|E\|_\psi = 0$. So for every T in $\mathcal{T}(\mathcal{N})$, ETE belong to I_ψ and $T-ETE$ belongs to I_ϕ. Hence $I_\psi + I_\phi = \mathcal{T}(\mathcal{N})$. Therefore, since π is not zero, $ker\pi$ does not contain I_ψ for $\psi \neq \phi$. ∎

6.12 Corollary. *The kernel of every topologically irreducible representation of $\mathcal{T}(\mathcal{N})$ contains the radical.*

6.13 Theorem. *The centre of each diagonal algebra $\mathcal{D}_\phi$ consists of the scalar multiples of the identity.*

Proof. If ϕ is evaluation at an atom A, then $\mathcal{D}_\phi = \mathcal{B}(A)$ which has trivial centre. So consider $\phi \equiv (N_0,0)$ when $(N_0)_- = N_0$. The case $\phi \equiv (N_0,1)$ is handled similarly. The proof is a more sophisticated version of Lemma 6.5. If $T+I_\phi$ is not scalar, then since $\lambda \rightarrow \|T-\lambda I\|_\phi$ is continuous and $\{\lambda : |\lambda| \leq 2\|T\|\}$ is compact,

$$\sigma = \inf\{\|T-\lambda I\|_\phi : |\lambda| \leq 2\|T\|\} > 0 \ .$$

Let $\{\lambda_j, j \geq 1\}$ be a dense subset of $\{\lambda : |\lambda| \leq 2\|T\|\}$. As in Lemma 6.5, if $N < N_0$ and n is an integer, one can find an $N' < N_0$ so that

$$\|\Phi_{N' \ominus N}(T-\lambda_j I)\| > \sigma/2 \ \text{ for } \ 1 \leq j \leq n \ .$$

Hence one may choose unit vectors x_j, y_j in $N' \ominus N$ so that

$$P(N' \ominus N)(T-\lambda_j I)x_j = \alpha_j y_j \ \text{ and } \ \alpha_j > \sigma/2 \ \text{ for } \ 1 \leq j \leq n \ .$$

Recursively choose $N_1 < N_2 < \dots$ in $\mathcal{N}$ with $N_0 = \bigvee_{k \geq 1} N_k$ and unit vectors $x_j^{(k)}$, $y_j^{(k)}$ in $N_{k+1} \ominus N_k$ such that

$$P(N_{k+1} \ominus N_k)(T - \lambda_j I)x_j^{(k)} = \alpha_j^{(k)} y_j^{(k)} \text{ and } \alpha_j^{(k)} > \sigma/2 \text{ for } 1 \leq j \leq k \ .$$

Let z_k be any unit vector in $N_{k+1} \ominus N_k$, and let λ be a cluster point of $\{(Tz_k, z_k), k \geq 1\}$. Drop to a subsequence k' so that $\lambda = \lim(Tz_{k'}, z_{k'})$. Relabel the subsequences as N_k, $x_j^{(k)}$, $y_j^{(k)}$, and z_k for $k \geq 1$. Choose a sequence $j_k \leq k$ so that λ_{j_k} converges to λ. Define $A = \sum_{k \geq 1} x_{j_k}^{(k)} \otimes z_{k+1}^*$. This belongs to $\mathcal{T}(\mathcal{N})$ since each $x_k \otimes z_{k+1}^*$ belongs by Lemma 2.8. Then A maps $N_{k+1} \ominus N_k$ into $N_k \ominus N_{k-1}$. Write E_k for $P(N_{k+1}) - P(N_k)$. One computes

$$\begin{aligned} E_k(AT - TA)z_{k+1} &= (Tz_{k+1}, z_{k+1})x_{j_k}^{(k)} - E_k T x_{j_k}^{(k)} \\ &= [(Tz_{k+1}, z_{k+1}) - \lambda_{j_k}]x_{j_k}^{(k)} - \alpha_{j_k}^{(k)} y_{j_k}^{(k)} \ . \end{aligned}$$

Now the first term tends to zero, and $\alpha_{j_k}^{(k)} > \sigma/2$ for all k. Thus for all k sufficiently large, $\|(AT - TA)z_{k+1}\| > \sigma/2$. Every test interval for ϕ contains most of the z_k. Hence $\|AT - TA\|_\phi > \sigma/2$, and T is not central. ∎

Let $Z(\mathcal{D}_\mathcal{N})$ denote the centre of $\mathcal{D}_\mathcal{N}$. This is characterized by the following result.

6.14 Theorem. *For T in $\mathcal{T}(\mathcal{N})$, the following are equivalent.*

1) *T belongs to $C^*(\mathcal{N}) + rad\, \mathcal{T}(\mathcal{N})$.*
2) *$\Phi(T)$ belongs to $Z(\mathcal{D}_\mathcal{N})$.*

2′) *$T + rad\, \mathcal{T}(\mathcal{N})$ belongs to $Z(\mathcal{T}(\mathcal{N})/rad\, \mathcal{T}(\mathcal{N}))$.*

3) *$\Phi_\phi(T)$ is scalar for every ϕ in $M_\mathcal{N}$.*
4) *$\Phi_\phi(T)$ is a continuous, scalar valued function of ϕ.*

Proof. Let $E = P(N_1) - P(N_2)$ for $N_2 < N_1$ in $\mathcal{N}$. Then if E belongs to $\mathcal{E}_\phi$, $\Phi_\phi(E) = \Phi_\phi(I) = I$. And if E does not belong to $\mathcal{E}_\phi$, there is a test interval in F in $\mathcal{E}_\phi$ orthogonal to E; so $\Phi_\phi(E) = 0$. Thus $\Phi(E)$ is scalar valued, and is the characteristic function of the closed and open set $\mathcal{O}_E$. Hence $\Phi_\phi(E)$ is a continuous function of ϕ. This extends to the closed linear span of these intervals, namely $C^*(\mathcal{N})$. Since Φ annihilates the radical, this proves 1) implies 4).

Clearly, 4) implies 3) and 3) implies 2). Theorem 6.10 shows that 2) and 2′) are equivalent. For $\Phi(T)$ to belong to the centre of $\mathcal{D}_\mathcal{N}$, it is necessary and sufficient that $\Phi_\phi(T)$ belong to the centre of $\mathcal{D}_\phi$ for each ϕ in $M_\mathcal{N}$. Thus Theorem 6.13 shows that 2) implies 3). Suppose 3) holds, let ϕ_0 belong to $M_\mathcal{N}$, and let $\lambda_0 I = \Phi_{\phi_0}(T)$. Then $\|T - \lambda_0 I\|_{\phi_0} = 0$. By Theorem 6.10,

$$\phi \rightarrow \|T-\lambda_0 I\|_\phi = |\Phi_\phi(T)-\Phi_{\phi_0}(T)|$$

is continuous at ϕ_0. Thus $\Phi_\phi(T)$ is a continuous function of ϕ.

If 4) holds, $t(\phi) = \Phi_\phi(T)$ is a continuous function on $M_{\mathcal{N}}$. Hence there is an operator A in $C^*(\mathcal{N})$ such that $\Phi_\phi(A) = \Phi_\phi(T)$ for all ϕ in $M_{\mathcal{N}}$. Hence $T-A$ belongs to $ker\Phi = rad\,\mathcal{T}(\mathcal{N})$. So 4) implies 1). ∎

Finally, we show that the algebras $\mathcal{D}_\phi$ can be represented isometrically as operator algebras on Hilbert space.

6.15 Theorem. *There is an isometric isomorphism of $\mathcal{D}_\phi$ into $\mathcal{B}(\mathcal{K})$ for some Hilbert space $\mathcal{K}$.*

Proof. If ϕ is evaluation at an atom A, then $\mathcal{D}_\phi = \mathcal{B}(A)$, and we are done. Suppose that $\phi \equiv (N_0,0)$ where $N_0 = (N_0)_-$. The case $\phi \equiv (N_0,1)$ with $N_0 = (N_0)_+$ is analogous. Choose an increasing sequence $N_1 < N_2 < \ldots$ in $\mathcal{N}$ with $N_0 = \bigvee_{k\geq 1} N_k$. Let $E_k = P(N_0)-P(N_k)$. Fix ω to be any non-zero multiplicative linear functional on ℓ^∞ which annihilates c_0.

Let X be the set of all bounded sequences $\{x_k\}$ such that x_k belongs to $N_0 \ominus N_k = E_k\mathcal{H}$. Define a sesquilinear form on X by

$$(\overline{x},\overline{y})_\omega = \omega(\{(x_k,y_k)\}) \ .$$

Let X_0 be the set of all $\overline{x}$ in X such that $(\overline{x},\overline{x})_\omega = 0$. Then X/X_0 is an inner product space. Let $\mathcal{K}$ be its Hilbert space completion. For T in $\mathcal{T}(\mathcal{N})$, define $T^{(\infty)}$ on X by $T^{(\infty)}\{x_k\} = \{E_kTx_k\}$. Then since ω is a positive linear functional,

$$\|T^{(\infty)}\overline{x}\|_\omega^2 = \omega(\|E_kTx_k\|^2) \leq \|T\|^2\omega(\|x_k\|^2) = \|T\|^2\|\overline{x}\|_\omega^2 \ .$$

Thus $T^{(\infty)}$ maps X_0 into itself, and hence determines an operator $\widetilde{T}$ on $\mathcal{K}$ with $\|\widetilde{T}\| \leq \|T\|$. Furthermore, for $k \geq n$,

$$\|E_kTx_k\| = \|E_kTE_kx_k\| \leq \|E_nTE_n\|\|x_k\| \ .$$

Thus

$$\|T^{(\infty)}\overline{x}\|_\omega \leq \|E_nTE_n\|\|\overline{x}\|_\omega$$

for all n. It follows that $\|T^{(\infty)}\overline{x}\|_\omega \leq \|T\|_\phi\|\overline{x}\|_\omega$. Hence $\|\widetilde{T}\| \leq \|T\|_\phi$.

On the other hand, let x_k be a unit vector in $N_0 \ominus N_k$ such that $\|Tx_k\| > (1-\frac{1}{k})\|E_kTE_k\|$. Then

$$\|T^{(\infty)}\{x_k\}\|_\omega^2 \geq \omega(\{(1-\frac{1}{k})^2\|E_kTE_k\|^2\}) = \|T\|_\phi^2 \ .$$

So $\|\widetilde{T}\| \geq \|T\|_\phi$. Consequently, the map taking $T+\mathcal{I}_\phi$ to $\widetilde{T}$ is isometric. It is clearly a homomorphism. Thus the theorem is established. ∎

This construction never produces an irreducible representation (except for an atom). Let x_k be a unit vector in $N_{k+1} \ominus N_k$, and let y_k be a unit vector in $N_{k+1}^{\perp}$. Then $\overline{x} = \{x_k\}$ and $\overline{y} = \{y_k\}$ determine unit vectors in $\mathcal{K}$. Furthermore, $(Tx_k,y_k) = 0$ for all k. Hence $(\widetilde{T}\overline{x},\overline{y}) = 0$ for all T in $\mathcal{T}(\mathcal{N})$. Thus $\mathcal{M} = \{\widetilde{T}\overline{x} : T \in \mathcal{T}(\mathcal{N})\}$ is a proper invariant subspace for this representation of $\mathcal{T}(\mathcal{N})$.

Notes and Remarks.

The characterization of the radical (Theorem 6.7) is due to Ringrose [3]. He did not explicitly identify each ϕ as an element of $Hom(\mathcal{N},2)$. This is done in Hopenwasser [1] and Hopenwasser-Larson [1]. The connection between $Hom(\mathcal{N},2)$ and $M_{\mathcal{N}}$ was known to a number of people. It is first used explicitly in Davidson [9] and Apostol-Davidson [1]. Ringrose also proved Theorems 6.10, 6.11 and 6.12. Theorems 6.13, 6.14 and 6.15 are due to Lance [1].

Exercises.

6.1 (Ringrose) Let $\phi_1,\dots,\phi_n$ be finitely many distinct points in $M_{\mathcal{N}}$. Given U_i in $\mathcal{D}_{\phi_i}$, $1 \le i \le n$, find T in $\mathcal{T}(\mathcal{N})$ such that $\Phi_{\phi_i}(T) = U_i$ for $1 \le i \le n$.

6.2 (Lance) Let J be a right (left, or 2-sided) ideal of $\mathcal{T}(\mathcal{N})$ containing the radical. Let $J_\phi = \Phi_\phi(J)$ for each ϕ in $M_{\mathcal{N}}$.

(i) Show that $J = \{T \in \mathcal{T}(\mathcal{N}) : \Phi_\phi(T) \in J_\phi$ for all ϕ in $M_{\mathcal{N}}\}$.

(ii) Deduce that T is invertible in $\mathcal{T}(\mathcal{N})$ if and only if $\Phi_\phi(T)$ is invertible for all ϕ in $M_{\mathcal{N}}$.

6.3 Let A be a positive invertible operator, and let $\mathcal{N}$ be the nest generated by the spectral subspaces $E_A[0,t]$, $t \ge 0$.

(a) (Deddens) Show that $\mathcal{T}(\mathcal{N})$ consists of all operators T such that

$$\sup_{n \ge 0} \| A^n T A^{-n} \| < \infty \ .$$

(b) (Erdos) Show that $rad\, \mathcal{T}(\mathcal{N})$ contains all operators T such that

$$\lim_{n \to \infty} \| A^n T A^{-n} \| = 0 \ .$$

(c) (Erdos) Show that $rad\, \mathcal{T}(\mathcal{N})$ is the closure of the union of the operators given in part (b) as A runs over all positive operators which generate $\mathcal{N}$ by spectral subspaces.

6.4* Is $\mathcal{D}_\phi$ primitive? That is, does $\mathcal{D}_\phi$ have a faithful irreducible representation?

6.5 (Hopenwasser, Larson) Let $\mathcal{L}$ be a complete lattice such that $\{P(L): L \in \mathcal{L}\}$ is commutative (a C.S.L.) and let $C^*(\mathcal{L})$ be the C^* algebra generated by these projections.

(i) Show that $Hom(\mathcal{L},\mathbf{2})$ is homeomorphic to the maximal ideal space of $C^*(\mathcal{L})$.

(ii) For each ϕ in $Hom(\mathcal{L},\mathbf{2})$, let $\mathcal{E}_\phi$ consist of all interval projections of $\mathcal{L}$ such that $\phi(E) = 1$. Define an ideal I_ϕ in $Alg\,\mathcal{L}$ as

$$I_\phi = \{T \in Alg\,\mathcal{L} : \inf_{\mathcal{E}_\phi} \|ETE\| = 0\} \ .$$

(iii) Show that for T in $Alg\,\mathcal{L}$ the following are equivalent:

1) T belongs to $\bigcap\{I_\phi : \phi \in Hom(\mathcal{L},\mathbf{2})\}$.

2) T belongs to the closure of $\bigcup\{rad\,\mathcal{T}(\mathcal{F}) : \mathcal{F}$ finite sublattice of $\mathcal{L}\}$.

3) For every $\epsilon > 0$, there is a finite sublattice $\mathcal{F}$ of $\mathcal{L}$ such that $\|\Delta_{\mathcal{F}}(T)\| < \epsilon$.

(iv) Show that for every non-zero irreducible representation π of $Alg\,\mathcal{L}$, there is exactly one ϕ such that $I_\phi \subseteq ker\,\pi$. Hence deduce that if T satisfies any of the conditions in (iii), then T belongs to $rad(Alg\,\mathcal{L})$.

6.6* Is $rad(Alg\,\mathcal{L}) = \bigcap\{I_\phi : \phi \in Hom(\mathcal{L},\mathbf{2})\}$ in question 6.5?

7. Unitary Invariants For Nests

7.1. We wish to distinguish nests by their spatial characteristics. Two nests $\mathcal{N}$ and $\mathcal{M}$ are *unitarily equivalent* if there is a unitary operator U such that $U\mathcal{N} = \{UN : N \in \mathcal{N}\}$ equals $\mathcal{M}$. The map ϕ_U taking N in $\mathcal{N}$ to UN in $\mathcal{M}$ is clearly an order isomorphism of $\mathcal{N}$ onto $\mathcal{M}$. Thus if $\mathcal{N}$ and $\mathcal{M}$ are unitarily equivalent, they must have the same order type. Now consider two order ω nests (as in Example 2.3), $\mathcal{P} = \{P_n, n \geq 1, \mathcal{H}\}$ with $dim P_n = n$ and $\mathcal{Q} = \{Q_n, n \geq 1, \mathcal{H}\}$ with $dim Q_n = 2n$. These fail to be unitarily equivalent because a unitary operator must preserve dimension. A somewhat more subtle example are the two continuous nests $\mathcal{N}$ on $L^2(0,1)$ and $\mathcal{M}$ on $L^2((0,1)^2)$ given in Examples 2.4 and 2.5. Any unitary operator U taking $\mathcal{N}$ onto $\mathcal{M}$ must carry $\mathcal{N}'$ onto $\mathcal{M}'$. But $\mathcal{N}' = \mathcal{N}''$ is abelian and $\mathcal{M}'$ is not abelian. Both these examples can be resolved by developing a notion of multiplicity for the abelian von Neumann algebras. This treatment is identical to the unitary invariants for self-adjoint operators. This is the content of Theorem 7.4.

We wish to define integrals of functions on a nest. If $\mathcal{N} = \{0 < N_1 < ... < N_k = \mathcal{H}\}$, then we require

$$\int_{\mathcal{N}} f(N)dN = \sum_{i=1}^{n} f(N_i)(P(N_i) - P(N_{i-1})) \equiv I_{\mathcal{N}}(f) \ .$$

If $\mathcal{N}$ is a cotinuous nest, then the appropriate notion is that of a Riemann sum. To combine the two notions, consider finite subsets $\mathcal{F} = \{0 < F_1 < ... < F_k = \mathcal{H}\}$ of $\mathcal{N}$. Say that $\int_{\mathcal{N}} f(N)dN$ exists if $\lim_{\mathcal{F}} I_{\mathcal{F}}(f)$ exists.

7.2 Lemma. *Let f be a continuous scalar valued function on $\mathcal{N}$. Then*

$$A = \int_{\mathcal{N}} f(N)dN \textit{ is well defined.}$$

Proof. We may restrict our attention to real valued functions. By Proposition 2.13, $\mathcal{N}$ is compact. So for every $\epsilon > 0$, there is a finite partition $\mathcal{F}$ consisting of $\{0\} = F_0 < F_1 < ... < F_n = \mathcal{H}$ so that

$$|f(N) - f(N')| < \epsilon \text{ for all } F_{i-1} < N \leq N' \leq F_i \ .$$

If $\mathcal{F}'$ is any refinement of $\mathcal{F}$, it follows that

$$\|I_{\mathcal{F}'}(f) - I_{\mathcal{F}}(f)\| \le \epsilon \ .$$

Thus $\lim_{\mathcal{F}} I_{\mathcal{F}}(f)$ exists. ∎

7.3. Lemma. *Let $\mathcal{N}$ be a nest on a separable Hilbert space. Then there is a self-adjoint operator A such that $\{A\}'' = \mathcal{N}''$ and $\mathcal{N} = \{E_A[0,t] : 0 \le t \le 1\}$ where E_A is the spectral measure of A.*

Proof. By Proposition 2.13, there is an order preserving homeomorphism Φ taking $\mathcal{N}$ into the unit interval such that $\Phi(0) = 0$ and $\Phi(\mathcal{H}) = 1$. Let

$$A = \int_{\mathcal{N}} \Phi(N) dN$$

where this is understood as a Reimann sum. By construction,

$$E_A[0,t] = \Phi^{-1}(t)$$

belongs to $\mathcal{N}$, and every element of the nest occurs in this way. Also notice that $E_A[0,t) = \Phi^{-1}(t)_-$ belongs to $\mathcal{N}$ also.

Since A belongs to $C^*(\mathcal{N})$, it belongs to $\mathcal{N}''$. On the other hand, $\{A\}''$ contains all the spectral projections of A. Thus $\{A\}'' = \mathcal{N}''$. ∎

7.4 Theorem. *Let $\mathcal{N}$ and $\mathcal{M}$ be nests in separable Hilbert spaces. Let Φ be an order preserving homeomorphism of $\mathcal{N}$ into $[0,1]$. Suppose that ψ is an order isomorphism of $\mathcal{N}$ onto $\mathcal{M}$. Then there is a unitary operator U implementing ψ if and only if*

$$A = \int_{\mathcal{N}} \Phi(N) dN \text{ is unitarily equivalent to } B = \int_{\mathcal{M}} \Phi \circ \psi^{-1}(M) dM \ .$$

Proof. If $UN = \psi(N)$ for all N in $\mathcal{N}$, then $B' = UAU^*$ is a self-adjoint operator such that

$$E_{B'}[0,t] = UE_A[0,t] = U\Phi^{-1}(t) = \psi \circ \Phi^{-1}(t) = E_B[0,t]$$

for all t. Thus $B' = B$. Conversely, if $B = UAU^*$, the same identity reversed shows that $UN = \psi(N)$ for all N in $\mathcal{N}$. ∎

7.5. Now we turn to the multiplicity theory of abelian von Neumann algebras. For convenience, we restrict our attention to separably acting algebras. These algebras are always generated by a single self-adjoint operator as in Lemma 7.3. To see this, note that since the unit ball of $\mathcal{M}$ is compact and metrizable in the weak*topology, one can find a countable dense subset of the self-adjoint part. Replace this by their spectral projections $\{E_n, n \ge 1\}$ corresponding to rational intervals. Let $A = \sum_{n \ge 1} 3^{-n} E_n$. Since $\sum_{n \ge 2} 3^{-n} = 1/6 < 1/3$, one obtains $E_1 = E_A[1/3, 1/2]$. Similarly, one finds that $C^*(A)$ contains $\{E_n, n \ge 1\}$. Thus, $\{A\}''$ equals $\mathcal{M}$.

Let X be the spectrum of A. Then $C^*(A)$ is isomorphic to $C(X)$ by the Gelfand theory. By the spectral theorem, there is a spectral measure $E_A(\cdot)$ on Borel sets of X, which is a countably additive, projection valued measure such that

$$f(A) = \int f(t)dE_A(t)$$

is defined for all bounded, Borel functions $Bor(X)$. This functional calculus is a continuous $*$-representation of $Bor(X)$ in that $fg(A) = f(A)g(A)$, $\overline{f}(A) = f(A)^*$ and $\|f(A)\| \le \|f\|_\infty$ In particular, we obtain the following immediate corollary of Lemma 7.3.

7.6 Corollary. *Let Φ be an order preserving homeomorphism of $\mathcal{N}$ onto a subset ω of $[0,1]$. Then there is a spectral measure $E_{\mathcal{N}}$ on the Borel subsets of ω such that $E_{\mathcal{N}}[0,t] = P(\Phi^{-1}(t))$.*

7.7 Lemma. *If f_n is a bounded sequence in $Bor(X)$ converging pointwise to f, then $f_n(A)$ converges weakly to $f(A)$. Consequently, $C^*(A)$ is weakly dense in the image of $Bor(X)$, which is therefore contained in $\mathcal{M}$.*

Proof. Let x, y be vectors in $\mathcal{H}$. Consider the scalar measure

$$\mu_{x,y}(\cdot) = (E_A(\cdot)x,y) \ .$$

By the Lebesgue Dominated Convergence Theorem,

$$\lim_{n\to\infty}(f_n(A)x,y) = \lim\int f_n(t)d\mu_{x,y}(t) = \lim\int f(t)d\mu_{x,y}(t) = (f(A)x,y) \ .$$

Hence $f_n(A)$ converges weakly to $f(A)$. Since every bounded Borel function on X is the pointwise limit of a bounded sequence of continuous functions, $C^*(A)$ is weakly dense in the image. But the weak operator closure of $C^*(A)$ is $\mathcal{M}$, concluding the proof. ■

In fact, $Bor(X)$ is taken *onto* $\mathcal{M}$. This will follow from Corollary 7.9.

An abelian von Neumann algebra is called *multiplicity free* if its commutant is also abelian. The following theorem characterizes these algebras.

7.8 Theorem. *Let $\mathcal{M}$ be an abelian von Neumann algebra, and let A be a self adjoint operator such that $\mathcal{M} = W^*(A)$. The following are equivalent:*

1) *$\mathcal{M}$ is multiplicity free.*
2) *$\mathcal{M}$ has a cyclic vector.*
3) *$\mathcal{M}$ is maximal abelian.*
4) *$\mathcal{M}$ is unitarily equivalent to the algebra $L^\infty(\mu)$ acting by multiplication on $L^2(\mu)$, where μ is a regular Borel measure on $X = \sigma(A)$.*

Proof. 1) implies 2) follows from Proposition 2.12 applied to $\mathcal{M}'$. Next, we show that 2) implies 4). Let x be a unit cyclic vector and define the state $\tau(M) = (Mx,x)$ for M in $\mathcal{M}$. The restriction of τ to $C^*(A)$ is a positive linear functional of norm one. By the Riesz Representation Theorem, there is a regular Borel measure μ (with $\|\mu\| = 1$) such that

$$\tau(f(A)) = \int f d\mu$$

for all f in $C(X)$.

Define a linear map $U{:}C(X)$ into $\mathcal{H}$ by $Uf = f(A)x$. Then

$$\begin{aligned}(Uf,Ug) &= (f(A)x,g(A)x) = ((\bar{g}f)(A)x,x) \\ &= \tau(\bar{g}f) = \int f\bar{g}d\mu \quad .\end{aligned}$$

Hence U extends to an isometry of $L^2(\mu)$ onto the closure of $\{f(A)x : f \in C(X)\}$. This contains, in particular, Mx for all M in $\mathcal{M}$ since $C^*(A)$ is strongly dense. But x is cyclic for $\mathcal{M}$, and hence U is unitary. Furthermore, for f and g in $C(X)$,

$$(UM_gU^*)f(A)x = UM_gf = Ugf = g(A)(f(A)x) \quad .$$

Hence $UM_gU^* = g(A)$ for all g in $C(X)$.

Since conjugation by U carries $C^*(M_x)$ onto $C^*(A)$, it carries the weak closure of $C^*(M_x)$ onto $\mathcal{M}$. It suffices to show that this algebra is multiplication by $L^\infty(\mu)$. By Lemma 7.7, M_f belongs to $C^*(M_x)''$ for every bounded Borel function. Since x is a separating vector for $\mathcal{M}$, $U^*x = 1$ is a separating vector for $C^*(M_x)''$. If $f = 0$ a.e. (μ), then

$$\|M_f1\|^2 = (M_{|f|^2}1,1) = \int |f|^2 d\mu = 0 \quad .$$

So the map taking g in $L^\infty(\mu)$ to M_g is well defined. On the other hand, suppose that T commutes with M_x. Let $g = T1$. For every f in $C(X)$,

$$Tf = TM_f1 = M_fT1 = fg = M_gf \quad .$$

But T is continuous, and $C(X)$ is dense in $L^2(\mu)$, so this extends to all of $L^2(\mu)$. If $r < \|g\|_\infty$, let y be the characteristic function of $E = \{x : |g(x)| > r\}$. Then

$$\|T\|^2 \geq \|Ty\|^2/\|y\|^2 = \int_E |g|^2 d\mu/\mu(E) \geq r^2 \quad .$$

Thus g belongs to $L^\infty(\mu)$. So $L^\infty(\mu) \subseteq C^*(M_x)'' \subseteq C^*(M_x)' \subseteq L^\infty(\mu)$.

This also proves that 4) implies 3). Finally, 3) implies 1) is trivial. ∎

7.9 Corollary. *Let $\mathcal{M}$ be an abelian von Neumann algebra on a separable Hilbert space. Then there is a regular Borel measure μ on $\mathbb{R}$ such that $\mathcal{M}$ is $*$-isomorphic to $L^\infty(\mu)$.*

Proof. By Proposition 2.12, $\mathcal{M}$ has a separating vector x. Let $\mathcal{M}_0$ be the restriction of $\mathcal{M}$ to the invariant subspace $\mathcal{H}_x$ which is the closed span of $\{Mx : M \in \mathcal{M}\}$. The restriction map taking M to $M|\mathcal{H}_x$ is a $*$-homomorphism. It is one to one since x is a separating vector, and thus it is a $*$-isomorphism. Clearly, x is a cyclic vector for $\mathcal{M}_0$. Thus by Theorem 7.8, there is a regular Borel measure μ such that $\mathcal{M}_0$ is unitarily equivalent to $L^\infty(\mu)$. ∎

7.10. It follows from this corollary that given a spectral measure E on the real line, one can find a scalar spectral measure mutually absolutely continuous to E. That is, if E is a spectral measure for $\mathcal{M}$ and μ is the measure constructed above, then $\mu(F) = 0$ if and only if $E(F) = 0$ for any Borel subset F of $\mathbb{R}$. The measure μ is determined up to its measure class $[\mu] = \{v : v \sim \mu\}$ since if $v \sim \mu$, then $L^\infty(v) = L^\infty(\mu)$. So given a nest $\mathcal{N}$ of order type X, one can talk about the measure class of the spectral measure.

7.11. For any cardinal $1 \leq n \leq \aleph_0$, a collection $\{U_{ij}, 0 \leq i,j < n\}$ of partial isometries are called $(n \times n)$ *matrix units* provided that $U_{ij}U_{k\ell} = \delta_{jk}U_{i\ell}$ for all i,j,k,ℓ and $\sum_{i<n} U_{ii} = I$. An abelian von Neumann algebra $\mathcal{M}$ has *uniform multiplicity* n if $\mathcal{M}'$ contains a system $\{U_{ij}, 0 \leq i,j < n\}$ of $n \times n$ matrix units such that $\mathcal{M}_0 = \mathcal{M}\,|U_{00}\mathcal{H}$ is multiplicity free. Thus, $\mathcal{M}$ is unitarily equivalent to

$$\mathcal{M}_0^{(n)} = \{A^{(n)} = A \oplus A \oplus \ldots \oplus A : A \in \mathcal{M}_0\}$$

acting on $\mathcal{H}_0^{(n)}$, the direct sum of n copies of $\mathcal{H}_0 = U_{00}\mathcal{H}$. The unitary that does this is $\sum_{0 \leq i < n} \oplus U_{i0}$. This same unitary carries $\mathcal{M}_0^{(n)\prime}$ onto $\mathcal{M}'$, and $\mathcal{M}_0^{(n)\prime}$ is $\mathcal{M}_0 \otimes M_n(\mathbb{C})$, the algebra of $n \times n$ matrices with coefficients in $\mathcal{M}_0$ acting on $\mathcal{H}_0^{(n)}$. If $n = \aleph_0$, replace $M_n(\mathbb{C})$ by $\mathcal{B}(\mathcal{H})$.

7.12 Lemma. *If $\mathcal{M}$ has uniform multiplicity, then the multiplicity is well defined.*

Proof. Suppose $\mathcal{M}$ has uniform multiplicity n and m with $n \leq m$ and $n < \aleph_0$. By the remarks above, $\mathcal{M}'$ is unitarily equivalent to $\mathcal{M}_0 \otimes \mathcal{M}_n(\mathbb{C})$ and M_0 is abelian. Thus if σ is any multiplicative linear functional on $\mathcal{M}_0$, $\sigma \otimes id_n$ yields a non-zero $*$-homomorphism of $\mathcal{M}'$ onto $M_n(\mathbb{C})$. Suppose that $\mathcal{M}'$ is also unitarily equivalent to $\mathcal{M}_1 \otimes \mathcal{M}_m(\mathbb{C})$. The restriction of $\sigma \otimes id$ to the subalgebra unitarily equivalent to $\mathbb{C}I \otimes M_m(\mathbb{C})$ is a unital $*$ homomorphism (thus not zero) of $M_m(\mathbb{C})$ (or $\mathcal{B}(\mathcal{H})$ if $m = \aleph_0$) onto $M_n(\mathbb{C})$. This forces $m = n$ (because M_m has no ideals and thus $dim M_m = m^2 = n^2$; and $\mathcal{B}(\mathcal{H})$ has only one ideal, $\mathcal{K}$, of infinite codimension). ∎

Let $\mathcal{M}$ be an abelian von Neumann algebra. Say that a non-zero projection P in $\mathcal{M}$ has multiplicity n if $\mathcal{M}\,|P\mathcal{H}$ has uniform multiplicity n.

7.13 Lemma. *If P is a projection of multiplicity n in an abelian von Neumann algebra $\mathcal{M}$, then so does every non-zero projection $P' \leq P$. If P_α are projections in $\mathcal{M}$ of multiplicity n, so is $P = \sup P_\alpha$.*

Proof. First, $\mathcal{M}\,|P\mathcal{H}$ is unitarily equivalent to $L^\infty(\mu)^{(n)}$. If $P' \leq P$, then P' is unitarily equivalent to $M_\chi^{(n)}$ where χ is the characteristic function of a subset X_0 of X. Let $\mu' = \mu\,|X_0$. Clearly, $\mathcal{M}\,|P'\mathcal{H}$ is unitarily equivalent to $L^\infty(\mu')^{(n)}$, and thus has multiplicity n.

Let $P = \sup P_\alpha$. It is not difficult to write P as an orthogonal direct sum $P = \sum_{k=1}^{\infty} P_k$ where $P_k \leq P_{\alpha_k}$ for some α_k (use separability). Each P_k has multiplicity n by the first paragraph. By Corollary 7.9, $\mathcal{M}$ is isomorphic to $L^\infty(\mu)$. So $\{P_k\}$ correspond to M_{χ_k} where χ_k are characteristic functions of pairwise disjoint sets X_k. Let $X_0 = \bigcup_{k=1}^{\infty} X_k$ and $\mu_k = \mu\,|X_k$ for $k \geq 0$. Then $\mathcal{M}\,|P_k\mathcal{H}$ is unitarily equivalent to $L^\infty(\mu_k)^{(n)}$ for all $k \geq 1$. By taking the direct sum, $\mathcal{M}\,|P\mathcal{H}$ is unitarily equivalent to $L^\infty(\mu_0)^{(n)}$, and thus has multiplicity n. ■

With this preparation, we can now state and prove the full classification theorem of (separably acting) abelian von Neumann algebras.

7.14 Theorem. *Let $\mathcal{M}$ be a separably acting, abelian von Neumann algebra. Then there are pairwise orthogonal projections P_n in $\mathcal{M}$, $1 \leq n \leq \aleph_0$ such that P_n is of multiplicity n in $\mathcal{M}$, and $\sum_{1 \leq n \leq \aleph_0} P_n = I$. Let A be a self-adjoint operator such that $\mathcal{M} = W^*(A)$. Then there are mutually singular measures μ_n on $X = \sigma(A)$, $1 \leq n \leq \infty$, so that $\mathcal{M}$ is unitarily equivalent to*

$$L^\infty(\mu_\infty)^{(\infty)} \oplus \sum_{n=1}^{\infty} \oplus\, L^\infty(\mu_n)^{(n)}$$

acting on $L^2(\mu_\infty)^{(\infty)} \oplus \sum_{n=1}^{\infty} \oplus\, L^2(\mu_n)^{(n)}$.

Proof. For each cardinal n, $1 \leq n \leq \aleph_0$, Lemma 7.13 shows that the supremum P_n of all projections P_α in $\mathcal{M}$ of multiplicity n is the largest projection of multiplicity n. For $n \neq m$, $P_n P_m$ is a projection less than both P_n and P_m. By Lemma 7.13, it has multiplicity both n and m. By Lemma 7.12, $P_n P_m = 0$. So $\{P_n, 1 \leq n \leq \aleph_0\}$ are pairwise orthogonal. It must be shown that $\sum_{1 \leq n \leq \aleph_0} P_n = I$. Let $Q = I - \sum_{1 \leq n \leq \aleph_0} P_n$.

Set $\mathcal{M}_0 = \mathcal{M}\,|Q\mathcal{H}$. If $Q \neq 0$, it suffices to find a projection R in $\mathcal{M}_0$ of uniform multiplicity. Recursively choose unit vectors x_n and projections Q_n onto $\{Mx_n : M \in \mathcal{M}_0\}$ such that $\{Q_n, n \geq 1\}$ are pairwise orthogonal and x_{n+1} is a separating vector for $\mathcal{M}\,|(\sum_{i=1}^{n} Q_i)^{\perp}\mathcal{H}$. The projections Q_n belong to $\mathcal{M}'_0$. Let $\overline{Q}_n$ denote the smallest projection in $\mathcal{M}_0$ such that $\overline{Q}_n \geq Q_n$. (This is known as the central cover of Q_n). Then $I = \overline{Q}_1 \geq \overline{Q}_n \geq \overline{Q}_{n+1}$ for all $n \geq 1$. There are two cases: (i) there is an integers n_0 such that $\overline{Q}_{n_0} = I > \overline{Q}_{n_0+1}$, and (ii) $\overline{Q}_n = I$ for all $n \geq 1$.

In case (i), let $R = I - \overline{Q}_{n_0+1}$. Let $\mathcal{M}_R = \mathcal{M}\,|R\mathcal{H}$ and let $A_R = A\,|R\mathcal{H}$. So $\mathcal{M}_R = W^*(A_R)$ is isomorphic to $L^\infty(\mu)$ where μ is a Borel measure on $X_R = \sigma(A_R)$ via the isomorphism $\Phi(f) = f(A_R)$. Since $Rx_{n_0+1} = 0$ and x_{n_0+1} is a separating vector for $\mathcal{M}_0\,|(\sum_{i=1}^{n_0} Q_i)^{\perp}\mathcal{H}$, it follows that $R \leq \sum_{i=1}^{n_0} Q_i$. Let $R_i = RQ_i$ and $y_i = Rx_i$ for $1 \leq i \leq n_0$. Then $\mathcal{M}_i = \mathcal{M}\,|R_i\mathcal{H}$ has a cyclic vector y_i. By Theorem 7.8, there is a regular Borel measure μ_i on X_R and a unitary U_i of $L^2(\mu_i)$ onto $R_i\mathcal{H}$ such that

$$U_i M_f U_i^* = f(A_R)\,|R_i\mathcal{H}$$

for every bounded Borel function f. This implements a * isomorphism Φ_i of $L^\infty(\mu_i)$ onto $\mathcal{M}_i$. Now y_i is a separating vector for $\mathcal{M}_R$. Thus, as in Corollary 7.9, the restriction map $\Psi_i(M) = M\,|R_i\mathcal{H}$ is a * isomorphism of $\mathcal{M}_R$ onto $\mathcal{M}_i$. Hence $\Phi_i^{-1} \circ \Psi_i \circ \Phi$ is a * isomorphism of $L^\infty(\mu)$ onto $L^\infty(\mu_i)$ which takes every Borel function f to itself. Hence μ_i and μ are mutually absolutely continuous, whence $L^\infty(\mu_i) = L^\infty(\mu)$ for $1 \leq i \leq n_0$. So $U = \sum_{i=1}^{n_0} \oplus U_i$ carries $L^2(\mu)^{(n_0)}$ onto $R\mathcal{H} = \sum \oplus R_i\mathcal{H}$ and $UM_f^{(n_0)}U^* = f(A_R)$ for every f in $L^\infty(\mu)$. Evidently, $\mathcal{M}_R$ has uniform multiplicity n, contradicting the maximality of P_{n_0}.

Case (ii) is dealt with in a similar manner. In this case, x_n is a separating vector for each n and $\{Q_n : n \geq 1\}$ are pairwise orthogonal. Extend this to a maximal family $\{y_n, n \geq 1\}$ of separating vectors such that the projections Q'_n onto $\{My_n : M \in \mathcal{M}\}^-$ are pairwise orthogonal. This family is countable since $\mathcal{H}$ is separable. Let $Q = (\sum_{n \geq 1} Q_n)^{\perp}$. Then $\overline{Q} \neq I$, for otherwise Proposition 2.12 applied to $\mathcal{M}\,|Q\mathcal{H}$ would yield a separating vector $y = Qy$, contradicting the maximality for $\{y_n\}$. Let $R = I - \overline{Q}$. Proceeding as in the previous paragraph, one shows that R has uniform infinite multiplicity. Again, a contradiction has been obtained.

For $1 \leq n \leq \aleph_0$, let $\mathcal{M}_n = \mathcal{M}|P_n\mathcal{H}$. Then $\mathcal{M}_n$ has uniform multiplicity n. By Corollary 7.9 and Remark 7.11, $\mathcal{M}_n$ is unitarily equivalent to $L^\infty(\mu_n)^{(n)}$. Furthermore, $\mathcal{M}$ is * isomorphic to $L^\infty(\mu)$ via $\Phi(f) = f(A)$. The projections $P_n = \Phi(\chi_n)$ where χ_n are the characteristic functions pairwise disjoint measurable sets E_n such that $\sum \mu(E_n) = 1$. Since P_n is the identity of $\mathcal{M}_n$, the measure μ_n is supported on E_n and may be taken to be the restriction of μ to E_n. In particular, the $\{\mu_n\}$ are mutually singular. Putting this all together yields a unitary equivalence between $\mathcal{M}$ and

$$L^\infty(\mu_\infty)^{(\infty)} \oplus \sum_{n=1}^{\infty} \oplus L^\infty(\mu_n)^{(n)} \ . \qquad \blacksquare$$

Let A be a self-adjoint operator. The complete unitary invariants of A can be read of from Theorem 7.14. First, there is the spectrum $X = \sigma(A)$. Then there is a spectral measure $E(\cdot)$. By 7.11, there is a class $[\mu]$ of scalar measures mutually absolutely continuous to E. Define a multiplicity function $m(x)$ by setting $m(x) = n$ if x belongs to E_n where E_n is the support of the multiplicity n part of $W^*(A)$. Then m is defined μ almost everywhere.

7.15 Theorem. *Two self-adjoint operators A and B are unitarily equivalent if and only if they have the same spectrum, scalar spectral measure, and multiplicity function.*

Proof. If $B = UAU^*$, the $f(B) = Uf(A)U^*$ for every bounded Borel function f. Thus, by Theorem 7.14, they have the same measure class and multiplicity function. Conversely, if they have the same measure class and multiplicity function, Theorem 7.14 provides a unitary U such that $f(B) = Uf(A)U^*$ for every bounded Borel function f. In particular, $B = UAU^*$. $\blacksquare$

7.16 Corollary. *Let $\mathcal{N}$ and $\mathcal{M}$ be two nests order isomorphic to a subset ω of $[0,1]$. Let Φ and Ψ be order isomorphisms of $\mathcal{N}$ and $\mathcal{M}$ onto ω. Then $\psi^{-1}\cdot\Phi$ is implemented by a unitary if and only if the two spectral measures for $\int_{\mathcal{N}}\Phi(N)dN$ and $\int_{\mathcal{M}}\Psi(M)dM$ have the same measure class and multiplicity function.*

Proof. Apply Theorems 7.4 and 7.15. $\blacksquare$

There is a certain difficulty with this Corollary. Namely, one must specify the order isomorphisms Φ and Ψ. These may not be unique. Thus to decide if $\mathcal{N}$ and $\mathcal{M}$ are unitarily equivalent, one must let Ψ run over all order isomorphisms of $\mathcal{M}$ onto ω (Φ may be fixed). The problem can be reduced to a measure theoretic problem. Consider the following special cases. A nest $\mathcal{N}$ is *multiplicity free* if $\mathcal{N}'$ is abelian.

7.17 Proposition. *Let $\mathcal{N}$ and $\mathcal{M}$ be two continuous, multiplicity free nests. Then $\mathcal{N}$ and $\mathcal{M}$ are unitarily equivalent.*

Proof. It suffices to show that $\mathcal{M}$ (and $\mathcal{N}$) is unitarily equivalent to the Volterra nest $\mathcal{N}_0$ given by $N_t = \{f \in L^2(0,1): supp(f) \subseteq [0,t]\}$. Let x be a unit cyclic and separating vector for $\mathcal{M}''$. The map $\Phi_x(M) = (Mx,x)$ carries $\mathcal{M}$ onto $[0,1]$. Let $E_{\mathcal{M}}(\cdot)$ be the spectral measure of $\mathcal{M}$, and define a scalar measure by $\mu(\cdot) = (E_{\mathcal{M}}(\cdot)x,x)$. It is clear that μ is mutually absolutely continuous with E. Furthermore

$$\mu(a,b] = (\Phi_x^{-1}(b)-\Phi_x^{-1}(a)x,x) = \Phi_x(\Phi_x^{-1}(b))-\Phi_x(\Phi_x^{-1}(a)) = b-a \ .$$

Hence μ equals Lebesgue measure. By Corollary 7.16, there is a unitary U of $L^2(0,1)$ onto $\mathcal{H}$ such that $UN_t = \Phi_x^{-1}(t)$ for all $0 \le t \le 1$. ∎

7.18 Example. It is implicit in 7.17 that $\mathcal{M}$ is parametrized by the spatially related function Φ_x, not some more arbitrary indexing. Consider a nonatomic measure μ on $(0,1)$ with $supp(\mu) = [0,1]$. Define $\mathcal{M}$ in $L^2(\mu)$ by setting

$$M_t = \{f \in L^2(\mu): supp(f) \subseteq [0,t]\} \ .$$

Then by Corollary 7.16, there is a unitary map U of $L^2(\mu)$ onto $L^2(0,1)$ such that $UM_t = N_t$ for $0 \le t \le 1$ if and only if $\mu \sim m$ (Lebesgue measure). On the other hand, if one sets $h(t) = \mu[0,t]$, then the proof of Proposition 7.17 shows that there is always a unitary U such that $UM_t = N_{h(t)}$ for $0 \le t \le 1$. More generally, if h is an order preserving homeomorphism of $[0,1]$ onto itself, then there is a unitary operator U such that $UN_t = N_{h(t)}$ if and only if the Lebesgue-Stieltjes measure dh is mutually absolutely continuous with Lebesgue measure m.

7.19 Example. Let $\mathcal{Q}$ be the Cantor nest of Example 2.6. This is a totally atomic, multiplicity free nest. Let $\mathcal{R}$ be the Cantor nest defined on $L^2(\mu) \oplus L^2(\mathbb{R})$ by $R_t^+ = \{f : supp(f) \subseteq (-\infty,t]\}$ and $R_t^- = \{f : supp(f) \subseteq (-\infty,t)\}$. So $R_t^{\pm} = Q_t^{\pm} \oplus N_t$ where N_t belongs to the canonical continuous nest on $L^2(\mathbb{R})$. The nest $\mathcal{R}$ is multiplicity free also. To see this, note that $\mathcal{R}''$ contains projections $A_t = (Q_t^+ - Q_t^-) \oplus 0$ onto all the atoms, and hence contains $I \oplus 0$. From this, one obtains that $\mathcal{R} \cong L^\infty(\mu) \oplus L^\infty(m) = L^\infty(\mu+m)$ which is multiplicity free. The map Φ taking $Q_t^{\pm}$ to $R_t^{\pm}$ is an order isomorphism of $\mathcal{Q}$ onto $\mathcal{R}$. However, no unitary operator can take $\mathcal{Q}$ onto $\mathcal{R}$ because $\mathcal{Q}''$ and $\mathcal{R}''$ are not $*$-isomorphic.($\mathcal{Q}''$ is atomic, isomorphic to ℓ^∞, and $\mathcal{R}''$ is isomorphic to $\ell^\infty \oplus L^\infty(\mathbb{R})$.) Thus Proposition 7.17 does not even extend to the general multiplicity free case.

7.20 Example. Let C be the Cantor set, and let h be a order preserving homeomorphism of $[0,1]$ such that $h(C)$ has positive measure. Let C_0 be a countable dense subset of C. Let μ be counting measure on C_0, and let υ be counting measure on $h(C_0)$. Define nests $\mathcal{L}$ and $\mathcal{M}$ in $L^2(\mu+m)$ and $L^2(\upsilon+m)$ respectively by

$$\begin{aligned} L_t^+ &= \{f \in L^2(\mu+m): supp(f) \subseteq [0,t]\} \\ L_t^- &= \{f \in L^2(\mu+m): supp(f) \subseteq [0,t)\} \\ M_t^+ &= \{f \in L^2(\upsilon+m): supp(f) \subseteq [0,t]\} \\ M_t^- &= \{f \in L^2(\upsilon+m): supp(f) \subseteq [0,t)\} \ . \end{aligned}$$

Both $\mathcal{L}$ and $\mathcal{M}$ are multiplicity free nests, and h is an order isomorphism of $\mathcal{L}$ onto $\mathcal{M}$. Furthermore, $\mathcal{L}'' \cong \ell^\infty \oplus L^\infty(m) \cong \mathcal{M}''$.

Suppose that there were a unitary operator U taking $\mathcal{L}$ onto $\mathcal{M}$. Let h' be the order isomorphism induced by U. By Corollary 7.16, h' carries the measure class of $\mu+m$ onto that of $\upsilon+m$. Now h' must take C_0 onto $h(C_0)$. Since h' is continuous, $h'(C) = h(C)$. But $E_{\mathcal{L}}(\cdot)\,|C$ is equivalent to μ which is atomic, whereas $E_{\mathcal{M}}(\cdot)\,|h(C)$ is equivalent to $\upsilon+m\ |h(C)$ which has a non trivial non-atomic part. This is impossible, so $\mathcal{L}$ and $\mathcal{M}$ are not unitarily equivalent.

7.21. Implicit in the proof of Proposition 7.17 is the fact that when x is a unit separating vector for $\mathcal{M}''$, the map $\Phi_x(M) = (P(M)x,x)$ is an order isomorphism of $\mathcal{M}$ which encodes the spectral measure E_x via Lebesgue measure in the sense that the measure μ_x on $X = \Phi_x(\mathcal{M})$ given by $\mu_x(X \cap [0,t]) = t$ is mutually absolutely continuous with E_x. In fact, let $[0,1]\backslash X = \bigcup_{n\geq 1}(\ell_n,r_n)$ is the disjoint union of open intervals. Then

$$\mu_x(S) = m(S \cap X) + \sum_{r_n \in S}(r_n - \ell_n) \ . \tag{7.21.1}$$

Let the *hull class* of $\mathcal{M}$ denote the set

$$h(\mathcal{M}) = \{\Phi_x(\mathcal{M}) : x \text{ is a unit separating vector for } \mathcal{M}''\} \ .$$

Say that an order preserving homeomorphism is absolutely continuous if f and f^{-1} preserve sets of measure zero. Let the set of all such maps be denoted by $AbsHom[0,1]$.

7.22 Proposition. *Let $\mathcal{M}$ be a multiplicity free nest, and let $X = \Phi_x(\mathcal{M})$ for some unit separating vector x. Then $h(\mathcal{M})$ consists of $\{f(X): f \in AbsHom[0,1]\}$.*

Proof. By Theorems 7.8 and 7.4, $\mathcal{M}$ is unitarily equivalent to the nest on $L^2(\mu_x)$ given by the subspaces M_t of functions supported in $[0,t]$ for t in $[0,1]$. Under this equivalence, x is taken to the constant function 1. We will assume that $\mathcal{M}$ is in this form.

If f and f^{-1} preserve sets of measure zero, then there is a strictly positive function h in $L^1(0,1)$ such that

$$f(t) = \int_0^t h(s)ds \quad .$$

Define a vector y in $L^2(\mu_x)$ by

$$y(r_x) = \left(\int_{\ell_n}^{r_n} h(s)ds\right)^{1/2} , \quad n \geq 1$$

$$y(t) = h(t)^{1/2} \quad , \quad t \in X \backslash \{r_n, n \geq 1\} \quad .$$

Then

$$\Phi_y(M_t) = \int_{X \cap [0,t]} |y(s)|^2 d\mu_x(s) = \int_0^t h(s)ds = f(t) \quad .$$

On the other hand, let y be a separating vector for $\mathcal{M}''$ and let $f(t) = (P(M_t)y,y)$. Then f carries X onto $\Phi_y(\mathcal{M})$. Extend f to a homeomorphism of $[0,1]$ by making it linear on each $[\ell_n, r_n]$. Let C be a subset of $[0,1]$ with measure zero, and let $C_0 = C \cap X \backslash \{r_n, n \geq 1\}$. Then $\mu_x(C_0) = m(C_0 \cap X) = 0$. Thus $M_{\chi_{C_0}} = 0$ in $L^\infty(\mu_x)$, so

$$m(f(C_0)) = \mu_y(C_0) = (M_{\chi_{C_0}} y, y) = 0 \quad .$$

As f is piecewise linear on $\bigcup [\ell_n, r_n]$, $m(f(C \backslash C_0)) = 0$. So f preserves sets of measure zero. Interchanging the roles of x and y shows that this also holds for f^{-1}. ■

7.23 Theorem. *Let $\mathcal{M}$ and $\mathcal{N}$ be multiplicity free nests. Then the following are equivalent:*

1) *$\mathcal{N}$ is unitarily equivalent to $\mathcal{M}$.*
2) *$h(\mathcal{N}) = h(\mathcal{M})$.*
3) *$h(\mathcal{N})$ intersects $h(\mathcal{M})$.*
4) *Given x, y unit separating vectors for $\mathcal{N}$ and $\mathcal{M}$, there is a function f in Abs Hom$[0,1]$ such that $f(\Phi_x(\mathcal{N})) = \Phi_y(\mathcal{M})$.*

Proof. Clearly, 1) implies 2) implies 3). But the first paragraph of Proposition 7.22 shows 3) implies that both $\mathcal{N}$ and $\mathcal{M}$ are unitarily equivalent to the same nest. Proposition 7.22 also shows the equivalence of 2) and 4).

∎

This leads to a slight improvement on Corollary 7.16. Given a nest $\mathcal{N}$, let x be a unit separating vector and define $\Phi_x(N) = (P(N)x,x)$. Let $X = \Phi_x(\mathcal{N})$. Define a multiplicity function $m_x(t)$ on $[0,1]$ defining it to be the multiplicity function of $A = \int_{\mathcal{N}} \Phi_x(N) dN$ for t in X, and on each interval (ℓ_n, r_n) corresponding to an atom A_n of $\mathcal{N}$, set $m_x(t) = -m_x(\{r_n\}) = -rank(A_n)$.

7.24 Theorem. *Let $\mathcal{N}$ and $\mathcal{M}$ be two nests with separating vectors x and y for $\mathcal{N}''$ and $\mathcal{M}''$ respectively. Then $\mathcal{N}$ and $\mathcal{M}$ are unitarily equivalent if and only if there is an absolutely continuous order homeomorphism f of $[0,1]$ such that $m_y(t) = m_x(f(t))$.*

Proof. The point is simply that the measure class of $\mathcal{N}$ on $\Phi_x(\mathcal{N})$ and $\Phi_y(\mathcal{M})$ are given as in (7.21.1) from Lebesgue measure on $[0,1]$. The homeomorphisms of $[0,1]$ which preserve this measure class are precisely $AbsHom[0,1]$. The condition $m_y = m_x \circ f$ shows that multiplicity is also preserved. Thus by Corollary 7.16, $\mathcal{N}$ and $\mathcal{M}$ are unitarily equivalent. The reverse implication is easy. ∎

7.25. Why is this not the whole answer? Because given $\mathcal{N}$, x, and m_x and likewise $\mathcal{M}$, y and m_y, how do you determine if an f in $AbsHom[0,1]$ exists such that $m_y = m_x \circ f$? The reason it is an improvement on 7.16 is that it replaces arbitrary measure classes by the more familiar Lebesgue measure. The resulting problem remaining is strictly a measure theoretic one. Given two partitions $\{A_n, n \in \mathbb{Z}\}$ and $\{B_n, n \in \mathbb{Z}\}$ of $[0,1]$ into Borel sets, determine when there is an f in $AbsHom[0,1]$ such that $m(f(A_n) \Delta B_n) = 0$ for all n.

Notes and Remarks.

The classification of self-adjoint operators is the classical theory of Hahn and Hellinger. The treatment used here was greatly influenced by Arvèson's book on C^*-algebras [8]. The corresponding theory for nests was developed by Erdos [2]. As he wished to deal with the non-separable case, he develops the theory from the beginning. Here, we have chosen to use Theorem 7.4 instead. Proposition 7.17 is due to Kadison and Singer [1], as is Theorem 7.23. Theorem 7.24 is new.

Exercises.

7.1 (Weyl-von Neumann) Show that every self-adjoint operator T on a separable Hilbert space is the sum of a diagonal operator and a compact one. Given $\epsilon > 0$, one can choose the compact operator with norm at most ϵ.

Hint: Reduce to the case of a cyclic vector x. Let $E_{j,n}$ be the spectral projections for the intervals $[2^{-n}j, 2^{-n}(j+1))$ and let P_n be the projection onto $span\{E_{j,n}x, j \in \mathbb{Z}\}$. Show that $rank P_n < \infty$, P_n increases strongly to I, and $\lim_{n \to \infty} \|P_n T - TP_n\| = 0$.

7.2 (a) (Halmos) Use the arguments of section 7.2 to show that if $\{N_k, k \geq 1\}$ is a family of commuting normal operators, then there is a single self-adjoint operator T such that $C^*(T)$ contains $\{N_k, k \geq 1\}$.

(b) (Berg) Deduce that if $\{N_k, k \geq 1\}$ is a family of commuting normal operators, then there is an atomic, maximal abelian von Neumann algebra $\mathcal{D}$ such that all N_k belong to $\mathcal{D} + \mathcal{K}$.

(c) (Berg) In particular, if N is normal and $\epsilon > 0$, show that $N = D + K$ where D is diagonal, K is compact, and $\|K\| < \epsilon$.

(d) (Berg) Suppose N and M are normal operators such that $\sigma(N) = \sigma_e(N) = \sigma_e(M) = \sigma(N)$. Given $\epsilon > 0$, show that there is a unitary U so that $N - UMU^*$ is compact of norm at most ϵ.

(e) Two operators S and T are *approximately unitarily equivalent* ($S \underset{a}{\sim} T$) if for every $\epsilon > 0$, there is a unitary U such that $\|S - UTU^*\| < \epsilon$. Show that two normal operators N and M on a seperable $\mathcal{H}$ have $N \underset{a}{\sim} M$ if and only if $\sigma(N) = \sigma(M)$ and the multiplicity of isolated eigenvalues are equal.

7.3 (Type I von Neumann Algebras) A von Neumann algebra $\mathcal{M}$ is type I if every central projection dominates a non-zero projection P in $\mathcal{M}$ such that $P\mathcal{M}P$ is abelian (an "abelian projection"). $\mathcal{M}$ is of type I_{nm}, $1 \leq n, m \leq \aleph_0$ if there is a maximal abelian von Neumann algebra $\mathcal{M}_0$ so that $\mathcal{M}$ is unitarily equivalent to $\mathcal{M}_0 \otimes M_n(\mathbb{C}) \otimes \mathbb{C}I_m$ on $\mathcal{H} \otimes \mathbb{C}^n \otimes \mathbb{C}^m$. Show that if $\mathcal{M}$ is type I, then there are pairwise orthogonal central projections E_{nm} with $\sum E_{nm} = I$ such that $\mathcal{M} | E_{nm}\mathcal{H}$ is type I_{nm}.

Outline:

i) Find a maximal abelian projection P. Show that the map of $\mathcal{M}'$ to $\mathcal{M}'_P$ is an isomorphism.

ii) Use Theorem 7.14 to decompose $\mathcal{M}'_P$. If P_m are the projections obtained, let E_m be the minimal central projections dominating them. $\mathcal{M}_{E_m}' \cong \mathcal{N}_m \otimes M_m$ where $\mathcal{N}_m$ is abelian.

iii) Apply Theorem 7.14 to each $\mathcal{N}_m$.

7.4* When are two nests unitarily equivalent? Or, by remark 7.25, given two partitions $\{A_n, n \in \mathbb{Z}\}$ and $\{B_n, n \in \mathbb{Z}\}$ of $[0,1]$ into Borel sets, determine when there is an order preserving homeomorphism f of $[0,1]$ with f and f^{-1} absolutely continuous such that $m(f(A_n) \Delta B_n) = 0$ for all $n \in \mathbb{Z}$.

8. Expectations

In this section, we consider the problem of projecting onto a von Neumann algebra, and its connection with triangular truncation.

An *expectation* of $B(\mathcal{H})$ onto a von Neumann subalgebra $\mathcal{A}$ is a norm one projection Φ of $B(\mathcal{H})$ onto $\mathcal{A}$.

8.1 Theorem. *Let Φ be an expectation of $B(\mathcal{H})$ onto a von Neumann subalgebra $\mathcal{A}$. Then*

i) *Φ is self-adjoint and order preserving.*

ii) *$\Phi(AXB) = A\Phi(X)B$ for all A, B in $\mathcal{A}$ and X in $B(\mathcal{H})$.*

iii) *$\Phi(X)^*\Phi(X) \leq \Phi(X^*X)$ for all X in $B(\mathcal{H})$.*

Proof. Let X be a self-adjoint operator of norm one, and write $\Phi(X) = A+iB$ where A and B are self-adjoint. Then

$$\begin{aligned} 1+n^2 = \|X \pm inI\|^2 &\geq \|\Phi(X \pm inI)\|^2 \\ &= \|A+i(B \pm nI)\|^2 \geq \|B \pm nI\|^2 \ . \end{aligned}$$

But the maximum of $\|B \pm nI\|^2$ is $(\|B\|+|n|)^2 \geq 2|n|\,\|B\|+n^2$. Thus $\|B\| \leq 1/2\,|n|$ for all n, and so $\Phi(X)$ is self-adjoint.

If $0 \leq X \leq I$, then $\Phi(X)$ is self-adjoint,

$$\|\Phi(X)\| \leq 1, \text{ and } \|I-\Phi(X)\| = \|\Phi(I-X)\| \leq 1 \ .$$

Hence $0 \leq \Phi(X) \leq I$, and so Φ is order preserving.

Let E be a projection in A, and X be a positive contraction. Since $EXE \leq E$, $\Phi(EXE) \leq E$. Thus

$$\Phi(EXE) = E\Phi(EXE)E \ .$$

By linearity, this holds for all operators X in $B(\mathcal{H})$. Next, consider an operator $X = EXE^{\perp}$ of norm one. It will be shown that $\Phi(X) = E\Phi(X)E^{\perp}$. Decompose $\mathcal{H}$ as $E\mathcal{H} \oplus E^{\perp}\mathcal{H}$. Then X has a matrix form $\begin{bmatrix} 0 & Y \\ 0 & 0 \end{bmatrix}$ and $\Phi(X)$ has the form (A_{ij}). Then if λ is any scalar

$$\begin{aligned} (1+|\lambda|^2)^{1/2} = \left\| \begin{bmatrix} \lambda & Y \\ 0 & 0 \end{bmatrix} \right\| &= \|\lambda E+X\| \geq \|\Phi(\lambda E+X)\| \\ &\geq \|E\Phi(\lambda E+X)E\| = \|\lambda I+A_{11}\| \geq \|\lambda I+ReA_{11}\| \ . \end{aligned}$$

However, $\max\|\pm nI+ReA_{11}\| = n+\|ReA_{11}\| > (1+n^2)^{1/2}$ for large n unless $ReA_{11} = 0$. Likewise, one obtains $ImA_{11} = 0$ so $A_{11} = 0$. Consideration of $\|\lambda E^{\perp}+X\|$ yields $A_{22} = 0$. Finally, $A = E^{\perp}\Phi(X)E = \begin{bmatrix} 0 & 0 \\ A_{21} & 0 \end{bmatrix}$ belongs to $\mathcal{A}$, so

$$\max\{\|X\|,n\|A_{21}\|\} = \left\| \begin{bmatrix} 0 & Y \\ nA_{21} & 0 \end{bmatrix} \right\| = \|nA+X\|$$

$$\geq \|\Phi(nA+X)\| = \|nA+\Phi(X)\|$$

$$\geq \|E^{\perp}(nA+\Phi(X))E\| = (n+1)\|A_{21}\| \ .$$

By letting n tend to infinity, one obtains $A_{21} = 0$.

It follows that

$$\Phi(EX) = \Phi(EXE)+\Phi(EXE^{\perp}) = E\Phi(EXE)+E\Phi(EXE^{\perp}) = E\Phi(EX) \ .$$

Since one therefore has $\Phi(E^{\perp}X) = E^{\perp}\Phi(E^{\perp}X)$ also, one obtains

$$E\Phi(X) = E\Phi(EX)+E\Phi(E^{\perp}X) = E\Phi(EX) = \Phi(EX) \ .$$

By the spectral theorem, every operator in $\mathcal{A}$ is the norm limit of a linear combination of projections in $\mathcal{A}$. Thus by the continuity of Φ, one obtains $\Phi(AX) = A\Phi(X)$ for all A in $\mathcal{A}$ and X in $\mathcal{B}(\mathcal{H})$. Taking adjoints yields $\Phi(XA) = \Phi(X)A$ as well. This proves (ii).

Let X belong to $\mathcal{B}(\mathcal{H})$. By (i) and (ii),

$$0 \leq \Phi((X-\Phi(X))^*(X-\Phi(X)))$$

$$= \Phi(X^*X-X^*\Phi(X)-\Phi(X)^*X+\Phi(X)^*\Phi(X))$$

$$= \Phi(X^*X)-\Phi(X^*)\Phi(X) \ . \qquad \blacksquare$$

8.2. A von Neumann algebra is called *approximately finite* (AF) if it is the weak operator closure of the union of an increasing net of finite dimensional C^* subalgebras. For separately acting algebras, a sequence will suffice.

Consider a finite dimensional von Neumann algebra $\mathcal{A}$. We wish to "average" over the unitary group $\mathcal{U}(\mathcal{A})$. As this group is compact, it has a Haar measure and integrating with respect to this measure will work nicely. However, a somewhat more elementary approach will be used here. By the Wedderburn Theorem for finite dimensional semi-simple rings, one knows that $\mathcal{A}$ is isomorphic to a direct sum of full matrix algebras $\mathcal{M}_n(\mathbb{C})$. The unitary group $\mathcal{U}_n$ of $\mathcal{M}_n$ contains a finite subgroup $\mathcal{G}_n$ such that $\mathcal{G}'_n = \mathbb{C}I$. For example, choose a basis $\{e_1,...,e_n\}$. Let $\mathcal{V}_n$ denote the group of diagonal symmetries with respect to this basis. (This is the group of diagonal matrices with diagonal entries equal to ± 1). And let $\mathcal{S}_n$ denote the group of permutation matrices given by $U_\sigma e_j = e_{\sigma(j)}$ for each permutation σ. Then $U_\sigma \mathcal{V}_n U_\sigma^* = \mathcal{V}_n$. Thus the group $\mathcal{G}_n$ generated by $\mathcal{V}_n$ and $\mathcal{S}_n$ is finite of order $2^n n!$. If T belongs to $\mathcal{M}_n$ and commutes with $\mathcal{V}_n$, then T is diagonal. So if T also commutes with $\mathcal{S}_n$, it must be scalar. Thus $\mathcal{G}'_n = \mathbb{C}I$.

Now if $\mathcal{A}$ is isomorphic to $\sum_{j=1}^{k} \oplus \mathcal{M}_{n_j}$, there is a finite subgroup $\mathcal{G}$ of $\mathcal{U}(\mathcal{A})$ isomorphic to $\mathcal{G}_{n_1} \times \mathcal{G}_{n_2} \times \dots \times \mathcal{G}_{n_k}$ such that $\mathcal{G}' = \mathcal{U}(\mathcal{A})' = \mathcal{A}'$. One may average an operator T over $\mathcal{G}$ by setting

$$m_{\mathcal{G}}(T) = \frac{1}{|\mathcal{G}|} \sum_{U \in \mathcal{G}} UTU^* \ .$$

8.3 Theorem. *Let $\mathcal{A}$ be a von Neumann algebra with AF commutant. Then there is an expectation Φ of $B(\mathcal{H})$ onto $\mathcal{A}$ such that for every X in $B(\mathcal{H})$, $\Phi(X)$ belongs to the weak* closed convex hull of $\{UXU^*: U \in \mathcal{U}(\mathcal{A}')\}$.*

Proof. First suppose $\mathcal{A}'$ is finite dimensional, and let $\mathcal{G}$ be a finite unitary group in $\mathcal{A}'$ such that $\mathcal{G}' = \mathcal{A}$. Define $\Phi(X) = m_{\mathcal{G}}(X)$. Then if V belongs to $\mathcal{G}$,

$$\begin{aligned} V\Phi(X) &= \frac{1}{|\mathcal{G}|} \sum_{U \in \mathcal{G}} VUXU^* \\ &= \frac{1}{|\mathcal{G}|} \sum_{VU \in \mathcal{G}} (VU)X(VU)^*V = \Phi(X)V \ . \end{aligned}$$

Thus $\Phi(X)$ belongs to $\mathcal{G}' = \mathcal{A}$. It is clearly in the convex hull of $\{UXU^*: U \in \mathcal{A}'\}$, and thus Φ is contractive. For any X in $\mathcal{A}$, $UXU^* = X$ for all U in $\mathcal{G}$, and hence $\Phi(X) = X$. So Φ is an expectation.

Now, let $\mathcal{A}'$ be the weak closure of the union of an increasing net $\mathcal{A}_n$ of finite dimensional subalgebras. Let Φ_n be the expectation onto $\mathcal{A}'_n$ constructed above. Think of Φ_n as elements of $\mathcal{L}(B(\mathcal{H}))$, the space of bounded linear maps of $B(\mathcal{H})$ into itself. Consider the point-weak operator topology on this space. This is the topology obtained from thinking of $\mathcal{L}(B(\mathcal{H}))$ as a subset of

$$\prod_{T \in B(\mathcal{H})} (B(\mathcal{H}), \mathrm{WOT})$$

where the embedding is given by sending Φ to $\{\Phi(T), T \in B(\mathcal{H})\}$. The unit ball of $B(\mathcal{H})$ is compact in the weak operator topology (WOT). And thus the set $\prod_{T \in B(\mathcal{H})} \{X \in B(\mathcal{H}): \|X\| \leq \|T\|\}$ is compact by Tychonoff's Theorem. So the unit ball of $\mathcal{L}(B(\mathcal{H}))$ is compact in the point-weak operator topology.

The net Φ_n is bounded, so it has a cluster point Φ in the point-weak operator topology. For any X in $B(\mathcal{H})$, $\Phi_n(X)$ belongs to $\mathcal{A}'_k$ for all $n \geq k$. Thus $\Phi(X)$ belongs to $\mathcal{A}'_k$ for all k, and hence belongs to their intersection $\mathcal{A}$. In particular, if X belongs to $\mathcal{A}$, $\Phi_n(X) = X$ for all n. So Φ is an expectation onto $\mathcal{A}$. Each $\Phi_n(X)$ belongs to the convex hull of $\{UXU^*: U \in \mathcal{U}(\mathcal{A}')\}$. Thus $\Phi(X)$ belongs to the weak operator closure. ■

8.4 Lemma. *Every von Neumann algebra $\mathcal{A}$ such that $\mathcal{A}$ or $\mathcal{A}'$ is abelian is AF.*

Proof. Let $\mathcal{A}$ be abelian. Associate to each partition $\mathcal{E} = \{E_1,...,E_n\}$ of the identity into the sum of projections E_j in $\mathcal{A}$, the subalgebra *span* $\mathcal{E}$. This is a net of finite subalgebras, ordered by inclusion whose union is dense in $\mathcal{A}$. So $\mathcal{A}$ is AF. If $\mathcal{A}$ acts on a separable Hilbert space, choose a countable sequence $\{E_j, j \geq 1\}$ of projections such that $span\{E_j, j \geq 1\}$ is weakly dense in $\mathcal{A}$. Then $\mathcal{A}_n = span\{E_1,...,E_n\}$ is an increasing sequence with the desired properties.

If $\mathcal{A}'$ is abelian, then by Theorem 6.14, there are central projections E_∞, E_1, E_2, ... summing strongly to the identity such that $\mathcal{A}'|E_n\mathcal{H}$ has multiplicity n. Thus there is a maximal abelian algebra $L^\infty(\mu_n)$ such that $\mathcal{A}|E_n\mathcal{H}$ is isomorphic to $M_n \otimes L^\infty(\mu_n)$ (or $\mathcal{B}(\mathcal{H}) \otimes L^\infty(\mu_n)$ if $n = \infty$). Since $L^\infty(\mu_n)$ is AF by the first paragraph, $M_n \otimes L^\infty(\mu_n)$ is AF. And since $\mathcal{B}(\mathcal{H})$ is also AF, this holds for $n = \infty$ as well. Putting it all together shows that $\mathcal{A}$ is AF. (See Exercises.) ∎

8.5 Corollary. *If $\mathcal{N}$ is a nest, there are expectations onto $\mathcal{N}'$ and $\mathcal{N}''$.*

One can ask when the expectation onto $\mathcal{A}$ is unique. An expectation is called *faithful* if $\Phi(X) > 0$ when $X > 0$. The expectation is called *normal* if Φ is weak* continuous. Even faithful, normal expectations need not be unique. For example, let $\mathcal{A}$ be the algebra of all scalars in $M_n(\mathbb{C})$. An expectation has the form $\Phi(X) = \tau(X)I$ where τ is a norm one linear functional on $M_n(\mathbb{C})$ such that $\tau(I) = 1$. Thus $\tau(X) = tr(XT)$ where $\|T\|_1 = 1 = trT$. This implies that T is positive. The expectation Φ is faithful if and only if T has rank n. For example, $T = n^{-1}I$ yields $\Phi_0(X) = tr(X)I$ which is "canonical". But there are many others, all of which are automatically normal. For example, on M_2, define $\Phi(X) = (\frac{3}{4}x_{11} + \frac{1}{4}x_{22})I$.

8.6 Theorem. *Let $\mathcal{A}$ be a von Neumann algebra with abelian commutant. If $\mathcal{A}'$ is atomic, then there is a unique expectation onto $\mathcal{A}$, and it is faithful and normal. If $\mathcal{A}'$ is not atomic, then there is no faithful or normal expectation onto $\mathcal{A}$, and the expectation is not unique.*

Proof. If $\mathcal{A}'$ is atomic, it contains minimal projections $\{E_n, n \geq 1\}$ and $\sum_{n\geq 1} E_n = I$. Thus if X belongs to $\mathcal{B}(\mathcal{H})$, E_nXE_n belong to $\mathcal{A}$ for all $n \geq 1$. Hence if Φ is any expectation onto $\mathcal{A}$,

$$\Phi(X) = \sum_{n\geq 1} \Phi(X)E_n = \sum_{n\geq 1} E_n \Phi(X) E_n$$

$$= \sum_{n\geq 1} \Phi(E_n X E_n) = \sum_{n\geq 1} E_n X E_n = \Delta(X) \ .$$

Thus the only expectation is the "compression to the diagonal", which is easily shown to be faithful and normal. (See Proposition 4.4).

If $\mathcal{A}'$ is not atomic, then $\mathcal{A}'$ contains a projection E such that $\mathcal{A}'|E\mathcal{H}$ is isomorphic to $L^\infty(0,1)$. (This is a classical theorem of measure theory that every finite nonatomic Borel measure is equivalent to Lebesgue measure (c.f. Royden [1], section 15.2). This fact is implicit in the proof of Proposition 7.17.) It follows as in Lemma 3.5 that if K is any compact operator in $\mathcal{B}(E\mathcal{H})$, and $\epsilon > 0$ is given, then there are projections $E_1,\ldots,E_n$ in $\mathcal{A}'$ such that $E = \sum_{i=1}^{n} E_i$ and $\|E_i K E_i\| < \epsilon$ for $1 \leq i \leq n$. Hence for any expectation Φ,

$$\Phi(K) = \sum \Phi(K)E_i = \sum E_i \Phi(K) E_i = \Phi(\sum E_i K E_i) \ .$$

Thus $\|\Phi(K)\| < \epsilon$ for all $\epsilon > 0$, and so $\Phi(K) = 0$. Therefore Φ is not faithful. As the compact operators on $E\mathcal{H}$ are weak* dense in $\mathcal{B}(E\mathcal{H})$, Φ is not normal either.

The non-uniqueness in general follows easily from the non-uniqueness in the multiplicity one case. This will follow from Corollary 8.10 below. ■

It follows from this proof that if K is compact, then $\Phi(K) = \Delta(K)$ is compact.

An *invariant mean* for a semigroup G is a state m on the von Neumann algebra $\ell^\infty(G)$ such that $m(F) = m(gF)$ where $gF(g') = F(gg')$ for all g in G and F in $\ell^\infty(G)$. A semigroup which has an invariant mean is called *amenable*.

8.7 Theorem. *Let G be an abelian semigroup. Then there is an invariant mean on G.*

Proof. Let $\mathcal{M}$ be the subspace spanned by $\{F - gF : g \in G, F \in \ell^\infty(G)\}$. It will be shown that $dist(1,\mathcal{M}) = 1$ where $1(g) = 1$ for all g. Let $H = \sum_{i=1}^{n} F_i - g_i F_i$ be an element of $\mathcal{M}$. Average over the set

$$S = \{\prod_{j=1}^{n} g_j^{k_j},\ 0 \leq k_j < p\} \ .$$

Let S_i be the subset of S such that $k_i = 0$. Then

$$\frac{1}{p^n}\sum_{g\in S} gH = \sum_{i=1}^{n}\frac{1}{p^n}\sum_{g\in S_i}\sum_{k=0}^{p-1} gg_i^kF_i - gg_i^{k+1}F_i$$
$$= \sum_{i=1}^{n}\frac{1}{p^{n-1}}\sum_{g\in S_i}\frac{1}{p}(F_i - g_i^pF_i) \ .$$

Hence

$$\|\frac{1}{p^n}\sum_{g\in S} gH\| \le \frac{2}{p}\sum_{i=1}^{n}\|F_i\|_\infty \ .$$

However, $g1 = 1$ for all g, so

$$\|1-H\| = \frac{1}{p^n}\sum_{g\in S}\|1-gH\| \ge \|1-\frac{1}{p^n}\sum_{g\in S} gH\| \ge 1-\frac{2}{p}\sum_{i=1}^{n}\|F_i\|_\infty \ .$$

Letting p tend to ∞ yields $\|1-H\| \ge 1$.

By the Hahn-Banach Theorem, there is a norm one linear functional m on $\ell^\infty(G)$ such that $m(1) = 1$ and m annihilates $\mathcal{M}$. Thus m is a positive linear functional, and

$$0 = m(F-gF) = m(F)-m(gF)$$

for all F in $\ell^\infty(G)$. So m is an invariant mean. ∎

This yields another proof of:

8.8 Corollary. *If $\mathcal{A}$ is a von Neumann algebra with abelian commutant, then there is an expectation Φ of $\mathcal{B}(\mathcal{H})$ onto $\mathcal{A}$ such that $\Phi(X)$ belongs to the weak* closed convex hull of $\{UXU^*: U \in \mathcal{U}(\mathcal{A}')\}$.*

Proof. Let m be an invariant mean on $\mathcal{U}(\mathcal{A}')$. For X in $\mathcal{B}(\mathcal{H})$ and T in C_1, define $F_{X,T}(U) = tr(UXU^*T)$. Define a linear functional on C_1 by $\phi(T) = m(F_{X,T})$. Since $C_1^* = \mathcal{B}(\mathcal{H})$ by Theorem 1.15, this determines a unique operator $\Phi(X)$ such that $\phi(T) = tr(\Phi(X)T)$. For V in $\mathcal{U}(\mathcal{A}')$,

$$F_{X,TV}(U) = tr(UXU^*TV) = tr(VUXU^*T)$$
$$= tr((VU)X(VU)^*VT) = F_{X,VT}(VU) = {}_VF_{X,VT}(U) \ .$$

Hence

$$tr(V\Phi(X)T) = tr(\Phi(X)TV) = m(F_{X,TV})$$
$$= m({}_VF_{X,VT}) = m(F_{X,VT})$$
$$= tr(\Phi(X)VT) \ .$$

This holds for every T in C_1. Hence $V\Phi(X) = \Phi(X)V$. Thus $\Phi(X)$ belongs to $\mathcal{U}(\mathcal{A}')' = \mathcal{A}$.

If $\Phi(X)$ were not in the weak* closed convex hull of $\{UXU^*:U \in \mathcal{U}(\mathcal{A}')\}$, then the Hahn-Banach Theorem provides a weak* continuous linear functional that separates them. Namely, there is a trace class operator T so that

$$\text{Re } m(F_{X,T}) = \text{Re } tr(\Phi(X)T) > \sup_U \text{Re } tr(UXU^*T) = \|\text{Re } F_{X,T}\|_\infty \ .$$

This is absurd since m is norm one. ■

Let $\mathcal{H}$ be $L^2(\mathbf{T},m)$ where $\mathbf{T}$ is the unit circle and m is normalized Lebesgue measure. The algebra $L^\infty = L^\infty(\mathbf{T},m)$ acts on L^2 by multiplication: $M_f h = fh$. This is a maximal abelian von Neumann algebra. Let $e_n(z) = z^n$ for n in $\mathbb{Z}$ be the standard orthonormal basis for $\mathcal{H}$. Let P_n be the projection onto $span\{e_k:k \geq n\}$; and let $\mathcal{P}$ be the nest $\{\{0\},\mathcal{H},P_n\mathcal{H},n \in \mathbb{Z}\}$. H^∞ is the subalgebra of L^∞ consisting of those functions h with analytic Fourier series $h \sim \sum_{n \geq 0} a_n z^n$.

8.9 Theorem. *There is an expectation Φ of $\mathcal{B}(L^2(\mathbf{T}))$ onto L^∞ such that $\Phi(P_n) = I$ for all $n \in \mathbb{Z}$, and Φ takes $\mathcal{T}(\mathcal{P})$ onto $\{M_h:h \in H^\infty\}$.*

Proof. Consider the semigroup of unitaries $\mathcal{S} = \{M_{\bar{z}^n},n \geq 0\}$. Let m be an invariant mean on $\mathcal{S}$, and repeat the proof of Corollary 8.8. Since $\mathcal{S}' = L^\infty$, one obtains an expectation Φ onto L^∞. Furthermore,

$$\begin{aligned}(\Phi(P_m)e_k,e_k) &= tr(\Phi(P_m)e_k \otimes e_k^*) \\ &= m(trM_{\bar{z}^n}P_m M_{z^n}e_k \otimes e_k^*) = m((P_m e_{k+n},e_{k+n})) = 1\end{aligned}$$

since $(P_m e_{k+n},e_{k+n}) = 1$ for n sufficiently large. Hence $\Phi(P_m) = I$. Finally, if X belongs to $\mathcal{T}(\mathcal{P})$, then for $k < \ell$ one has $(Xe_k,e_\ell) = 0$. By a similar computation to the one above,

$$(\Phi(X)e_k,e_\ell) = m((Xe_{k+n},e_{\ell+n})) = 0 \ .$$

Now $\Phi(X) = M_h$ and $h = \Phi(X)e_0$ belongs to $span\{e_k,k \geq 0\}$. Hence h belongs to H^∞. ■

8.10 Corollary. *The expectation onto $L^\infty(\mathbf{T})$ is not unique.*

Proof. In place of the semigroup $\mathcal{S}$, use $\mathcal{S}^*$. This yields another expectation Ψ. But a reworking of the previous proof yields $\Psi(P_n) = 0$ for all n. So $\Psi \neq \Phi$. ■

Now we return to the problem considered in Chapter 4 of the convergence of triangular integrals.

8.11 Theorem. *Let* $\mathcal{N}$ *be a nest, and let* $\{\mathcal{F}\}$ *be the finite subnests of* $\mathcal{N}$ *ordered by inclusion. Then* $\lim_{\mathcal{F}} \mathcal{D}_{\mathcal{F}}(X)$ *converges in norm if and only if* X *belongs to the closure of*

$$\mathcal{D}(\mathcal{N}) + rad\,\mathcal{T}(\mathcal{N}) + rad\,\mathcal{T}(\mathcal{N})^* \ .$$

In this case, the limit $\mathcal{D}_{\mathcal{N}}(X)$ *agrees with* $\Phi(X)$ *for every expectation onto* $\mathcal{D}(\mathcal{N})$.

Proof. Suppose $X = D + R + S^*$ where D commutes with $\mathcal{N}$ and R and S belong to the radical of $\mathcal{T}(\mathcal{N})$. For every $\epsilon \geq 0$, there is, by Theorem 6.7, a finite subnest $\mathcal{F}$ such that $\|\Delta_{\mathcal{F}}(R)\| + \|\Delta_{\mathcal{F}}(S)\| < \epsilon$. So

$$\Delta_{\mathcal{F}}(X) = D + \Delta_{\mathcal{F}}(R) + \Delta_{\mathcal{F}}(S)^*$$

converges to D. If Φ is any expectation, and E_i are the intervals of $\mathcal{F}$,

$$\Phi(R) = \sum \Phi(R) E_i = \sum E_i \Phi(R) E_i = \Phi(\sum E_i R E_i) \ .$$

Thus $\Phi(R) = 0 = \Phi(S^*)$. So $\Phi(X) = \mathcal{D}_{\mathcal{N}}(X)$. As $\mathcal{D}_{\mathcal{F}}$ and Φ are continuous, these statements extend to the norm closure.

On the other hand, if $\lim_{\mathcal{F}} \mathcal{D}_{\mathcal{F}}(X)$ exists, let D be this limit. For every $\epsilon > 0$, there is a finite subnest $\mathcal{F}$ such that $\|\Delta_{\mathcal{F}}(X - D)\| < \epsilon$. Then

$$X - \Delta_{\mathcal{F}}(X - D) = D + \mathcal{L}_{\mathcal{F}}(X) + \mathcal{L}_{\mathcal{F}}(X^*)^* \ .$$

The operator $\mathcal{L}_{\mathcal{F}}(X)$ belongs to $rad\,\mathcal{T}(\mathcal{N})$ and $\mathcal{L}_{\mathcal{F}}(X^*)^*$ belongs to $rad\,\mathcal{T}(\mathcal{N})^*$. So X belongs to the closure of $\mathcal{D}(\mathcal{N}) + rad\,\mathcal{T}(\mathcal{N}) + rad\,\mathcal{T}(\mathcal{N})^*$. ■

8.12 Theorem. *Let* $\mathcal{N}$ *be a nest. Then* $\mathcal{U}_{\mathcal{N}}(X) = \lim_{\mathcal{F}} \mathcal{U}_{\mathcal{F}}(X)$ *exists if and only if* X *belongs to* $\mathcal{T}(\mathcal{N}) + rad\,\mathcal{T}(\mathcal{N})^*$. *Likewise,* $\mathcal{L}_{\mathcal{N}}(X)$ *exists if and only if* X *belongs to* $\mathcal{T}(\mathcal{N})^* + rad\,\mathcal{T}(\mathcal{N})$. *Any two of* $\mathcal{U}_{\mathcal{N}}(X)$, $\mathcal{L}_{\mathcal{N}}(X)$ *and* $\mathcal{D}_{\mathcal{N}}(X)$ *exist if and only if all three exist if and only if* X *belongs to* $\mathcal{D}(\mathcal{N}) + rad\,\mathcal{T}(\mathcal{N}) + rad\,\mathcal{T}(\mathcal{N})^*$.

Proof. First we remark that the closures of these sets are not used because truncation is not continuous. Let $X = T + R^*$ where T belongs to $\mathcal{T}(\mathcal{N})$ and R belongs to $rad\,\mathcal{T}(\mathcal{N})$. Then for $\epsilon > 0$, there is a finite nest $\mathcal{F}$ so that $\|\Delta_{\mathcal{F}}(R)\| < \epsilon$ by the Ringrose condition (Theorem 6.7). Hence

$$\mathcal{U}_{\mathcal{F}}(X) = T + \mathcal{U}_{\mathcal{F}}(R^*) = T + \Delta_{\mathcal{F}}(R)^*$$

converges to T. On the other hand, if $T = \lim_{\mathcal{F}} \mathcal{U}_{\mathcal{F}}(X)$, then

$$R = X - T = \lim_{\mathcal{F}} X - \mathcal{U}_{\mathcal{F}}(X) = \lim_{\mathcal{F}} \mathcal{L}_{\mathcal{F}^\perp}(X) \ .$$

Each $\mathcal{L}_{\mathcal{F}^\perp}(T)$ belongs to $rad\,\mathcal{T}(\mathcal{N})^*$, which is closed. So R belongs to $rad\,\mathcal{T}(\mathcal{N})^*$.

Now as $\mathcal{L}_{\mathcal{N}}(X) = X - \mathcal{U}_{\mathcal{N}^\perp}(X)$, the lower truncation result is immediate. Also $\mathcal{D}_{\mathcal{F}}(X) = \mathcal{U}_{\mathcal{F}}(X) - \mathcal{L}_{\mathcal{F}}(X)$, so if two limits exist, so does the third. If X belongs to $\mathcal{D}(\mathcal{N}) + rad\,\mathcal{T}(\mathcal{N}) + rad\,\mathcal{T}(\mathcal{N})^*$, all the limits exist as above. Conversely, if all the integrals converge, then

$$X = \mathcal{D}_{\mathcal{N}}(X) + \mathcal{L}_{\mathcal{N}}(X) + \mathcal{L}_{\mathcal{N}^{\perp}}(X)$$

is written as a sum in $\mathcal{D}(\mathcal{N}) + rad\,\mathcal{T}(\mathcal{N}) + rad\,\mathcal{T}(\mathcal{N})^*$. ∎

Let $\mathcal{N}$ be a nest. A *partition* is a possibly infinite collection $\mathcal{E}$ of intervals $E_\alpha = P(N_\alpha) - P(M_\alpha)$ for $M_\alpha < N_\alpha$ in $\mathcal{N}$ such that $\sum E_\alpha = I$. Define a homomorphism of $\mathcal{T}(\mathcal{N})$ into itself by $\Delta_{\mathcal{E}}(T) = \sum E_\alpha T E_\alpha$. Larson's $\mathcal{R}^\infty(\mathcal{N})$ ideal is the set of operators T in $\mathcal{T}(\mathcal{N})$ such that for every $\epsilon > 0$, there is a partition $\mathcal{E}$ such that $\|\Delta_{\mathcal{E}}(T)\| < \epsilon$. This is a generalized Ringrose condition. If $\mathcal{N}$ is atomic, one may take the maximal partition $\{A_\alpha\}$ of atoms of $\mathcal{N}$. In this case, $\mathcal{R}^\infty(\mathcal{N})$ is the strictly upper triangular operators, which coincides with the kernel of the diagonal map.

8.13 Proposition. *$\mathcal{R}^\infty(\mathcal{N})$ is a closed, two sided ideal of $\mathcal{T}(\mathcal{N})$. The net $\mathcal{D}_{\mathcal{F}}(T)$ converges strongly to 0 for every T in $\mathcal{R}^\infty(\mathcal{N})$, and $\Phi(T) = 0$ for every expectation onto the diagonal. The set $\mathcal{D}(\mathcal{N}) + \mathcal{R}^\infty(\mathcal{N})$ is a norm closed subalgebra of $\mathcal{T}(\mathcal{N})$ on which $\mathcal{D}_{\mathcal{N}}(\cdot)$ exists, and agrees with every expectation onto $\mathcal{D}(\mathcal{N})$.*

Proof. Let R belong to $\mathcal{R}^\infty(\mathcal{N})$ and let T belong to $\mathcal{T}(\mathcal{N})$. Then $\Delta_{\mathcal{E}}(RT) = \Delta_{\mathcal{E}}(R)\Delta_{\mathcal{E}}(T)$ and $\Delta_{\mathcal{E}}(TR) = \Delta_{\mathcal{E}}(T)\Delta_{\mathcal{E}}(R)$. Thus it is immediate that $\mathcal{R}^\infty(\mathcal{N})$ is a 2-sided ideal. Since each $\Delta_{\mathcal{E}}$ is contractive, it is easy to verify that $\mathcal{R}^\infty(\mathcal{N})$ is closed. Let x be a vector, and let $\epsilon \geq 0$ be given. Choose $\mathcal{E}$ so that $\|\Delta_{\mathcal{E}}(R)\| < \epsilon$. There is a finite subset $E_1,\ldots,E_n$ of $\mathcal{E}$ such that $\|(\sum_{i=1}^{n} E_i)^{\perp} x\| < \epsilon$. Let $\mathcal{F}$ be any finite subnest of $\mathcal{N}$ containing the end points of $E_1,\ldots,E_n$. Then

$$\begin{aligned}
\|\mathcal{D}_{\mathcal{F}}(R)x\| &= \|\mathcal{D}_{\mathcal{F}}(\sum_{i=1}^{n} E_i R E_i + (\sum_{i=1}^{n} E_i)^{\perp} R (\sum_{i=1}^{n} E_i)^{\perp})x\| \\
&\leq \|\sum E_i R E_i\|\,\|x\| + \|\mathcal{D}_{\mathcal{F}}((\sum_{i=1}^{n} E_i)^{\perp} R (\sum_{i=1}^{n} E_i)^{\perp})\|\,\|(\sum_{i=1}^{n} E_i)^{\perp} x\| \\
&< \epsilon(\|x\| + \|R\|) \ .
\end{aligned}$$

Thus $\mathcal{D}_{\mathcal{F}}(R)$ converges strongly to 0. (Note that $\mathcal{D}_{\mathcal{F}}(R)^* x$ tends to zero also.)

Next, if Φ is any expectation onto $\mathcal{D}(\mathcal{N})$, then

$$\Phi(R) = \sum \Phi(R) E_\alpha = \sum E_\alpha \Phi(R) E_\alpha = \Phi(\sum E_\alpha R E_\alpha) \ .$$

As the right hand side can be made arbitrarily small, $\Phi(R) = 0$. The last sentence of the theorem is trivial except for the closure of $\mathcal{D}(\mathcal{N}) + \mathcal{R}^\infty(\mathcal{N})$.

If $T_n = D_n + R_n$ converges to T, then

$$\|D_n - D_m\| = \|\Phi(T_n - T_m)\| \leq \|T_n - T_m\|$$

for any expectation Φ. Thus D_n converges to some D in $\mathcal{D}(\mathcal{N})$. Whence R_n converges to $T-D$. Since $\mathcal{R}^\infty$ is closed, $T-D$ belongs to $\mathcal{R}^\infty$ as desired. ∎

Notes and Remarks.

Theorem 8.1 is due to Tomiyama [1]. Schwartz [1] introduced *property P* for a von Neumann algebra. A von Neumann algebra $\mathcal{A}$ has this property if the weak closed convex hull of $\{UXU^*: U \in \mathcal{U}(\mathcal{A}')\}$ always intersects $\mathcal{A}$ for each X in $\mathcal{B}(\mathcal{H})$. Theorem 8.3 is essentially due to Schwartz, and shows that AF implies property P. This in turn implies that there is an expectation of $\mathcal{B}(\mathcal{H})$ onto $\mathcal{A}$. Such algebras are called *injective* since they are precisely the von Neumann algebras such that every completely positive map ϕ of a von Neumann algebra $\mathcal{B}_0 \subset \mathcal{B}$ into $\mathcal{A}$ extends to a completely positive map of $\mathcal{B}$ into $\mathcal{A}$. This follows from a theorem of Arveson [4] (see Theorem 20.10) which shows that $\mathcal{B}(\mathcal{H})$ is injective followed by the expectation onto $\mathcal{A}$. The big theorem in this connection is a combination of work due to Connes [1], Choi and Effros [1,2], and Effros and Lance [1].

Theorem. *Let $\mathcal{A}$ be a von Neumann algebra. Then the following are equivalent.*

(i) *$\mathcal{A}$ is AF.*

(ii) *$\mathcal{A}$ has property P.*

(iii) *$\mathcal{A}$ is injective.*

(iv) *$\mathcal{A}$ is semidiscrete.*

A von Neumann algebra is *semidiscrete* if there is a nest of finite rank, completely positive, weak* continuous contractions of $\mathcal{A}$ into itself converging pointwise to the identity map in the weak* topology. In particular, every type I von Neumann algebra is injective (see Exercises) and $\mathcal{A}$ is AF if and only if $\mathcal{A}'$ is AF.

Theorem 8.6 is due to Arveson [2] and Theorem 8.9 and Corollary 8.10 are also due to Arveson [7]. Theorems 8.11 and 8.12 are due to Erdos and Longstaff [1]. Proposition 8.13 is due to Larson [4].

Exercises.

8.1 (a) Show that the von Neumann algebra direct sum of AF algebras is AF.

(b) If $\mathcal{A}$ and $\mathcal{B}$ are von Neumann algebras on $\mathcal{H}$ and $\mathcal{K}$ respectively, let $\mathcal{A} \underset{\min}{\otimes} \mathcal{B}$ denote the weakly closed subalgebra of $\mathcal{B}(\mathcal{H} \otimes \mathcal{K})$ generated by $\{A \otimes B : A \in \mathcal{A}, B \in \mathcal{B}\}$. Show that if $\mathcal{A}$ and $\mathcal{B}$ are AF, then so is $\mathcal{A} \underset{\min}{\otimes} \mathcal{B}$.

8.2 Use Exercise 7.3 to prove that type I von Neumann algebras are AF.

8.3 Generalize the Hahn-Banach Theorem to maps into $L^\infty(\mu)$ as follows:

(i) Show that the real valued functions $L^\infty_{\mathbb{R}}(\mu)$ is a complete lattice under ess.sup and ess.inf.

(ii) Let Φ be a norm one linear map of a real Banach space X into $L^\infty_{\mathbb{R}}(\mu)$. Let X be a subspace of Y, and let y belong to Y. In order to extend Φ to a contraction of *span*$\{X,y\}$ into $L^\infty_{\mathbb{R}}(\mu)$ it suffices to find an f_0 in $L^\infty_{\mathbb{R}}(\mu)$ such that

$$\underset{x \in X}{\text{ess.sup}} - \| x+y \| - \Phi(x) \leq f_0 \leq \underset{x \in X}{\text{ess.inf}} \| x+y \| - \Phi(x) \ .$$

(iii) Show that this is always possible. Then use Zorn's lemma.

(iv) Complexify.

(v) Obtain an expectation onto any abelian von Neumann algebra.

8.4 Let $\mathcal{N}$ be a nest, and $\{\mathcal{F}\}$ its net of finite subnests. Let

$$J(\mathcal{N}) = \{T \in \mathcal{T}(\mathcal{N}) : \mathcal{D}_{\mathcal{F}}(T) \text{ converges strongly to zero}\} \ .$$

(a) Show that $J(\mathcal{N})$ is a closed left ideal of $\mathcal{T}(\mathcal{N})$ containing $\mathcal{R}^\infty(\mathcal{N})$.

(b)* Characterize $J(\mathcal{N})$.

9. Distance Formulae

In this section, two important estimates for the distance to von Neumann algebras and nest algebras are obtained. These results are crucial to later developments. A number of related estimates will be developed as well.

9.1 Lemma. *Let A, B, and C be operators in $\mathcal{B}(\mathcal{H}_1)$, $\mathcal{B}(\mathcal{H}_2,\mathcal{H}_1)$, and $\mathcal{B}(\mathcal{H}_1,\mathcal{H}_2)$ respectively. Then there is an operator X_0 in $\mathcal{B}(\mathcal{H}_2)$ such that*

$$\left\| \begin{bmatrix} A & B \\ C & X_0 \end{bmatrix} \right\| = \inf_X \left\| \begin{bmatrix} A & B \\ C & X \end{bmatrix} \right\| = \max\left\{ \| [A \ B] \|, \left\| \begin{bmatrix} A \\ C \end{bmatrix} \right\| \right\} .$$

If A is compact, then X_0 may be taken to be compact.

Proof. If $A = 0$, take $X_0 = 0$. Otherwise, scale the operators A, B, C so that $\max\left\{ \| [A \ B] \|, \left\| \begin{bmatrix} A \\ C \end{bmatrix} \right\| \right\} = 1$. Then

$$I \geq \begin{bmatrix} A \\ C \end{bmatrix}^* \begin{bmatrix} A \\ C \end{bmatrix} = A^*A + C^*C .$$

So $C^*C \leq I - A^*A$. Hence for any vector x,

$$\|Cx\|^2 = (C^*Cx, x) \leq (1 - A^*Ax, x) = \|(1-A^*A)^{1/2}x\|^2 .$$

Thus one may define an operator L in $\mathcal{B}(\mathcal{H}_1,\mathcal{H}_2)$ by

$$L(1-A^*A)^{1/2}x = Cx$$

for vectors in the range of $(1-A^*A)^{1/2}$, and $Ly = 0$ if y is orthogonal to this range. The inequality shows that $\|L\| \leq 1$, and hence L extends by continuity to all of $\mathcal{H}_1$. So $C = L(1-A^*A)^{1/2}$. Similarly, one obtains $BB^* \leq 1 - AA^*$, and hence $B^* = K(1-AA^*)^{1/2}$ where K is a contraction in $\mathcal{B}(\mathcal{H}_1,\mathcal{H}_2)$. So $B = (1-AA^*)^{1/2}K^*$.

Let $X_0 = -LA^*K^*$. One has

$$\begin{bmatrix} A & B \\ C & X_0 \end{bmatrix} = \begin{bmatrix} I & 0 \\ 0 & L \end{bmatrix} \begin{bmatrix} A & (1-AA^*)^{1/2} \\ (1-A^*A)^{1/2} & -A^* \end{bmatrix} \begin{bmatrix} I & 0 \\ 0 & K^* \end{bmatrix} = L'UK' .$$

The two terms L' (in $\mathcal{B}(\mathcal{H}_1^{(2)}, \mathcal{H}_1 \oplus \mathcal{H}_2)$) and K' (in $\mathcal{B}(\mathcal{H}_1 \oplus \mathcal{H}_2, \mathcal{H}_1^{(2)})$) are contractions. The middle term U is unitary since

$$UU^* = \begin{bmatrix} I & A(1-AA^*)^{1/2}-(1-A^*A)^{1/2}A^* \\ (1-A^*A)^{1/2}A^*-A(1-A^*A)^{1/2} & I \end{bmatrix} .$$

Now $A(A^*A)^k = (A^*A)^k A^*$ for all $k \geq 0$, so $Ap(A^*A) = p(AA^*)A$ for every polynomial p. The identity $UU^* = I^{(2)}$ follows from approximating $x^{1/2}$ by polynomials uniformly on $[0,1]$. Similarly, $U^*U = I^{(2)}$. Therefore, $\left\| \begin{bmatrix} A & B \\ C & X_0 \end{bmatrix} \right\| \leq 1$.

It is clear that $\|[A\ B]\|$ and $\left\| \begin{bmatrix} A \\ C \end{bmatrix} \right\|$ are lower bounds for $\left\| \begin{bmatrix} A & B \\ C & X_0 \end{bmatrix} \right\|$ and hence X_0 minimizes this quantity. If A is compact, so is A^* and hence X_0 is compact. ■

9.2 Corollary. *Let A_{ij} be operators in $B(\mathcal{H}_j,\mathcal{H}_i)$ for $j < i$. Let X_{ij} be operators in $B(\mathcal{H}_j,\mathcal{H}_i)$ for $i \leq j$. Then*

$$\min_{\{X_{ij}\}} \left\| \begin{bmatrix} X_{11} & X_{12} & & X_{1n} \\ A_{21} & X_{22} & & \\ A_{31} & A_{32} & & \\ & & & \\ A_{n1} & & A_{nn-1} & X_{nn} \end{bmatrix} \right\| = \max_{2 \leq k \leq n} \left\| \begin{bmatrix} A_{k1} & A_{kk-1} \\ & \\ A_{n1} & A_{nk-1} \end{bmatrix} \right\| .$$

Proof. Let β be the right hand side. Set $X_{11},\ldots,X_{1n}$, and $X_{2n},\ldots,X_{n,n}$ all equal to zero. Choose X_{22} by Lemma 9.1 so that

$$\left\| \begin{bmatrix} A_{21} & X_{22} \\ A_{31} & A_{32} \\ & \\ A_{n1} & A_{n2} \end{bmatrix} \right\| \leq \beta .$$

With X_{22} fixed, one can now use Lemma 9.1 again to choose $\begin{bmatrix} X_{23} \\ X_{33} \end{bmatrix}$ so that

$$\left\| \begin{bmatrix} A_{21} & X_{22} & X_{23} \\ A_{31} & A_{32} & X_{33} \\ A_{41} & A_{42} & A_{43} \\ & & \\ A_{n1} & A_{n2} & A_{n3} \end{bmatrix} \right\| \leq \beta .$$

In this way, all X_{ij} are defined after $(n-1)$ steps so that the norm is at most β. ■

9.3 Corollary. *Let $\mathcal{F}$ be a finite nest $\{0\} = N_0 < N_1 < \ldots < N_n = \mathcal{H}$. Then for every A in $B(\mathcal{H})$,*

$$dist(A, \mathcal{T}(\mathcal{F})) = \max_{0 \le k \le n} \| P(N_k)^{\perp} A P(N_k) \| .$$

Proof. Decompose $\mathcal{H}$ as the direct sum of the n blocks $N_k \ominus N_{k-1}$, $1 \le k \le n$. Then A becomes an $n \times n$ matrix (A_{ij}). The operator T in $\mathcal{T}(\mathcal{F})$ have the form (T_{ij}) where $T_{ij} = 0$ for $j < i$. Thus

$$\inf\{\|A - T\| : T \in \mathcal{T}(\mathcal{F})\} = \inf \|(X_{ij})\|$$

where $X_{ij} = A_{ij}$ for $j < i$. By Corollary 9.2, this is precisely the quantity $\max_{0 \le k \le n} \|P(N_k)^{\perp} A P(N_k)\|$. ■

9.4 Lemma. *Let Λ be a net, and suppose that $\{\mathcal{A}_\lambda : \lambda \in \Lambda\}$ is a collection of weak*-closed linear subspaces of $B(\mathcal{H})$ such that $\mathcal{A}_\mu \subseteq \mathcal{A}_\lambda$ if $\mu \ge \lambda$. Let $\mathcal{A}$ be the intersection of $\{\mathcal{A}_\lambda\}$. Then for T in $B(\mathcal{H})$,*

$$dist(T, \mathcal{A}) = \sup_{\Lambda} dist(T, \mathcal{A}_\lambda) = \lim_{\Lambda} dist(T, \mathcal{A}_\lambda) .$$

Proof. Since $\mathcal{A}_\lambda$ is decreasing in λ, it is clear that

$$dist(T, \mathcal{A}) \ge \sup_{\Lambda} dist(T, \mathcal{A}_\lambda) = \lim_{\Lambda} dist(T, \mathcal{A}_\lambda) .$$

Let the $\sup_{\Lambda} dist(T, \mathcal{A}_\lambda) = L$. For each λ, let

$$C_\lambda = \{A \in \mathcal{A}_\lambda : \|T - A\| \le L\} = \mathcal{A}_\lambda \cap \{A : \|T - A\| \le L\} .$$

Then C_λ is the intersection of a weak* closed and a weak* compact set, thus is weak*-compact. Moreover, $\{C_\lambda : \lambda \in \Lambda\}$ has the finite intersection property. For if $\lambda_1, \ldots, \lambda_n$ belong to Λ, there is some λ_0 in Λ such that $\lambda_0 \ge \lambda_i$ for $1 \le i \le n$. So $\bigcap_{i=1}^{n} C_\lambda$ contains C_{λ_0}. By a standard compactness argument, there is an operator A in the intersection of all the C_λ. Clearly, A belongs to $\mathcal{A}$ and $\|T - A\| \le L$. ■

9.5 Theorem (Arveson's Distance Formula). *Let $\mathcal{N}$ be a nest. For every A in $B(\mathcal{H})$, $dist(A, \mathcal{T}(\mathcal{N})) = \sup_{\mathcal{N}} \|P(N)^{\perp} A P(N)\|$.*

Proof. The set $\{\mathcal{T}(\mathcal{F})\}$ as $\mathcal{F}$ runs over finite subnests of $\mathcal{N}$ ordered by inclusion is a net of weakly closed subspaces with intersection $\mathcal{T}(\mathcal{N})$. By Lemma 9.5 and Corollary 9.3,

$$dist(A, \mathcal{T}(\mathcal{N})) = \sup_{\mathcal{F}} dist(A, \mathcal{T}(\mathcal{F})) = \sup_{\mathcal{N}} \|P(N)^{\perp} T P(N)\| .$$ ■

Now, we turn our attention to von Neumann algebras. If $\mathcal{A}$ is a von Neumann algebra, then $\mathcal{A} = \mathcal{A}'' = Alg\ Lat\,\mathcal{A}$ where $Lat\,\mathcal{A}$ consists of the ranges of projections in $\mathcal{A}'$. So, for T in $B(\mathcal{H})$, one might hope for a connection between $dist(T,\mathcal{A})$, and the quantities

$$\sup\{\|P^{\perp}TP\| : P \in Proj(\mathcal{A}')\}$$

and

$$\sup\{\|TA-AT\| : A \in \mathcal{A}', \|A\| \leq 1\} \ .$$

Define the derivation operation δ_T by $\delta_T(A) = TA-AT$. Note that

$$\delta_T(AB) = TAB-ATB+ATB-ABT = \delta_T(A)B+A\delta_T(B) \ .$$

9.6 Theorem. *Let $\mathcal{A}$ be a von Neumann algebra such that $\mathcal{A}'$ is AF. Then for T in $B(\mathcal{H})$,*

$$\sup_{P \in Lat\,\mathcal{A}} \|P^{\perp}TP\| \leq \frac{1}{2}\|\delta_T | \mathcal{A}'\| \leq dist(T,\mathcal{A}) \leq \|\delta_T | \mathcal{A}'\| \leq 4 \sup_{P \in Lat\,\mathcal{A}} \|P^{\perp}TP\| \ .$$

Proof. Let Φ be an expectation onto $\mathcal{A}$ provided by Theorem 8.3. Then since $\Phi(T)$ belongs to the weak* closed convex hull of $\{UTU^* : U \in \mathcal{U}(\mathcal{A}')\}$,

$$\begin{aligned} dist(T,\mathcal{A}) &\leq \|T-\Phi(T)\| \leq \sup\{\|T-UTU^*\| : U \in \mathcal{U}(\mathcal{A}')\} \\ &= \sup\{\|\delta_T(U)\| : U \in \mathcal{U}(\mathcal{A}')\| = \|\delta_T | \mathcal{A}'\| \ . \end{aligned}$$

On the other hand, if $\|T-A\| = dist(T,\mathcal{A})$ for some A in $\mathcal{A}$, then

$$\begin{aligned} \|\delta_T | \mathcal{A}'\| &= \sup\{\|(T-A)B-B(T-A)\| : B \in \mathcal{A}', \|B\| \leq 1\} \\ &\leq 2\|T-A\| = 2\,dist(T,\mathcal{A}) \ . \end{aligned}$$

When P is a projection in $Lat\,\mathcal{A}$, $2P-I$ is a unitary in $\mathcal{A}'$. So

$$\begin{aligned} \max\{\|PTP^{\perp}\|, \|P^{\perp}TP\|\} &= \|TP-PT\| \\ &= \frac{1}{2}\|T(2P-I)-(2P-I)T\| \leq \frac{1}{2}\|\delta_T | \mathcal{A}'\| \ . \end{aligned}$$

On the other hand, the extreme points of the unit ball of the *self-adjoint* operators in $\mathcal{A}'$ are precisely $2P-I$ for P in $Lat\,\mathcal{A}$. Thus

$$\sup \|P^{\perp}TP\| = \frac{1}{2}\|\delta_T | \mathcal{A}'_{s.a.}\| \ .$$

Every norm one element of $\mathcal{A}'$ can be written $X = A+iB$ when A, B are self-adjoint contractions. So

$$\|\delta_T | \mathcal{A}'\| \leq 2\|\delta_T | \mathcal{A}_{s.a.}'\| \leq 4 \sup_{P \in Lat\,\mathcal{A}} \|P^{\perp}TP\| \ . \qquad \blacksquare$$

9.7 Example. Consider the 3×3 diagonal algebra $\mathcal{D}$ and the operator $T = \begin{bmatrix} 1 & 1 & 1 \\ 1 & 1 & 1 \\ 1 & 1 & 1 \end{bmatrix}$. As $\mathcal{D} = \mathcal{D}'$, one finds $Lat\,\mathcal{D} = Proj(\mathcal{D})$ consists of $\{0,P_i,P_i^\perp,I,i{=}1,2,3\}$ where P_i is the projection onto the ith basis vector. So $\sup_{P \in Lat\,\mathcal{D}} \|P^\perp TP\| = \|[1\ \ 1]\| = \sqrt{2}$. Suppose that D is diagonal and $\|T-D\| = dist(T,\mathcal{D})$. Now $\|T-UDU^*\| = \|T-D\|$ for every permutation matrix U in S_3. Averaging over this group leaves T fixed, but converts D to a scalar. So

$$dist(T,\mathcal{D}) = \inf\{\|T-\lambda I\| : \lambda \in \mathbb{C}\} \ .$$

Now T is self-adjoint with $\sigma(T) = \{0,3\}$. So

$$\|T-\lambda I\| = \max\{|\lambda|, |3-\lambda|\} \geq 3/2 \ ;$$

and $dist(T,\mathcal{D}) = \|T-3/2I\| = 3/2$. Thus $\|\delta_T|\mathcal{D}\| \leq 3$. Let $W = \begin{bmatrix} 1 & 0 & 0 \\ 0 & \omega & 0 \\ 0 & 0 & \overline{\omega} \end{bmatrix}$ where $\omega = -\frac{1}{2}+\frac{\sqrt{3}}{2}i$. Then W is unitary, so

$$\|\delta_T|\mathcal{D}\| \geq \|TW-WT\| = \|W^*TW-T\|$$

$$\geq \left\| \begin{bmatrix} 0 & 1-\overline{\omega} & 1-\omega \\ 1-\omega & 0 & 1-\overline{\omega} \\ 1-\overline{\omega} & 1-\omega & 0 \end{bmatrix} \begin{bmatrix} 1/\sqrt{3} \\ 1/\sqrt{3} \\ 1/\sqrt{3} \end{bmatrix} \right\| = \left\| \begin{bmatrix} \sqrt{3} \\ \sqrt{3} \\ \sqrt{3} \end{bmatrix} \right\| = 3 \ .$$

So three of the four inequalities in Theorem 9.6 are strict in this example.∎

With some more work, the constant 4 can be reduced to 2 if $\mathcal{A}$ or $\mathcal{A}'$ is abelian.

9.8 Lemma. *If $\mathcal{A}$ is a von Neumann algebra with abelian commutant, then*

$$dist(T,\mathcal{A}) \leq 2\sup\{\|P^\perp TP\| : P \in Lat\,\mathcal{A}\}$$

for T in $\mathcal{B}(\mathcal{H})$.

Proof. Let P and Q belong to $Lat\,\mathcal{A}$. Then $P+Q-PQ$ is the projection in $Lat\,\mathcal{A}$ onto the span of $P\mathcal{H}$ and $Q\mathcal{H}$. So

$$(2P-I)(2Q-I) = I-2(P+Q-PQ) = 2(P+Q-PQ)^\perp - I \ .$$

Thus $\mathcal{G} = \{2P-I : P \in Lat\,\mathcal{A}\}$ is a subgroup of $\mathcal{U}(\mathcal{A}')$. The projections span $\mathcal{A}'$, hence $\mathcal{G}' = \mathcal{A}$. The proof of Corollary 8.8 yields an expectation Φ of $\mathcal{B}(\mathcal{H})$ onto $\mathcal{A}$ such that $\Phi(X)$ belongs to the weak* closed convex hull of $\{UXU^* : U \in \mathcal{G}\}$. Hence

$$dist(T,\mathcal{A}) \leq \|T-\Phi(T)\| \leq \sup_{P \in Lat\,\mathcal{A}} \|T-(2P-I)T(2P-I)\|$$

$$= \sup_{P \in Lat\,\mathcal{A}} \|T(2P-I)-(2P-I)T\| = 2 \sup_{P \in Lat\,\mathcal{A}} \|P^{\perp}TP\| \ .$$

The last equality comes from the proof of Theorem 9.6. ■

9.9 Lemma. *If 0 is in the convex hull of a subset Λ of $\mathbb{C}$, then 0 is in the convex hull of three points of Λ. In particular, if $\Lambda = \{\lambda_n, n \geq 1\}$ and $\sum_{n=1}^{\infty} \lambda_n = 0$, then 0 is in the convex hull of Λ.*

Proof. The convex hull of any finite subset of $\mathbb{C}$ is a polygon with vertices $\lambda_0,\dots,\lambda_n$ in order around the boundary. It is the union of the triangles with vertices $\{\lambda_0,\lambda_j,\lambda_{j+1}\}$ for $1 \leq j \leq n-1$. Thus the convex hull of Λ coincides with the convex combinations of 3 point subsets of Λ.

If 0 is not in the convex hull of Λ, then there is a scalar $e^{i\theta}$ such that $Re(e^{i\theta}\lambda_n) \geq 0$ for all λ_n in Λ. But $0 = Re(e^{i\theta}\sum_{n\geq 1}\lambda_n) = \sum_{n \geq 1} Re(e^{i\theta}\lambda_n)$ forces λ_n to be on the line $\mathrm{Re}(e^{i\theta}z) = 0$. So $\sum_{n\geq 1}\lambda_n = 0$ implies $\mathrm{Im}(e^{i\theta}\lambda_n)$ takes both signs, and so 0 is in the convex hull. ■

9.10 Lemma. *If $K \in C_1$, and $tr(K) = 0$, then there is a unit vector e such that $(Ke,e) = 0$.*

Proof. Choose an orthonormal basis $\{e_n, n \geq 1\}$ and let $\lambda_n = (Ke_n,e_n)$. Then $0 = trK = \sum \lambda_n$. By the preceding lemma, it may be supposed that 0 is in the convex hull of λ_1, λ_2 and λ_3. Each λ_i belongs to the *numerical range* $W(K) = \{(Kx,x): \|x\| = 1\}$. So it suffices to prove that $W(K)$ is convex. To this end, consider (Kx_1,x_1) and (Kx_2,x_2). They belong to $W(K_0)$ also, where K_0 is the compression of K to $span\{x_1,x_2\}$. Since $W(K_0)$ is a subset of $W(K)$, it suffices to show that $W(K_0)$ is convex. It is clear that if a, b are scalars and U is unitary,

$$W(aUK_0U^*+bI) = aW(K_0)+b \ .$$

So put K_0 in upper triangular form and scale it so that $trK_0 = 0$, and $\sigma(K_0) = \{0\}$ or $\{-1,+1\}$, and $(K_0e_2,e_1) \geq 0$. This reduces our problem to computing $W\left(\begin{bmatrix} 0 & 1 \\ 0 & 0 \end{bmatrix}\right)$ and $W\left(\begin{bmatrix} 1 & 2r \\ 0 & -1 \end{bmatrix}\right)$ where $r \geq 0$. The first is the closed disc of radius 1/2, centre 0 and the latter is the set

$$\{\cos 2\theta + re^{i\alpha}\sin 2\theta : 0 \le \theta, \alpha \le 2\pi\}$$

which can be shown to be the ellipse $\{x+iy : \frac{x^2}{1+r^2} + \frac{y^2}{r^2} \le 1\}$ (See Exercise 9.2). Both are convex as claimed. ∎

9.11 Lemma. *Let $\mathcal{A}$ be an abelian von Neumann algebra. Then for T in $B(\mathcal{H})$,*

$$dist(T,\mathcal{A}) = \sup dist(T,\mathcal{M})$$

where the sup *is taken over all maximal abelian von Neumann algebras $\mathcal{M}$ containing $\mathcal{A}$.*

Proof. Since $\mathcal{A}$ is weak* closed, the Hahn-Banach Theorem provides a weak* continuous linear function ϕ of norm one which annihilates $\mathcal{A}$ and

$$\phi(T) = dist(T,\mathcal{A}) \ .$$

By Theorem 1.15, $\phi(X) = tr(XK)$ for some trace class operator K with $\|K\| = 1$. Using Zorn's Lemma, one can find a maximal, abelian self-adjoint algebra $\mathcal{M}$ in $ker\phi$. It suffices to show that $\mathcal{M}$ is maximal abelian in $B(\mathcal{H})$.

If not, then by Theorem 7.14, $\mathcal{M}$ contains a projection P of uniform multiplicity $n \ge 2$. If P dominates a minimal projection, we may assume that P itself is minimal. In this case, $\mathcal{M}_0 = \mathcal{M}\,|P\mathcal{H}$ consists of the scalars in $\mathcal{M}_n$ (or $B(\mathcal{H})$ if $n = \aleph_0$). Otherwise, $\mathcal{M}_0\,|P\mathcal{H}$ is unitarily equivalent to $L^\infty(\mu)^{(n)}$ acting on $L^2(\mu)^{(n)}$ for a non-atomic measure μ. Let ϕ_0 be the restriction of ϕ to $B(P\mathcal{H})$ given by $\phi_0(X) = tr(XK_0)$, and note that it annihilates $\mathcal{M}_0$.

In the first case, $trK_0 = 0$. By Lemma 9.10, there is a rank one projection $E < P$ such that $trEK_0 = 0$. It is easy to check that the abelian von Neumann algebra generated by $\mathcal{M}$ and E is just $\{A+\lambda E : A \in \mathcal{M}, \lambda \in \mathbb{C}\}$. So ϕ annihilates this larger algebra, contrary to hypothesis.

In the continuous case, consider the restriction of ϕ_0 to the $n \times n$ matrices with coefficients in L^∞. Let ϕ_{ij} be the restriction of ϕ_0 to the copy of $L^\infty(\mu)$ in the ij component of the matrix. Since ϕ_{ij} is weak*-continuous and $L^\infty(\mu) = L^1(\mu)^*$, there is an $L^1(\mu)$ function f_{ij} such that $\phi_{ij}(h) = \int h f_{ij} d\mu$. Since $\phi_0\,|\mathcal{M}_0 = 0$, $\sum\limits_{i \ge 1} f_{ii} = 0$. The atomic case can be repeated "pointwise" provided care is taken to keep all the functions Borel. First choose Borel representatives $f_{ij}(x)$. Then $\sum\limits_{i\ge 1} f_{ii}(x) = 0$, and converges absolutely, except on a set of measure zero. By Lemma 9.9, for each x, 0 is in the convex hull of some triple $f_{i_1 i_1}(x)$, $f_{i_2 i_2}(x)$, $f_{i_3 i_3}(x)$. The convex hull of a finite set is a Borel function of the set, so

$$\{x : 0 \text{ belongs to } conv\{f_{i_1i_2}(x), f_{i_2i_2}(x), f_{i_3i_3}(x)\}\}$$

is Borel for each triple. Thus there is one triple for which this has positive measure. It may be assumed to be 1,2,3 (or a pair 1,2 if $n = 2$.). By making P smaller, it may be supposed that this holds for all x.

By Lemma 9.10, there are scalars $\alpha_1(x)$, $\alpha_2(x)$ and $\alpha_3(x)$ such that $\sum_{i=1}^{3} |\alpha_i(x)|^2 = 1$ and $\sum_{i=1}^{3}\sum_{j=1}^{3} f_{ij}(x)\alpha_i(x)\overline{\alpha_j(x)} = 0$. To make this choice Borel, we insist that $\alpha_1(x) \geq 0$ and maximal, $\alpha_2(x) = r_2(x)e^{i\theta_2(x)}$ where $0 \leq \theta_2(x)$ is minimal and $r_2(x)$ is maximal subject to earlier restraints, and likewise for $\alpha_3(x)$. (Note that if $n = 2$, then α_3 does not appear.) Let E be the projection in $M_n \otimes L^\infty(\mu)$ given by $E_{ij} = M_{\alpha_i\bar{\alpha}_j}$ for $1 \leq i,\ j \leq 3$ and $E_{ij} = 0$ otherwise. This has been constructed so that if $P_\Omega^{(n)}$ is the diagonal matrix in $\mathcal{M}_0$ corresponding to the characteristic function of any Borel set Ω, then

$$\begin{aligned}\phi_0(P_\Omega^{(n)}E) &= tr(P_\Omega^{(n)}EK_0)\\ &= \int_\Omega \sum_{i=1}^{3}\sum_{j=1}^{3} f_{ij}(x)\alpha_i(x)\overline{\alpha_j(x)}d\mu(x) = 0 \ .\end{aligned}$$

Thus ϕ_0 annihilates the von Neumann algebra generated by $\mathcal{M}_0$ and E, and hence ϕ annihilates $W^*(\mathcal{M},E)$. This contradicts the maximality of $\mathcal{M}$.

Thus $\mathcal{M}$ is maximal abelian, and since ϕ annihilates $\mathcal{M}$,

$$dist(T,\mathcal{A}) \geq dist(T,\mathcal{M}) \geq \phi(T) = dist(T,\mathcal{A}) \ . \qquad \blacksquare$$

9.12 Theorem. *Let $\mathcal{A}$ be an abelian von Neumann algebra. For T in $\mathcal{B}(\mathcal{H})$,*

$$dist(T,\mathcal{A}) \leq 2\sup\{\|P^\perp TP\| : P \in Lat\,\mathcal{A}\} \ .$$

Proof. By Lemma 9.11 and Lemma 9.8, there is a m.a.s.a. $\mathcal{M}$ such that

$$dist(T,\mathcal{A}) = dist(T,\mathcal{M}) \leq 2 \sup_{P \in Lat\,\mathcal{M}} \|P^\perp TP\| \leq 2 \sup_{P\in Lat\,\mathcal{A}} \|P^\perp TP\| \ . \qquad \blacksquare$$

We complete this chapter with a sharp formula for the distance to the scalar operators. Define the *maximal numerical range* of an operator T to be

$$W_0(T) = \{\lambda \in \mathbb{C} : \exists x_n \in \mathcal{H}, \|x_n\| = 1, \lim(Tx_n, x_n) = \lambda \text{ and } \lim\|Tx_n\| = \|T\|\} \ .$$

9.13 Lemma. *$W_0(T)$ is a closed convex subset of $\overline{W(T)}$.*

Proof. One need only verify that $W_0(T)$ is convex. Normalize so that $\|T\| = 1$. Let λ and μ be distinct points in $W_0(T)$ determined by sequences $\{x_n\}$ and $\{y_n\}$ respectively, and let $\lambda_n = (Tx_n, x_n)$ and $\mu_n = (Ty_n, y_n)$. Choose ϕ so that $(x_n, e^{i\phi}y_n) \geq 0$, and set $z_n = e^{i\phi}y_n - x_n$. Then

$$\mu_n = (Tx_n + z_n, x_n + z_n) = \lambda_n + (Tx_n, z_n) + (Tz_n, e^{i\phi}y_n) \ .$$

Hence $|\mu_n - \lambda_n| \leq 2\|z_n\|$. Now

$$|(x_n, y_n)| = \frac{1}{2}(\|x_n\|^2 + \|y_n\|^2 - \|z_n\|^2)$$
$$\leq 1 - \frac{1}{8}|\mu_n - \lambda_n|^2 \leq \theta < 1$$

for n large. Let $\epsilon_n = \max\{1 - \|Tx_n\|^2, 1 - \|Ty_n\|^2\}$. Then

$$\|(I - T^*T)x_n\|^2 = \|x_n\|^2 - 2Re(T^*Tx_n, x_n) + \|T^*Tx_n\|^2$$
$$\leq 2(1 - \|Tx_n\|^2) \leq 2\epsilon_n \ .$$

Similarly, $\|(I - T^*T)y_n\|^2 \leq 2\epsilon_n$. If u_n is a unit vector in $span\{x_n, y_n\}$, write $u_n = a_n x_n + b_n y_n$. Then

$$1 = \|u_n\|^2 = |a_n|^2 + |b_n|^2 + 2Re\ a_n\overline{b_n}(x_n, y_n)$$
$$\geq |a_n|^2 + |b_n|^2 - 2|a_n|\,|b_n|\theta$$
$$= (1-\theta)(|a_n|^2 + |b_n|^2) + \theta(|a_n| - |b_n|)^2 \ .$$

Hence $|a_n|^2 + |b_n|^2 \leq (1-\theta)^{-1}$. So by the Cauchy-Schwartz inequality,

$$\|(I - T^*T)u_n\| \leq |a_n|\,\|(I - T^*T)x_n\| + |b_n|\,\|(I - T^*T)y_n\|$$
$$\leq 2(1-\theta)^{-1/2}\epsilon_n^{1/2} \ .$$

Hence

$$\|Tu_n\|^2 = \|u_n\|^2 - ((I - T^*T)u_n, u_n) \geq 1 - 2(1-\theta)^{-1/2}\epsilon_n^{1/2} \ .$$

Now let η belong to the convex hull of λ and μ, and let η_n belong to the convex hull of λ_n and μ_n with $\eta = \lim \eta_n$. Let P_n be the projection onto $span\{x_n, y_n\}$, and set $T_n = P_n T | P_n \mathcal{H}$. Then λ_n and μ_n belong to $W(T_n)$, which is convex (Lemma 9.10 and Exercise 9.2). So there is a unit vector u_n in $span\{x_n, y_n\}$ such that $(Tu_n, u_n) = (T_n u_n, u_n) = \eta_n$. By the previous paragraph, $\|Tu_n\|$ tends to 1. So η belongs to $W_0(T)$. Thus $W_0(T)$ is convex. ∎

9.14 Lemma. *If* $\|T\| = dist(T,\mathbb{C}I)$, *then* 0 *belongs to* $W_0(T)$.

Proof. Assume $W_0(T)$ does not contain 0. As $W_0(T)$ is convex and closed, one can replace T by a scalar multiple so that $ReW_0(T) \geq 1$. Let

$$S = \{x \in \mathcal{H}: \|x\| = 1 \text{ and } Re(Tx,x) \leq 1/2\} \ ,$$

and let $r = \sup\{\|Tx\|: x \in S\}$. Then $r < \|T\|$, so

$$\mu = \min\{1/2, (\|T\|-r)/2\} > 0 \ .$$

Consider $T-\mu I$. For x in S,

$$\|(T-\mu I)x\| \leq \|Tx\|+\mu \leq r+(\|T\|-r)/2 < \|T\| \ .$$

For any unit vector x not in S, $Re(Tx,x) > 1/2$, so

$$\|(T-\mu I)x\|^2 = \|Tx\|^2+\mu^2-2\mu Re(Tx,x) < \|T\|^2-\mu(1-\mu) \ .$$

Hence $\|T-\mu I\| < \|T\|$. This contradiction establishes the lemma. ∎

9.15 Theorem. *For all* T *in* $B(\mathcal{H})$, $dist(T,\mathbb{C}I) = \frac{1}{2}\|\delta_T\|$.

Proof. By compactness, there is a scalar λ so that $\|T-\lambda I\| = dist(T,\mathbb{C}I)$. Without loss of generality, $\lambda = 0$ and $\|T\| = 1$. So by Lemma 9.14, 0 belongs to $W_0(T)$. Let $\{x_n\}$ be unit vectors such that $\lim(Tx_n,x_n) = 0$, and $y_n = Tx_n$ have $\lim\|y_n\| = 1$. Note that $(T^*y_n,x_n) = (y_n,y_n)$ tends to 1. Let $A_n = x_n \otimes x_n^*-y_n \otimes y_n^*$. Clearly, $\|A_n\| \leq 1$. Moreover

$$\begin{aligned}|(\delta_T(A_n)x_n,y_n)| &= |((y_n \otimes x_n^* - Ty_n \otimes y_n^* - x_n \otimes (T^*x_n)^* \\ &\qquad + y_n \otimes (T^*y_n)^*)x_n,y_n)| \\ &\geq \|y_n\|^2-2|(x_n,y_n)|+\|y_n\|^4 \ .\end{aligned}$$

So

$$\|\delta_T\| \geq \sup\|\delta_T(A_n)\| \geq 2 = 2dist(T,\mathbb{C}I) \ .$$

On the other hand, for all A

$$\|\delta_T(A)\| \leq \|TA\|+\|AT\| \leq 2\|T\| \, \|A\| = 2dist(T,\mathbb{C}I) \, \|A\| \ .$$

Thus $\|\delta_T\| = 2dist(T,\mathbb{C}I)$. ∎

Notes and Remarks.

Lemma 9.1 was proved by Parrott [2] using a lifting theorem of Sz. Nagy and Foias [1]. It was proved independently by Davis, Kahane, Weinburger [1] and it is their proof provided here. They explicitly give all solutions X satisfying $\left\|\begin{bmatrix} A & B \\ C & X\end{bmatrix}\right\| = \gamma$. Theorem 9.5 is due to Arveson [7]. The proof given here is due to Power [1]. Another proof due to Lance [2] will be given in Chapter 16. Theorem 9.6 is due to Christensen [4].

Example 9.7 has been credited to Choi in Davidson [1]. Theorem 9.12 is due to Rosenoer [1]. The essence of Lemma 9.10 is that the numerical range of an operator is convex. This is due to Hadamard (c.f. Halmos [2]). Theorem 9.15 is due to Stampfli [1].

Exercises.

9.1 (Arveson)

(i) Let ϕ be a weak* continuous linear functional on $B(\mathcal{H})$. Show that there are vectors x and y in $\mathcal{H}^{(\infty)}$ such that

$$\phi(A) = (A^{(\infty)}x,y) \ .$$

(ii) If $\mathcal{A}$ is a weak* closed, unital subalgebra of $B(\mathcal{H})$, show that for all T in $B(\mathcal{H})$,

$$dist(T,\mathcal{A}) = \sup\{\|P^{\perp}T^{(\infty)}P\| : P \in Lat\,\mathcal{A}^{(\infty)}\} \ .$$

(iii) Let $\mathcal{F}$ be a finite nest $\{0\} = F_0 < \dots < F_n = \mathcal{H}$ with atoms $A_k = F_k \ominus F_{k-1}$. Let P belong to $Lat(\mathcal{T}(\mathcal{F})^{(\infty)})$. Show that this belongs to $\mathcal{F}'' \otimes B(\mathcal{H})$ and thus has the form $\sum_{i=1}^{n} P(A_i) \otimes P_i$ where P_i are projections. Show that $P_k \geq P_{k+1}$ for $1 \leq k \leq n-1$.

(iv) Rewrite P as $\sum_{j=1}^{n} P(F_j) \otimes Q_j$ where Q_j are pairwise orthogonal. Combine this with (ii) to prove Corollary 9.3.

9.2 Show that $x+iy$ belongs to $W\left(\begin{bmatrix} 1 & 2r \\ 0 & -1 \end{bmatrix}\right)$ if and only if $(x-c)^2+y^2 = r^2(1-c^2)$ has a solution c in $[-1,1]$. Compute the discriminant of this quadratic in c and thereby deduce that the numerical range is the ellipse $\{x+iy : \frac{x^2}{1+r^2}+\frac{y^2}{r^2} \leq 1\}$.

9.3 (Davidson-Power) Let $\mathcal{N}$ be a nest of order type $\mathbb{N} \cup \{\infty\}$, $\{-\infty\} \cup -\mathbb{N}$, or $\{-\infty\} \cup \mathbb{Z} \cup \{+\infty\}$. Given K compact, find T in $\mathcal{T}(\mathcal{N}) \cap \mathcal{K}$ such that $dist(K,\mathcal{T}(\mathcal{N})) = \|K-T\|$. *Hint*: Reduce to the finite nest case by finding $N_1 < N_2$ in $\mathcal{N}$ so that $\|KP(N_1)\| \leq ½\, dist(K,\mathcal{T}(\mathcal{N}))$ and $\|P(N_2)^{\perp}K\| \leq ½\, dist(K,\mathcal{T}(\mathcal{N}))$.

9.4 (Davidson-Power)

(i) Let $\mathcal{N}$ be the Volterra nest on $L^2(0,1)$ and let $K = 1 \otimes 1^*$. Show that $½ = dist(K,\mathcal{T}(\mathcal{N})) < \|K-T\|$ for every compact T in $\mathcal{T}(\mathcal{N})$. *Hint*: Let $x = \sqrt{2}\chi_{[½,1]}$. Prove that if $\|K-T\| = ½$, then $TP(N_t)x = ½P(N_t)x$ for all t in $[½,1]$.

(ii) $dist(K, \mathcal{T}(\mathcal{N})) = dist(V^*, \mathcal{T}(\mathcal{N}))$ where V is the Volterra operator of Chapter 5. Let D be the diagonal operator which is multiplication by $f(x) = \min\{x, 1-x\}$. Show that $\|V^*-D\| = \frac{1}{2}$.

9.5 (Davidson-Power) Let $\mathcal{N}$ be given by an increasing sequence of finite dimensional subspaces sense in $\mathcal{H}$. Prove that for A in $B(\mathcal{H})$,

$$dist(A, \mathcal{T}(\mathcal{N}) \cap K) = \max\{\|A\|_e, dist(T, \mathcal{T}(\mathcal{N}))\} .$$

9.6 (i) Let $\mathcal{A}$ be a reflexive algebra. For T in $B(\mathcal{H})$, show that

$$dist(T, \mathcal{A}) \geq \sup\{\|P^{\perp}TP\| : P \in Lat\ \mathcal{A}\} .$$

(ii) Let $\mathcal{A}_\theta$ be the algebra of 2×2 matrices with $e_1 = \begin{bmatrix} 1 \\ 0 \end{bmatrix}$ and $e_\theta = \begin{bmatrix} \cos\theta \\ \sin\theta \end{bmatrix}$ as eigenvectors. Find $Lat\ \mathcal{A}_\theta$. Let $E = \begin{bmatrix} 1 & 0 \\ 0 & 0 \end{bmatrix}$. Compare $dist(E, \mathcal{A}_\theta)$ to $\max\{\|P^{\perp}EP\| : P \in Lat\ \mathcal{A}_\theta\}$.

(iii) (Kraus-Larson) Let $\mathcal{A} = \sum_{n=1}^{\infty} \oplus \mathcal{A}_{1/n}$. Prove that $\mathcal{A}$ is reflexive. Find operators E_n such that

$$\lim_{n \to \infty} \frac{\max\{\|P^{\perp}E_nP\| : P \in Lat\ \mathcal{A}\}}{dist(E_n, \mathcal{A})} = 0 .$$

10. Derivations of C*-Algebras

A *bimodule* of a Banach algebra $\mathcal{A}$ is a Banach space $\mathcal{E}$ which is a 2-sided module over $\mathcal{A}$ satisfying $\|AE\| \le \|A\|\,\|E\|$ and $\|EA\| \le \|E\|\,\|A\|$ for all A in $\mathcal{A}$ and E in $\mathcal{E}$. A *derivation* δ of an algebra $\mathcal{A}$ into an $\mathcal{A}$-bimodule $\mathcal{E}$ is a linear map δ satisfying $\delta(AB) = \delta(A)B + A\delta(B)$ for all A and B in $\mathcal{A}$. The derivation is called *inner* if there is an element E in $\mathcal{E}$ such that $\delta(A) = \delta_E(A) \equiv AE - EA$. For our purposes, the bimodules of interest will be $\mathcal{A}$, $\mathcal{B}(\mathcal{H})$, and $\mathcal{K}$.

10.1 Lemma. *Let A and B belong to a C^*-algebra $\mathcal{A}$. If B is positive, $\|B\| \le 1$, and $AA^* \le B^4$, then there is an element C in $\mathcal{A}$ such that $A = BC$ and $\|C\| \le 1$.*

Proof. If $\mathcal{A}$ has no unit, then let $\mathcal{A}_1$ be the C^*-algebra with unit adjointed. Define $C_t = (B+tI)^{-1}A$ and note that this belongs to $\mathcal{A}$, and

$$C_tC_t^* = (B+tI)^{-1}AA^*(B+tI)^{-1} \le (B+tI)^{-1}B^4(B+tI)^{-1} \le B^2 \le I \ .$$

So $\|C_t\| \le 1$. It will be shown that $C = \lim_{t\to 0} C_t$ exists.

$$\begin{aligned} C_s - C_t &= (B+sI)^{-1}(B+tI)^{-1}(B+tI-B-sI)A \\ &= (t-s)(B+sI)^{-1}(B+tI)^{-1}A \ . \end{aligned}$$

Hence

$$\begin{aligned} \|C_s - C_t\|^2 &= (t-s)^2\|(B+sI)^{-1}(B+tI)^{-1}AA^*(B+tI)^{-1}(B+sI)^{-1}\| \\ &\le (t-s)^2\|(B+sI)^{-1}(B+tI)^{-1}B^4(B+tI)^{-1}(B+sI)^{-1}\| \\ &\le (t-s)^2 \ . \end{aligned}$$

So C_t is Cauchy, and C exists as claimed. Furthermore,

$$BC = \lim_{t\to 0} B(B+tI)^{-1}A = \lim_{t\to 0} A - tC_t = A \ . \qquad \blacksquare$$

10.2 Lemma. *Let $\mathcal{A}$ be an infinite dimensional C^*-algebra. Then $\mathcal{A}$ contains a self-adjoint element A with infinite spectrum.*

Proof. It clearly does not hurt to adjoin a unit, so we assume that $\mathcal{A}$ contains an identity I. Let $\mathcal{M}$ be a maximal abelian subalgebra. Then $\mathcal{M}$ is isomorphic to $C(X)$ for a compact Hausdorff space X. If X is infinite, then $\mathcal{M}$ contains an element A corresponding to a real function with infinite range, and thus infinite spectrum. If X is finite, $\mathcal{M}$ contains a finite set $E_1,\ldots,E_n$ of minimal projections with $\sum_{j=1}^{n} E_j = I$. Now, $E_j \mathcal{A} E_j$ is a C^*-algebra in which $\mathbb{C}E_j$ is a maximal abelian subalgebra. Hence $E_j \mathcal{A} E_j = \mathbb{C}E_j$. If $E_i \mathcal{A} E_j$ is not zero, choose $A = E_i A E_j$ of norm one. Then $A^*A = E_j$ and $AA^* = E_i$. Let $B = E_i B E_j$. Then there is a scalar λ so that $BA^* = \lambda AA^*$. Thus

$$B - \lambda A = (B - \lambda A)E_j = (BA^* - \lambda AA^*)A = 0 \ .$$

So $E_i \mathcal{A} E_j$ is one dimensional, and hence $\mathcal{A}$ is at most n^2 dimensional. ∎

10.3 Theorem. *Every derivation δ of a C^*-algebra into a bimodule $\mathcal{E}$ is continuous.*

Proof. Let $\mathcal{J}$ be the set of elements J in $\mathcal{A}$ such that the map D_J taking T to $\delta(JT)$ is continuous. Clearly $\mathcal{J}$ is a right ideal. Also

$$\delta(AJT) = A\delta(JT) + \delta(A)JT$$

is continuous in T for every A in $\mathcal{A}$ and J in $\mathcal{J}$. So $\mathcal{J}$ is also a left ideal. To show that $\mathcal{J}$ is closed, consider J_n in $\mathcal{J}$ converging to J. Then

$$\begin{aligned} D_J(T) = \delta(JT) &= J\delta(T) + \delta(J)T \\ &= \lim_{n \to \infty} J_n \delta(T) + \delta(J)T \\ &= \lim_{n \to \infty} D_{J_n}(T) - \delta(J_n)T + \delta(J)T \ . \end{aligned}$$

So D_J is the pointwise limit of continuous maps. Thus D_J is continuous by the Uniform Boundedness Principle.

Next, we show that δ is continuous on $\mathcal{J}$. For otherwise, there are J_n in $\mathcal{J}$, $n \geq 1$, such that $\|J_n\|^2 \leq 2^{-n}$ and $\|\delta(J_n)\| \geq n$. Thus $B = (\sum_{n \geq 1} J_n J_n^*)^{1/4}$ is a positive element of $\mathcal{J}$ with $\|B\| \leq 1$. By Lemma 10.1, there are elements C_n in $\mathcal{J}$ with $\|C_n\| \leq 1$ such that $J_n = BC_n$. Hence

$$\|D_B(C_n)\| = \|\delta(BC_n)\| = \|\delta(J_n)\| \geq n$$

contradicting the continuity of D_B. Let $M = \|\delta\ |\mathcal{J}\|$.

Suppose that $\mathcal{A}/\mathcal{J}$ is infinite dimensional. Then by Lemma 10.2, it contains a self-adjoint element $\dot{A}$ so that $C^*(\dot{A})$ is isomorphic to $C_0(X)$ where $X \cup \{0\}$ is an infinite, compact subset of $[-1,1]$. There are countably many positive functions f_j, $j \geq 1$ in $C_0(X)$ such that $f_j f_k = 0$ if $j \neq k$ and $f_j(\dot{A}) \neq 0$. Let A_j be positive, norm one elements of $\mathcal{A}$ such that $A_j + \mathcal{J}$ is a scalar multiple of $f_j(\dot{A})$. Then $A_j A_k$ belongs to $\mathcal{J}$ for $j \neq k$, but $A_j^2 + \mathcal{J} = f_j(\dot{A})^2 \neq 0$. Since A_j^2 is not in $\mathcal{J}$, D_{A_j} is unbounded.

Thus one can choose T_j in $\mathcal{A}$ with $\|T_j\| \le 2^{-j}$ and $\|\delta(A_j^2T_j)\| \ge \|\delta(A_j)\|+j$. Let $X = \sum_{j\ge1} A_jT_j$. Then $A_kX = A_k^2T_k+J_k$ where $J_k = \sum_{j\ne k} A_kA_jT_j$ belongs to J, and $\|J_k\| \le 1$. So

$$\begin{aligned}\|\delta(X)\| \ge \|A_k\delta(X)\| &= \|\delta(A_kX)-\delta(A_k)X\| \\ &\ge \|\delta(A_k^2T_k)\|-\|\delta(J_k)\|-\|\delta(A_k)\|\ \|X\| \\ &\ge \|\delta(A_k)\|+k-M-\|\delta(A_k)\| = k-M \ .\end{aligned}$$

But k is arbitrary, and so this is impossible.

Hence $\mathcal{A}/J$ is finite dimensional. As δ is continuous on J, it must be continuous on all of $\mathcal{A}$. ∎

Let $\mathcal{A}$ be a C^*-algebra acting on a Hilbert space $\mathcal{H}$. An $\mathcal{A}$-bimodule $\mathcal{E}$ which is the dual of a Banach space $\mathcal{E}_*$, is a *normal dual bimodule* if the maps $A \to AE$ and $A \to EA$ are continuous from the weak* topology of $\mathcal{B}(\mathcal{H})$ restricted to $\mathcal{A}$ to the weak* topology on $\mathcal{E}$. It is easy to extend the module action of $\mathcal{A}$ to all of $\mathcal{A}''$ by the weak* continuity. To study the weak* continuity of derivations, we first need some basic facts about the double dual of a C^*-algebra.

10.4 Theorem. *Let $\mathcal{A}$ be a C^*-algebra. Then the second dual $\mathcal{A}^{**}$ is isometrically isomorphic to a von Neumann algebra $\mathcal{M}$, and this isomorphism is a homeomorphism of $\mathcal{A}^{**}$ with its weak* topology onto $\mathcal{M}$ with its weak* topology.*

Proof. For each state f on $\mathcal{A}$, the GNS construction provides a representation $(\mathcal{H}_f,\pi_f)$ with a cyclic vector ξ_f such that $(\pi_f(A)\xi_f,\xi_f) = f(A)$ for all A in $\mathcal{A}$. The representation $\pi = \sum \oplus \pi_f$ on $\mathcal{H}_\pi = \sum \oplus \mathcal{H}_f$ is the *universal atomic representation*. Let $\mathcal{M}$ denote the weak operator closure of $\pi(\mathcal{A})$. This is known as the *universal enveloping von Neumann algebra* of $\mathcal{A}$. Every positive linear functional on $\mathcal{A}$ is given by a vector state, and thus extends to be a weak* continuous functional on $\mathcal{M}$. Conversely, if ϕ is a weak* continuous functional on $\mathcal{M}$, then ϕ is determined by its restriction to $\mathcal{A}$. Because the unit ball of $\mathcal{A}$ is weak* dense in the ball for $\mathcal{M}$ (Kaplansky's Density Theorem), $\|\phi\| = \|\phi\,|\mathcal{A}\|$. So $\mathcal{A}^*$ is isometrically isomorphic to $\mathcal{M}_*$, the space of weak* continuous functionals on $\mathcal{M}$. By a standard theorem of functional analysis, the weak* closed subspace $\mathcal{M}$ of the dual space $\mathcal{B}(\mathcal{H}_\pi)$ is the dual of $\mathcal{M}_*$. Thus $\mathcal{M}$ is isometrically isomorphic to $\mathcal{A}^{**}$, and the weak* topology on $\mathcal{A}^{**}$ coincides with the weak* topology on $\mathcal{M}$. ∎

10.5 Corollary. *Let ρ be a representation of a C^*-algebra $\mathcal{A}$ on a Hilbert space $\mathcal{H}$. Then there is a unique normal representation $\tilde{\rho}$ of $\mathcal{A}^{**}$ extending ρ, and this takes $\mathcal{A}^{**}$ onto $\rho(\mathcal{A})''$. Furthermore, there is a central projection P in $\mathcal{A}^{**}$ such that the restriction of $\tilde{\rho}$ to $P\mathcal{A}^{**}$ is a $*$-isomorphism onto $\rho(\mathcal{A})''$.*

Proof. Every representation ρ is the direct sum of cyclic representations ρ_α (easy application of Zorn's lemma), and every cyclic representation (ρ_α, x_α) equivalent to π_{f_α} where f_α is the state $f_\alpha(A) = (\rho_\alpha(A)x_\alpha, x_\alpha)$. One maps $\mathcal{A}^{**}$ into $\pi_{f_\alpha}(\mathcal{A})''$ by restriction to $\mathcal{H}_\alpha$, and this map is normal. Since $\mathcal{A}^{**}$ is weakly closed, it maps onto $\pi_{f_\alpha}(\mathcal{A})''$. The direct sum of these maps is a normal representation $\tilde{\rho}$ of $\mathcal{A}^{**}$ onto $\rho(\mathcal{A})''$. But $\mathcal{A}$ is weak* dense in $\mathcal{A}^{**}$, so this extension must be unique.

Let P be a maximal projection in $J = ker(\tilde{\rho})$. This is the identity for J as this is a weak* closed C^*-algebra. Moreover, P is central in $\mathcal{A}^{**}$ since if A belongs to $\mathcal{A}$, then PA and AP are in J. Thus $PA = PAP = AP$. The restriction map of $\tilde{\rho}$ to $P^{\perp}\mathcal{A}^{**}$ is thus a $*$-isomorphism. ■

10.6 Theorem. *Let ρ be a representation of a C^*-algebra $\mathcal{A}$ on a Hilbert space $\mathcal{H}$. Let δ be a derivation of $\rho(\mathcal{A})$ into a dual normal bimodule $\mathcal{E}$. Then δ is weak* continuous, and thus extends to a derivation of $\rho(\mathcal{A})''$ into $\mathcal{E}$ with the same norm.*

Proof. Let $\tilde{\rho}$ be the extension of ρ to a normal representation of $\mathcal{A}^{**}$ onto $\rho(\mathcal{A})''$. By Corollary 10.5, there is a central projection P in $\mathcal{A}^{**}$ such that the restriction σ of $\tilde{\rho}$ to $P\mathcal{A}^{**}$ is a $*$-isomorphism. Make $\mathcal{E}$ into an $\mathcal{A}^{**}$ bimodule by

$$A \cdot E = \tilde{\rho}(A)E \quad \text{and} \quad E \cdot A = E\tilde{\rho}(A)$$

for A in $\mathcal{A}^{**}$ and E in $\mathcal{E}$. Since $\tilde{\rho}$ is normal, and $\mathcal{E}$ is a dual normal $\rho(\mathcal{A})''$ bimodule, it becomes a dual normal $\mathcal{A}^{**}$ bimodule.

Let ϕ be a functional in $\mathcal{E}_*$. By Theorem 10.3, δ is continuous, so $f(A) = \phi(\delta(\rho(A))$ is a continuous functional (in $\mathcal{A}^*$). Thus f is weak* continuous on $\mathcal{A}^{**}$. Therefore, $\delta \circ \rho$ is a normal map of $\mathcal{A}$ into $\mathcal{E}$. Thus it extends to a linear map $\tilde{\delta}$ of $\mathcal{A}^{**}$ onto $\mathcal{E}$ which is weak*-weak* continuous. Since the unit ball of a Banach space is dense in the ball of its double dual, $\|\tilde{\delta}\| = \|\delta\|$. Furthermore, $\tilde{\delta}$ is a derivation. To see this, fix A in $\mathcal{A}$ and let B_β be a net in $\mathcal{A}$ converging weak* to B in $\mathcal{A}^{**}$. Then

$$\begin{aligned}\tilde{\delta}(AB) &= \lim \delta \circ \rho(AB_\beta) \\ &= \lim A\delta(\rho(B_\beta)) + \delta(\rho(A))B_\beta = A\tilde{\delta}(B) + \tilde{\delta}(A)B\ .\end{aligned}$$

Now, fix B and take a net A_α in $\mathcal{A}$ converging weak* to A. By the same argument, we obtain $\tilde{\delta}(AB) = A\tilde{\delta}(B)+\tilde{\delta}(A)B$. Also, $\tilde{\delta}(P) = \tilde{\delta}(P^2) = \tilde{\rho}(P)\tilde{\delta}(P)+\tilde{\delta}(P)\tilde{\rho}(P) = 2\tilde{\delta}(P)$. Thus $\tilde{\delta}(P) = 0$.

Define a map $\hat{\delta}$ of $\rho(\mathcal{A})''$ into $\mathcal{E}$ by $\hat{\delta}(X) = \tilde{\delta}(\sigma^{-1}(X))$. This is a weak*-weak* continuous map since both σ^{-1} and $\tilde{\delta}$ are normal, and it is easily seen to be a derivation. Furthermore, if A belong to $\mathcal{A}$,

$$\begin{aligned}\hat{\delta}(\rho(A)) &= \tilde{\delta}(\sigma^{-1}(\rho(A))) = \tilde{\delta}(PA) \\ &= \tilde{\delta}(P)\rho(A)+\tilde{\rho}(P)\tilde{\delta}(A) \\ &= \delta(\rho(A))\ .\end{aligned}$$

So $\hat{\delta}$ is an extension of δ as desired. ∎

10.7 Lemma. *Every derivation δ of a finite dimensional C^*-algebra is inner, and $\delta = \delta_E$ for E in $\mathcal{E}$ with $\|E\| \le \|\delta\|$.*

Proof. Let $\mathcal{G}$ be a finite group of unitaries in the C^*-algebra $\mathcal{A}$ as in section 8.2 so that $\mathcal{A}$ is the span of $\mathcal{G}$. Let

$$E = \frac{1}{|\mathcal{G}|}\sum_{U\in\mathcal{G}} \delta(U)U^*\ .$$

Then if V belongs to $\mathcal{G}$,

$$\begin{aligned}\delta_E(V) = EV-VE &= \frac{1}{|\mathcal{G}|}\sum_{U\in\mathcal{G}} \delta(U)U^*V - V\delta(U)U^* \\ &= \frac{1}{|\mathcal{G}|}\sum_{U\in\mathcal{G}} (\delta(U)U^* - \delta(VU)(VU)^*)V + \delta(V) = \delta(V)\ .\end{aligned}$$

So $\delta = \delta_E$, and $\|E\| \le \max\|\delta(U)U^*\| \le \|\delta\|$. ∎

10.8 Theorem. *Let $\mathcal{A}$ be an AF von Neumann algebra. Then every derivation δ of $\mathcal{A}$ into a dual normal bimodule is inner, and $\delta = \delta_E$ with $\|E\| \le \|\delta\|$.*

Proof. Let $\mathcal{A}_\alpha$ be an increasing net of finite dimensional subalgebras of $\mathcal{A}$ whose union is weak* dense. By Theorem 10.3, δ is bounded. By Lemma 10.7, for each α there is an element E_α in $\mathcal{E}$ so that $\delta\,|\mathcal{A}_\alpha = \delta_{E_\alpha}$ and $\|E_\alpha\| \le \|\delta\|$. By the Banach-Alaoglu Theorem, the ball of radius $\|\delta\|$ in $\mathcal{E}$ is weak* compact. Hence there is a cofinal subnet Λ such that $E = \omega^*\!-\lim_\Lambda E_\lambda$ exists. For every α, there is a λ in Λ so that $\alpha \le \lambda$. Thus if $\mu \ge \lambda$, $\delta_{E_\mu}|\mathcal{A}_\alpha = \delta\,|\mathcal{A}_\alpha$. Hence $\delta_E\,|\mathcal{A}_\alpha = \delta\,|\mathcal{A}_\alpha$ for every α. By Theorem 10.6, δ is weak* continuous (as is δ_E, clearly) so $\delta = \delta_E$. ∎

Now, we turn to the consideration of derivations into the compact operators. The main result will be very important for its applications to nest algebras. We begin with the abelian case.

10.9 Theorem. *Let $\mathcal{M}$ be an abelian von Neumann algebra. Every derivation δ of $\mathcal{M}$ into the compact operators $\mathcal{K}$ is inner, and $\delta = \delta_K$ for some K in $\mathcal{K}$ with $\|K\| \le \|\delta\|$.*

Proof. By considering δ as a derivation into $\mathcal{B}(\mathcal{H})$ which is a dual normal $\mathcal{M}$-module, one has by Theorem 10.8 that $\delta = \delta_T$ for some T in $\mathcal{B}(\mathcal{H})$. Let Φ be an expectation onto $\mathcal{M}'$ given by Theorem 8.3. Set $K = T - \Phi(T)$. Clearly, $\delta_K = \delta_T$. It suffices to prove that K is compact.

Suppose that $\|K\|_e = \gamma \neq 0$. Let $\mathcal{P}$ be a maximal chain of projections in $\mathcal{M}$ such that $\|PKP\|_e = \gamma$ for all P in $\mathcal{P}$. If P belongs to $\mathcal{P}$ and $P = P_1 + P_2$, then

$$\begin{aligned} \gamma = \|PKP\|_e &= \|P_1KP_1 + P_2KP_2 + P_1KP_2 + P_2KP_1\|_e \\ &= \|P_1KP_1 + P_2KP_2 + P_1\delta(P_2) + P_2\delta(P_1)\|_e \\ &= \|P_1KP_1 + P_2KP_2\|_e \\ &= \max\{\|P_1KP_1\|, \|P_2KP_2\|\} \ . \end{aligned} \tag{10.9.1}$$

Let $P_0 = \inf\{P : P \in \mathcal{P}\}$. If P_0 belongs to $\mathcal{P}$, then the computation (10.9.1) along shows that P_0 must be minimal in $\mathcal{P}$ (for otherwise some smaller projection also belongs to $\mathcal{P}$). But if P_0 is minimal, then since $P_0\Phi(T) = P_0TP_0$,

$$P_0KP_0 = P_0TP_0 - P_0\Phi(T)P_0 = 0 \ .$$

Thus, P_0 does not belong to $\mathcal{P}$. By equation (10.9.1), $\|(P-P_0)K(P-P_0)\| = \gamma$ for every P in $\mathcal{P}$.

Let $\mathcal{P} = \{P_\alpha\}$ ordered by itself in decreasing order. Since the norm is lower semi-continuous in the weak* topology,

$$\gamma = \|(P-P_0)K(P-P_0)\| = \lim_\alpha \|(P-P_\alpha)K(P-P_\alpha)\| \ .$$

Hence one may choose a sequence $\alpha_1 > \alpha_2 > \dots$ such that $Q_k = P_{\alpha_k} - P_{\alpha_{k+1}}$ satisfies $\|Q_kKQ_k\| > \gamma/2$. By Theorem 8.3,

$$Q_kKQ_k = Q_k(K - \Phi(K))Q_k$$

belongs to the weak* closed convex hull of

$$\{Q_k(K - UKU^*)Q_k : U \in \mathcal{U}(\mathcal{M})\} = \{Q_k\delta(U)U^*Q_k : U \in \mathcal{U}(\mathcal{M})\} \ .$$

Choose a unitary U_k in $\mathcal{M}$ such that $\|Q_k\delta(U_k)Q_k\| > \gamma/2$. Now

$$\begin{aligned} Q_k\delta(Q_k^\perp U_k)Q_k &= Q_k\delta(Q_k^\perp U_k Q_k^\perp)Q_k \\ &= Q_k Q_k^\perp \delta(U_k Q_k^\perp) + Q_k\delta(Q_k^\perp)U_k Q_k^\perp Q_k = 0 \ . \end{aligned}$$

So $\gamma/2 < \|Q_k\delta(Q_kU_k)Q_k\|$. Let $A = \sum_{k=1}^{\infty} Q_kU_k$. This belongs to $\mathcal{M}$, but

$$\begin{aligned} \|Q_k\delta(A)Q_k\| &= \|Q_k\delta(Q_kU_k)Q_k + Q_k\delta(Q_k^\perp A Q_k^\perp)Q_k\| \\ &= \|Q_k\delta(Q_kU_k)Q_k\| \geq \gamma/2 \ . \end{aligned}$$

This contradicts the compactness of $\delta(A)$. Thus K is compact.

Since $K = K - \Phi(K)$, it belongs to the weak* closed convex hull of $K - UKU^* = \delta(U)U^*$ and hence $\|K\| \leq \|\delta\|$. ∎

10.10 Lemma. *Let $\mathcal{A}$ be the algebra of all $n \times n$ matrices with coefficients in an abelian von Neumann algebra $\mathcal{M}$. ($\mathcal{A} = M_n \otimes \mathcal{M}$). Then every derivation δ of $\mathcal{A}$ into the compact operators is inner, and $\delta = \delta_K$ with K in $\mathcal{K}$ and $\|K\| \leq \|\delta\|$.*

Proof. Again by Theorem 10.8, it suffices to consider $\delta = \delta_T$ for T in $\mathcal{B}(\mathcal{H})$. By Lemma 10.7, the restriction of δ to the $n \times n$ matrices with scalar entries is inner, so equals δ_C for some compact operator C. Let $\delta' = \delta_{T-C}$. Let E_{ij} be the standard matrix units of M_n. Then for M in $\mathcal{M}$,

$$\delta'(E_{ij}M) = \delta'(E_{i1}(E_1ME_1)E_{1j}) = E_{i1}\delta'(E_1ME_1)E_{1j} \ .$$

The derivation δ'' obtained by restricting δ' to the 1–1 entry is then a derivation into $E_{11}\mathcal{K}E_{11}$. By Theorem 10.8, $\delta'' = \delta_L$ where L is compact. Let $K = L^{(n)}$ be the direct sum of n copies of L. One has

$$\delta'(E_{ij}M) = E_{i1}(LM - ML)E_{1j} = K(E_{ij}M) - (E_{ij}M)K \ .$$

So $\delta' = \delta_K$ and thus $\delta = \delta_{C+K}$. Replace $C+K$ by $K' = C+K-\Phi(C+K)$ where Φ is any expectation onto $\mathcal{A}'$ (Theorem 8.3). Then by the remark following Theorem 8.6, $\Phi(C+K)$ is compact. Furthermore, K' belongs to the convex hull of $\{C+K-U(C+K)U^* : U \in \mathcal{U}(\mathcal{A})\}$ which equals $\{\delta(U)U^* : U \in \mathcal{U}(\mathcal{A})\}$. Hence $\|K'\| \leq \|\delta\|$. ∎

10.11 Lemma. *Let $\mathcal{A} = \mathcal{B}(\mathcal{H}) \otimes \mathcal{M}$ be the algebra of all infinite matrices $(A_{ij})_{i,j\in\mathbb{Z}}$ with coefficients in a von Neumann algebra $\mathcal{M}$ which are bounded operators. Every derivation of $\mathcal{A}$ into the compacts is inner, and $\delta = \delta_K$ for K in $\mathcal{K}$ with $\|K\| \leq \|\delta\|$.*

Proof. First note that

$$\mathcal{A} = \mathcal{B}(\mathcal{H}) \otimes \mathcal{M} \cong \mathcal{B}(\mathcal{H} \otimes \mathcal{H}) \otimes \mathcal{M} \cong \mathcal{B}(\mathcal{H}) \otimes (\mathcal{B}(\mathcal{H}) \otimes \mathcal{M})$$

so $\mathcal{A}$ can be thought of as infinite matrices with coefficients in $\mathcal{A}$. Let U be the bilateral shift of infinite multiplicity given by the matrix $(\delta_{ij+1}I)$.

Then $\mathcal{U} = \{U\}''$ is an abelian von Neumann algebra. By Theorem 10.9, $\delta\,|\mathcal{U} = \delta_K$ with K compact. Let $\delta' = \delta - \delta_K$. For any operator A in $\mathcal{U}' \cap \mathcal{A} = \{(A_{ij}) : A_{ij} = A_{i+1,j+1} \in \mathcal{A}\}$,

$$\delta'(A)U = \delta'(AU) - A\delta'(U) = \delta'(UA) = U\delta'(A) \ .$$

So $\delta'(A)$ belongs to $\mathcal{U}' \cap \mathcal{K} = \{0\}$. Hence $\delta\,|\mathcal{U}' \cap \mathcal{A} = \delta_K$.

Now considering $\mathcal{U}$ as a subalgebra of $\mathcal{A}$, the algebra $\mathcal{D}$ of diagonal matrices with entries in $\mathcal{U}$ is an abelian von Neumann subalgebra of $\mathcal{A}$ such that $\mathcal{D}'$ contains no non-zero compact operators. Again use Theorem 10.9 to obtain a compact operator C so that $\delta'\,|\mathcal{D} = \delta_C$. Write $C = (C_{ij})$, and note that each C_{ij} is compact. The constant diagonal matrix $U^{(\infty)}$ belongs to $\mathcal{D} \cap \mathcal{U}'$, so

$$0 = \delta'(U^{(\infty)}) = \delta_C(U^{(\infty)}) = (C_{ij}U - UC_{ij}) \ .$$

Hence each C_{ij} belongs to $\mathcal{U}' \cap \mathcal{K} = \{0\}$, whence $C = 0$.

This shows that $\delta = \delta_K$ on the algebras $\mathcal{U}'$ and $\mathcal{D}$. But these generate $\mathcal{A}$ as a weak* closed algebra. So by Theorem 10.6, $\delta = \delta_K$ on all of $\mathcal{A}$. As K was obtained from $\delta\,|\mathcal{U}$, one has $\|K\| \leq \|\delta\,|\mathcal{U}\| \leq \|\delta\|$. ∎

10.12 Johnson-Parrott Theorem. *If $\mathcal{A}$ is a type I von Neumann algebra, then every derivation δ into $\mathcal{K}$ is inner, and $\delta = \delta_K$ for K in $\mathcal{K}$ with $\|K\| \leq \|\delta\|$.*

Proof. We prove this theorem when $\mathcal{A}'$ is abelian. The general case is essentially the same, but relies on the extension of Theorem 7.14 given in Exercise 7.3 to classify all type I von Neumann algebras. By that theorem, $\mathcal{A}$ is unitarily equivalent to a direct sum $\sum_{n=1}^{\infty} \oplus (M_n \otimes \mathcal{R}_n) \oplus (\mathcal{B}(\mathcal{H}) \otimes \mathcal{R}_\infty)$ where $\mathcal{R}_n$ are abelian von Neumann algebras. By theorem 10.9, the restriction of δ to the centre $\mathcal{A} \cap \mathcal{A}'$ has the form δ_K where K is compact. Let $\delta' = \delta - \delta_K$. Then $\delta'(E_nA) = E_n\delta'(A)E_n$ for every central projection E_n, and in particular, δ' can be restricted to each summand $M_n \otimes \mathcal{R}_n$. By Lemmas 9.10 and 9.11, $\delta' = \sum_{n=1}^{\infty} \oplus \delta_{K_n} \oplus \delta_{K_\infty}$ where each K_n is compact, and $\|K_n\| \leq \|\delta'\,|M_n \otimes \mathcal{R}_n\|$. Choose U_n unitary in $M_n \otimes \mathcal{R}_n$, $1 \leq n \leq \infty$, such that $\|\delta'_{K_n}(U_n)\| \geq ½\|K_n\|$. Then $U = \sum \oplus U_n$ belongs to $\mathcal{A}$, and $\delta'(U) = \sum \oplus \delta_{K_n}(U_n)$ is compact. Thus

$$0 = \lim_{n \to \infty} 2\|\delta_{K_n}(U_n)\| \geq \lim_{n \to \infty} \|K_n\|$$

and hence $K' = \sum_{n=1}^{\infty} \oplus K_n \oplus K_\infty$ is compact. So $\delta = \delta_{K+K'}$.

Let Φ be the expectation of $\mathcal{B}(\mathcal{H})$ onto $\mathcal{A}$ given in Theorem 8.3. Then $\delta = \delta_{K+K'-\Phi(K+K')} = \delta_L$. As in the proof of Lemma 10.10, L is compact and $\|L\| \le \|\delta\|$. ■

Notes and Remarks.

Sakai [1] proved that derivations of C^*-algebras into themselves are norm continuous. Kadison [1] showed that derivations of C^*-algebras into themselves are weak* continuous and hence extend to their weak* closures. Sakai [2] building on Kadison [1], proved that every derivation of a von Neumann algebra into itself is inner. The generalizations Theorems 10.3 and 10.6 for modules are due to Ringrose [6] and the proofs given here are his. Theorem 10.4 is due to Grothiendieck [1], Sherman [1] and Takeda [1]. Theorem 10.8 is considerably easier than the Kadison-Sakai Theorem. Note that one needs some condition such as weak closure, for δ_T is a derivation of the compact operators $\mathcal{K}$ into itself for every bounded operator T. Theorems 10.9 - 10.12 are due to Johnson and Parrott [1]. They prove Theorem 10.12 for purely infinite von Neumann algebras as well. This was extended to all von Neumann algebras by Popa [1].

Exercises.

10.1 Show that there are no non-zero derivations of $C(X)$ into itself. *Hint.* Show $\delta(1) = 0$. Then if $\delta(f) = f'$ and $x \in X$, write $f - f(x) = f_1^2 - f_2^2$ with $f_i(x) = 0$. Compute $f'(x)$.

10.2 (i) Let δ be a derivation of a C^*-algebra $\mathcal{A}$ into itself. Let $\alpha_t = \exp(t\delta)$. Show that α_t is a one parameter group of automorphisms such that $t \to \alpha_t$ is norm continuous for all t.

(ii) If α is an automorphism of $\mathcal{A}$ such that $\|\alpha - id\| < 1$, show that $\delta = \log\alpha$ is a derivation.

(iii) Let $t \to \alpha_t$ be a norm continuous, one parameter group of automorphisms, and let δ be its infinitesimal generator. So $\alpha_t = \exp(t\delta)$ for some bounded map δ of $\mathcal{A}$ into itself. Show that δ is a derivation.

10.3 Let $\mathcal{A}$ be a C^*-algebra containing the compact operators $\mathcal{K}$, and let α be an $*$-automorphism of $\mathcal{A}$.

(i) Show that α maps $\mathcal{K}$ onto itself.

(ii) Show that there is a unitiary U such that $\alpha(K) = UKU^*$ for all K in $\mathcal{K}$.

(iii) Prove that $\alpha(A) = UAU^*$ for all A in $\mathcal{A}$.

10.4 (Johnson-Parrot) Let $\mathcal{A}$ be an AF von Neumann algebra, and let δ be a derivation of $\mathcal{A} + \mathcal{K}$ into itself. Let Φ be an expectation of $\mathcal{B}(\mathcal{H})$ onto $\mathcal{A}$.

(i) Show that $\Phi\delta \mid \mathcal{A}$ is a derivation of $\mathcal{A}$ into itself.

(ii) Show that $\delta - \Phi\delta \mid \mathcal{A}$ is a derivation of $\mathcal{A}$ into $\mathcal{K}$.

(iii) Prove that $\delta = \delta_T$ for some operator T in $\mathcal{A} + \mathcal{A}' + \mathcal{K}$.

11. M-Ideals

In this section, a method is developed for computing the distance to certain ideals. The important applications to nest algebra will be given in the next chapter. The key idea is a Banach space notion introduced by Alfsen and Effros. A closed subspace $\mathcal{M}$ of a Banach space $\mathcal{X}$ is an *M-ideal* if there is a linear projection

$$\eta : \mathcal{X}^* \longrightarrow \mathcal{M}^\perp$$

of the dual space $\mathcal{X}^*$ onto the annihilator $\mathcal{M}^\perp$ of $\mathcal{M}$ in $\mathcal{X}^*$ such that for all Φ in $\mathcal{X}^*$,

$$\|\Phi\| = \|\eta\Phi\| + \|\Phi - \eta\Phi\| \ .$$

In this case, $\mathcal{M}^\perp$ is an *L-summand* of $\mathcal{X}^*$. The range $\mathcal{N}$ of $(1-\eta)$ is a complementary subspace of $\mathcal{M}^\perp$, and $\mathcal{X}^* = \mathcal{M}^\perp \oplus_{\ell_1} \mathcal{N}$, meaning that the norm on $\mathcal{X}^*$ is the ℓ_1 sum of the norms on the two subspaces.

11.1 Lemma. *Let $\mathcal{M}$ be an M-ideal in $\mathcal{X}$. Then every ϕ in $\mathcal{M}^*$ has a unique Hahn-Banach extension $\tilde{\phi}$ in $\mathcal{X}^*$. When $\mathcal{M}^*$ is identified as a subspace of X^* in this way, one obtains a decomposition $\mathcal{X}^* = \mathcal{M}^* \oplus_{\ell_1} \mathcal{M}^\perp$ with $\|\tilde{\phi}+\psi\| = \|\phi\| + \|\psi\|$ for ϕ in $\mathcal{M}^*$ and ψ in $\mathcal{M}^\perp$.*

Proof. Let Φ be any Hahn-Banach extension of ϕ to $\mathcal{X}^*$. Then $\eta\Phi$ annihilates $\mathcal{M}$, so $\Phi - \eta\Phi$ is an extension of ϕ. Furthermore

$$\|\phi\| \le \|\Phi - \eta\Phi\| \le \|\Phi - \eta\Phi\| + \|\eta\Phi\| = \|\Phi\| = \|\phi\| \ .$$

Thus $\eta\Phi = 0$. The most general extension of ϕ is $\Phi + \psi$ where ψ belongs to $\mathcal{M}^\perp$. However, since $\eta(\Phi+\psi) = \psi$,

$$\|\Phi+\psi\| = \|\Phi\| + \|\psi\| = \|\phi\| + \|\psi\| \ .$$

So the only Hahn-Banach extension of ϕ is Φ, which we denote by $\tilde{\phi}$. Identify $\mathcal{M}^*$ with $\{\tilde{\phi} : \phi \in \mathcal{M}^*\}$.

Let η be the projection onto $\mathcal{M}^\perp$. Suppose that $\eta\Phi = 0$. Then $\phi = \Phi\,|\,\mathcal{M}$ belongs to $\mathcal{M}^*$ and the above argument shows that $\Phi = \tilde{\phi}$. Thus $\mathcal{M}^* = \ker\eta$. For any Φ in $\mathcal{X}^*$, one has $\Phi = (\Phi - \eta\Phi) + \eta\Phi$ is a sum of an element of $\mathcal{M}^*$ and an element of $\mathcal{M}^\perp$ such that

$$\|\Phi\| = \|\Phi - \eta\Phi\| + \|\eta\Phi\| \ .$$

Furthermore, $\mathcal{M}^* \cap \mathcal{M}^\perp = \{0\}$. So this is a direct sum. ∎

A basic and elementary property of M-ideals is that they are *proximinal*, meaning that for every x in $\mathcal{X}$, there is an m in $\mathcal{M}$ such that $\|x-m\| = dist(x,\mathcal{M})$.

11.2 Lemma. *Let $\mathcal{M}$ be an M-ideal in $\mathcal{X}$. Let y be a vector in $\mathcal{X}$ such that $dist(y,\mathcal{M}) = 1$ and $\|y\| = 1+r$, and let $\epsilon > 0$. Then there is an m in $\mathcal{M}$ with $\|m\| \leq r$ and $\|y-m\| < 1+\epsilon$.*

Proof. Suppose, to the contrary, the set $S = \{y-m : \|m\| \leq r, m \in \mathcal{M}\}$ is disjoint from the set $(1+\epsilon)D$ where D is the open unit ball of $\mathcal{X}$. By the Hahn-Banach Theorem, there is a non-zero linear functional Φ in $\mathcal{X}^*$ so that

$$\sup\{Re\Phi(x) : x \in (1+\epsilon)D\} \leq \inf\{Re\Phi(y-m) : \|m\| \leq r, m \in \mathcal{M}\} \ .$$

Evaluating these expressions yields

$$(1+\epsilon)\|\Phi\| \leq Re\Phi(y) - r\|\Phi\,|\mathcal{M}\| \ .$$

Write $\Phi = \tilde{\phi}+\psi$, where $\phi = \Phi\,|\mathcal{M}$ and $\psi = \eta\Phi$. Then

$$Re\Phi(y) = Re(\tilde{\phi}(y)+\psi(y)) \leq \|\tilde{\phi}\|\ \|y\| + \|\psi\| dist(y,\mathcal{M}) \ .$$

Hence

$$(1+\epsilon)\|\Phi\| \leq \|\tilde{\phi}\|(1+r) + \|\psi\| - r\|\tilde{\phi}\| = \|\Phi\| \ .$$

This contradiction establishes the claim. ■

11.3 Theorem. *Let $\mathcal{M}$ be an M-ideal in $\mathcal{X}$. Then $\mathcal{M}$ is proximinal in $\mathcal{X}$.*

Proof. Let x_0 belong to $\mathcal{X}$, normalized so that $dist(x_0,\mathcal{M}) = 1$. Fix $\epsilon > 0$. Choose m_0 in $\mathcal{M}$ with $\|m_0\| \leq \|x_0\|-1$ so that $\|x_0-m_0\| < 1+\epsilon/2$. Recursively, choose m_j in $\mathcal{M}$, $j \geq 1$ so that $\|m_j\| \leq \epsilon/2^j$ and $\|x_0 - \sum_{j=0}^{n} m_j\| < 1+\epsilon/2^{n+1}$. Then

$$\sum_{j=0}^{\infty}\|m_j\| \leq \|x_0\|-1+\sum_{j=1}^{\infty}\frac{\epsilon}{2^j} = \|x_0\|-1+\epsilon \ .$$

Thus $m = \sum_{j=0}^{\infty} m_j$ belongs to $\mathcal{M}$. Furthermore,

$$\|x_0-m\| = \limsup_{n\to\infty}\|x_0 - \sum_{j=0}^{n} m_j\| \leq 1 \ .$$

Hence $\|x_0-m\| = 1$. (Plus, $\|x_0-m\|+\|m\| < \|x_0\|+\epsilon$.) ■

The important examples of M-ideals are ideals of C^*-algebras. To prove this, we need the material on the double dual of $\mathcal{A}$ from section 10.4.

11.4 Theorem. *Every (closed, two sided) ideal of a C^*-algebra is an M-ideal.*

Proof. Let $\mathcal{J}$ be an ideal in C^*-algebra $\mathcal{A}$. By Theorem 10.4, $\mathcal{A}^{**}$ is a von Neumann algebra. The weak* closure of $\mathcal{J}$ in $\mathcal{A}^{**}$ is $\mathcal{J}^{**}$, and this is an ideal in $\mathcal{A}^{**}$. Let P be the largest projection in $\mathcal{J}^{**}$. For any A in $\mathcal{A}^{**}$, PA and AP belong to $\mathcal{J}^{**}$. So $PA = (PA)P = P(AP) = AP$. Thus P is a central projection in $\mathcal{A}^{**}$.

Define a mapping η on $\mathcal{A}^*$ by

$$(\eta\Phi)(A) = \Phi(P^{\perp}A) \ .$$

Then since $P^{\perp}\mathcal{J}^{**} = \{0\}$, $\eta\Phi$ belongs to $\mathcal{J}^{\perp}$. Conversely, if Φ belongs to $\mathcal{J}^{\perp}$, then Φ annihilates $\mathcal{J}^{**}$. Thus

$$\Phi(A) = \Phi(A-PA) = \eta\Phi(A) \ .$$

Thus η is a projection onto $\mathcal{J}^{\perp}$. Furthermore, since P is central, $\mathcal{A}^{**}$ is the direct sum of $\mathcal{J}^{**}$ and $P^{\perp}\mathcal{A}^{**}$. The functional $\eta\Phi$ attains its norm on $P^{\perp}\mathcal{A}^{**}$, and $(\Phi-\eta\Phi)(A) = \Phi(PA)$ attains its norm on $\mathcal{J}^{**}$. So if Φ belongs to $\mathcal{A}^*$, choose J in $\mathcal{J}^{**}$ and A in $P^{\perp}\mathcal{A}^{**}$ with $\|J\| = \|A\| = 1$ such that

$$\eta\Phi(A) = \|\eta\Phi\| \quad \text{and} \quad \Phi-\eta\Phi(J) = \|\Phi-\eta\Phi\| \ .$$

Then $\|J+A\| = 1$, and

$$\begin{aligned} \|\Phi\| &\geq \Phi(J+A) = \eta\Phi(A)+(\Phi-\eta\Phi)(J) \\ &= \|\eta\Phi\|+\|\Phi-\eta\Phi\| \geq \|\Phi\| \ . \end{aligned}$$

Hence $\mathcal{J}^*$ is an L-summand of $\mathcal{A}^*$, and $\mathcal{J}$ is an M-ideal. ∎

Although no use will be made of the converse to Theorem 11.4, we include a proof for completeness.

11.5 Theorem. *Every M-ideal of a C^*-algebra is a two sided ideal.*

Proof. Let $\mathcal{M}$ be an M-ideal in a C^*-algebra $\mathcal{A}$. Then $\mathcal{A}^* \cong \mathcal{M}^* \oplus_{\ell_1} \mathcal{M}^{\perp}$ and $\mathcal{A}^{**} \cong \mathcal{M}^{**} \oplus_{\ell_\infty} \mathcal{M}^{*\perp}$, where $\mathcal{M}^{**}$ is identified with $\mathcal{M}^{\perp\perp}$. By Theorem 10.4, $\mathcal{A}^{**}$ can be represented as a von Neumann algebra containing $\mathcal{A}$ which is the weak* closure of $\mathcal{A}$. Moreover, every state on $\mathcal{A}$ is a vector state in this universal representation. Let the identity I of $\mathcal{A}^{**}$ decompose as $P \oplus Q$.

Let f be any state on $\mathcal{A}$, decomposed as $f = f_1 \oplus f_2$. Then

$$\|f\| = 1 = f(I) = f_1(I)+f_2(I) \leq \|f_1\|+\|f_2\| = \|f\| \ .$$

So $h_i = \|f_i\|^{-1}f_i$ are states, $i = 1,2$. Moreover,

$$1 = h_1(I) = h_1(P)+h_1(Q) = h_1(P)$$

and

$$1 = h_2(I) = h_2(P)+h_2(Q) = h_2(Q) \ .$$

Hence $f(P) = \|f_1\|$ belongs to $[0,1]$ for every state, and thus $0 \leq P \leq I$. This also shows that every state is a convex combination of a state in $\mathcal{M}^*$ and a state in $\mathcal{M}^{\perp}$; thus, together they span all of $\mathcal{A}^*$.

Now suppose the state f is given by $f(A) = (Ax,x)$. For f in $\mathcal{M}^*$,

$$1 = f(P) = (Px,x) = \|P^{1/2}x\|^2 \ .$$

Hence x is an eigenvector for P, and $Px = x$. Thus if A belongs to $\mathcal{A}^{**}$, one has

$$\begin{aligned} f(A) = (Ax,x) &= (APx,x) = (PAx,x) \\ &= f(AP) = f(PA) \ . \end{aligned}$$

Likewise, if f belongs to $\mathcal{M}^{\perp}$, then $Qx = x$, whence $Px = 0$. So

$$0 = (APx,x) = (PAx,x) = f(AP) = f(PA) \ .$$

Consequently, $g(AP) = g(PA)$ for every functional g in $\mathcal{A}^*$. So P is central. Moreover, if A belongs to $\mathcal{M}^{**}$ and f belongs to $\mathcal{M}^{\perp}$,

$$f(A) = 0 = f(AP) \ .$$

So $g(A) = g(AP)$ for all g in $\mathcal{A}^*$ and thus P is the identity for $\mathcal{M}^{**}$. In particular, P is a projection.

Hence $\mathcal{M}^{**} = P\mathcal{A}^{**}$ and $\mathcal{M}^{*\perp} = P^{\perp}\mathcal{A}^{**}$ are ideals of $\mathcal{A}^{**}$. Therefore $\mathcal{M} = \mathcal{M}^{**} \cap \mathcal{A}$ is an ideal in $\mathcal{A}$. ■

Let $\mathcal{M}$ be a M-ideal in a Banach space $\mathcal{X}$. The $\mathcal{M}^*$ topology on $\mathcal{X}$ is the weakest topology such that for every for ϕ in $\mathcal{M}^*$, $\tilde{\phi}$ is continuous on $\mathcal{X}$. In particular, a net m_α of elements of $\mathcal{M}$ converges $\mathcal{M}^*$ to x in $\mathcal{X}$ if and only if

$$\lim_\alpha \phi(m_\alpha) = \tilde{\phi}(x)$$

for every ϕ in $\mathcal{M}^*$. This topology is not in general Hausdorff. In the case of an ideal $\mathcal{J}$ of a C^*-algebra $\mathcal{A}$, the $\mathcal{J}^*$ topology is Hausdorff exactly when $\mathcal{J}$ is essential, meaning that $A\mathcal{J} = 0$ implies $A = 0$.

An important special case is the ideal of compact operators $\mathcal{K}$, which is an M-ideal in $\mathcal{B}(\mathcal{H})$. In this case, $\mathcal{K}^*$ is identified with $\mathcal{C}_1$ by Theorem 1.13. So the $\mathcal{K}^*$ topology is precisely the weak* topology by Theorem 1.15.

The main result of this chapter is the following:

11.6 Theorem. *Let $\mathcal{M}$ be an M-ideal in a Banach space $\mathcal{X}$. Suppose that S is a subspace of $\mathcal{X}$ such that $S \cap \mathcal{M}$ is $\mathcal{M}^*$ dense in S. Then*

i) *$S \cap \mathcal{M}$ is an M-ideal in S. Furthermore, $S+\mathcal{M}$ is closed and the quotient maps*

$$\sigma : S/S \cap \mathcal{M} \to S+\mathcal{M}/\mathcal{M}$$

and

$$\tau : \mathcal{M}/S \cap \mathcal{M} \to S+\mathcal{M}/S$$

are isometric.

ii) *$S+\mathcal{M}/S$ is an M-ideal in $\mathcal{X}/S$. Thus if S is proximinal in $\mathcal{X}$, so is $S+\mathcal{M}$.*

Proof. Let η be the L-projection of $\mathcal{X}^*$ onto $\mathcal{M}^{\perp}$. First, we show that $\eta S^{\perp}$ is contained in $S^{\perp}$. Let Φ belong to $S^{\perp}$, and let s belong to S. Let m_{α} be a net in $S \cap \mathcal{M}$ converging $\mathcal{M}^*$ to s. Then since $\Phi - \eta\Phi$ belongs to $\mathcal{M}^*$,

$$\eta\Phi(s) = \Phi(s) - (\Phi - \eta\Phi)(s) = -\lim_{\alpha}(\Phi - \eta\Phi)(m_{\alpha}) = -\lim_{\alpha}\Phi(m_{\alpha}) = 0 \ .$$

Thus $\eta\Phi$ belongs to $S^{\perp}$. Therefore, $S^{\perp}$ splits as the ℓ^1 direct sum

$$S^{\perp} = \eta S^{\perp} \oplus_{\ell_1} (1-\eta) S^{\perp}$$

in $\mathcal{X}^* = \mathcal{M}^{\perp} \oplus_1 \mathcal{M}^*$. Thus, S^* splits as an ℓ^1 direct sum

$$S^* \cong \mathcal{X}^*/S^{\perp} \cong \mathcal{M}^{\perp}/\eta S^{\perp} \oplus_{\ell_1} \mathcal{M}^*/(1-\eta)S^{\perp} \ .$$

It also follows that

$$S^{\perp} + \mathcal{M}^{\perp} = \mathcal{M}^{\perp} \oplus_{\ell_1} (1-\eta) S^{\perp} \ .$$

So $S^{\perp}+\mathcal{M}^{\perp}$ is closed. It is obviously contained in $(S \cap \mathcal{M})^{\perp}$. Conversely, suppose ϕ belongs to $(S \cap \mathcal{M})^{\perp}$. Then $\eta\phi$ belongs to $\mathcal{M}^{\perp}$, and $\phi - \eta\phi$ belongs to $\mathcal{M}^* \cap (S \cap \mathcal{M})^{\perp}$. As $S \cap \mathcal{M}$ is $\mathcal{M}^*$ dense in S, it follows that $\phi - \eta\phi$ annihilates S. So $(S \cap \mathcal{M})^{\perp} = S^{\perp} + \mathcal{M}^{\perp}$. Thus, thinking of $S \cap \mathcal{M}$ as a subspace of S, its annihilator in S^* is

$$(S \cap \mathcal{M})^{\perp}/S^{\perp} \cong \mathcal{M}^{\perp}/\eta S^{\perp} \oplus_{\ell_1} 0 \ .$$

This is an L-summand of S^*. Hence $S \cap \mathcal{M}$ is an M-ideal in S.

Next, to compute the quotient norms, let s belong to S and ψ belong to $(S \cap \mathcal{M})^{\perp}$. So $\psi = \eta\psi \oplus (1-\eta)\psi$, and $(1-\eta)\psi$ annihilates S. Also, $\eta\psi$ is an arbitrary element of $\mathcal{M}^{\perp}$. Hence

$$dist(s,S \cap M) = \sup\{\psi(s):\psi \in (S \cap M)^{\perp},\|\psi\| \leq 1\}$$
$$= \sup\{\eta\psi(s):\eta\psi \in M^{\perp},\|\eta\psi\| \leq 1\} = dist(s,M) \ .$$

Therefore, the natural map σ of $S/S \cap M$ onto $S+M/M$ is isometric. Consequently, $S+M$ is complete and hence closed in $\mathcal{X}$.

Similarly, if m belongs to M,

$$dist(m,S \cap M) = \sup\{\psi(m):\psi \in (S \cap M)^{\perp},\|\psi\| \leq 1\}$$
$$= \sup\{(\psi-\eta\psi)(m):\psi \in (S \cap M)^{\perp},\|\psi\| \leq 1\}$$
$$= \sup\{(\psi-\eta\psi)(m):\psi \in S^{\perp},\|\psi\| \leq 1\}$$
$$= \sup\{\psi(m):\psi \in S^{\perp},\|\psi\| \leq 1\} = dist(m,S) \ .$$

So, the map of $M/S \cap M$ onto $S+M/S$ also isometric.

For part ii),

$$(S+M)^{\perp} = M^{\perp} \cap S^{\perp} = \eta S^{\perp} \oplus_{\ell_1} 0 \ .$$

But $(\mathcal{X}/S)^*$ is isomorphic to $S^{\perp} = \eta S^{\perp} \oplus_1 (1-\eta)S^{\perp}$. Thus $(S+M/S)^{\perp} = (S+M)^{\perp}$ as a subset of $S^{\perp}$ is an L-summand. So $S+M/S$ is an M-ideal in $\mathcal{X}/S$. Now, suppose S is proximinal in $\mathcal{X}$. For x in $\mathcal{X}$, apply Lemma 11.3 to $S+M/S$ and choose m in M so that

$$dist(x-m,S) = dist(x,S+M) \ .$$

Since S is proximinal, there is an s in S so that

$$\|x-m-s\| = dist(x-m,S) = dist(x,S+M) \ .$$

Hence $S+M$ is proximinal. ∎

11.7 Corollary. *Let S be a weak* closed subspace of $B(\mathcal{H})$ such that $S \cap K$ is weak* dense in S. Then $S+K$ is closed and the map σ of $S/S \cap K$ onto $S+K/K$ is isometric. Furthermore, $S \cap K$ is proximinal in S and $S+K$ is proximinal in $B(\mathcal{H})$.*

Proof. It was observed that the K^* topology is the weak* topology. Every weak* closed subspace of $B(\mathcal{H})$ is proximinal. For if x belongs to $B(\mathcal{H})$, let s_n belong to S such that $\|x-s_n\|$ converges to $dist(x,S)$. Let s be any weak* cluster point of the sequence s_n. Then by the lower semicontinuity of the norm,

$$\|x-s\| \leq \limsup\|x-s_n\| = dist(x,S) \ .$$

Now, apply Theorem 11.6. ∎

In many cases of interest, the M^* density of $S \cap M$ can be assured by the use of approximate identities. A *left approximate identity* for an algebra B is a net $\{e_\lambda:\lambda \in \Lambda\}$ such that $b = \lim_\Lambda e_\lambda b$ for every b in B.

11.8 Lemma. *Let J be an ideal in a C^*-algebra $\mathcal{A}$. Suppose that S is a (possibly non self-adjoint) subalgebra of $\mathcal{A}$ and that $S \cap J$ contains a bounded approximate identity for J. Then $S \cap J$ is J^* dense in S.*

Proof. Let $\{e_\lambda : \lambda \in \Lambda\}$ be the bounded approximate identity. Let P be any weak* limit point of this net in $\mathcal{A}^{**}$. Then

$$Pj = w^* - \lim_\Lambda e_\lambda j = j$$

for all j in J. Thus P is the identity on J^{**}, and hence it is the central support projection of J^{**}. Since P is the unique limit point, e_λ converges weak* to P. Let η be the L-projection of $\mathcal{A}^*$ onto $J^\perp$. If s belongs to S and ψ belongs to $\mathcal{M}^*$,

$$\lim_\Lambda \psi(e_\lambda s) = \psi(Ps) = \psi(s) - \eta\psi(s) = \psi(s) \ .$$

Thus $e_\lambda s$ converges J^* to s. But S is an algebra and J is an ideal, so $e_\lambda s$ belongs to $J \cap S$. Thus $J \cap S$ is J^* dense in S. ■

We conclude this section with a useful method for improving approximate identities in a given subspace of a C^*-algebra. First, we include a standard C^*-algebra fact.

11.9 Lemma. *Let $\mathcal{A}$ be a C^*-algebra. Then $\mathcal{A}$ contains a two-sided approximate identity $\{e_\lambda : \lambda \in \Lambda\}$ such that $0 \le e_\lambda \le 1$. If $\mathcal{A}$ is separable, $\{e_\lambda\}$ may be taken to be a sequence.*

Proof. Let Λ be the net of finite subsets of $\mathcal{A}_{s.a.}$ ordered by inclusion. For $\lambda = \{a_1, \dots, a_n\}$, let

$$e_\lambda = f_n(a_1^2 + \dots + a_n^2)$$

where f_n is the continuous function

$$f_n(x) = \begin{cases} 0 & x \le 0 \\ n^2 x & 0 \le x \le 1/n^2 \\ 1 & x \ge 1/n^2 \end{cases} .$$

Since $f_n(0) = 0$, e_λ belongs to $\mathcal{A}$, and $0 \le e_\lambda \le 1$. Furthermore, if one represents $\mathcal{A}$ on a Hilbert space $\mathcal{H}$, one sees that e_λ dominates the spectral projection $E_{a_i}\{t : |t| \ge 1/n\}$. Hence

$$\|a_i - e_\lambda a_i\| \le \|1 - e_\lambda\| \, \|E_{a_i}[-\frac{1}{n}, \frac{1}{n}] a_i\| \le \frac{1}{n} \ .$$

Thus $e_\lambda a$ converges to a for all self-adjoint elements of $\mathcal{A}$, and thus for all of $\mathcal{A}$. Taking adjoints yields that it is a right approximate identity as well.

If $\mathcal{A}$ is separable, let $\{a_n, n \geq 1\}$ be a dense subset of $\mathcal{A}_{s.a.}$. Let $\lambda_n = \{a_1,\ldots,a_n\}$ and $e_n = e_{\lambda_n}$ constructed above. Then if a belongs to $\mathcal{A}_{s.a.}$ and $\epsilon > 0$, choose i so that $\|a-a_i\| < \epsilon$ and an $n \geq i$ so that $n\epsilon > 1$. Then

$$\|a-e_n a\| \leq \|1-e_n\| \, \|a-a_i\|+\|(1-e_n)a_i\| < 2\epsilon \ .$$

So, as above, e_n is a two sided approximate identity for $\mathcal{A}$. ∎

11.10 Lemma. *Let $\mathcal{A}$ be a separable C^*-algebra. Suppose that $\{b_k\}$ is a bounded left approximate identity for $\mathcal{A}$. Then there exist convex combinations e_n of $\{b_k\}$ such that*

$$\lim_{n\to\infty}\|e_n\| = \lim_{n\to\infty}\|1-e_n\| = 1$$

and e_n is a two sided approximate identity for $\mathcal{A}$.

Proof. Let f_j, $j \geq 1$, be a fixed two sided approximate identity for $\mathcal{A}$ such that $0 \leq f_j \leq 1$ for all j. Let $C = \sup\|b_k\|$, and let N be an integer. Choose an integer $M \geq C^2N^2$. Let $j_1 = N$, and recursively choose j_i and k_i, $1 \leq i \leq M$ such that

1) $\|f_{j_i}-b_{k_i}f_{j_i}\| < 1/N$.
2) $\|b_{k_i}-b_{k_i}f_{j_{i+1}}\| < 1/N$.
3) $\|f_{j_i}-f_{j_i}f_{j_{i+1}}\| < 1/M$.

This is possible because k_i is chosen to satisfy 1) since $\{b_k\}$ is a left approximate identity. Then j_{i+1} is chosen to satisfy 2) and 3) since $\{f_j\}$ is an approximate identity. Define

$$e_N = \frac{1}{M}\sum_{i=1}^{M} b_{k_i} \text{ and } h_N = \frac{1}{M}\sum_{i=1}^{M} f_{j_i} \ .$$

It is routine to verify that h_N is a two sided approximate identity with $0 \leq h_N \leq 1$. Now compute

$$\begin{aligned}\|e_N-h_N\| &= \|\frac{1}{M}\sum_{i=1}^{M} b_{k_i}-f_{j_i}\| \\ &= \|\frac{1}{M}\sum_{i=1}^{M}(b_{k_i}-b_{k_i}f_{j_{i+1}})+b_{k_i}(f_{j_{i+1}}-f_{j_i})+(b_{k_i}f_{j_i}-f_{j_i})\| \\ &\leq \frac{2}{N}+\|\frac{1}{M}\sum_{i=1}^{M} b_{k_i}\Delta_i\|\end{aligned}$$

where $\Delta_i = f_{j_{i+1}}-f_{j_i}$. When $|i-j| \geq 2$, one has by 3) that $\|\Delta_i\Delta_j\| \leq 4/M$. Split the sum into two sums over the even and odd terms respectively.

$$\|\frac{1}{M}\sum_{i\ even} b_{k_i}\Delta_i\|^2 = \|\frac{1}{M^2}\sum_{i=1}^{M/2}\sum_{j=1}^{M/2} b_{k_{2i}}\Delta_{2i}\Delta_{2j}b_{k_{2j}}^*\|$$

$$\leq \frac{1}{M^2}\left[\sum_{i=1}^{M/2}\|b_{k_{2i}}\Delta_{2i}\|^2 + \frac{M^2}{4}\cdot C^2\cdot\frac{4}{M}\right]$$

$$\leq \frac{1}{M^2}\left[\frac{M}{2}C^2 + MC^2\right] = \frac{3C^2}{2M} < \frac{3}{2N^2} \ .$$

The sum over the odd terms is treated in the same way. Thus

$$\|\frac{1}{M}\sum_{i=1}^{M} b_{k_i}\Delta_i\| \leq 2\left[\frac{3}{2}\right]^{1/2}/N < 3/N \ .$$

Hence $\|e_N - h_N\| < 5/N$, from which it follows that e_N is also a two sided approximate identity with

$$\lim_{n\to\infty}\|e_N\| = \lim_{N\to\infty}\|1-e_N\| = 1 \ .$$

By construction, e_N is in the convex hull of $\{b_k\}$. ■

11.11 Proposition. *Let $\mathcal{A}$ be a C^*-algebra with an ideal J. Let S be a subalgebra of $\mathcal{A}$ such that $S \cap J$ contains an approximate identity e_n satisfying*

$$\lim_{n\to\infty}\|e_n\| = \lim_{n\to\infty}\|1-e_n\| = 1 \ .$$

Then for any a in $\mathcal{A}$,

$$dist(a, S+J) = \lim_{n\to\infty} dist(a-ae_n, S) = \lim_{n\to\infty} dist(a-e_n a, S) \ .$$

Proof. Since ae_n belongs to J, $dist(a-ae_n, S) \geq dist(a, S+J)$. So $dist(a, S+J) \leq \lim_{n\to\infty} dist(a-ae_n, S)$. Conversely, if J belongs to J and s is in S, then

$$(a-s-j)(1-e_n) = (a-ae_n)+(s-se_n)+(j-je_n) \ .$$

Since $\|j-je_n\|$ tends to zero and $(s-se_n)$ belongs to s,

$$\lim_{n\to\infty} dist(a-ae_n, S) \leq \lim_{n\to\infty}\|a-s-j\|\ \|1-e_n\| = \|a-s-j\| \ .$$

Taking the infimum over all s and j yields the desired equality.

The "left" version is proved in the same manner. ■

Notes and Remarks.

M-ideals were introduced by Alfsen and Effros [1]. They say that a subspace $\mathcal{M}$ of a Banach space $\mathcal{X}$ has the n-ball property if given n open balls $B_1,\ldots,B_n$ such that $\bigcap_{i=1}^{n} B_i \neq \emptyset$ and $B_i \cap M \neq \emptyset$ for $1 \leq i \leq n$, then

$\mathcal{M} \cap \bigcap_{i=1}^{n} B_i \neq \emptyset$. Their main result is

Theorem. *Let $\mathcal{M}$ be a subspace of a Banach space $\mathcal{X}$. The following are equivalent*

(1) *$\mathcal{M}$ is an M-ideal.*

(2) *$\mathcal{M}$ has the n-ball property for all n.*

(3) *$\mathcal{M}$ has the 3-ball property.*

Theorem 11.3 is due to them, but the simple proof given here is taken from Gamelin, Marshall, Younis and Zame [1]. Theorems 11.4 and 11.5 are due to Alfsen and Effros in the real case, and to Smith and Ward [1] in the complex case.

The remainder of this chapter is taken from Davidson and Power [2] except for the standard result Lemma 11.9 on approximate identities in C^*-algebras. This paper also contains a constructive approach generalizing a method due to Axler,Berg,Jewell and Shields [1].

Exercises.

11.1 A subspace $\mathcal{M}$ of a Banach space has the strong n-ball property if given n open balls $B_1,\ldots,B_n$ such that $\bigcap_{i=1}^{n} B_i \neq \emptyset$ and $\overline{B_i} \cap \mathcal{M} \neq \emptyset$, $1 \leq i \leq n$, then $\mathcal{M} \cap \bigcap_{i=1}^{n} \overline{B_i} \neq \emptyset$.

(i) Show that the strong n ball property implies the n ball property, which in turn implies the strong $(n-1)$ ball property.

(ii) Show that the 2 ball property implies that $\mathcal{M}$ is proximinal.

(iii) (Holmes, Scranton, Ward) Suppose that $\mathcal{M}$ has the strong two ball property. Show that for each $x \notin \mathcal{M}$, the set of closest approximants in $\mathcal{M}$ spans all of $\mathcal{M}$.

Hint: Say $dist(x,\mathcal{M}) = 1$. For each unit vector m in $\mathcal{M}$, consider $B_1(x)$ and $B_1(x+m/2)$.

11.2 (i) Use approximate identities to show that if A is a (two sided) ideal in a C^*-algebra B, and B is an ideal in a C^*-algebra C, then A is an ideal in C.

(ii) Show that if $\mathcal{M}$ is an M-ideal in $\mathcal{X}$, and $\mathcal{X}$ is an M-ideal in $\mathcal{Y}$, then $\mathcal{M}$ is an M-ideal in $\mathcal{Y}$.

11.3 (Davidson-Power) Let $\mathcal{J}$ be an ideal in a C^*-algebra $\mathcal{A}$. The $\mathcal{J}$-weak topology is the weakest topology in $\mathcal{A}$ such that $\phi(AJ)$ is continuous for every J in $\mathcal{J}$ and ϕ in $\mathcal{J}^*$. Similarly, the $\mathcal{J}$-strong (and $\mathcal{J}$-strong*) are the weakest topologies such that $A \rightarrow AJ$ (and also $A \rightarrow JA$) are continuous for all J in $\mathcal{J}$.

(i) Think of $\mathcal{A}$ as acting on $\mathcal{J}$ by left multiplication, and thus as a subalgebra of $\mathcal{B}(\mathcal{J})$. Show that the $\mathcal{J}$-weak, strong, strong* topologies are induced by the corresponding operator topologies. Hence deduce that the $\mathcal{J}$-weak, $\mathcal{J}$-strong and $\mathcal{J}$-strong* topologies have the same closed convex sets.

(ii) Show that the $\mathcal{J}$-weak topology coincides with the $\mathcal{J}^*$ topology.

Hint: Use the Cohen factorization theorem to show that every ψ in $\mathcal{J}^*$ has the form $\psi(X) = \phi(XJ)$ for some J in $\mathcal{J}$, ϕ in $\mathcal{J}^*$.

11.4 (Davidson-Power) Let $\mathcal{M}$, $\mathcal{X}$ and $\mathcal{S}$ be as in Theorem 11.5. Define a map τ of $\mathcal{S}$ into $\mathcal{M}^{**}$ by $\tau(s)(\phi) = \tilde{\phi}(s)$.

(i) Show that $\tau(\mathcal{S}) \cap \mathcal{M}$ is weak* dense in $\tau(\mathcal{S})$, and hence show that $\tau(\mathcal{S})$ is contained in $(\tau(\mathcal{S}) \cap \mathcal{M})^{**}$.

(ii) Deduce that the unit ball of $\tau(\mathcal{S}) \cap \mathcal{M}$ is weak* dense in the ball of $\tau(\mathcal{S})$, and hence that the ball of $\mathcal{S} \cap \mathcal{M}$ is $\mathcal{M}^*$ dense in ball of $\mathcal{S}$.

(iii) If $\mathcal{X}$ is a C^*-algebra and $\mathcal{M}$ is an ideal, show that the ball of $\mathcal{S} \cap \mathcal{M}$ is $\mathcal{M}$-strong* dense in the ball of $\mathcal{S}$.

11.5 (Axler-Berg-Jewell-Shields)

(i) Let K be compact, and let B_n be a sequence of operators converging strong* to 0. Show that for $\epsilon > 0$, there is an integer n_0 so that $n \geq n_0$ implies $\|K+B_n\| < \max\{\|K\|,\|B_n\|\}+\epsilon$.

(ii) Let B_n be as above, and let A be an operator. Given $\epsilon > 0$, find n_0 so that $n \geq n_0$ implies

$$\|A+B_n\| \leq \max\{\|A\|,\|A\|_e+\|B_n\|\}+\epsilon \ .$$

Hint: Write $A = A_0+K$ where $\|A_0\| = \|A\|_e$, $\|K\| = \|A\|-\|A\|_e$, and K is compact.

(iii) Let A be the weak* limit of a sequence K_n of compact operators. Let $B_n = A-K_n$ and $C = \sup\|B_n\|$. Let $\epsilon_n = C^{n-1}/(C+1)^n$. Note that

$$\sum_{n=K+1}^{\infty} \epsilon_n = \left(\frac{C}{C+1}\right)^k = (C+1)\epsilon_{k+1} \ .$$

Show that one may choose integers n_k by induction so that $\|\sum_{k=1}^{m} \epsilon_k B_{n_k}\| < \|A\|_e$ for all $m \geq 1$.

(Hint: Apply (ii) to the partial sum and appropriate ϵ.)

(iv) With A, K_n as above, find a convex combination K of $\{K_n, n \geq 1\}$ so that $\|A-K\| = \|A\|_e$.

12. Quasitriangular Algebras

An operator T in $\mathcal{B}(\mathcal{H})$ is *quasitriangular* if there is an increasing sequence P_n of finite rank projections such that $\lim_{n\to\infty} \|P_n^\perp T P_n\| = 0$. This notion was used in Theorem 3.1 to obtain invariant subspaces for compact operators. It is easy to find a compact perturbation of T which is upper triangular with respect to some basis. In fact, it follows from Theorem 12.2 below that T belongs to $\mathcal{T}(\mathcal{P})+\mathcal{K}$ where $\mathcal{P}$ is the nest $\{P_n\mathcal{H}, n \geq 1\}$. It is convenient to consider the more general family of algebras $\mathcal{Q}\mathcal{T}(\mathcal{N}) = \mathcal{T}(\mathcal{N})+\mathcal{K}$ for an arbitrary nest. This will be called the quasitriangular algebra of $\mathcal{N}$. However, $\mathcal{Q}\mathcal{T}(\mathcal{N})$ NEED NOT consist of quasitriangular operators. Precisely which kinds of operators occur in $\mathcal{Q}\mathcal{T}(\mathcal{N})$ will be surveyed in chapter 21.

12.1 Theorem. *$\mathcal{Q}\mathcal{T}(\mathcal{N})$ is norm closed. For each T in $\mathcal{T}(\mathcal{N})$, there is a compact operator K in $\mathcal{T}(\mathcal{N}) \cap \mathcal{K}$ such that $\|T-K\| = \|T\|_e$. Hence the quotient map π of $\mathcal{T}(\mathcal{N})/\mathcal{T}(\mathcal{N}) \cap \mathcal{K}$ onto $\mathcal{Q}\mathcal{T}(\mathcal{N})/\mathcal{K}$ is isometric. Also,*

$$dist(K,\mathcal{T}(\mathcal{N}) \cap \mathcal{K}) = dist(K,\mathcal{T}(\mathcal{N}))$$

for all K compact.

Proof. By the Erdos Density Theorem 3.11, $\mathcal{T}(\mathcal{N}) \cap \mathcal{K}$ is weak* dense in $\mathcal{T}(\mathcal{N})$. Thus by Theorem 11.5 and Corollary 11.6, $\mathcal{T}(\mathcal{N}) \cap \mathcal{K}$ is an M-ideal in $\mathcal{T}(\mathcal{N})$ and the quotient maps π defined above and σ taking $\mathcal{K}/\mathcal{T}(\mathcal{N})\cap\mathcal{K}$ onto $\mathcal{Q}\mathcal{T}(\mathcal{N})/\mathcal{T}(\mathcal{N})$ are isometric. By Theorem 11.3, $\mathcal{T}(\mathcal{N}) \cap \mathcal{K}$ is proximinal in $\mathcal{T}(\mathcal{N})$. Thus there is a K in $\mathcal{K}$ such that

$$\|T-K\| = dist(T,\mathcal{T}(\mathcal{N}) \cap \mathcal{K}) = dist(T,\mathcal{K}) = \|T\|_e \ .$$

The other estimate is a restatement of the fact that σ is isometric. ∎

12.2 Theorem. *Let $\mathcal{P}$ be a nest $\{P_n\mathcal{H}, n \geq 1\}$ where P_n is an increasing sequence of finite rank projections. Then for A in $\mathcal{B}(\mathcal{H})$,*

$$dist(A,\mathcal{Q}\mathcal{T}(\mathcal{P})) = \limsup_{n\to\infty} \|P_n^\perp A P_n\| \ .$$

Consequently, $\mathcal{Q}\mathcal{T}(\mathcal{P})$ consists of all operators which are quasitriangular with respect to the sequence $\{P_n, n \geq 1\}$.

Proof. $\mathcal{T}(\mathcal{P}) \cap \mathcal{K}$ contains the sequence $\{P_n\}$ which is a two sided approximate identity for $\mathcal{K}$ satisfying $\|P_n\| = \|P_n^\perp\| = 1$. Hence by Proposition 11.11,

$$dist(A, Q\,\mathcal{T}(\mathcal{P})) = \limsup_{n \to \infty} dist(P_n^\perp A, \mathcal{T}(\mathcal{P})) \ .$$

But by Arveson's distance formula (Theorem 9.5),

$$dist(P_n^\perp A, \mathcal{T}(\mathcal{N})) = \sup_{k \geq 1} \|P_k^\perp (P_n^\perp A) P_k\| = \sup_{k \geq n} \|P_k^\perp A P_k\| \ .$$

This is decreasing in n, thus

$$dist(A, Q\,\mathcal{T}(\mathcal{P})) = \limsup_{n \to \infty} \|P_n^\perp A P_n\| \ .$$

If this limit is zero, then A belongs to $Q\,\mathcal{T}(\mathcal{P})$ by Theorem 12.1. ∎

This can be generalized in a certain sense, but Proposition 12.3 below is not particularly satisfying. A better result is Theorem 12.6.

12.3 Proposition. *Let $\mathcal{N}$ be a nest on a separable space. There is an approximate identity E_n for $\mathcal{K}$ in $\mathcal{T}(\mathcal{N}) \cap \mathcal{K}$ such that for all A in $\mathcal{B}(\mathcal{H})$,*

$$dist(A, Q\,\mathcal{T}(\mathcal{N})) = \lim_{n \to \infty} \sup_{N \in \mathcal{N}} \|P(N)^\perp A (I - E_n) P(N)\| \ .$$

Proof. By the Erdos Density Theorem 3.11, $\mathcal{T}(\mathcal{N}) \cap \mathcal{K}$ contains a bounded approximate identity for $\mathcal{K}$. Thus by Lemma 11.9, it contains an approximate identity E_n such that

$$\lim_{n \to \infty} \|E_n\| = \lim_{n \to \infty} \|I - E_n\| = 1 \ .$$

So by Proposition 11.10 and the distance formula Theorem 9.5,

$$\begin{aligned} dist(A, Q\,\mathcal{T}(\mathcal{N})) &= \lim_{n \to \infty} dist(A(I - E_n), \mathcal{T}(\mathcal{N})) \\ &= \lim_{n \to \infty} \sup_{N \in \mathcal{N}} \|P(N)^\perp A (I - E_n) P(N)\| \ . \end{aligned}$$ ∎

12.4 Let $\mathcal{N}$ be a nest. For A in $\mathcal{B}(\mathcal{H})$, define a function Φ_A on $\mathcal{N}$ by $\Phi_A(N) = P(N)^\perp A P(N)$. This map is continuous in a certain sense. For if N_α converges to N in the order topology, then $P(N_\alpha)$ converges to $P(N)$ in the strong operator topology by Proposition 2.13. Thus $\Phi_A(N_\alpha)$ converges to $\Phi_A(N)$ in the strong operator topology. Likewise, $\Phi_A(N_\alpha)^*$ converges strongly to $\Phi_A(N)^*$. So Φ_A is a strong* continuous, bounded function. The set $C_{s^*}(\mathcal{N}, \mathcal{B}(\mathcal{H}))$ is the space of all bounded, strong* continuous functions of $\mathcal{N}$ into $\mathcal{B}(\mathcal{H})$ endowed with the norm $\|\Phi\| = \sup \|\Phi(N)\|$. This is a C^*-algebra under point-wise multiplication because multiplication is jointly continuous in the strong* topology on *bounded* sets. In this formulation, Arveson's distance formula becomes

$$dist(A,\mathcal{T}(\mathcal{N})) = \|\Phi_A\| \ .$$

Let $C_n(\mathcal{N},\mathcal{K})$ denote the subalgebra of norm continuous functions into the compact operators. This is in fact a two sided ideal in $C_{s*}(\mathcal{N},\mathcal{B}(\mathcal{H}))$. Clearly, the product of a compact valued function with a bounded one is compact valued. So it suffices to show that if a bounded net A_α converges to A strong* and K_α converges to K in norm, then $A_\alpha K_\alpha$ converges to AK and $K_\alpha A_\alpha$ converges to KA. But

$$\|A_\alpha K_\alpha - AK\| \leq \|A_\alpha\| \, \|K_\alpha - K\| + \|(A_\alpha - A)K\| \ .$$

By Proposition 1.18, the second term tends to zero, and the first clearly does. The multiplication on the right is dealt with by taking adjoints.

12.5 Theorem. *A belongs to $Q\mathcal{T}(\mathcal{N})$ if and only if Φ_A belongs to $C_n(\mathcal{N},\mathcal{K})$.*

Proof. If $A = T+K$ where T belongs to $\mathcal{T}(\mathcal{N})$ and K is compact, then $\Phi_A(N) = \Phi_K(N) = P(N)^\perp KP(N)$. The constant map $\Phi(N) = K$ is in $C_n(\mathcal{N},\mathcal{K})$, and $\Pi(N) = P(N)$ belongs to $C_{s*}(\mathcal{N},\mathcal{K})$. Since $C_n(\mathcal{N},\mathcal{K})$ is an ideal in $C_{s*}(\mathcal{N},\mathcal{K})$, $\Phi_A = \Pi^\perp \Phi \Pi$ also belongs to $C_n(\mathcal{N},\mathcal{K})$.

On the other hand, suppose Φ_A belongs to $C_n(\mathcal{N},\mathcal{K})$. Let $\epsilon > 0$, and use the continuity of Φ_A to select a finite subnest

$$\mathcal{F}: \{0\} = N_0 < N_1 < \ldots < N_n = \mathcal{H}$$

such that

$$\|\Phi_A(N) - \Phi_A(N_j)\| < \epsilon \ \text{ for } \ N_{j-1} < N \leq N_j, \ 1 \leq j \leq n \ .$$

Let $P_j = P(N_j)$ and define the compact operator $K (= \mathcal{L}_{\mathcal{F}^\perp}(A))$ by

$$K = \sum_{j=1}^{n-1} P_j^\perp A(P_j - P_{j-1}) = \sum_{j=1}^{n-1} \Phi_A(N_j) P_{j-1}^\perp \ .$$

Thus

$$A - K = \mathcal{U}_{\mathcal{F}}(A) = \sum_{j=1}^{n} P_j A(P_j - P_{j-1}) \ .$$

So if $N_{j-1} < N \leq N_j$ and $P = P(N)$, then $P^\perp P_i = 0$ for $i < j$ and $(P_i - P_{i-1})P = 0$ for $i > j$. Hence

$$\Phi_{A-K}(N) = P^\perp P_j A(P_j - P_{j-1})P = (P_j - P_{j-1})\Phi_A(N)(P_j - P_{j-1}) \ .$$

But $(P_j - P_{j-1})\Phi_A(N_j) = 0$, hence

$$\|\Phi_{A-K}(N)\| = \|(P_j-P_{j-1})(\Phi_A(N)-\Phi_A(N_j))(P_j-P_{j-1})\| < \epsilon \ .$$

By the distance formula, $dist(A-K,\mathcal{T}(\mathcal{N})) \le \epsilon$. But $\epsilon > 0$ is arbitrary, and thus A belongs to the closure of $\mathcal{T}(\mathcal{N})+\mathcal{K}$. By Theorem 12.1, A belongs to $Q\mathcal{T}(\mathcal{N})$. ∎

Now this result is combined with the results of Chapter 11 to obtain a distance formula. For Φ in $C_{s^*}(\mathcal{N},\mathcal{K})$, let

$$\|\Phi\|_e = dist(\Phi, C_n(\mathcal{N},\mathcal{K})) \ .$$

12.6 Theorem. *Let $\mathcal{N}$ be a nest, and let A belong to $\mathcal{B}(\mathcal{H})$. Then*

$$dist(A,Q\mathcal{T}(\mathcal{N})) = \|\Phi_A\|_e \ .$$

Proof. Let $\mathcal{S}$ consist of $\{\Phi_A : A \in \mathcal{B}(\mathcal{H})\}$. From the distance formula, one has that $\mathcal{S}$ is isometrically isomorphic to $\mathcal{B}(\mathcal{H})/\mathcal{T}(\mathcal{N})$ and thus it is closed. The subset $\{\Phi_A : A \in Q\mathcal{T}(\mathcal{N})\}$ is the intersection $\mathcal{S} \cap C_n(\mathcal{N},\mathcal{K})$ by Theorem 12.5. So

$$dist(A,Q\mathcal{T}(\mathcal{N})) = \inf\{\|\Phi_{A-K}\| : K \in \mathcal{K}\} = dist(\Phi_A, \mathcal{S} \cap C_n(\mathcal{N},\mathcal{K})) \ .$$

We wish to show that $\mathcal{S} \cap C_n(\mathcal{N},\mathcal{K})$ is $C_n(\mathcal{N},\mathcal{K})^*$ dense in $\mathcal{S}$. By the Erdos Density Theorem 3.11, there is a bounded approximate identity E_α for $\mathcal{K}$ in $\mathcal{T}(\mathcal{N}) \cap \mathcal{K}$. Let Ψ_α be the function on $\mathcal{N}$ with constant value E_α. If Φ belongs to $C_n(\mathcal{N},\mathcal{K})$, then the range of Φ is the continuous image of a compact set, and thus is compact. Hence $\Phi(N)E_\alpha$ converges to $\Phi(N)$ uniformly on $\mathcal{N}$. So Ψ_α is a bounded approximate identity for $C_n(\mathcal{N},\mathcal{K})$. From the proof of Lemma 11.7, it follows that $\Phi\Psi_\alpha$ converges to Φ in the $C_n(\mathcal{N},\mathcal{K})^*$ topology for every Φ in $C_{s^*}(\mathcal{N},\mathcal{K})$. Now let Π be the projection in $C_{s^*}(\mathcal{N},\mathcal{K})$ given by $\Pi(N) = P(N)$. For each ϕ in $C_n(\mathcal{N},\mathcal{K})^*$, define another functional $\phi_\Pi(\Phi) = \phi(\Phi\Pi)$. Then

$$\lim_\alpha \phi(\Phi\Psi_\alpha\Pi) = \lim_\alpha \phi_\Pi(\Phi\Psi_\alpha) = \phi_\Pi(\Phi) = \phi(\Phi\Pi) \ .$$

Hence $\Phi\Psi_\alpha\Pi$ converges $C_n(\mathcal{N},\mathcal{K})^*$ to $\Phi\Pi$. For Φ_A in $\mathcal{S}$, $\Phi_A = \Phi_A\Pi$ and

$$\Phi_A\Psi_\alpha\Pi(N) = P(N)^\perp AP(N)E_\alpha P(N) = P(N)^\perp AE_\alpha P(N) = \Phi_{AE_\alpha}(N) \ .$$

Thus, $\mathcal{S} \cap C_n(\mathcal{N},\mathcal{K})$ is $C_n(\mathcal{N},\mathcal{K})^*$ dense in $\mathcal{S}$.

By Theorem 11.5, one now has

$$\begin{aligned} dist(A,Q\mathcal{T}(\mathcal{N})) &= dist(\Phi_A, \mathcal{S} \cap C_n(\mathcal{N},\mathcal{K})) \\ &= dist(\Phi_A, C_n(\mathcal{N},\mathcal{K})) = \|\Phi_A\|_e \ . \end{aligned}$$

∎

Note that $\|\Phi_A\|_e$ is NOT just $\sup\|P(N)^\perp AP(N)\|_e$. For example, in the special case of Theorem 12.2, Φ_A is always compact valued. It may however fail to be continuous at the end point $\mathcal{H}$. That is precisely what is measured by $\lim_{n\to\infty}\sup\|P_n^\perp AP_n\|$.

12.7. Now we turn to the major problem solved in this section: When are $Q\ \mathcal{T}(\mathcal{N})$ and $Q\ \mathcal{T}(\mathcal{M})$ equal? One simple way this can occur is to take a finite dimensional subspace R such that $P(R)$ belongs to $\mathcal{N}'$. The nest $\mathcal{N}^R = \{\{0\}, R\vee N : N \in \mathcal{N}\}$ is another nest; and such a nest will be called a *finite perturbation of* $\mathcal{N}$. For any operator T in either $\mathcal{T}(\mathcal{N})$ or $\mathcal{T}(\mathcal{N}^R)$, the compact perturbation $P(R)^\perp TP(R)^\perp$ belongs to both $\mathcal{T}(\mathcal{N})$ and $\mathcal{T}(\mathcal{N}^R)$. Thus $Q\ \mathcal{T}(\mathcal{N}^R) = Q\ \mathcal{T}(\mathcal{N})$. An order isomorphism θ of $\mathcal{N}$ onto $\mathcal{M}$ *preserves dimension* if $dim\ \theta(N')/\theta(N) = dim\ N'/N$ for all $N < N'$ in $\mathcal{N}$. Two nests $\mathcal{N}$ and $\mathcal{M}$ will be called *compactly perturbed* if there are finite perturbations $\mathcal{N}^N$ and $\mathcal{M}^M$ and a dimension preserving order isomorphism θ of $\mathcal{N}^N$ onto $\mathcal{M}^M$ such that $\theta - id(L) = P(\theta(L)) - P(L)$ is continuous and compact valued (i.e. belongs to $C_n(\mathcal{N}^N, \mathcal{K})$).

Two nests $\mathcal{N}$ and $\mathcal{M}$ are *similar* if there is an invertible operator S such that $S\mathcal{N} \equiv \{SN : N \in \mathcal{N}\}$ equals $\mathcal{M}$. A similarity induces an order isomorphism θ_S of $\mathcal{N}$ onto $\mathcal{M}$. One has the identity

$$\mathcal{T}(S\mathcal{N}) = S\ \mathcal{T}(\mathcal{N})S^{-1}\ .$$

For if T belongs to $\mathcal{T}(\mathcal{N})$, and N belongs to $\mathcal{N}$,

$$STS^{-1}(SN) = ST(N) \subseteq SN\ .$$

So STS^{-1} belongs to $\mathcal{T}(\mathcal{M})$. And the reverse inclusion follows by reversing the roles of $\mathcal{N}$ and $\mathcal{M}$. If S has the form $I+K$ with K compact, then clearly $Q\ \mathcal{T}(S\mathcal{N}) = Q\ \mathcal{T}(\mathcal{N})$.

Two nests $\mathcal{N}$ and $\mathcal{M}$ *approximately unitarily equivalent* if there is a sequence U_n of unitaries and an order isomorphism θ of $\mathcal{N}$ onto $\mathcal{M}$ such that $\Phi_n(N) = P(U_nN) - P(\theta(N))$ is compact valued and converges uniformly to 0. If all the U_n are compact perturbations of the identity, it will be shown that $Q\ \mathcal{T}(\mathcal{N}) = Q\ \mathcal{T}(\mathcal{M})$.

The main theorem can now be stated.

12.8 Theorem. *Let $\mathcal{N}$ and $\mathcal{M}$ be nests on a separable Hilbert space. The following are equivalent:*

1) $Q\ \mathcal{T}(\mathcal{N}) = Q\ \mathcal{T}(\mathcal{M})$.

2) *$\mathcal{N}$ and $\mathcal{M}$ are compactly perturbed nests.*

3) *$\mathcal{N}$ and $\mathcal{M}$ have finite perturbations which are approximately unitarily equivalent by compact perturbations of the identity.*

4) *$\mathcal{N}$ and $\mathcal{M}$ have finite perturbations which are similar by a compact perturbation of the identity.*

The proof of 1) implies 2) follows from Lemmas 12.9-12.14 and Theorem 12.15. Theorem 12.16 shows 2) implies 3). And 3) implies 4) is a consequence of Lemma 12.17 and Theorem 12.18. The implication 4) implies 1) has already been done.

12.9 Lemma. *Let $\mathcal{N}$ and $\mathcal{M}$ be nests with $Q\,\mathcal{T}(\mathcal{N}) = Q\,\mathcal{T}(\mathcal{M})$. Let Φ be an expectation of $\mathcal{B}(\mathcal{H})$ onto $\mathcal{M}''$. Then the map $N \to P(N)-\Phi(P(N))$ is norm continuous and compact valued.*

Proof. First, we show that $\delta = \delta_{P(N)}$ is a derivation of $\mathcal{M}'$ into $\mathcal{K}$. For $\mathcal{M}'$ belongs to $Q\,\mathcal{T}(\mathcal{N}) \cap Q\,\mathcal{T}(\mathcal{N})^*$, and thus any M in $\mathcal{M}'$ can be written as $M = A+K = B^*+L$ where A and B belong to $\mathcal{T}(\mathcal{N})$, and K and L are compact. Thus

$$\begin{aligned}\delta(M) &= P(N)(B^*+L)P(N)^{\perp}-P(N)^{\perp}(A+K)P(N)\\ &= P(N)LP(N)^{\perp}-P(N)^{\perp}KP(N)\ .\end{aligned} \tag{12.9.1}$$

This is compact valued. Hence by the Johnson-Parrott Theorem 10.12, $\delta = \delta_K$ where $K = P(N)-\Phi(P(N))$ is compact, and $\|K\| \le \|\delta_{P(N)}\|$.

Let N_k be a sequence in $\mathcal{N}$ converging to N. Then $P_k = P(N_k)$ converges to $P = P(N)$ in the strong operator topology by Proposition 2.13. Let $\delta_k = \delta_{P_k}$. By (12.9.1),

$$\begin{aligned}(\delta-\delta_k)(M) &= (PLP^{\perp}-P^{\perp}KP)-(P_kLP_k^{\perp}-P_k^{\perp}KP_k)\\ &= (P-P_k)LP^{\perp}-P_kL(P-P_k)\\ &\quad +(P-P_k)KP-P_kK(P-P_k)\ .\end{aligned}$$

By Proposition 1.18, $(\delta-\delta_k)(M)$ converges to 0 in norm as k tends to infinity. Now consider $\delta_{\sum\oplus P-P_k} = \sum\oplus\delta-\delta_k$ which is derivation of $(\mathcal{M}')^{(\infty)}$ into the compacts because of the previous computation. By the Johnson-Parrott Theorem 10.12, there is a compact operator $K' = \sum\oplus K'_k$ such that $\delta-\delta_k = \delta_{K'_k}$. Now $K_k = P(N_k)-\Phi(P(N_k))$ satisfies $\delta-\delta_k = \delta_{K-K_k}$ and $\|K-K_k\| \le \|\delta-\delta_k\| \le 2\|K'_k\|$. This tends to zero as k tends to infinity. Thus the map taking N to $P(N)-\Phi(P(N))$ is norm continuous. ∎

12.10 Lemma. *Let P be a projection such that $P^{\perp}TP$ is compact for every T in $\mathcal{T}(\mathcal{M})$. Then there is an M on $\mathcal{M}$ such that $P-P(M)$ is compact.*

Proof. In particular, P is in the essential commutant of $\mathcal{M}'$, since $PM-MP = (P^{\perp}M^*P)^*-P^{\perp}MP$ is compact for every M in $\mathcal{M}'$. So by the Johnson-Parrott Theorem 10.12, $P = A+K$ where K is compact and A belongs to $\mathcal{M}''$. Now, $P = P^* = ReA+ReK$, so we may suppose that $A = A^*$. Also, $A-A^2$ is compact, and thus $\sigma_e(A)$ is contained in $\{0,1\}$. So using the functional calculus of A, one obtains a projection Q in $\mathcal{M}''$ such that $P-Q$ is compact. Define

$$M_0 = \sup\{M \in \mathcal{M} : P(M)Q^\perp \text{ is compact}\}$$

and

$$M_1 = \inf\{M \in \mathcal{M} : P(M_-)^\perp Q \text{ is compact}\} .$$

For each M in $\mathcal{M}$, either $P(M)Q^\perp$ or $P(M_-)^\perp Q$ is compact; for otherwise, there is an infinite rank partial isometry

$$U = P(M)Q^\perp UQP(M_-)^\perp .$$

This belongs to $\mathcal{T}(\mathcal{M})$ by Lemma 2.8. But $Q^\perp UQ = U$ is not compact, contradicting the essential invariance of Q (and P). It follows that $M_1 \leq M_0$.

If there is an M in $\mathcal{M}$ with $M_1 < M < M_0$, then $P(M)Q^\perp$ and $P(M_-)^\perp Q$ are both compact. Hence

$$P(M)-Q = P(M)Q^\perp - P(M)^\perp Q$$

is compact. Likewise, if $M_1 = (M_0)_- < M_0$, then $P(M_1)Q^\perp$ and $P(M_1)^\perp Q = P((M_0)_-)^\perp Q$ are both compact. So as above, $P(M_1)-Q$ is compact.

In the remaining case, $M_0 = M_1$ and one of $P(M_0)Q^\perp$ or $P(M_0)^\perp Q$ is compact. It must be shown that both are compact, implying that $Q-P(M_0)$ is compact. Suppose that $P(M_0)Q^\perp$ is compact but $P(M_0)^\perp Q$ is not. For every $M > M_0$, $P(M_-)^\perp Q$ is a compact projection. Thus it is finite rank and is therefore the sum of minimal projections in $\mathcal{M}''$. These minimal projections are precisely the atoms of $\mathcal{M}$. Let M_α be a decreasing net with limit M_0. Then $P(M_\alpha)^\perp Q$ converges strongly to $P(M_0)^\perp Q$, and thus $P(M_0)^\perp Q = \sum_{n\geq 1} A_n$ where A_n are finite rank atoms of $\mathcal{M}$. Also, if $M > M_0$, then $(P(M)-P(M_0))Q^\perp$ is not compact. This projection is the strong limit of $(P(M)-P(M_\alpha))Q^\perp$ and thus this net is eventually non-zero. Hence one may construct a decreasing sequence M_j, $j \geq 2$ so that $P(M_j)A_j = 0$ and $R_j \equiv (P(M_j)-P(M_{j+1}))Q^\perp \neq 0$. Define a partial isometry

$$U_j = R_j U_j A_j = P(M_j) R_j U_j A_j P(M_j)^\perp ,$$

for $j \geq 2$. This belongs to $\mathcal{T}(\mathcal{M})$ by Lemma 2,8, and so the infinite rank partial isometry $U = \sum_{j\geq 2} U_j$ belongs to $\mathcal{T}(\mathcal{M})$. But $U = Q^\perp UQ$ contradicts the essential invariance of Q. So $P(M_0)^\perp Q$ must have been compact as claimed.

One can repeat the proof in the case $P(M_0)^\perp Q$ is compact but $P(M_0)Q^\perp$ is not by reversing the order considerations. A contradiction is obtained as above. In all these cases, one has found an M in $\mathcal{M}$ so that

$P(M)-Q$ and thus $P(M)-P$ is compact. ∎

12.11 Corollary. *Let Θ be a map of intervals $E = P(N_1)-P(N_2)$ onto $\mathcal{M}''$ defined by $\Theta(E) = \chi(\Phi(E))$ where χ is the characteristic function of $[½,1]$. Then $\Theta(E)$ belongs to $C^*(\mathcal{M})$, $\Theta(E)-E$ is compact for every E, and when E_α converges strongly to zero, $\Theta(E_\alpha)-E_\alpha$ converges to zero in norm.*

Proof. By Lemmas 12.9 and 12.10, $\Phi(P(N_i))$ differs from some $P(M_i)$ by a compact operator. Thus $\Phi(E)$ differs from an interval of $\mathcal{M}$ by a compact operator in $\mathcal{M}''$. Every compact operator in $\mathcal{M}''$ is normal, and thus is a norm convergent sum $\sum \lambda_n A_n$ where A_n are minimal, finite rank projections in $\mathcal{M}''$, and thus are atoms of $\mathcal{M}$. Thus $\mathcal{M}'' \cap \mathcal{K}$ is contained in $C^*(\mathcal{M})$. Furthermore, χ is a continuous function of $\sigma(\Phi(E))$ which is a countable subset of $[0,1]$ with $\{0,1\}$ as its only possible limit points. So $\Theta(E)$ belongs to $C^*(\mathcal{M})$ and $\Phi(E)-\Theta(E)$ is compact. By Lemma 12.9, the map taking E to $E-\Phi(E)$ is continuous from the strong operator topology on E (induced from the order topology on $\mathcal{N}$) to the norm topology on $\mathcal{K}$. The map of E to $\Phi(E)-\chi(\Phi(E))$ is continuous at E provided that $\sigma(\Phi(E))$ does not contain ½. In particular, this is the case if $\|E-\Phi(E)\| < ½$. This holds eventually for E_α if E_α converges strongly to zero. So $\Theta(E)-E$ is continuous at zero. ∎

Lemma 12.10 shows that $\mathcal{N}$ and $\mathcal{M}$ are equal modulo the compact operators. It must now be shown that the atoms of $\mathcal{N}$ and $\mathcal{M}$ essentially match up, and fit into the nests in the "same places". To locate the atoms, we introduce the maximal ideal space $M^e_{\mathcal{N}}$ of $C^*(\widetilde{\mathcal{N}}) = C^*\{\tilde{P}(N):N \in \mathcal{N}\} = C^*(\mathcal{N})+\mathcal{K}/\mathcal{K}$. This is an invariant of $\mathcal{Q}\,\mathcal{T}(\mathcal{N})$ since Lemma 12.10 shows that $\widetilde{\mathcal{N}}$ is determined as the essentially invariant projections of $\mathcal{Q}\,\mathcal{T}(\mathcal{N})$. The set $M^e_{\mathcal{N}}$ sits naturally as a closed subset of $M_{\mathcal{N}}$, for if ϕ belongs to $M^e_{\mathcal{N}}$, then $\phi \circ \pi$ belongs to $M_{\mathcal{N}}$. (Recall sections 6.1-6.2 about $M_{\mathcal{N}}$.)

12.12 Proposition. *The complement $M_{\mathcal{N}} \backslash M^e_{\mathcal{N}}$ consists of the evaluations δ_A at finite rank atoms of $\mathcal{N}$.*

Proof. Every compact operator in $C^*(\mathcal{N})$ is normal, and thus is a norm convergent sum $\sum \lambda_n A_n$ where A_n are minimal, finite rank projections (i.e. atoms) in $C^*(\mathcal{N})$. The maximal ideals in $M^e_{\mathcal{N}}$ correspond in a bijective way with the maximal ideals of $C^*(\mathcal{N})$ which contain all of $\mathcal{K} \cap C^*(\mathcal{N})$. Clearly $ker\,\delta_A$ does not contain A, so evaluation at finite rank atoms are not in $M^e_{\mathcal{N}}$. On the other hand, if ϕ in $M_{\mathcal{N}}$ is not evaluation at a finite rank atom, then $\phi(A) = 0$ for every finite atom A and hence $ker\,\phi$ contains all $\mathcal{K} \cap C^*(\mathcal{N})$. So, ϕ belongs to $M^e_{\mathcal{N}}$. ∎

Put a partial order $\ll$ on the intervals of $\mathcal{N}$ by setting $E \ll F$ if and only if $E\mathcal{T}(\mathcal{N})F = E\mathcal{B}(\mathcal{H})F$. In particular, if $\mathcal{F} = \{0 = F_0 < F_1 < \ldots < F_n = \mathcal{H}\}$ is a finite subnest of $\mathcal{N}$, then $E_j = P(F_j) - P(F_{j-1})$ for $1 \le j \le n$ is a partition of $\mathcal{N}$ into intervals such that $E_1 \ll E_2 \ll \ldots \ll E_n$.

The space $M_{\mathcal{N}}$ is linearly ordered (Chapter 6), and $M_{\mathcal{N}} \backslash M^e_{\mathcal{N}}$ consists of the disjoint union of countably many maximal open intervals of finite rank atoms. The end points of these intervals belong to $M^e_{\mathcal{N}}$, and serve to mark the place of a finite rank atom in $M^e_{\mathcal{N}}$. So define the *markers* $M(\delta_A)$ to be the end points of the maximal interval of finite rank atoms containing δ_A. It may occur that one open interval in $M_{\mathcal{N}} \backslash M^e_{\mathcal{N}}$ is an initial or terminal segment. So in this case one end point may be missing. The typical example is a nest $\mathcal{P} = \{P_n, n \ge 1\}$ of finite projections. In this case $M_{\mathcal{P}} = \{\delta_n, n \ge 1, \phi_0\}$ where δ_n is evaluation at the atom $P_n - P_{n-1}$, and $\phi_0(P) = 0$ if $P \ne I$. $M^e_{\mathcal{P}} = \{\phi_0\}$, so the left end point of $M_{\mathcal{P}} \backslash M^e_{\mathcal{P}}$ is not defined in $M^e_{\mathcal{P}}$.

Suppose that A and B are atoms of $\mathcal{N}$ with $M(A) < M(B)$. That is, $M(A) = \{\phi_0, \phi_1\}$ and $M(B) = \{\psi_0, \psi_1\}$, and $\phi_0 < \psi_0$ or equivalently $\phi_1 \le \psi_0$ since ϕ_1 is the immediate successor of ϕ_0 in $M^e_{\mathcal{N}}$. Let N be the least element of $\mathcal{N}$ containing A, and let M be the greatest element of $\mathcal{N}$ orthogonal to B. It follows that $\phi_1(N) = 0$ and $\phi_1(M) = 1$. So $E = P(M) - P(N)$ is a test interval in $\mathcal{E}_{\phi_1}$ (i.e. $\phi_1(E) = 1$) with $A \ll E \ll B$. Since ϕ_1 annihilates the compact operators, E has infinite rank.

12.13 Lemma. *Suppose that $Q\mathcal{T}(\mathcal{N}) = Q\mathcal{T}(\mathcal{M})$, and let* $\mathbf{A}$ *and* $\mathbf{B}$ *be the sets of atoms of $\mathcal{N}$ and $\mathcal{M}$ respectively. There are finite subsets* $\mathbf{A}_0$ *and* $\mathbf{B}_0$ *of* $\mathbf{A}$ *and* $\mathbf{B}$ *such that for every A, A' in* $\mathbf{A} \backslash \mathbf{A}_0$:

i) $\Theta(A)$ *is an atom in* $\mathbf{B} \backslash \mathbf{B}_0$.

ii) $\|\Theta(A) - A\| < \frac{1}{2}$ *and rank* $\Theta(A) =$ *rank* A.

iii) $M(\Theta(A)) = M(A)$.

iv) $A \ll A'$ *implies* $\Theta(A) \ll \Theta(A')$.

Furthermore, Θ is a bijection between $\mathbf{A} \backslash \mathbf{A}_0$ *and* $\mathbf{B} \backslash \mathbf{B}_0$.

Proof. Let $\mathbf{A} = \{A_n, n \ge 1\}$. By Corollary 12.11, $\|A_n - \Theta(A_n)\| < \frac{1}{2}$ for all except finitely many atoms. After deleting these, *rank* $A_n =$ *rank* $\Theta(A_n)$. Also, $\Theta(A_m) = \Theta(A_n)$ implies $\|A_n - A_m\| < 1$, and so $A_n = A_m$ and Θ is one to one. If $\Theta(A_n)$ is not an atom, then there is a partial isometry $U_n = \Theta(A_n) U_n \Theta(A_n)$ such that $dist(U_n, \mathcal{T}(\mathcal{M})) = 1$. Suppose this occurs for infinitely many atoms A_n. Chose an infinite subsequence Λ so that

$$\sum_{n\in\Lambda} \|A_n - \Theta(A_n)\| < \infty \ .$$

The operator $U = \sum_{n\in\Lambda} U_n$ is not in $Q\,\mathcal{T}(\mathcal{M})$. However,

$$U - \sum_{n\in\Lambda} A_n U_n A_n = \sum_{n\in\Lambda} (\Theta(A_n) - A_n)U_n + A_n U_n(\Theta(A_n) - A_n) \ .$$

The right hand side is a norm convergent sum of compact operators, and thus U is a compact perturbation of $\sum_{n\in\Lambda} A_n U_n A_n$ which belongs to $\mathcal{T}(\mathcal{N})$. This contradicts the equality $Q\,\mathcal{T}(\mathcal{N}) = Q\,\mathcal{T}(\mathcal{M})$. So i) and ii) hold.

Suppose that (A_n, A'_n) are disjoint pairs in $\mathbf{A}$ with $A_n \ll A'_n$ and $\Theta(A_n) \gg \Theta(A'_n)$. Then there is a partial isometry of the form $U_n = \Theta(A_n)U_n\Theta(A'_n)$, and it follows that $\|U_n\| = dist(U_n, \mathcal{T}(\mathcal{M})) = 1$. As above, one obtains a contradiction if the set of pairs is infinite. Thus after deleting finitely many more atoms, iv) holds.

Similarly, suppose that $M(\Theta(A_n)) \neq M(A_n)$ infinitely often. For definiteness, suppose $M(A_n) \lll M(\Theta(A_n))$ infinitely often. This means that there is a ϕ_n in $M^e_{\mathcal{N}} = M^e_{\mathcal{M}}$ and intervals E_n of $\mathcal{N}$ and F_n of $\mathcal{M}$ such that $\phi_n(E_n) = \phi_n(F_n) = 1$, and $A_n \lll E_n$ and $F_n \lll B_n$. Since ϕ_n annihilates the compact operators, $\phi_n(\Theta(E_n)) = 1$ and hence $P_n = \Theta(E_n)F_n$ is an infinite rank projection. Repeating once more the argument of the first paragraph, drop to a subsequence Λ so that $\sum_{n\geq 1} \|A_n - \Theta(A_n)\| < \infty$ and construct rank one partial isometrices $U_n = B_n U_n P_n$. Then $U = \sum_{n\in\Lambda} U_n$ is not in $Q\,\mathcal{T}(\mathcal{M})$, but does belong to $Q\,\mathcal{T}(\mathcal{N})$. So iii) holds.

Let $\mathbf{A}_0$ be the finite collection of atoms deleted from $\mathbf{A}$ in order to satisfy i)-iv). Let $\mathbf{B}_0 = \mathbf{B}\backslash\Theta(\mathbf{A}\backslash\mathbf{A}_0)$. It must be shown that $\mathbf{B}_0$ is finite. By reversing the role of $\mathcal{M}$ and $\mathcal{N}$, one obtains a one to one map Θ^* of $\mathbf{B}$ into $C^*(\mathcal{N})$ such that $\Theta^*(B)$ is an atom except finitely often and $\lim_{n\to\infty} \|\Theta^*(B_n) - B_n\| = 0$. So for n sufficiently large,

$$\|B_n - \Theta(\Theta^*(B_n))\| < 1$$

and hence $\Theta(\Theta^*(B_n)) = B_n$. Hence Θ maps onto all but finitely many atoms of $\mathbf{B}$. So $\mathbf{B}_0$ is finite. ∎

12.14 Lemma. *Let E_j and F_j, $1 \leq j \leq n$, be two families of pairwise orthogonal projections such that $E_j - F_j$ is compact and $\|E_j - F_j\| < 1$, $1 \leq j \leq n$. Then $codim \sum_{j=1}^n E_j = codim \sum_{j=1}^n F_j$.*

Proof. Since $\|E_j(E_j - F_j)E_j\| < 1$, $E_jF_jE_j \geq \epsilon E_j$ for some $\epsilon > 0$, $1 \leq j \leq n$. Similarly, $F_jE_jF_j \geq \epsilon F_j$. Let $T = \sum_{j=1}^n E_jF_j$. Notice that

$$\sum_{j=1}^{n} F_j \geq T^*T = \sum_{j=1}^{n} F_j E_j F_j \geq \epsilon \sum_{j=1}^{n} F_j$$

and

$$\sum_{j=1}^{n} E_j \geq TT^* = \sum_{j=1}^{n} E_j F_j E_j \geq \epsilon \sum_{j=1}^{n} E_j \quad .$$

So T is bounded below on its domain $(\sum_{j=1}^{n} F_j)\mathcal{H}$ which is mapped onto its range $(\sum_{j=1}^{n} E_j)\mathcal{H}$. Now T is a compact perturbation of $\sum_{j=1}^{n} F_j$ which is self-adjoint, and thus T is Fredholm of index zero, or is not semi Fredholm. So if $\sum_{j=1}^{n} F_j$ has finite codimension, then

$$0 = ind\ T = codim \sum_{j=1}^{n} F_j - codim \sum_{j=1}^{n} E_j \quad .$$

Likewise for $\sum_{j=1}^{n} E_j$. In the remaining case, both have infinite codimension. ∎

12.15 Theorem. *If $Q\mathcal{T}(\mathcal{N}) = Q\mathcal{T}(\mathcal{M})$, then $\mathcal{N}$ and $\mathcal{M}$ are compactly perturbed nests.*

Proof. Let Θ be the map given by Corollary 12.11. The first paragraph of this proof modifies $\mathcal{N}$ and $\mathcal{M}$ so that $\|\Theta(A)-A\| < 1$ for every infinite rank atom of $\mathcal{N}$. Consider any infinite rank atom A in $\mathbf{A}_0$, the finite set of atoms provided by Lemma 12.13. Now $\tilde{A}$ is a minimal projection in $C^*(\widetilde{\mathcal{N}})$. This C^*-algebra equals $C^*(\widetilde{\mathcal{M}})$ by Lemma 12.10, so $\tilde{A}$ is a minimal projection in $C^*(\widetilde{\mathcal{M}})$. Thus there is an infinite atom B of $\mathcal{M}$ so that $A-B$ is compact. Choose projections A' and B' of finite codimension in A and B respectively so that $\|A'-B'\| < ½$. Let R be the sum of the finite atoms in $\mathbf{A}_0$ and of $(A-A')$ for the infinite atoms. Likewise, define S to be the corresponding sum for $\mathbf{B}_0$. Extend the definition of Θ to intervals of $\mathcal{N}^R$ onto $\mathcal{M}^S$ by setting $\Theta(A') = B'$ and $\Theta(R) = S$.

For each ϕ in $M_{\mathcal{N}}$ (except δ_R), there is a test interval E_ϕ such that for every interval $E \leq E_\phi$, one has $\|E-\Theta(E)\| < ½$. This follows since it is true for atoms by Lemma 12.13 ii), and holds at other points by the continuity at zero guaranteed by Corollary 12.11. Thus there is a partition of $\mathcal{N}^R$ into intervals $R = E_0 \ll E_1 \ll ... \ll E_n$ so that $\|\Theta(E)-E\| < ½$ for all intervals $E \leq E_j$, $1 \leq j \leq n$. Let $F_j = \Theta(E_j)$, $0 \leq j \leq n$. For $1 \leq i < j \leq n$,

$$\|F_i F_j\| < \|(F_i - E_i)E_j\| + \|F_i(E_j - F_j)\| < 1 \quad .$$

Since F_i and F_j commute, $F_i F_j = 0$ for $1 \leq i < j \leq n$. By Lemma 12.14, $rank\, R = rank\, S$.

Define a map θ of $\mathcal{N}^R$ into $\mathcal{M}^S$ as follows. If $\sum_{j<k} E_j < N \leq \sum_{j\leq k} E_j$, set

$$P(\theta(N)) = \sum_{j<k} F_j + \Theta(P(N)E_k) \ .$$

Now $E = P(N)E_k$ is an interval and thus $\Theta(E) = F \leq F_k$ differs from an interval by a finite linear combination of finite atoms. However by Lemma 12.13 (iii) and (iv), F must be an interval. In fact, F must be an initial segment of F_k for the same reasons. So $\theta(N)$ belongs to $\mathcal{M}^S$. By Lemma 12.13 (ii), and the previous paragraph, θ preserves the dimension of atoms. Thus, θ is a dimension preserving order isomorphism of $\mathcal{N}^R$ into $\mathcal{M}^S$ such that $\theta(\mathcal{N}^R)$ and $\mathcal{M}^S$ have the same atoms. Both are complete nests, and hence θ is surjective.

With N, E and F as above, $P(N)-P(\theta(N)) = \sum_{j<k} E_j - F_j + E - F$ is compact for all N in $\mathcal{N}$. Suppose that $\sum_{j<k} E_j \leq P(N_\alpha) \leq \sum_{j\leq k} E_j$, and N_α converges to N. Set $E_\alpha = P(N_\alpha)E_k$. Then

$$(P(N)-P(\theta(N)))-(P(N_\alpha)-P(\theta(N_\alpha))) = (E-\Theta(E))-(E_\alpha-\Theta(E_\alpha)) \ .$$

Since E_α converges strongly to E, this term converges to zero in norm by Corollary 12.11. Hence it follows easily that the map of N to $P(N)-P(\theta(N))$ is norm continuous on $\mathcal{N}$ (with the order topology). So $\mathcal{N}$ and $\mathcal{M}$ are compactly perturbed nests. ■

12.16 Theorem. *Let $\mathcal{N}$ and $\mathcal{M}$ be nests, and let θ be a dimension preserving order isomorphism of $\mathcal{N}$ onto $\mathcal{M}$ such that $\theta - id$ belongs to $C_n(\mathcal{N},\mathcal{K})$. Then one can finite rank projections E in $\mathcal{N}'$ and F in $\mathcal{M}'$ dominated by the infinite dimensional atoms such that $\mathcal{N}^E$ and $\mathcal{M}^E$ are approximately unitarily equivalent by a sequence U_n of unitaries such that $U_n - I$ is compact.*

Proof. Extend θ to a map Θ of intervals $E = P(N_1)-P(N_2)$ by $\Theta(E) = P(\theta(N_1))-P(\theta(N_2))$. We first show how to modify $\mathcal{N}$ so that $\|\Theta(A)-A\| < 1$ for all infinite rank atoms. The continuity of $\theta - id$ implies that $\Theta - id$ is continuous from the strong operator topology on the intervals to the norm topology on $\mathcal{K}$. Since $\mathcal{N}$ is compact, this continuity implies there is a partition $E_1 < E_2 < \ldots < E_n$ of $\mathcal{N}$ so that E_j is either an atom or $\|\Theta(E)-E\| < 1$ for every interval $E \leq E_j$. Let $F_j = \Theta(E_j)$. Consider the infinite rank atoms E_j such that $\|E_j - F_j\| = 1$ and choose finite rank projections $A_j < E_j$ and $B_j < F_j$ such that $\|(E_j - A_j)-(F_j - B_j)\| < 1$. Let $E'_j = E_j - A_j$, $F'_j = F_j - B_j$, and let E_0 be the sum of the A_j and let F_0 be the sum of the B_j. One can use Lemma 12.14 to show that $rank\, E_0 = rank\, F_0$ provided one notes that if E_j is a finite rank atom and $\|E_j - F_j\| = 1$, then $rank\, E_j = rank\, F_j$ because θ preserves dimension. So, replace $\mathcal{N}$ by $\mathcal{N}^{E_0}$ and $\mathcal{M}$ by $\mathcal{M}^{F_0}$ and modify θ to

θ' in the obvious way. It is easy to verify that θ' is now a dimension preserving order isomorphism of $\mathcal{N}^{E_0}$ onto $\mathcal{M}^{F_0}$ such that $\theta' - id'$ belongs to $C_n(\mathcal{N}^{E_0}, \mathcal{K})$. Moreover, the extended map Θ' has the property that $\|\Theta'(A) - A\| < 1$ for every infinite rank atom. For convenience of notation, we will assume that our original nests have this property, so that the "$'$"s can be dropped.

Let $0 < \epsilon < 1$ be given. Since $\mathcal{N}$ is compact and $\Theta - id$ is continuous into $\mathcal{K}$, one can obtain as before a partition $E_1 \ll E_2 \ll \ldots \ll E_n$ of $\mathcal{N}$ such that either $\|E - \Theta(E)\| < \epsilon$ for every interval $E \leq E_j$ or E_j is an atom. Let $F_j = \Theta(E_j)$. If $\|E_j - F_j\| < 1$, let U_j be the partial isometry in the polar decomposition of F_jE_j. Since $F_j - E_j$ is compact and $\|E_jF_j\| < 1$, U_j takes $E_j\mathcal{H}$ onto $F_j\mathcal{H}$ and $U_j - E_j$ is compact. If $\|E_j - F_j\| = 1$, then E_j is a finite rank atom and $rank\, E_j = rank\, F_j$. So one can define a partial isometry U_j of $E_j\mathcal{H}$ onto $F_j\mathcal{H}$. Let $U = \sum_{j=1}^{n} U_j$. This is a unitary operator, and $U - I = \sum_{j=1}^{n} U_j - E_j$ is compact. Moreover, $UE_jU^* = F_j$ for $1 \leq j \leq n$.

Let N belong to $\mathcal{N}$, and choose k so that

$$\sum_{j<k} E_j < P(N) \leq \sum_{j\leq k} E_j \ .$$

Let $E = P(N)E_k$ and $F = \Theta(E)$. Then $P(\theta(N)) = \sum_{j<k} F_j + F$ and

$$UP(N)U^* = \sum_{j<k} F_j + U_kEU_k^* \ .$$

So

$$\|P(\theta(N)) - UP(N)U^*\| = \|F - U_kEU_k^*\| = \|FU_k - U_kE\| \ .$$

If E_k is an atom, then $E = E_k$ and so this term is zero. Otherwise $\|E_k - F_k\| < \epsilon$ and hence $E_kF_kE_k \geq (1-\epsilon)E_k$. Let $R = (E_k^\perp + E_kF_kE_k)^{-1/2}$. Then $U_k = F_kE_kR$ and $\|R\| \leq (1-\epsilon)^{-1/2}$. So

$$\begin{aligned}
\|FU_k - U_kE\| &\leq \|FF_kE_k - F_kE_kE\| \, \|R\| \\
&= \|FE_kE^\perp - F^\perp F_kE\| \, \|R\| \\
&\leq (1-\epsilon)^{-1/2} \max\{\|F(F-E)E_kE^\perp\|, \|F^\perp F_k(F-E)E\|\} \\
&< \epsilon(1-\epsilon)^{-1/2} \ .
\end{aligned}$$

Hence $\|\theta - AdU\| < \epsilon(1-\epsilon)^{-1/2}$.

It remains to choose a sequence U_n with ϵ_n tending to 0. ∎

12.17 Lemma. *Let $\mathcal{N}$ and $\mathcal{M}$ be approximately unitarily equivalent via unitaries U_n, and let θ be the induced order isomorphism. Let $\delta = \|\theta - id\|$. Given $\epsilon > 0$, there is an invertible operator T such that $\|T-I\| \leq 10\delta$, and the map $\theta_T(N) = TN$ satisfies $\|\theta_T - \theta\| < \epsilon$. If $U_n - I$ is compact for all $n \geq 1$, then $T-I$ is compact also.*

Proof. As in the previous lemma, extend θ to a map Θ of intervals. Then $\|\Theta - id\| \leq 2\|\theta - id\| \leq 2\delta$. Let $U = U_n$ for n so large that $\|\theta_{U_n} - \theta\| < \epsilon \leq \delta$. If $\delta \geq 1/5$, take $T = U$. Otherwise, we can assume that $\delta < 1/5$. Compute $dist(U, \mathcal{T}(\mathcal{N}))$:

$$\begin{aligned}\|P(N)^{\perp}UP(N)\| &= \|P(N)^{\perp}P(UN)UP(N)\| \\ &\leq \|P(UN) - P(\theta(N))\| + \|P(\theta(N)) - P(N)\| \\ &\leq \|\theta_U - \theta\| + \|\theta - id\| < \epsilon + \delta \leq 2\delta \ .\end{aligned}$$

By Arveson's distance formula, $dist(U, \mathcal{T}(\mathcal{N})) < 2\delta$. So, choose A in $\mathcal{T}(\mathcal{N})$ such that $\|U - A\| < 2\delta$. If $U = I + K$ and K is compact, then by Theorem 12.1

$$dist(U, \mathcal{T}(\mathcal{N})) = dist(K, \mathcal{T}(\mathcal{N})) = dist(K, \mathcal{T}(\mathcal{N}) \cap \mathcal{K}) \ .$$

So one may take $A - I$ to be compact in this case.

Since $\|U - A\| < 2/5$, A is invertible in $\mathcal{B}(\mathcal{H})$, and $\|A^{-1}\| \leq 5/3$. It must be shown that A^{-1} belongs to $\mathcal{T}(\mathcal{N})$, or equivalently, that $AN = N$ for every N in $\mathcal{N}$. If it were otherwise, then since AN is a proper subspace of N, there would be a unit vector x in N orthogonal to AN. Now

$$\|P(N) - P(UN)\| < \|id - \theta\| + \|\theta - \theta_U\| \leq 2/5 \ ,$$

and $y = P(UN)x = Uz$ for some vector z in N with $\|z\| \leq 1$. So

$$\|x - Az\| \leq \|x - y\| + \|(U - A)z\| \leq 4/5 < 1 \ .$$

This contradicts the assumption that $AN \neq N$. So A^{-1} belongs to $\mathcal{T}(\mathcal{N})$.

Let $T = UA^{-1}$. Then

$$\|T - I\| = \|(U - A)A^{-1}\| < 2\delta(5/3) < 4\delta \ .$$

And if N belongs to $\mathcal{N}$, $TN = U(A^{-1}N) = UN$ so $\theta_T = \theta_U$. Hence $\|\theta_T - \theta\| < \epsilon$. ∎

12.18 Theorem. *Let $\mathcal{N}$ and $\mathcal{M}$ be approximately unitarily equivalent via unitaries U_n, and let θ be the induced order isomorphism. Then $\mathcal{N}$ and $\mathcal{M}$ are similar via an invertible operator S such that $\theta_S = \theta$. Given $\epsilon > 0$, S may be taken to be $U + L$ where U is unitary and $\|L\| < \epsilon$. If furthermore, $U_n - I$ is compact for all $n \geq 1$, $S - I$ may be taken to be compact, and also of the form $U + K$ where U is unitary in $I + \mathcal{K}$, K is compact, and $\|K\| < \epsilon$.*

Proof. Let $\epsilon > 0$ be given, and choose a decreasing sequence ϵ_n, $n \geq 1$ such that $\prod_{n \geq 1}(1+10\epsilon_n) < 1+\epsilon$. Choose a unitary $U = U_{n_0}$ so that $\|\theta_U - \theta\| < \epsilon_1$. Recursively apply Lemma 12.17 to obtain operators T_n with $\|T_n - I\| < 10\epsilon_n$ so that $\|\theta_{T_n T_{n-1} \cdots T_1 U} - \theta\| < \epsilon_{n+1}$. If each $U_n - I$ is compact, then make $T_n - I$ compact also. The infinite product $T = \lim_{n \to \infty} T_n T_{n-1} \ldots T_1 U$ converges in norm, and

$$\|T-U\| \leq \prod_{n \geq 1} \|T_n - I\| - 1 < \epsilon$$

(see Exercise 12.4). If furthermore, each $T_n - I$ is compact, then $T-U$ is compact.

For any operator X, $\|P(XN) - P(N)\| \leq 2\|X-I\|$. This is trivial if $\|X-I\| \geq 1/2$. On the other hand, if $\|X-I\| < 1$ and y is any unit vector in N, then y and Xy meet at an angle θ, $0 \leq \theta < \pi/2$. Thus

$$\|y - \frac{Xy}{\|Xy\|}\| = 2\sin(\theta/2) \leq 2\sin\theta \leq 2\|X-I\|$$

and conversely, every unit vector in XN has the form $Xy/\|Xy\|$ for some unit vector y in N. Thus

$$\|\theta_T - \theta\| \leq \|\theta_T - \theta_{T_n \ldots T_1 U}\| + \|\theta_{T_n \ldots T_1 U} - \theta\| \leq 2\|T - T_n \ldots T_1 U\| + \epsilon_{n+1} \ .$$

This tends to zero as N increases, so $\theta_T = \theta$ as desired. ∎

12.19 Example. It is possible that $\mathcal{Q}\,\mathcal{T}(\mathcal{N}) = \mathcal{Q}\,\mathcal{T}(\mathcal{M})$, and that $\mathcal{N}$ is unitarily equivalent to $\mathcal{M}$, but no *unitary* of the form $I+K$ with K compact will work. In our example, it will also be seen that the isomorphism θ of $\mathcal{N}$ onto $\mathcal{M}$ may be arbitrarily close to *id*, and the unitary implementing it is always far from the identity.

Let $\{e_n, n \geq 1\}$ be an orthonormal basis for $\mathcal{H}$, and let P_n be the $span\{e_k, 1 \leq k \leq n\}$. Let $\mathcal{P} = \{\{0\}, P_n, n \geq 1, \mathcal{H}\}$. Any other nest $\mathcal{Q} = \{\{0\}, Q_n, n \geq 1, \mathcal{H}\}$ with $dim Q_n = n$ for all n is unitarily equivalent to $\mathcal{P}$. It follows from Theorem 12.2, $\mathcal{Q}\,\mathcal{T}(\mathcal{Q}) = \mathcal{Q}\,\mathcal{T}(\mathcal{P})$ provided $\lim_{n \to \infty} \|P_n - Q_n\| = 0$. The converse follows from Theorem 12.8 or directly by more elementary means (see Exercise 12.2).

Recall Example 4.1. Let $t_{ij} = 0$ if $i \geq j$, and $t_{ij} = \frac{1}{j-i}$ if $j > i$. Let T_n be the matrix

$$T_n = [t_{ij}] = \begin{bmatrix} 0 & 1 & \frac{1}{2} & \frac{1}{3} & & & \frac{1}{n-1} \\ 0 & 0 & 1 & \frac{1}{2} & & & \frac{1}{n-2} \\ 0 & 0 & 0 & 1 & & & \frac{1}{n-3} \\ 0 & 0 & 0 & 0 & & & \\ & & & & & & \\ & & & & 0 & 1 & \frac{1}{2} \\ & & & & 0 & 0 & 1 \\ 0 & & & & 0 & 0 & 0 \end{bmatrix}$$

Then $\|T_n\| > .8\log n$ but $\|T_n - T_n^*\| \leq \pi$. Let

$$R_n = \frac{T_n + T_n^*}{\|T_n + T_n^*\|} \quad \text{and} \quad S_n = \frac{2T_n}{\|T_n + T_n^*\|} .$$

Since $2\|T_n\| - \pi \leq \|T_n + T_n^*\| \leq 2\|T_n\| + \pi$, one has $\lim_{n\to\infty} \|S_n\| = 1$, and $\lim_{n\to\infty} \|R_n - S_n\| = 0$. Let N_k^n be the span of the first k basis vectors, and let $\mathcal{T}_n$ be the $n \times n$ matrices which are upper triangular with respect to $\{N_k^n, 1 \leq k \leq n\}$. Since S_n belongs to $\mathcal{T}_n$, $\lim_{n\to\infty} dist(R_n, \mathcal{T}_n) = 0$. Let $U_n = e^{iR_n}$ and let $V_n = e^{iS_n}$. Then U_n is unitary, V_n belongs to $\mathcal{T}_n$, and the diagonal $\Delta(V_n) = I$. Moreover,

$$\begin{aligned} \|U_n - V_n\| &= \|\sum_{k\geq 0} \frac{(iR_n)^k - (iS_n)^k}{k!}\| \leq \sum_{k\geq 1} \frac{1}{k!} \|R_n^k - S_n^k\| \\ &\leq \sum_{k\geq 1} \frac{1}{k!} (\sum_{j=0}^{k-1} \|S_n\|^j) \|R_n - S_n\| \\ &= \sum_{j\geq 0} \|S_n\|^j (\sum_{k\geq j} \frac{1}{k!}) \|R_n - S_n\| \\ &< (e-1)e^{\|S_n\|} \|R_n - S_n\| . \end{aligned}$$

The estimate $\|R_n^k - S_n^k\| \leq \sum_{j=0}^{k-1} \|S_n\|^j \|R_n - S_n\|$ is easily established by induction. So $\lim_{n\to\infty} \|U_n - V_n\| = 0$.

Let $M_k^n = U_n N_k^n$. Then $M_k^n = U_n V_n^{-1} N_k^n$, and since $\lim_{n\to\infty} \|U_n V_n^{-1} - I\| = 0$, it follows that

$$\lim_{n\to\infty} \sup_{1\le k\le n} \|P(M_k^n)-P(N_k^n)\| = 0 \ .$$

Let W_n be another unitary carrying each N_k^n onto M_k^n, $1 \le k \le n$. Then $U_n^* W_n$ takes each N_k^n onto itself, as does $W_n^* U_n$. So $U_n^* W_n$ belongs to $\mathcal{T}_n \cap \mathcal{T}_n^* = \mathcal{D}_n$, the diagonal algebra (i.e. $W_n = UD$ where D is a diagonal unitary). So

$$\|W_n - I\| = \|U_n - D^*\| \ge \|\Delta(U_n) - D^*\|$$
$$\ge \|\Delta(V_n) - D^*\| - \|U_n - V_n\| = \|I - D^*\| - \|U_n - V_n\| \ .$$

But since $\|R_n\| = 1$, the spectrum of U_n contains one of the $e^{\pm i}$, and so $\|U - I\| = |e^i - 1| = 2\sin\frac{1}{2}$. So

$$\|U_n - D^*\| \ge \|U_n - I\| - \|D^* - I\| = 2\sin\tfrac{1}{2} - \|D^* - I\| \ .$$

Combining these two inequalities yields

$$\|W_n - I\| \ge \sin(\frac{1}{2}) - \|U_n - V_n\| \ .$$

Identify $\mathcal{H}$ with $\sum_{n\ge n_0} \oplus \mathbb{C}^n$. Let $\mathcal{P}$ be identified with the nest obtained by concatenating the nests $\{N_k^n, 1 \le k \le n\}$ for $n \ge n_0$. Let $U = \sum_{n\ge n_0} \oplus U_n$, and $\mathcal{Q} = U\mathcal{P}$. By construction, $\mathcal{Q}$ is asymptotic to $\mathcal{P}$ and

$$\sup_{k\ge 1} \|P(Q_k) - P(P_k)\| \le \sup_{n\ge n_0} 2\|U_n V_n^{-1} - I\| \ .$$

So for n_0 large, $\mathcal{Q}$ and $\mathcal{P}$ are both close and compactly perturbed. Suppose that W is a unitary carrying $\mathcal{P}$ onto $\mathcal{Q}$. Then W carries each block $\mathbb{C}^n$ onto itself. So $W = \sum_{n\ge n_0} \oplus W_n$, and W_n carries $\{N_k^n, 1 \le k \le n\}$ onto $\{M_k^n, 1 \le k \le n\}$. As we have just seen,

$$\lim_{n\to\infty} \|W_n - I\| = \sin(\frac{1}{2}) > 0 \ .$$

Consequently, $W - I$ is neither compact nor of small norm.

Notes and Remarks.

The notion of quasitriangularity was introduced by Halmos [3]. It is a key idea in the proof of a number of invariant subspace theorems. An important theorem of Apostol, Foias, and Voiculescu [1] characterizes quasitriangular operators as those operators T such that $ind(T-\lambda I) \ge 0$ whenever $T-\lambda I$ is semi Fredholm. Most of Theorem 12.1 is due to Fall, Arveson and Muhly [1] with the exception that the quotient map π is isometric. The proof given here is from Davidson and Power [2]. Theorem 12.2 is due to Arveson [7]. Theorem 12.5 is in Fall, Arveson and Muhly. Theorem 12.6 is in Davidson and Power. Another proof that $\mathcal{Q}\,\mathcal{T}(\mathcal{N})$ is proximinal in $\mathcal{B}(\mathcal{H})$ is given by Feeman [1,2].

Theorem 12.8 is due to Andersen [1] and Davidson [6]. Andersen proved the equivalence of 1), 2) and 3). Davidson added 4). The approach given here is new in several ways. In particular, the use of $M^e_{\mathcal{N}}$ to locate the atoms is taken from Apostol and Davidson [1]. Lemma 12.10 is a special case of a theorem in Davidson [1].

Exercises.

12.1 Let T be a quasitriangular operator and let $\epsilon > 0$ be given. Show that there is a compact operator K with $\|K\| < \epsilon$ and an orthonormal basis $\{e_n, n \geq 1\}$ so that $T-K$ is triangular with respect to this basis.

12.2 Use Theorem 12.2 to prove that if $\mathcal{P} = \{P_n, n \geq 1\}$ and $\mathcal{Q} = \{Q_n, n \geq 1\}$ are nests of finite rank projections with $\sup P_n = I = \sup Q_n$ and $\lim\|P_n - Q_n\| = 0$, then $Q\mathcal{T}(\mathcal{P}) = Q\mathcal{T}(\mathcal{Q})$.

12.3 (i) Let X be a compact Hausdorff space. Use the Riesz Representation Theorem to show that the dual of $C_n(X,\mathcal{K})$ is $\mathcal{M}(X, C_1)$, the space of bounded regular Borel measures on X with values in C_1.

(ii) Show that a *bounded* net Φ_α in $C_{s^*}(X, \mathcal{B}(\mathcal{H}))$ converges to Φ in the $C_n(X,\mathcal{K})^*$ topology if and only if $\Phi_n(x)$ converges weak* to $\Phi(x)$ pointwise.

12.4 This example is designed to show that Lemma 12.14 fails if $E_j - F_j$ is not compact. Let $n \geq 4$ be a positive integer, and let $\mathcal{H} = \mathcal{H}_1 \oplus \ldots \oplus \mathcal{H}_n$, where $\mathcal{H}_j$ are isomorphic Hilbert spaces. Let E_j be the projection onto $\mathcal{H}_j$. Let $Q(x_1,\ldots,x_n) = (\bar{x}, \bar{x}, \ldots, \bar{x})$ where $\bar{x} = \frac{1}{n}\sum_{i=1}^n x_i$. Then Q is the orthogonal projection onto $\mathcal{Q} = \{(x,\ldots,x) : x \in \mathcal{H}_1\}$. Let S be the unilateral shift acting on $\mathcal{Q}$, and let $V = Q^\perp + SQ$. Define $F_j = VE_jV^*$. Then $\sum_{j=1}^n F_j = VV^*$ has codimension 1 while $\sum_{j=1}^n E_j = I$. Prove that $\|E_j - F_j\| \leq (\frac{4}{n})^{1/2}$ for $1 \leq j \leq n$.

12.5 Let ϵ_n be positive numbers such that $\prod_{n=1}^\infty (1+\epsilon_n) = 1+\epsilon$ exists. Let A_n be operators with $\|A_n\| \leq \epsilon_n$. Define $T_0 = I$ and $T_n = (1+A_n)T_{n-1}$. Prove that $\|T_n - I\| \leq \prod_{k=1}^n (1+\epsilon_k) - 1$. Hence establish that T_n is Cauchy with a limit T such that $\|T-I\| \leq \epsilon$.

12.6 Let $\mathcal{N}$ be a nest and let A be an operator such that the range of $\Phi_A(N) = P(N)^\perp AP(N)$ is a norm compact subset of $\mathcal{K}$. Prove that A belongs to $\mathcal{Q}\,\mathcal{T}(\mathcal{N})$. Hint: Compare the norm and strong* topologies on $Ran\,\Phi_A$.

12.7 Let $\mathcal{N}$ and $\mathcal{M}$ be nests. Suppose that α is an algebra isomorphism of $\mathcal{Q}\,\mathcal{T}(\mathcal{N})$ onto $\mathcal{Q}\,\mathcal{T}(\mathcal{M})$.

(i) Prove that α carries the set of finite rank operators onto itself.

(ii) Show that there is an invertible operator S so that $\alpha(F) = SFS^{-1}$ for all finite rank operators.

(iii) Deduce that $\alpha(T) = STS^{-1}$ for all T. Hence α is continuous.

(iv) Show that $\mathcal{N}$ and $\mathcal{M}$ have finite perturbations which are similar.

(v) (Plastiras) If $\mathcal{N} = \{N_k, k \geq 1, \mathcal{H}\}$ and $\mathcal{M} = \{M_k, k \geq 1, \mathcal{H}\}$ consist of finite dimensional subspaces with dense union and $\mathcal{Q}\,\mathcal{T}(\mathcal{N})$ and $\mathcal{Q}\,\mathcal{T}(\mathcal{M})$ are isomorphic, show that there are integers n_0 and k_0 so that

$$dim N_n = dim M_{n+k_0}$$

for all $n \geq n_0$.

13. Similarity and Approximate Unitary Equivalence

The main result of this chapter is the Similarity Theorem 13.20 which shows that two nests are similar if and only if there is a dimension preserving order isomorphism of one nest onto the other. This is quite different from the unitary invariants of chapter 7; for neither multiplicity nor the measure class need be preserved in general. Several examples will be given at the end to demonstrate some of the unusual phenomena. In this chapter, *all Hilbert spaces will be separable.*

13.1. The first big step is to show that all continuous nests are approximately unitarily equivalent (Anderson's Theorem 13.9). Let $\mathcal{N}$ be a continuous nest $\{N_t, 0 \leq t \leq 1\}$ parametrized by $[0,1]$. Let θ be an order isomorphism of $\mathcal{N}$ onto another nest $\mathcal{M}$. Then $\mathcal{M}$ may be parametrized as $\{M_t, 0 \leq t \leq 1\}$ so that $\theta(N_t) = M_t$. Say that $\mathcal{N}$ and $\mathcal{M}$ are ϵ-unitarily equivalent if there is a unitary U such that $P(M_t)U - UP(N_t)$ is compact for all $0 \leq t \leq 1$, and

$$\sup_{0 \leq t \leq 1} \|P(M_t)U - UP(N_t)\| < \epsilon \ .$$

Thus $\mathcal{N}$ and $\mathcal{M}$ are approximately unitarily equivalent (via θ) if they are ϵ-unitarily equivalent for all $\epsilon > 0$.

Now consider a method for decomposing a nest $\mathcal{N}$. Let P be a projection in $\mathcal{N}'$. Let $\mathcal{N}_1$ and $\mathcal{N}_2$ be the nests

$$\mathcal{N}_1 = \{N_t^{(1)} = N_t \cap P\mathcal{H} : 0 \leq t \leq 1\}$$

and

$$\mathcal{N}_2 = \{N_t^{(2)} = N_t \cap P^{\perp}\mathcal{H} : 0 \leq t \leq 1\} \ .$$

It is quite possible that the parametrization of $\mathcal{N}_1$ or $\mathcal{N}_2$ may be redundant. Indeed, $N_s^{(1)} = N_t^{(1)}$ precisely if $N_t \ominus N_s$ is orthogonal to P. We will write $\mathcal{N} \cong \mathcal{N}_1 \oplus \mathcal{N}_2$. More generally, if $\mathcal{N}_k$ are nests parametrized in $[0,1]$, the direct sum $\sum_{k \geq 1} \oplus \mathcal{N}_k$ will mean the nest $\mathcal{N}$ with elements $N_t \cong \sum_{k \geq 1} \oplus N_t^{(k)}$. The nests $\mathcal{N}_k$ are called *summands* of $\mathcal{N}$.

Let $E_{\mathcal{N}}$ denote the spectral measure of $\mathcal{N}$ introduced in Corollary 7.6. Given a vector x in $\mathcal{H}$, define the *support* of x (supp(x)) to be the smallest *closed* subset C of $[0,1]$ such that $E_{\mathcal{N}}(C)x = x$. If $\mathcal{N}_1$ is a summand of $\mathcal{N}$, then supp($\mathcal{N}_1$) is the smallest closed set C such that $E_{\mathcal{N}}(C)x = x$ for all x in $N_1^{(1)}$. In particular, if $\mathcal{N}_1$ has a cyclic vector x_1 (meaning that

$\underset{0 \le t \le 1}{span} P(N_t^{(1)})x_1 = N_1^{(1)})$, then $\text{supp}(\mathcal{N}_1) = \text{supp}(x_1)$.

13.2 Lemma. *Let $\mathcal{N}$ be a continuous nest, and let J_i be open subsets of $J_0 = (0,1)$ for $i \ge 0$. Then $\mathcal{N}$ has pairwise orthogonal summands $\mathcal{N}_i$ with* support $\overline{J_i}$ *such that $\mathcal{N} \cong \sum_{i=0}^{\infty} \oplus \mathcal{N}_i$. If $\mathcal{N}$ is cyclic, then $\mathcal{N}_i$ has a cyclic vector $x_i = E_{\mathcal{N}}(J_i)x_i$.*

Proof. Let μ be a finite regular Borel measure on $[0,1]$ mutually absolutely continuous to the spectral measure $E_{\mathcal{N}}$. (For example, $\mu(A) = (E_{\mathcal{N}}(A)x,x)$ where x is a separating vector for $\mathcal{N}''$.) Since $\mathcal{N}$ is continuous, μ is non-atomic. It is an exercise in measure theory to produce disjoint Borel sets A_i contained in J_i such that $J_0 = \bigcup_{i \ge 0} A_i$ and $\mu(A_i \cap I) > 0$ for every open subset I of J_i (see Exercise 13.2). Set $\mathcal{N}_i = \mathcal{N} | E_{\mathcal{N}}(A_i)\mathcal{H}$. It follows that the support of $\mathcal{N}_i$ is $\overline{A_i} = \overline{J_i}$, and $\mathcal{N} \cong \sum_{i \ge 0} \oplus \mathcal{N}_i$. If x is cyclic for $\mathcal{N}$, then $x_i = E_{\mathcal{N}}(A_i)x$ is cyclic for $\mathcal{N}_i$. A fortiori, $x_i = E_{\mathcal{N}}(J_i)x_i$. ∎

The next lemma shows that every continuous nest has "approximate infinite multiplicity" in a certain sense.

13.3 Lemma. *Let $\mathcal{N}$ be a continuous nest. Then $\mathcal{N} \cong \sum_{i=1}^{\infty} \oplus \mathcal{N}_i$ where each $\mathcal{N}_i$ is cyclic with* support $[0,1]$.

Proof. For each non-zero vector x, let $\mathcal{H}_x$ denote the closed span of $\{P(N)x : N \in \mathcal{N}\}$. The projection onto $\mathcal{H}_x$ commutes with $\mathcal{N}$, and $\mathcal{N}_x = \mathcal{N} | \mathcal{H}_x$ has x as a cyclic vector. By Proposition 2.12, there is a separating vector x_0 for $\mathcal{N}''$. Clearly, $\text{supp}(x_0) = [0,1]$. Extend $\{x_0\}$ to a maximal (countable or finite) family $\{x_n, 0 \le n < \alpha\}$ of non-zero vectors such that $\mathcal{H}_{x_n}$ are pairwise orthogonal. The maximality guarantees that $\mathcal{H}$ is the direct sum $\sum \oplus \mathcal{H}_{x_n}$. Let $C_n = \text{supp}(x_n)$.

Let $J_n = (0,1) \backslash C_n$ for $1 \le n < \alpha$. Since $C_0 = [0,1]$, one may apply Lemma 13.2 to $\mathcal{N}_{x_0}$ to obtain a decomposition

$$\mathcal{N}_{x_0} \cong \sum_{1 \le n < \alpha} \oplus \mathcal{M}_n \oplus \sum_{j=1}^{\infty} \oplus \mathcal{L}_j$$

where $\text{supp}(\mathcal{M}_n) = \overline{J_n}$ and $\text{supp}(\mathcal{L}_j) = [0,1]$ for $j \ge 1$. Since $\mathcal{N}_{x_0}$ has a cyclic vector x_0, each summand is cyclic. Let y_n be a cyclic vector for $\mathcal{M}_n$, $1 \le n < \alpha$ and z_j be a cyclic vector for $\mathcal{L}_j$, $j \ge 1$. Also, $E_{\mathcal{N}}(J_n)y_n = y_n$.

Notice that if x and y have supports S_1 and S_2 satisfying $E_{\mathcal{N}}(S_1 \cap S_2) = 0$, then $\mathcal{H}_{x+y} = \mathcal{H}_x \oplus \mathcal{H}_y$ and $\mathrm{supp}(x+y) = S_1 \cup S_2$. Let $w_n = x_n + y_n$ for $1 \leq n < \alpha$. By construction, $\mathcal{N}_{w_n} \cong \mathcal{N}_{x_n} \oplus \mathcal{M}_n$ and $\mathrm{supp}(w_n) = [0,1]$. So

$$\begin{aligned}\mathcal{N} &\cong \mathcal{N}_{x_0} \oplus \sum_{n=1}^{\alpha} \oplus \mathcal{N}_{x_n} \\ &\cong \sum_{n=1}^{\alpha} \mathcal{M}_n \oplus \mathcal{N}_{x_n} \oplus \sum_{j=1}^{\infty} \oplus \mathcal{L}_j \\ &\cong \sum_{n=1}^{\alpha} \oplus \mathcal{N}_{w_n} \oplus \sum_{j=1}^{\infty} \oplus \mathcal{N}_{z_j} \ .\end{aligned}$$

Consequently, $\mathcal{N}$ is the direct sum of countably many cyclic nests with support $[0,1]$. ∎

Next, we wish to construct a sequence F_n of positive finite rank contractions converging strongly to the identity operator such that $\sup\{\|F_n P(N) - P(N) F_n\| : N \in \mathcal{N}\}$ tends to zero as n increases. This will be called a *quasi-central approximate identity*.

13.4 Lemma. *Let y_j, z_j, $1 \leq j < L$ be vectors such that $\|y_i\| \leq 1$, $\|z_j\| \leq 1$, and $\max_{i<j}\{|(y_i,y_j)|, |(z_i,z_j)|\} < 1/L^2$. Then $\|\sum_{j=1}^{L} y_j \otimes z_j^*\| \leq 2$.*

Proof. Let $e_1,\ldots,e_L$ be an orthnormal set. Let $A = \sum_{j=1}^{L} y_j \otimes e_j^*$ and $B = \sum_{j=1}^{L} e_j \otimes z_j^*$. Then $AB = \sum_{j=1}^{L} y_j \otimes z_j^*$. Then

$$\begin{aligned}\|A\|^2 = \|A^*A\| &= \|\sum_{i=1}^{L}\sum_{j=1}^{L} (y_j,y_i) e_i \otimes e_j^*\| \\ &\leq \max_{1 \leq j \leq L} |(y_j,y_j)| + \sum_{i \neq j}\sum |(y_j,y_i)| \leq 1 + L^2/L^2 = 2 \ .\end{aligned}$$

Similarly, $\|B\|^2 \leq 2$. So $\|\sum_{j=1}^{L} y_j \otimes z_j^*\| \leq \|A\| \, \|B\| \leq 2$. ∎

13.5 Lemma. *Let $\mathcal{N}$ be a continuous nest with cyclic vector x. Let $E_{\mathcal{N}}$ be the spectral measure for $\mathcal{N}$, and set $E_{k,n} = E_{\mathcal{N}}\left(\frac{k-1}{2^n}, \frac{k}{2^n}\right)$ for $1 \leq k \leq 2^n$. Let $x_{k,n} = \|E_{k,n}x\|^{-1} E_{k,n} x$, and let P_n be the projection onto span$\{x_{k,n} : 1 \leq k \leq 2^n\}$. Given $n \geq 1$ and $\epsilon > 0$, there is a finite rank positive contraction $F \geq P_n$ which is a convex combination of $\{P_m, m \geq n\}$ such that $\sup_{0 \leq t \leq 1} \|P(N_t)F - FP(N_t)\| < \epsilon$.*

Proof. Let $f(t) = \|P(N_t)x\|^2$, and note that this is a strictly increasing continuous function of [0,1] onto itself. From the uniform continuity of f, one obtains

$$\lim_{n\to\infty} \max_{1\le k\le 2^n} \|E_{k,n}x\|^2 \le \lim_{n\to\infty} \sup_{|s-t|\le 2^{-n}} |f(s)-f(t)| = 0 \ .$$

For $m = n+p$, one has

$$x_{k,n} = \sum_{j=2^p(k-1)+1}^{2^p k} a_j x_{j,m}$$

where

$$a_j = (x_{k,n}, x_{j,m}) = \|E_{k,n}x\|^{-1}\|E_{j,m}x\|^{-1}\|E_{j,m}x\|^2 = \|E_{k,n}x\|^{-1}\|E_{j,m}x\| \ .$$

Let

$$\delta(n,m) = \max_{1\le j\le 2^m} a_j \le \max_{1\le k\le 2^n} \|E_{k,n}x\|^{-1} \cdot \max_{1\le j\le 2^m} \|E_{j,m}x\| \ .$$

Then $\lim_{m\to\infty} \delta(n,m) = 0$ for any fixed n.

Let $L = 2[\epsilon^{-1}+1]$, and choose integers $n = m_1 < m_2 < \dots < m_L$ so that $\delta(m_j, m_{j+1}) < 1/L^2$ for $1 \le j \le L$. Set $F = L^{-1}\sum_{j=1}^{L} P_{m_j}$. Fix t in [0,1] and compute $[P(N_t),F] = P(N_t)F - FP(N_t)$. For each j, there is an integer k_j such that $k_j - 1 \le 2^{m_j}t \le k_j$. Now

$$P_{m_j} = \sum_{k=1}^{2^{m_j}} x_{k,m_j} \otimes x^*_{k,m_j} \ .$$

If $k \ne k_j$, then $P(N_t)$ commutes with $x_{k,m_j} \otimes x^*_{k,m_j}$. So

$$[P(N_t),P_{m_j}] = [P(N_t), x_{k_j,m_j} \otimes x^*_{x_j,m_j}] = y_j \otimes z_j^* - z_j \otimes y_j^*$$

where $y_j = P(N_t)x_{k_j,m_j}$ and $z_j = P(N_t)^\perp x_{k_j,m_j}$. So

$$[P(N_t),F] = L^{-1}\sum_{j=1}^{L} y_j \otimes z_j^* - z_j \otimes y_j^* \ .$$

If $i < j$, then

$$|(y_i,y_j)| + |(z_i,z_j)| = (x_{k_i,m_i}, x_{k_j,m_j}) \le \delta(m_i,m_j) < 1/L^2 \ .$$

Hence by Lemma 13.4,

$$\|[P(N_t),F]\| = L^{-1}\max\{\|\sum_{j=1}^{L} y_j \otimes z_j^*\|, \|\sum_{j=1}^{L} z_j \otimes y_j^*\|\} \le 2L^{-1} < \epsilon \ . \quad \blacksquare$$

Now, this quasi-central approximate unit will be used to approximately embed $\mathcal{N}$ as a summand of another continuous nest $\mathcal{M}$.

13.6 Lemma. *Let $\mathcal{N}$ and P_n be as in Lemma 13.5. Let $\mathcal{M}$ be another continuous nest with cyclic vector y. Given an integer n_0, and an $\epsilon > 0$, there is a unitary operator U such that*

$$\sup_{0 \le t \le 1} \|(P(M_t)U - UP(N_t))P_{n_0}\| < \epsilon .$$

Proof. Let $x_{k,n}$ be defined as in Lemma 13.5. Similarly, define unit vectors $y_{k,n}$ for $\mathcal{M}$. Choose $n > n_0$ sufficiently large so that $\delta(n_0,n) < \epsilon/2$. Define U by setting $Ux_{k,n} = y_{k,n}$ for $1 \le k \le 2^n$ and extending it in such a way that

$$UE_{\mathcal{N}}\left(\frac{k-1}{2^n},\frac{k}{2^n}\right) = E_{\mathcal{M}}\left(\frac{k-1}{2^n},\frac{k}{2^n}\right), \quad 1 \le k \le 2^n .$$

Fix t in $[0,1]$, and choose integers k and i so that

$$k-1 \le 2^{n_0}t \le k \quad \text{and} \quad i-1 \le 2^n t \le i .$$

Then $P(M_t)Ux_{j,n} = UP(N_t)x_{j,n}$ for $j \ne i$. So in particular, $P(M_t)U - UP(N_t)x_{k',n_0} = 0$ for $k' \ne k$. Write $x_{k,n_0} = \sum a_j x_{j,n}$. Then

$$\begin{aligned} \|(P(M_t)U - UP(N_t))P_{n_0}\| &= \|(P(M_t)U - UP(N_t))x_{k,n_0}\| \\ &= |a_i| \, \|(P(M_t)U - UP(N_t))x_{i,n}\| \\ &\le 2\delta(n_0,n) < \epsilon . \end{aligned}$$

■

13.7 Lemma. *Let E be a positive operator, and let T be an operator with $\|T\| \le 1$. Then $\|[E^{1/2},T]\| \le \sqrt{2}\|[E,T]\|^{1/2}$.*

Proof. Note that for any operators A and B that

$$\|[A^n,B]\| = \|\sum_{k=0}^{n-1} A^k[A,B]A^{n-1-k}\| \le n\|A\|^{n-1}\|[A,B]\| .$$

Now consider the power series

$$f(t) = 1-(1-t)^{1/2} = \sum_{n \ge 1} a_n t^n, \quad |t| < 1$$

where $a_n = \dfrac{(-1)^{n-1}}{n!}\prod_{k=0}^{n-1}(\frac{1}{2}-k) > 0$ for all $n \ge 1$. Differentiating yields

$$f'(t) = 2^{-1}(1-t)^{-1/2} = \sum_{n \ge 1} n a_n t^{n-1} .$$

Normalize E so that $\|E\| = 1$. (This doesn't affect the relationship in the lemma.) For $0 < \delta < 1$, set $A = (1-\delta)I - E = I-(E+\delta I)$. Then $(E+\delta I)^{1/2} = I - f(A)$, whence

$$\|[(E+\delta I)^{1/2},T]\| = \|[\sum_{n\geq 1} a_n A^n,T]\| \leq \sum_{n\geq 1} a_n n\|A\|^{n-1}\|[A,T]\|$$
$$= f'(\|A\|)\|[E,T]\| \leq f'(1-\delta)\|[E,T]\|$$
$$= \frac{\|[E,T]\|}{2\delta^{1/2}} .$$

On the other hand,

$$0 \leq (E+\delta I)^{1/2}-E^{1/2} = \delta((E+\delta I)^{1/2}+E^{1/2})^{-1} \leq \delta^{1/2}I .$$

So

$$\|[E^{1/2},T]\| \leq \|[(E+\delta I)^{1/2},T]\|+\|[(E+\delta I)^{1/2}-E^{1/2}-\tfrac{1}{2}\delta^{1/2}I,T]\|$$
$$\leq \|[E,T]\|/2\delta^{1/2}+2\|(E+\delta I)^{1/2}-E^{1/2}-\tfrac{1}{2}\delta^{1/2}I\|\ \|T\|$$
$$\leq \|[E,T]\|/2\delta^{1/2}+\delta^{1/2}\|T\| .$$

Set δ so that $(2\delta)^{1/2} = \|[E,T]\|^{1/2}\|T\|^{-1/2}$. Then

$$\|[E^{1/2},T]\| \leq \sqrt{2}\|[E,T]\|^{1/2}\|T\|^{1/2} .$$ ■

13.8 Lemma. *Let $\mathcal{M}$ be a continuous nest, and let $\mathcal{N}$ be a continuous nest with cyclic vector. Given $\epsilon > 0$, there is an isometry U such that $f(t) = P(M_t)U-UP(N_t)$ is norm continuous, and compact of norm at most ϵ for all $0 \leq t \leq 1$ (i.e. $f \in C_n(\mathcal{N},K)$ and $\|f\| \leq \epsilon$.)*

Proof. By Lemma 13.3, $\mathcal{M}$ can be decomposed as $\mathcal{M} \cong \sum_{j=1}^{\infty} \oplus \mathcal{M}_j$ on $\mathcal{H} = \sum_{j=1}^{\infty} \oplus \mathcal{H}_j$ where each M_j has a cyclic vector y_j with support $[0,1]$.

Let P_n be defined as in Lemma 13.5 for $\mathcal{N}$. Recursively choose integers n_k, positive finite rank contractions F_k, and unitaries U_k of $\mathcal{H}$ into $\mathcal{H}_k$ so that $n_0 = 0$, $n_1 = 1$, $F_0 = 0$ and

i) $P_{n_k} \leq F_k \leq P_{n_{k+1}}$.

ii) $\sup_{0\leq t\leq 1}\|P(N_t)F_k-F_kP(N_t)\| < (\epsilon/2^{k+1})^2/10$.

iii) $\sup_{0\leq t\leq 1}\|(P(M_t)U_k-U_kP(N_t))P_{n_{k+1}}\| < 2^{-k-1}\epsilon$.

This is done as follows: given n_k, F_{k-1} and U_{k-1}, use Lemma 13.5 to obtain F_k satisfying $F_k \geq P_{n_k}$ and ii). Then n_{k+1} is chosen to satisfy i). Now Lemma 13.6 is used to obtain U_k satisfying iii).

Define $E_k = (F_k-F_{k-1})^{1/2}$ for $k \geq 1$, and $U = \sum_{k=1}^{\infty} U_kE_k$. Since the U_k have orthogonal ranges,

$$U^*U = \sum_{k=1}^{\infty} E_k^2 = I$$

and thus U is an isometry. Also, by Lemma 13.7

$$\begin{aligned}\|E_kP(N_t)-P(N_t)E_k\| &\le \sqrt{2}\|[F_k-F_{k-1},P(N_t)]\|^{1/2} \\ &\le \sqrt{2}((\epsilon/2^k)^2/10+(\epsilon/2^{k+1})^2/10)^{1/2} \\ &= \epsilon/2^{k+1} \ .\end{aligned}$$

Thus

$$\begin{aligned}\|P(M_t)U-UP(N_t)\| &= \|\sum_{k=1}^{\infty}(P(M_t)U_k-U_kP(N_t))E_k+U_k[P(N_t),E_k]\| \\ &\le \sum_{k=1}^{\infty}\|(P(M_t)U_k-U_kP(N_t))P_{n_{k+1}}\|+\|[P(N_t),E_k]\| \\ &< \sum_{k=1}^{\infty} 2^{-k-1}\epsilon+2^{-k-1}\epsilon = \epsilon \ .\end{aligned}$$

Furthermore, this is a convergent sum of finite rank operators, and thus is compact. The strong continuity of $t \to P(M_t)$ and $t \to P(N_t)$ guarantees that each term is norm continuous. The limit is uniform in t, and hence is also norm continuous. ∎

13.9 Andersen's Theorem. *Let $\mathcal{N}$ and $\mathcal{M}$ be continuous nests on a separable Hilbert space, and let θ be an order isomorphism of $\mathcal{N}$ onto $\mathcal{M}$. Given $\epsilon > 0$, there is a unitary operator W such that $f(N) = P(\theta(N))-WP(N)W^*$ is continuous, compact valued and of norm at most ϵ for all N in $\mathcal{N}$.*

Proof. Parametrize $\mathcal{N}$ as $\{N_t, 0 \le t \le 1\}$ and $\mathcal{M}$ as $\{M_t, 0 \le t \le 1\}$ so that $\theta(N_t) = M_t$. Apply Lemma 13.3 to decompose $\mathcal{N}^{(\infty)}$, the direct sum of countably many copies of $\mathcal{N}$, as $\mathcal{N}^{(\infty)} \cong \sum_{i=1}^{\infty} \oplus \mathcal{N}_i$ on $\mathcal{H}^{(\infty)} = \sum_{i=1}^{\infty} \oplus \mathcal{H}_i$ where each $\mathcal{N}_i$ is cyclic with support $[0,1]$. Likewise, decompose $\mathcal{M} \cong \sum_{i=1}^{\infty} \oplus \mathcal{M}_i$ on $\mathcal{H}' = \sum_{j=1}^{\infty} \oplus \mathcal{H}'_i$. By Lemma 13.8, there are isometries U_i of $\mathcal{H}_i$ into $\mathcal{H}'_i$ so that $f_i(t) = P(M_t)U_i-U_iP(N_t^{(i)})$ belongs to $C_n(\mathcal{N},\mathcal{K})$ and $\|f_i\| \le 2^{-i}\epsilon$. Thus $U = \sum_{i=1}^{\infty} \oplus U_i$ is an isometry of $\mathcal{H}$ into $\mathcal{H}'$ such that $f(t) = P(M_t)U-UP(N_t^{(\infty)})$ belongs to $C_n(\mathcal{N},\mathcal{K})$ and $\|f\| \le \sum_{i\ge 1}\|f_i\| \le \epsilon$.

Let P be the projection $P = UU^*$, and compute:

$$P(M_t)P-PP(M_t) = (P(M_t)U-UP(N_t^{(\infty)}))U^*-U(P(M_t)U-UP(N_t^{(\infty)}))^* \ .$$

Thus $t \to [P(M_t),P]$ is continuous, and compact of norm at most 2ϵ for all $0 \le t \le 1$. This produces an "approximate summand" of $\mathcal{M}$ unitarily equivalent to $\mathcal{N}^{(\infty)}$. If one proceeds formally, one obtains a "complement" $\mathcal{L}$ of $\mathcal{N}^{(\infty)}$ in $\mathcal{M}$. A formal computation yields

$$\mathcal{M} \sim \mathcal{L} \oplus \mathcal{N}^{(\infty)} \cong (\mathcal{L} \oplus \mathcal{N}^{(\infty)}) \oplus \mathcal{N} \sim \mathcal{M} \oplus \mathcal{N} \ .$$

To make this precise, let S be the shift on $\mathcal{H}^{(\infty)}$ given by

$$S(x_1,x_2,...) = (0,x_1,x_2,...)$$

and let J be the isometry of $\mathcal{H}$ into $\mathcal{H}^{(\infty)}$ given by $Jx = (x,0,0,...)$. Define a unitary W of $\mathcal{H}' \oplus \mathcal{H}$ onto $\mathcal{H}'$ by $[P^{\perp}+USU^* \ UJ]$. Use the facts that S commutes with $P(N_t^{(\infty)})$ and $P(N_t^{(\infty)})J = JP(N_t)$ to compute $P(M_t)W-W(P(M_t) \oplus P(N_t))$. The first matrix entry is

$$\begin{aligned} P(M_t)P^{\perp}+P(M_t)USU^*-P^{\perp}P(M_t)-USU^*P(M_t) &= \\ = [P(M_t),P^{\perp}]+(P(M_t)U-UP(N_t^{(\infty)}))SU^* & \\ +US(UP(N_t^{(\infty)})-P(M_t)U)^* & \end{aligned}$$

which is continuous in t, compact, and has norm at most 4ϵ. The second matrix entry is

$$P(M_t)UJ-UJP(N_t) = (P(M_t)U-UP(N_t^{(\infty)}))J$$

which is continuous and compact with norm at most ϵ. Hence $P(M_t)W-W(P(M_t) \oplus P(N_t))$ is continuous and compact with norm at most 5ϵ.

Reversing the roles of $\mathcal{N}$ and $\mathcal{M}$ yields another unitary W' such that $(P(M_t) \oplus P(N_t))W'-W'P(N_t)$ is continuous and compact valued and has norm at most 5ϵ. So $t \to P(M_t)WW'-WW'P(N_t)$ is continuous, compact and of norm at most 10ϵ. ∎

Combining this result with Theorem 12.8 yields:

13.10 Theorem. *Let $\mathcal{N}$ and $\mathcal{M}$ be continuous nests on a separable Hilbert space. Let θ be any order isomorphism of $\mathcal{N}$ onto $\mathcal{M}$, and let $\epsilon > 0$ be given. Then there is an invertible operator $S = U+K$ where U is unitary, K is compact, and $\|K\| < \epsilon$ such that $S\mathcal{N} = \mathcal{M}$ and $\theta_S = \theta$.*

Proof. By Theorem 13.9, there is a unitary U such that $U\mathcal{N}$ and $\mathcal{M}$ are compactly perturbed nests via the isomorphism $\theta \circ \theta_U^{-1}$. So by Theorem 12.8 and the norm estimate of Theorem 12.18, the invertible operator S exists as required. ∎

13.11 Example. It is an immediate consequence of this theorem that neither multiplicity nor measure class is preserved by similarity. For example, let $\mathcal{N}$ be the Volterra nest $\{N_t : 0 \le t \le 1\}$ of functions in $L^2(0,1)$ supported on $[0,t]$, $0 \le t \le 1$. Let $\mathcal{M}$ be the nest of Example 1.6 on $L^2(0,1)^2$ given by M_t, the subspace of functions supported on $[0,t] \times [0,1]$ for $0 \le t \le 1$. Then $\mathcal{M}$ has infinite multiplicity since $\mathcal{M}''$ has uniform infinite multiplicity, whereas $\mathcal{N}''$ is maximal abelian and so $\mathcal{N}$ has multiplicity one. By Theorem 13.10, these two nests are similar.

13.12 Example. Now let h be an order preserving homeomorphism of $[0,1]$ onto itself, and consider the automorphism of $\mathcal{N}$ given by $\theta(N_t) = N_{h(t)}$. The scalar spectral measure for $\theta(\mathcal{N})$ is equivalent to the Lebesgue-Stieltjes measure dh. By Theorem 7.16, there is a unitary operator U such that $UN_t = N_{h(t)}$ for every t in $[0,1]$ precisely when dh is mutually absolutely continuous with Lebesgue measure. However, even if dh is singular to Lebesgue measure, there is an invertible operator S so that $SN_t = N_{h(t)}$.

13.13 Example. Let V be the Volterra operator. By Theorem 5.5, $Lat(V)$ equals $\mathcal{N}$. Let S be the invertible operator guaranteed in Example 13.11 such that $S\mathcal{N} = \mathcal{M}$. Then SVS^{-1} is a compact operator such that $Lat(SVS^{-1}) = \mathcal{M}$. So, although SVS^{-1} has a maximal nest of invariant subspaces, it does not have a multiplicity free nest of invariant subspaces.

An operator with a multiplicity free nest of invariant subspaces has been called *hyperintransitive*. The example above shows that not every compact operator has this property. However:

13.14 Corollary. *Every compact operator K is similar to an operator with a multiplicity free nest of invariant subspaces.*

Proof. By Corollary 3.2, K has a maximal nest $\mathcal{N}$ of invariant subspaces. Thus all the atoms of $\mathcal{N}$ are one dimensional. Let P be the projection onto the *span* of all atoms of $\mathcal{N}$. Then $\mathcal{N}|P^{\perp}\mathcal{H}$ is a continuous nest. By Theorem 13.10, there is an invertible operator S of $P^{\perp}\mathcal{H}$ onto $L^2(0,1)$ such that $S(\mathcal{N}|P^{\perp}\mathcal{H})$ is the Volterra nest. Hence $T = P + SP^{\perp}$ is an invertible operator of $\mathcal{H}$ onto $P\mathcal{H} \oplus L^2(0,1)$ such that $T\mathcal{N}$ is multiplicity free. So TKT^{-1} is hyperintransitive.

A refinement of this is to take any $\epsilon > 0$ and choose $S = U + L$ where U is unitary, L is compact, and $\|L\| < \epsilon$. Then $T_\epsilon = P + U^*SP^{\perp}$ satisfies $\|T_\epsilon - I\| < \epsilon$ and $T_\epsilon K T_\epsilon^{-1}$ is hyperintransitive. So K is the limit of hyperintransitive operators similar to K. ∎

13.15. Now we turn to arbitrary nests. A typical nest to keep in mind is Example 2.6 of the canonical atomic nest $\mathcal{Q}$ on $\ell^2(\mathbb{Q})$. This nest is order isomorphic to the Cantor set (section 2.14) and hence has been called the Cantor nest. Let $\mathcal{N}$ be the usual continuous nest on $L^2(\mathbb{R})$ given by the subspaces N_t of functions supported on $(-\infty,t]$. The nest $\mathcal{R}$ given by $Q_t^\pm \oplus N_t$ for $t \in \mathbb{R}$ is also order isomorphic to the Cantor set. The map θ of $\mathcal{Q}$ onto $\mathcal{R}$ sending $Q_t^\pm$ to $Q_t^\pm \oplus N_t$ is a dimension preserving, order isomorphism. But $\mathcal{Q}$ and $\mathcal{R}$ are not unitarily equivalent since $\mathcal{Q}$ is atomic and $\mathcal{R}$ has a continuous part.

An invertible operator T is said to act *absolutely continuously* on a nest $\mathcal{N}$ if the spectral measure of $E_{T\mathcal{N}}$ is absolutely continuous with respect to $E_{\mathcal{N}}$ (where the order isomorphism between $\mathcal{N}$ and $T\mathcal{N}$ is taken to be θ_T). Otherwise, it is said to act *discontinuously*, or even *singularly* in the event that $E_{T\mathcal{N}}$ is singular to $E_{\mathcal{N}}$. Recall from section 8.13 that a *partition* of $\mathcal{N}$ is a possibly infinite collection of intervals $\{E_\alpha\}$ of $\mathcal{N}$ and that $\sum E_\alpha = I$.

13.16 Lemma. *Let $\mathcal{N}$ be a nest, and let T be an invertible operator. If $E = P(N)-P(N')$ is an interval of $\mathcal{N}$, let $\hat{E} = P(TN)-P(TN')$. Then T acts absolutely continuously on $\mathcal{N}$ if and only if for every partition $\{E_\alpha\}$, the corresponding intervals $\{\hat{E}_\alpha\}$ form a partition of $T\mathcal{N}$.*

Proof. Suppose that T acts absolutely continuously. Fix a homomorphism of $\mathcal{N}$ onto a subset ω of $[0,1]$, and let $E_{\mathcal{N}}$ be the associated spectral measure on $[0,1]$. For each α, there are $N'_\alpha < N_\alpha$ in $\mathcal{N}$ and $a_\alpha < b_\alpha$ in $[0,1]$ such that

$$E_\alpha = P(N_\alpha)-P(N'_\alpha) = E_{\mathcal{N}}(a_\alpha,b_\alpha] \ .$$

Also,

$$\hat{E}_\alpha = P(TN_\alpha)-P(N'_\alpha) = E_{T\mathcal{N}}(a_\alpha,b_\alpha] \ .$$

Let $\delta = [0,1]\backslash\bigcup_\alpha(a_\alpha,b_\alpha]$. Since $\{E_\alpha\}$ is a partition, $\sum E_\alpha = I$ and thus $E_{\mathcal{N}}(\delta) = 0$. By absolute continuity, $E_{T\mathcal{N}}(\delta) = 0$. So

$$\sum \hat{E}_\alpha = E_{T\mathcal{N}}([0,1]\backslash\delta) = I \ .$$

Conversely, suppose that T preserves partitions. Let δ be a closed set such that $E_{\mathcal{N}}(\delta) = 0$. The complement $[0,1]\backslash\delta$ is a countable union of intervals (a_n,b_n). Each $E_n = E_{\mathcal{N}}(a_n,b_n)$ is an interval of $\mathcal{N}$. Also, $\{E_n\}$ is a partition since $\sum E_n = I - E_{\mathcal{N}}(\delta) = I$. Since T preserves partitions,

$$E_{T\mathcal{N}}(\delta) = I - \sum E_{T\mathcal{N}}(a_n,b_n) = I - \sum \hat{E}_n = 0 \ .$$

The spectral measures $E_{\mathcal{N}}$ and $E_{T\mathcal{N}}$ are regular Borel measures, thus $E_{\mathcal{N}}(\delta) = 0$ implies $E_{T\mathcal{N}}(\delta) = 0$ for every Borel set δ. Thus T acts absolutely continuously. ∎

13.17 Lemma. *Let $\mathcal{N}$ be an uncountable nest order isomorphic to a subset ω of $[0,1]$. Then $\mathcal{N}$ has a partition $E_1, E_2, \ldots$ into infinite dimensional intervals $E_n = E_{\mathcal{N}}(a_n, b_n)$ such that $\omega \setminus \bigcup_{n=1}^{\infty} (a_n, b_n)$ is uncountable.*

Proof. First, it will be shown that any subset of K of $[0,1]$ homeomorphic to the Cantor set contains a Cantor set C such that every component of $[0,1] \setminus C$ has uncountable intersection with K. There is no loss of generality in letting K be the usual Cantor set. Let f be the Cantor function - a continuous non-decreasing function of K onto $[0,1]$ which is one to one except on a countable set. Let K' be a Cantor subset of $[1/3, 2/3]$ which does not meet any of the countable many multiple image points. Then $C = f^{-1}(K')$ is a Cantor subset of K. The complement $[0,1] \setminus K'$ is a disjoint union $\bigcup_{n=1}^{\infty} J_n$ of open intervals. The complement of C in $[0,1]$ is the disjoint union of intervals I_n such that $I_n \cap K = f^{-1}(J_n)$ is uncountable.

Let μ be a scalar spectral measure for $\mathcal{N}$ on ω. Since ω is uncountable, it contains a Cantor set K such that $\mu(K) = 0$. (For example, let L be any Cantor subset of ω disjoint from the (countable) set of atoms of μ. If $\mu(L) \neq 0$, then $\mu \,|L$ is a continuous measure. The usual "middle thirds" argument using the function $g(x) = \mu([0,x] \cap L)/\mu(L)$ yields a Cantor subset of L of μ-measure zero.) Inside K, choose a Cantor set C as in the first paragraph. Let $[0,1] \setminus C$ be $\bigcup_{n=1} I_n$ where I_n are the components of the complement. Then $E_n = E_{\mathcal{N}}(I_n)$ forms a partition of $\mathcal{N}$ since $E_{\mathcal{N}}(C) = 0$, and since $I_n \cap \omega$ is uncountable, each E_n is infinite dimensional. ■

13.18 Lemma. *Let $\mathcal{N}$ be an uncountable nest. Given $\epsilon > 0$, there is a positive, invertible operator T such that $T - I$ is compact, $\|T - I\| < \epsilon$, and T acts discontinuously on $\mathcal{N}$.*

Proof. Let E_n be the partition given by Lemma 13.17, where $E_n = E_{\mathcal{N}}(a_n, b_n)$ and $C = \omega \setminus \bigcup_{n=1} (a_n, b_n)$ is uncountable. By Lemma 13.16 it suffices to obtain an invertible operator T which does not preserve the partition E_n. We construct a continuous nest $\mathcal{M}$ which also has E_n as a partition. To this end, let m be Lebesgue measure on $\bigcup_{n=1} (a_n, b_n)$. Let $\mathcal{M}'$ be the continuous nest in $L^2(m)$ given by subspaces M'_t of functions supported on $[0,t]$. Let U be a unitary of $L^2(m)$ onto $\mathcal{H}$ which takes $M'_{b_n} \ominus M'_{a_n} = L^2(a_n, b_n)$ onto $E_n \mathcal{H}$. Then $\mathcal{M} = U\mathcal{M}'$ is a continuous nest, and E_n is a partition of $\mathcal{M}$.

Since C is uncountable, it supports a non-zero non-atomic measure ν. Let $\mathcal{Q}$ be the continuous nest on $\mathcal{H} \oplus L^2(\nu)$ given by

$$Q_t = M_t \oplus \{f \in L^2(\nu):\text{supp}(f) \subseteq [0,t]\} \ .$$

The map taking M_t to Q_t is an order isomorphism. By Theorem 13.10, there is an invertible operator $S = U+K$ where U is unitary, K is compact and $\|K\| < \epsilon/3$ such that $SM_t = Q_t$ for all t in ω. Since $E_Q(C) = L^2(\mu)$, S does not preserve the partition $\{E_n\}$. Thus the same applies to $U^*S = I+U^*K$. Let

$$T = |I+U^*K| = ((I+U^*K)^*(I+U^*K))^{1/2} \ .$$

Since $T = W(I+U^*K)$ where $I+U^*K = W^*T$ is the polar decomposition, T also acts discontinuously. Clearly, T is positive and $T-I$ is compact. Furthermore

$$\|T^2-I\| = \|K^*U+U^*K+K^*K\| < \epsilon \ .$$

So $\|T-I\| < \epsilon\|T+I\|^{-1} < \epsilon$. ∎

Say that an interval E of $\mathcal{N}$ has a *non-atomic part* if E is greater than the sum of the atoms it dominates. The interval E will be called *countable* or *uncountable* depending on the cardinality of the restriction $\mathcal{N}|E\mathcal{H}$. Lemma 13.18 will be used repeatedly to ensure that a nest similar to $\mathcal{N}$ has a maximal non-atomic part.

13.19 Lemma. *Let $\mathcal{N}$ be a nest, and let $\epsilon > 0$ be given. There is a positive operator such that $T-I$ is compact and $\|T-I\| < \epsilon$ such that $T\mathcal{N}$ has non-atomic part on every uncountable interval.*

Proof. Choose $\epsilon_k > 0$ such that $\prod_{k=1}^{\infty}(1+\epsilon_k) < 1+\epsilon$. Fix a subset ω of $[0,1]$ order isomorphic to $\mathcal{N}$, and let $E_{\mathcal{N}}$ be the spectral measure on ω. Positive operators S_n will be constructed so that S_n-I is compact and $\|S_n-I\| < \epsilon_n$. Define $T_n = S_nS_{n-1}...S_1$, and let $\mathcal{N}_n = T_n\mathcal{N}$. If T_{n-1} has been defined, divide ω into 2^n diadic intervals $(k2^{-n},(k+1)2^{-n}]$. Let E_k be the corresponding intervals of $\mathcal{N}_{n-1}$. If E_k is countable, or if E_k has a non-atomic part, define $S_n|E_k = I|E_k$. If E_k is uncountable and atomic, use Lemma 13.18 define $S_n|E_k = E_kS_nE_k$ so that $E_kS_nE_k$ acts discontinuously on $\mathcal{N}_{n-1}|E_k\mathcal{H}$. Thus $S_n\mathcal{N}_{n-1}|E_k\mathcal{H}$ must have a non-atomic part. Thus $S_n = \sum_{k=1}^{2^n} S_nE_k$ is defined. The nest $\mathcal{N}_n = T_n\mathcal{N} = S_n\mathcal{N}_{n-1}$ has the property that every diadic interval of length 2^{-n} is countable, or has non-atomic part.

Write $S_n = I+K_n$. One readily verifies by induction that

$$\|(\prod_{j=1}^{n} I+K_j)-I\| < (\prod_{j=1}^{n} 1+\epsilon_j)-1 < \epsilon$$

and

$$\|\prod_{j=1}^{m} I+K_j-\prod_{j=1}^{n} I+K_j\| < (1+\epsilon)((\prod_{j=n+1}^{\infty} 1+\epsilon_j)-1) \ .$$

So $T = \lim_{t\to\infty} T_n$ converges on norm, and $T-I$ is compact with $\|T-I\| < \epsilon$. Let E be an interval of $T\mathcal{N}$. If E is uncountable, then E contains an uncountable diadic interval E_0 corresponding to $(k2^{-n},(k+1)2^{-n}]$. This interval has a non-atomic part in $\mathcal{N}_n$. Each operator S_k for $k > n$ has been constructed so that it is the identity on the non-atomic part of $\mathcal{N}_{k-1}$. So E_0 still has a non-atomic part in the limit nest $T\mathcal{N}$. ■

Let P be the sum of all atoms of $T\mathcal{N}$. Then $T\mathcal{N}|P^{\perp}\mathcal{H}$ is a non-atomic (i.e. continuous) nest. The supportω_0 of this subnest intersects every uncountable interval. Since ω_0 is closed, its complement consists of countably many open intervals, each of which must be countable. Indeed, ω_0 is the set of points x in ω such that every neighbourhod of x meets ω in an uncountable set. Let ω' denote the set of all limit points of ω. For each ordinal α, define $\omega^{(\alpha)} = \omega^{(\beta)\prime}$ if $\alpha = \beta+1$ and $\omega^{(\alpha)} = \bigcap_{\beta<\alpha} \omega^{(\beta)}$ if α is a limit ordinal. Then $\omega_0 = \bigcap \omega^{(\alpha)}$ is the maximal perfect subset of ω.

13.20 Similarity Theorem. *Let $\mathcal{N}$ and $\mathcal{M}$ be nests on a separable Hilbert space. Then $\mathcal{N}$ and $\mathcal{M}$ are similar if and only if there is an dimension preserving order isomorphism of $\mathcal{N}$ onto $\mathcal{M}$. Given $\epsilon > 0$ and any dimension preserving order isomorphism θ, there is a invertible operator $S = U+K$ with U unitary, K compact, and $\|K\| < \epsilon$ such that $S\mathcal{N} = \mathcal{M}$ and $\theta_S = \theta$.*

Proof. Consider the non-trivial direction. Let $\epsilon > 0$ and θ be given. Let ω be a subset of $[0,1]$ order isomorphic to $\mathcal{N}$, and to $\mathcal{M}$ also via θ. Using Lemma 13.19, one obtains operators X and Y such that $X-I$ and $Y-I$ are compact, and have norms at most $\epsilon/4$, and such that $\mathcal{N}' = X\mathcal{N}$ and $\mathcal{M}' = Y\mathcal{M}$ both have non-atomic parts with supportω_0, the maximal perfect subset of ω. The map $\theta' = \theta_Y \theta \theta_X^{-1}$ is an order preserving isomorphism of $\mathcal{N}'$ onto $\mathcal{M}'$. Let P and Q be the projections onto the atomic parts of $\mathcal{N}'$ and $\mathcal{M}'$. Since θ' preserves dimension, there is a unitary operator U_a from $P\mathcal{H}$ onto $Q\mathcal{H}$ which carries each atom A of $\mathcal{N}$ onto $\theta'(A)$. The restriction of θ' to $\mathcal{N}'|P^{\perp}\mathcal{H}$ is an order isomorphism onto $\mathcal{M}'|Q^{\perp}\mathcal{H}$. By Theorem 13.10, there is an invertible operator $S_0 = U_c+K_c$ where U_c is unitary, K_c is compact, and $\|K_c\| < \epsilon/4$ such that S_c takes $\mathcal{N}'|P^{\perp}\mathcal{H}$ onto $\mathcal{M}'|Q^{\perp}\mathcal{H}$ and implements θ'. Let $S' = U_a \oplus S_c = (U_a \oplus U_c)+(0 \oplus K_c)$. It is clear that S' takes $\mathcal{N}'$ onto $\mathcal{M}'$ and implements θ'. Thus $S = Y^{-1}S'X$ is the required similarity of $\mathcal{N}$ onto $\mathcal{M}$ with $\theta_S = \theta$. Clearly, $S-I$ is compact of norm at most

$$(1+\epsilon/4)^2(1-\epsilon/4)^{-1}-1 < \epsilon$$

provided $\epsilon < .05$. ■

13.21 Corollary. *For two nests $\mathcal{N}$ and $\mathcal{M}$ on separable Hilbert spaces, the following are equivalent.*

1) *There is an order isomorphism of $\mathcal{N}$ onto $\mathcal{M}$ which preserves dimension.*

2) *$\mathcal{N}$ and $\mathcal{M}$ are similar.*

3) *$\mathcal{N}$ and $\mathcal{M}$ are approximately unitarily equivalent.*

4) *$Q\ \mathcal{T}(\mathcal{N})$ and $Q\ \mathcal{T}(\mathcal{M})$ are unitarily equivalent.*

2′) *There is a unitary operator U and a compact operator K such that $(U+K)\mathcal{N} = \mathcal{M}$.*

3′) *There is an approximate unitary equivalence $\{U_n\}$ of $\mathcal{N}$ onto $\mathcal{M}$ such that $U_n - U_0$ is compact for all $n \geq 1$.*

Proof. 1) implies 2′) by Theorem 13.20. 2′) easily implies 4). By Theorem 12.8, 4) implies 3′) which in turn implies 3). Theorem 12.18 shows 3) implies 2) which readily implies 1). ■

13.22 Example. Consider the nests $\mathcal{Q}$ and $\mathcal{R}$ mentioned in section 13.15. The natural isomorphism θ of $\mathcal{Q}$ onto $\mathcal{R}$ preserves dimension since all atoms are one dimensional. Thus, by Theorem 13.20, $\mathcal{Q}$ and $\mathcal{R}$ are similar. This is in spite of the fact that $\mathcal{Q}$ is atomic and $\mathcal{R}$ has a maximal continuous part. This is what Lemma 13.19 does to $\mathcal{Q}$ in a much less precise way.

Recall from section 8.13 the ideal $\mathcal{R}^\infty(\mathcal{N})$. It consists of those operators T in $\mathcal{T}(\mathcal{N})$ such that for every $\epsilon > 0$, there is a countable partition $\mathcal{E}$ of $\mathcal{N}$ such that $\|\Delta_{\mathcal{E}}(T)\| < \epsilon$. It has a striking connection with absolute continuity.

13.23 Theorem. *Let $\mathcal{N}$ be a nest on a separable space, and let T be an invertible operator. The following are equivalent:*

1) *T acts absolutely continuously on $\mathcal{N}$.*

2) *$\mathcal{R}^\infty(T\mathcal{N})$ contains $T\mathcal{R}^\infty(\mathcal{N})T^{-1}$.*

3) *$\mathcal{R}^\infty(\mathcal{N})$ does not contain a non-zero indempotent P such that TPT^{-1} is self-adjoint.*

Proof. 1) $\Rightarrow$ 2). Let A belong to $\mathcal{R}^\infty(\mathcal{N})$. Let $\epsilon > 0$ be given, and let $\mathcal{E}$ be a partition such that $\|\Delta_{\mathcal{E}}(A)\| < \epsilon$. If T acts absolutely continuously, then by Lemma 13.16 the image $\hat{\mathcal{E}}$ is a partition of $T\mathcal{N}$. Let E be an interval of $\mathcal{E}$, and let $\hat{E}$ be the corresponding interval in $T\mathcal{N}$. Write $E = P(N_1) - P(N_0)$. Then $T^{-1}\hat{E} = P(N_1)T^{-1}\hat{E}$ and $\hat{E}T = \hat{E}TP(N_0)^\perp$. So

$$\hat{E}TAT^{-1}\hat{E} = \hat{E}TP(N_0)^{\perp}AP(N_1)T^{-1}\hat{E} = \hat{E}TEAET^{-1}\hat{E} \ .$$

Thus $\|\hat{E}TAT^{-1}\hat{E}\| \le \|T\| \, \|T^{-1}\|\epsilon$, and hence TAT^{-1} belongs to $\mathcal{R}^\infty(T\mathcal{N})$.

2) ⇒ 3). By Proposition 8.13, $\mathcal{R}^\infty(T\mathcal{N}) \cap \mathcal{D}(T\mathcal{N}) = \{0\}$. So $\mathcal{R}^\infty(T\mathcal{N})$ contains no non-zero self-adjoint operator, and in particular cannot contain TPT^{-1} if this is a projection.

3) ⇒ 1). If T does not act absolutely continuously, then Lemma 13.16 provides a partition $\mathcal{E}$ so that the corresponding intervals $\hat{\mathcal{E}}$ do not partition $T\mathcal{N}$. Let $Q = (\sum \hat{E}:\hat{E} \in \hat{\mathcal{E}})^{\perp}$ and let $P = T^{-1}QT$. For each $E = P(N_1) - P(N_0)$ in $\mathcal{E}$,

$$EPE = ET^{-1}P(TN_0)^{\perp}QP(TN_1)TE = ET^{-1}\hat{E}Q\hat{E}TE = 0 \ .$$

Thus P belongs to $\mathcal{R}^\infty(\mathcal{N})$ and $TPT^{-1} = Q$ is self-adjoint. ∎

13.24 Corollary. *Let $\mathcal{N}$ be a nest on a separable space. The following are equivalent:*

1) *$\mathcal{N}$ is uncountable.*
2) *There is an invertible operator T which acts discontinuously on $\mathcal{N}$.*
3) *$\mathcal{R}^\infty(\mathcal{N})$ contains a non-zero idempotent.*

Proof. 1) implies 2) is Lemma 13.18. 2) implies 3) follows from Theorem 13.23. If P is a non-zero indempotent, then $T = [P^*P+(I-P^*)(I-P)]^{1/2}$ is invertible and TPT^{-1} is self-adjoint. So 3) implies 2), again by Theorem 13.23. Finally, if $\mathcal{N}$ is countable and T is invertible, then $T\mathcal{N}$ is countable. So both nests are atomic and θ_T implements a bijective correspondence between the atoms of $\mathcal{N}$ and the atoms of $T\mathcal{N}$. So the spectral measures are atomic with masses corresponding exactly, and thus T acts absolutely continuously on $\mathcal{N}$. So 2) implies 1). ∎

Notes and Remarks.

The main results of this chapter are due to Andersen, Larson, and Davidson. Theorem 13.9 is from Andersen [1]. The proof given here is due to Davidson [8]. A third proof was given by Arveson [10] generalizing his approach [9] to Voiculescu's Theorem [2]. The proof of Lemma 13.7 was provided by G. Pedersen. Larson [4] proved that any two continuous nests are similar. The generalization, Theorem 13.10, is due to Davidson [6]. The examples and Corollary 13.14 are due to Larson. Larson [6] also proved Lemmas 13.16 and 13.18. The proofs given here are from Davidson [6], where Theorem 13.20 is proven. Larson's Theorem solved an old problem due to Ringrose by showing that neither multiplicity nor measure class is preserved by similarity of nests. Andersen [2] showed that a weaker version follows immediately from his results and those of Lance [2]. Theorem 13.23 is also due to Larson.

Exercises.

13.1 Let A and B be self-adjoint operators on a separable Hilbert space. Theorem 7.15 shows that A and B are unitarily equivalent if and only if $\sigma(A) = \sigma(B)$ and they have the same measure class and multiplicity function.

(i) Show that if A and B are similar, then they are unitarily equivalent. *Hint.* If $AS = SB$ and $S = UP$ is the polar decomposition, then $(U^*AU)P^k = P^kB$ for all $k \geq 1$.

(ii) Use the Weyl-von Neumann Theorem (Exercise 7.1) to show that A and B are approximately unitarily equivalent (i.e. there are a U_n unitary such that $\|A - U_n^*BU_n\|$ tends to zero) if and only if $\sigma(A) = \sigma(B)$ and the multiplicities of isolated eigenvalues are equal.

(iii) Note the parallels and differences between the three equivalences (unitary, similarity, and approximate unitary) for self-adjoint operators and nests.

13.2 Prove the measure theoretic fact needed in Lemma 13.2.

Hint: Let $\nu_0, \nu_1, \ldots$ be mutually singular, non-atomic probability measures with support $[0,1]$. Let $\nu = \sum_{n=0}^{\infty} 2^{-n-1}\nu_n$. Let $h_\nu(x) = \nu[0,x]$ and $h_\mu(x) = \mu[0,x]$. Let E_k be disjoint Borel sets with union $[0,1]$ such that $\nu_n(E_n) = 1$. Define $F_n = h_\mu^{-1} \circ h_\nu(E_i)$. Let $A_n = F_n \cap J_n$ for $n \geq 1$ and $A_0 = J_0 \backslash \bigcup_{n=1}^{\infty} A_n$. Prove that A_n satisfy the criterion.

13.3 (i) Let $\mathcal{G}$ be a finite group of operators under multiplication. Let $T = (\sum_{G \in \mathcal{G}} G^*G)^{1/2}$. Show that $T^2G = (G^{-1})^*T^2$ for all G in $\mathcal{G}$. Hence deduce that TGT^{-1} is unitary for all G in $\mathcal{G}$.

(ii) Let $\mathcal{A}$ be a unital, norm closed algebra of operators. Suppose that P, A, B belong to $\mathcal{A}$ such that $P^2 = P$, $A = PA(I-P)$, $B = (I-P)AP$, and $(A+B)^2 - I$ is compact. Show that there are $A' = PA'(1-P)$ and $B' = (1-P)B'P$ in $\mathcal{A}$ so that $A'B' = P'$ and $B'A' = Q'$ are idempotent, and $P - P'$ and $I - P - Q'$ are finite rank.

(iii) Let $\mathcal{A}$ be as in (ii). Suppose that $\mathcal{A}/\mathcal{A} \cap K$ contains elements α, β generating a group isomorphic to S_3 by satisfying $\alpha^2 = 1$, $\beta^3 = 1$, and $\alpha\beta = \beta^2\alpha$. Use the Riesz functional calculus to obtain a preimage B of β in $\mathcal{A}$ of the form $B = P_0 + \omega P_1 + \omega^2 P_2$ where $\omega = e^{2\pi i/3}$ and P_0, P_1, P_2 are commuting idempotents summing to I. Then obtain a preimage A' of α in $\mathcal{A}$ of the form $A' = P_0A'P_0 + P_1A'P_2 + P_2A'P_1$. Use (ii) to modify A' so that $P_1A'P_2A'P_1 = P'_1$ and $P_2A'P_1A'P_2 = P'_2$ are idempotents differing from P_1 and P_2 by a finite rank operator. Finally, adjust A' to A by another compact operator in $\mathcal{A} \cap \mathcal{K}$ so that $A^2 = I = B^3$ and $AB = B^2A$.

13.4 (Larson) Let $\mathcal{N}$ be the Volterra nest and let $\mathcal{M}$ be a continuous nest of uniform multiplicity 3. Then $\mathcal{M}$ contains a finite unitary group isomorphic to S_3. Use Andersen's Theorem to obtain a copy of S_3 in $Q\mathcal{T}(\mathcal{N})/\mathcal{K}$. Then using Exercise 13.3, obtain a copy of S_3 in $\mathcal{T}(\mathcal{N})$. Use Exercise 13.3 (i) to obtain a positive operator T that unitarizes this group. Show that the nest $T\mathcal{N}$ is not multiplicity free.

13.5 (Larson) If $\mathcal{M}$ is a continuous nest of multiplicity 2, then $\mathcal{M}'$ contains a partial isometry U so that UU^*-U^*U is invertible. Use Andersen's Theorem to show that if $\mathcal{N}$ is any continuous nest, then $\mathcal{T}(\mathcal{N})/\mathcal{T}(\mathcal{N}) \cap \mathcal{K}$ contains elements α and β so that $\alpha\beta-\beta\alpha = 1$. Use the fact that $\mathcal{T}(\mathcal{N}) \cap \mathcal{K}$ belongs to the radical to lift this to $\mathcal{T}(\mathcal{N})$. Conclude that the commutator ideal of $\mathcal{T}(\mathcal{N})$ (i.e. the ideal generated by all operators $AB-BA$) is all of $\mathcal{T}(\mathcal{N})$. Deduce this directly from the first sentence and Theorem 13.10.

13.6 (Larson)

(a) Show that $\mathcal{R}^\infty(\mathcal{N})$ is contained in the strong operator closure of the radical of $\mathcal{T}(\mathcal{N})$.

(b) If $\mathcal{N}$ and $\mathcal{M}$ are nests and $\mathcal{N}$ is contained in $\mathcal{M}$, show that $\mathcal{R}^\infty(\mathcal{N})$ is contained in $\mathcal{R}^\infty(\mathcal{M})$.

13.7* (i) Let $\mathcal{N}$ be the Volterra net. Find an *explicit* operator T so that $T\mathcal{N}$ is not multiplicity free.

(ii) Let $\mathcal{Q}$ be the Cantor nest. Find an *explicit* operator T so that $T\mathcal{Q}$ is not purely atomic.

III. ADDITIONAL TOPICS

14. Factorization

In this chapter, we consider the following factorization questions and a number of variations.

Q1. Given a positive invertible operator Q and a nest $\mathcal{N}$, when is there an invertible element A of $\mathcal{T}(\mathcal{N})$ so that $Q = A^*A$?

Q2. Given an operator T in $\mathcal{T}(\mathcal{N})$ which is invertible in $B(\mathcal{H})$, when can T be factored as $T = UA$ where U is a unitary in $\mathcal{T}(\mathcal{N})$ and A is invertible in $\mathcal{T}(\mathcal{N})$?

14.1 Lemma. *Let T be an invertible operator, and let $\mathcal{N}$ be a nest. The following are equivalent:*

1) *there is an A in $\mathcal{T}(\mathcal{N})^{-1}$ such that $A^*A = T^*T$,*

2) *there is an A in $\mathcal{T}(\mathcal{N})^{-1}$ and a unitary operator U such that $T = UA$,*

3) *there is a unitary operator U so that $TN = UN$ for all N in $\mathcal{N}$,*

4) *T preserves the multiplicity and measure class of $\mathcal{N}$.*

Proof. Suppose 1) holds. Then $|T| = |A|$ and hence by the polar decompositions $T = V|T|$ and $A = W|T|$, one has $T = UA$ where $U = VW^*$. Given 2), one has AN contained in N and $A^{-1}N$ contained in N, whence $AN = N$ for all N in $\mathcal{N}$. Hence $TN = U(AN) = UN$ for all N in $\mathcal{N}$. Given 3), $A = U^*T$ carries N onto N for every N in $\mathcal{N}$. Hence A and A^{-1} belong to $\mathcal{T}(\mathcal{N})$. Clearly $A^*A = T^*T$ so 1) holds. The equivalence of 3) and 4) is Corollary 7.16. ∎

14.2 Theorem. *Let $\mathcal{N}$ be a nest on a separable Hilbert space.*

1) *If $\mathcal{N}$ is countable, then every positive invertible operator Q factors as $Q = A^*A$ for some A in $\mathcal{T}(\mathcal{N})^{-1}$.*

2) *If $\mathcal{N}$ is uncountable and $\epsilon > 0$, then there is a positive operator $Q = I+K$ where K is compact and $\|K\| < \epsilon$ such that Q does not factor as A^*A for any A in $\mathcal{T}(\mathcal{N})^{-1}$.*

Proof. Let $\mathcal{N}$ be countable, and consider the nest $Q^{1/2}\mathcal{N}$ and the order isomorphism $\theta = \theta_{Q^{1/2}}$. Countable nests are atomic, and θ preserves dimension. So θ preserves multiplicity and measure class. By Lemma 14.1, Q has the desired factorization.

If $\mathcal{N}$ is uncountable, then by Lemma 13.18, there is a positive operator $T = I+K$ with K compact and $\|K\| < \epsilon/3$ so that T fails to act absolutely continuously on $\mathcal{N}$. By Lemma 14.1, $Q = T^*T$ fails to have the desired factorization. ∎

14.3 Corollary. *Let $\mathcal{N}$ be a countable nest and suppose that T is an invertible operator in $\mathcal{T}(\mathcal{N})$. Then T factors as $T = UA$ where U is a unitary in $\mathcal{T}(\mathcal{N})$ and A belongs to $\mathcal{T}(\mathcal{N})^{-1}$.*

Proof. By Theorem 14.2, there is an A in $\mathcal{T}(\mathcal{N})^{-1}$ such that $T^*T = A^*A$. Let $U = TA^{-1}$. Then for each N in $\mathcal{N}$,

$$UN = T(A^{-1}N) = TN \subseteq N \ .$$

So U is a unitary in $\mathcal{T}(\mathcal{N})$ as desired. ∎

However, one can always get a "weak factorization".

14.4 Lemma. *Let $\mathcal{N}$ be a nest on a separable space $\mathcal{H}$, and let T be an invertible operator in $\mathcal{B}(\mathcal{H})$. Then T factors as $T = US$ where U is unitary and S belongs to $\mathcal{T}(\mathcal{N})$.*

Note: S is invertible, but in general, S^{-1} does not belong to $\mathcal{T}(\mathcal{N})$.

Proof. The countable case follows from Theorem 14.2 and the equivalence of 1) and 2) in Lemma 14.1. Assume that $\mathcal{N}$ is uncountable, and define

$$N_{-\infty} = \bigvee\{N \in \mathcal{N} : \mathcal{N} \wedge N \text{ is countable}\}$$

$$N_{+\infty} = \bigwedge\{N \in \mathcal{N} : \mathcal{N} \vee N \text{ is countable}\} \ .$$

The sets $\{N \in \mathcal{N} : N \leq N_{-\infty}\}$ and $\{N \in \mathcal{N} : N \geq N_{+\infty}\}$ are countable and the intervals $[N_{-\infty}, N]$ and $[N, N_{+\infty}]$ are uncountable whenever $N_{-\infty} < N < N_{+\infty}$. Hence one may choose a sequence N_n in $\mathcal{N}$, $n \in \mathbb{Z}$ such that $N_n < N_{n+1}$ and $N_{n+1} \ominus N_n$ is infinite dimensional for all n in $\mathbb{Z}$, and $\bigwedge_n N_n = N_{-\infty}$ and $\bigvee_n N_n = N_{+\infty}$. Let $\mathcal{N}_0$ be the countable nest given by $\{N \in \mathcal{N} : N \leq N_{-\infty}\} \cup \{N_n, n \in \mathbb{Z}\} \cup \{N \in \mathcal{N} : N \geq N_{+\infty}\}$.

Let T be a given invertible operator. By the countable case, there is a unitary W such that W^*T belongs to $\mathcal{T}(\mathcal{N}_0)^{-1}$. Let V be a unitary operator which is the identity on $N_{-\infty} \oplus N_{+\infty}^{\perp}$ and carries $N_{n+1} \ominus N_n$ onto $N_n \ominus N_{n-1}$ for all n in $\mathbb{Z}$. Then $A = VW^*T$ takes N_{n+1} into N_n for every N in $\mathbb{Z}$. It also takes N onto itself if $N \leq N_{-\infty}$ or $N \geq N_{+\infty}$. So A belongs to $\mathcal{T}(\mathcal{N})$. Thus $T = (WV^*)A$ is the desired factorization. ∎

The corresponding version for positive operators is immediate, and leads to a complement of Corollary 14.3.

14.5 Corollary. *Let $\mathcal{N}$ be a nest on a separable space. For every positive invertible operator Q, there is an operator T in $\mathcal{T}(\mathcal{N})$ so that $Q = T^*T$.*

Proof. By Lemma 14.4, factor $Q^{1/2} = UT$ where T belongs to $\mathcal{T}(\mathcal{N})$ and U is unitary. ∎

14.6 Corollary. *Let $\mathcal{N}$ be an uncountable nest on a separable Hilbert space. Then there is an invertible operator T in $\mathcal{T}(\mathcal{N})$ which does not factor as $T = UA$ for any A in $\mathcal{T}(\mathcal{N})^{-1}$ and U unitary.*

Proof. Let Q be a positive, invertible operator provided by Theorem 14.2 which does not factor as A^*A for any A in $\mathcal{T}(\mathcal{N})^{-1}$. Let T be an operator in $\mathcal{T}(\mathcal{N})$ such that $T^*T = Q$ as provided by Corollary 14.5. The equivalence of 1) and 2) of Lemma 14.1 shows that T fails to have the desired factorization. ∎

Now we consider the class of invertible, positive operators of the form $I+A$ where A belongs to the Macaev ideal C_ω. It will be shown that these operators have a good factorization with respect to any nest. The key is the boundedness of triangular truncation.

14.7 Lemma. *Let $\mathcal{N}$ be a nest, and let A belong to C_ω with $\|A\|_\omega < 1$. Then $I+A$ has a unique factorization of the form*

$$I+A = (I+K_-)(I+D)(I+K_+)$$

where K_+ belongs to $\mathcal{K} \cap \operatorname{rad} \mathcal{T}(\mathcal{N})$, K_- belongs to $\mathcal{K} \cap \operatorname{rad} \mathcal{T}(\mathcal{N}^\perp)$, and D belongs to $\mathcal{K} \cap \mathcal{D}(\mathcal{N})$. If A is self-adjoint, then $K_- = K_+^*$ and $D = D^*$.

Proof. First we prove uniqueness without any condition on $\|A\|_\omega$. Suppose $(I+L_-)(I+E)(I+L_+)$ is another factorization. Then

$$(I+L_-)^{-1}-I = \sum_{n\geq 1} L_-^n$$

belongs to $\operatorname{rad} \mathcal{T}(\mathcal{N}^\perp)$ and $(I+K_+)^{-1}-I$ belongs to $\operatorname{rad} \mathcal{T}(\mathcal{N})$. Hence

$$(I+L_-)^{-1}(I+K_-) = (I+E)(I+L_+)(I+K_+)^{-1}(I+D)^{-1} .$$

The left hand side belongs to $I+\operatorname{rad} \mathcal{T}(\mathcal{N}^\perp)$ whereas the right hand side belongs to $\mathcal{T}(\mathcal{N})$. Hence both are equal to I. So $L_- = K_-$, and likewise $L_+ = K_+$. Hence $E = D$ and uniqueness is obtained. If $A = A^*$, then

$$I+A = (I+K_+^*)(I+D^*)(I+K_-^*)$$

and uniqueness implies that $K_- = K_+^*$ and $D = D^*$.

So now consider existence. It is enough to obtain a factorization $I+A = (I+B_-)(I+B_+)$ where B_- belongs to $\mathcal{K} \cap rad\, \mathcal{T}(\mathcal{N}^\perp)$ and B_+ belongs to $\mathcal{K} \cap \mathcal{T}(\mathcal{N})$. Then if $D = \Delta(B_+)$, one sets $I+K_+ = (I+D)^{-1}(I+B_+)$. Since Δ is a homomorphism on $\mathcal{T}(\mathcal{N})$, one obtains $\Delta(K_+) = 0$ and hence by Corollary 6.8, K_+ belongs to $\mathcal{K} \cap rad\, \mathcal{T}(\mathcal{N})$. Recall Theorem 4.16 which shows that the triangular projection $\mathcal{P} = \mathcal{U}_\mathcal{N}$ maps $\mathcal{C}_\omega$ into $\mathcal{K} \cap \mathcal{T}(\mathcal{N})$ and has norm one. Likewise, $\mathcal{P}^\perp = I - \mathcal{P} = \mathcal{L}_{\mathcal{N}^\perp}$ maps $\mathcal{C}_\omega$ into $\mathcal{K} \cap rad\, \mathcal{T}(\mathcal{N}^\perp)$ and has norm one.

Suppose that the factorization $I+A = (I+B_-)(I+B_+)$ exists and define $C_\pm$ by $I+C_+ = (I+B_+)^{-1}$ and $I+C_- = (I+B_-)^{-1}$ and note that C_+ belongs to $\mathcal{K} \cap \mathcal{T}(\mathcal{N})$ and C_- belongs to $\mathcal{K} \cap rad\, \mathcal{T}(\mathcal{N}^\perp)$. One obtains

$$I+B_+ = (I+C_-)(I+A) = I+C_-+A+C_-A \ .$$

Cancelling and applying $\mathcal{P}^\perp$ to both sides yields

$$0 = C_- + \mathcal{P}^\perp A + \mathcal{P}^\perp(C_-A) \ .$$

Let $\mathcal{R}_A$ denote the operator taking $\mathcal{B}(\mathcal{H})$ into $\mathcal{C}_\omega$ by $\mathcal{R}_A(X) = XA$. This has norm $\|\mathcal{R}_A\| = \|A\|_\omega < 1$. One obtains

$$(I+\mathcal{P}^\perp \mathcal{R}_A)(C_-) = -\mathcal{P}^\perp(A) \ .$$

Since $\mathcal{P}^\perp \mathcal{R}_A$ maps $\mathcal{B}(\mathcal{H})$ into itself (in fact into $\mathcal{K} \cap rad\, \mathcal{T}(\mathcal{N}^\perp)$) and $\|\mathcal{P}^\perp \mathcal{R}_A\| \le \|A\|_\omega < 1$, $I+\mathcal{P}^\perp\mathcal{R}_A$ is invertible. Hence

$$C_- = -(I+\mathcal{P}^\perp \mathcal{R}_A)^{-1}\mathcal{P}^\perp(A) \ .$$

Similarly, if $\mathcal{L}_A(X) = AX$ is left multiplication, one obtains

$$C_+ = -(I+\mathcal{P}\mathcal{L}_A)^{-1}\mathcal{P}(A) \ .$$

So, given A, define C_+ and C_- by these formulae and let $B_+ = \mathcal{P}(A+C_-A)$ and $B_- = \mathcal{P}^\perp(A+AC_+)$. Now,

$$\begin{aligned}(I+C_-)(I+A) &= I+C_-+A+C_-A \\ &= I+C_-+\mathcal{P}^\perp(A)+\mathcal{P}^\perp(C_-A)+\mathcal{P}(A+C_-A) \\ &= I+(I+\mathcal{P}^\perp\mathcal{R}_A)C_-+\mathcal{P}^\perp(A)+B_+ \\ &= I+B_+ \ ,\end{aligned}$$

and B_+ belongs to $\mathcal{K} \cap \mathcal{T}(\mathcal{N})$. Likewise, one obtains

$$(I+A)(I+C_+) = I+\mathcal{P}^\perp(A+AC_+) = I+B_-$$

and B_- belongs to $\mathcal{K} \cap rad\, \mathcal{T}(\mathcal{N}^\perp)$. So

$$(I+C_-)(I+B_-) = (I+C_-)(I+A)(I+C_+) = (I+B_+)(I+C_+) \ .$$

This product belongs to $\{I+rad\,\mathcal{T}(\mathcal{N}^\perp)\} \cap \{I+\mathcal{K} \cap \mathcal{T}(\mathcal{N})\} = \{I\}$. Thus, $I+B_- = (I+C_-)^{-1}$ and $I+B_+ = (I+C_+)^{-1}$. (These operators are Fredholm of index zero, so one sided inverses imply invertibility.) Hence

$$(I+B_-)(I+B_+) = (I+B_-)[(I+C_-)(I+A)(I+C_+)](I+B_+) = I+A \ . \quad ■$$

14.8 Lemma. *Let $\mathcal{N}$ be a nest, and let F be a finite rank operator such that $I+F$ is positive and invertible. Then $I+F$ factors as $(I+K)^*(I+D)(I+K)$ where K belongs to $\mathcal{K} \cap rad\,\mathcal{T}(\mathcal{N})$ and D belongs to $\mathcal{K} \cap \mathcal{D}(\mathcal{N})$.*

Proof. This will be proven by induction on $k = rank\,F$. Consider $k = 1$. If $rank\,F = 1$ and $\|F\| < 1$, then $\|F\|_\omega = \|F\|$ so the previous lemma provides the desired factorization. If $\|F\| \geq 1$, then F must be positive of the form $\alpha x \otimes x^*$ where $\|x\| = 1$ and $\alpha \geq 1$. So $(I+F)^{-1} = I - \dfrac{\alpha}{1-\alpha} x \otimes x^*$. Since $|\alpha(1+\alpha)^{-1}| < 1$, apply Lemma 14.7 to $(I+F)^{-1}$ and $\mathcal{N}^\perp$. So

$$(I+F)^{-1} = (I+L^*)(I+E)(I+L)$$

where E belongs to $\mathcal{D}(\mathcal{N}) \cap \mathcal{K}$ and L belongs to $\mathcal{K} \cap rad\,\mathcal{T}(\mathcal{N}^\perp)$. Thus

$$I+F = (I+L)^{-1}(I+E)^{-1}(I-L^*)^{-1} = (I+K^*)(I+D)(I+K)$$

where $K = (I+L^*)^{-1}-I$ belongs to $\mathcal{K} \cap rad\,\mathcal{T}(\mathcal{N}^\perp)^* = \mathcal{K} \cap rad\,\mathcal{T}(\mathcal{N})$, and $D = (I+E)^{-1}-I$ belongs to $\mathcal{D}(\mathcal{N}) \cap \mathcal{K}$.

Now suppose that the lemma holds for all eligible F of rank less than k, for $k \geq 2$. For F of rank k, write $F = F_1+F_2$ where F_1 is rank one and F_2 has rank $k-1$. Then $I+F_1$ is positive and invertible, so factors as $(I+K_1^*)(I+D_1)(I+K_1) = (I+L^*)(I+L)$ where $I+L = (I+D_1)^{1/2}(I+K_1)$ and L belongs to $\mathcal{T}(\mathcal{N}) \cap \mathcal{K}$. Then

$$(I+L^*)^{-1}(I+F)(I+L)^{-1} = I+(I+L^*)^{-1}F_2(I+L)^{-1} = I+G \ .$$

Furthermore, $rank\,G = k-1$. So $I+G$ factors as $(I+M^*)(I+M)$ for some M in $\mathcal{K} \cap \mathcal{T}(\mathcal{N})$. Hence

$$I+F = (I+L^*)(I+M^*)(I+M)(I+L) = (I+S^*)(I+S)$$

where $S = L+M+ML$ belongs to $\mathcal{T}(\mathcal{N}) \cap \mathcal{K}$. Let

$$I+D = (I+\Delta(S^*))(I+\Delta(S))$$

and

$$I+K = (I+\Delta(S))^{-1}(I+S) \ .$$

Then $I+F = (I+K^*)(I+D)(I+K)$. Clearly, D belongs to $\mathcal{D}(\mathcal{N}) \cap \mathcal{K}$, and K belongs to $\mathcal{K} \cap \mathcal{T}(\mathcal{N})$ and satisfies $\Delta(K) = 0$. By Corollary 6.8, K belongs to $rad\,\mathcal{T}(\mathcal{N})$. ■

14.9 Theorem. *Let $\mathcal{N}$ be any nest, and let A belong to $\mathcal{C}_\omega$ be such that $I+A$ is positive and invertible. Then $I+A$ factors uniquely as*

$$I+A = (I+K)^*(I+D)(I+K) = (I+C)^*(I+C)$$

where D belongs to $\mathcal{D}(\mathcal{N}) \cap \mathcal{K}$ and K belongs to $\mathcal{K} \cap \operatorname{rad} \mathcal{T}(\mathcal{N})$, and C belongs to $\mathcal{K} \cap \mathcal{T}(\mathcal{N})$.

Proof. Let E be a finite rank spectral projection of A such that $\|E^\perp A\|_\omega < 1$. By Lemma 14.7, $I+E^\perp A$ factors as $(I+C_1^*)(I+C_1)$ for C_1 in $\mathcal{K} \cap \mathcal{T}(\mathcal{N})$. So

$$(I+C_1^*)^{-1}(I+A)(I+C_1)^{-1} = I+(I+C_1^*)EA(I+C_1) = I+F \ .$$

This operator satisfies the hypotheses of Lemma 14.8 and hence factors as $(I+C_2^*)(I+C_2)$. Thus

$$I+A = (I+C_1^*)(I+C_2^*)(I+C_2)(I+C_1) = (I+C)^*(I+C)$$

where $C = C_1+C_2+C_2C_1$. As in the previous lemma, one can factor out the diagonal part to obtain $I+A = (I+K)^*(I+D)(I+K)$. The uniqueness follows as in Lemma 14.7. ■

14.10 Corollary. *Let $\mathcal{N}$ be a nest and let A belong to $\mathcal{C}_\omega$ and let U be a unitary operator such that $U+A$ is invertible. Then the order isomorphism θ_{U+A} of $\mathcal{N}$ onto $(U+A)\mathcal{N}$ is implemented by a unitary operator.*

Proof. By Lemma 14.1, this reduces to showing that $(U+A)^*(U+A) = I+B$ factors as T^*T for some T in $\mathcal{T}(\mathcal{N})^{-1}$. But $I+B$ is positive and invertible, and B belongs to $\mathcal{C}_\omega$. So this is a consequence of Theorem 14.9. ■

14.11. Next, we wish to consider factorizations of operators which are not invertible. A replacement is needed for the elements of $\mathcal{T}(\mathcal{N})^{-1}$. As these operators are characterized as those operators A such that $AN = N$ for all N in $\mathcal{N}$, the following definition is an appropriate generalization. An operator A in $\mathcal{T}(\mathcal{N})$ is called *outer* if the range projection $P(R_A)$ commutes with $\mathcal{N}$ and AN is dense in $N \cap R_A$ for every N in $\mathcal{N}$. A partial isometry U in $\mathcal{T}(\mathcal{N})$ is called *inner* if U^*U commutes with $\mathcal{N}$. One can now ask if a positive operator P factors as A^*A for A outer, and if every T in $\mathcal{T}(\mathcal{N})$ factors as $T = UA$ with A outer and U inner. These definitions parallel well known factorizations of L^∞ functions with respect to the subalgebra H^∞ on the unit circle. This connection is made explicit in Exercise 14.2.

Recall that a set is *well ordered* if every subset has a least element. Such sets are order isomorphic to some ordinal, and are characterized by the fact that every element has an immediate successor. So $\mathcal{N}$ is well ordered if and only if $N \neq N_+$ for every N in $\mathcal{N}$ (except $N = \mathcal{H}$).

14.12 Proposition. *Let $\mathcal{N}$ be a nest which is not well ordered. Then there is a positive operator Q which cannot be factored as A^*A for any outer operator A in $\mathcal{T}(\mathcal{N})$.*

Proof. Pick N_0 in $\mathcal{N}$ such that $N_0 = (N_0)_+$. Let x be a unit vector in $N_0^\perp$ which is not in $N^\perp$ for any $N > N_0$. Set $Q = x \otimes x^*$. Suppose $Q = A^*A$ and A is in $\mathcal{T}(\mathcal{N})$. Then A is rank one, so $A = y \otimes x^*$ for some y in $\mathcal{H}$. By Lemma 3.7, y must belong to N_0. Then y belongs to R_A so $AN_0 = \{0\}$ is not dense in $R_A \cap N_0$. Thus A is not outer. ■

We wish to prove that such factorizations always exist in the well ordered case. This takes some preparation.

14.13 Lemma. *Let $\begin{bmatrix} A & B \\ B^* & D \end{bmatrix}$ represent a positive operator on $\mathcal{H}_1 \oplus \mathcal{H}_2$. Then there is a bounded operator X so that $B = A^{1/2}X$ and R_X is contained in $\overline{R_A}$. On the other hand, if B has this form, then $\begin{bmatrix} A & B \\ B^* & D \end{bmatrix}$ is positive if and only if $D \geq X^*X$.*

Proof. Let $\delta \geq \|D\|$. Since

$$0 \leq \begin{bmatrix} A & B \\ B^* & D \end{bmatrix} \leq \begin{bmatrix} A & B \\ B^* & \delta I \end{bmatrix}$$

one has

$$0 \leq \begin{bmatrix} \delta I & -B \end{bmatrix} \begin{bmatrix} A & B \\ B^* & \delta I \end{bmatrix} \begin{bmatrix} \delta I \\ -B^* \end{bmatrix} = \delta(\delta A - BB^*) \ .$$

For $\delta > 0$, one obtains $BB^* \leq \delta A$. So if $D = 0$, then $B = 0$. Otherwise, one has $|B^*| \leq \delta^{1/2} A^{1/2}$. So there is an operator X in $\mathcal{B}(\mathcal{H}_2, \mathcal{H}_1)$ with $\|X\| \leq \delta^{1/2}$ so that $B^* = X^*A^{1/2}$ and $B = A^{1/2}X$. This X is unique provided that it is stipulated that X^* is zero on $R_A^\perp$. Thus R_X is contained in $\overline{R_A}$.

Now suppose that $B = A^{1/2}X$ and $\begin{bmatrix} A & B \\ B^* & D \end{bmatrix} \geq 0$. For $t > 0$, let E_t denote the spectral projection $E_A[t,\infty)$. From the normal functional calculus, there is a positive operator T_t in $W^*(A)$ such that $T_tA^{1/2} = A^{1/2}T_t = E_t$. So

$$0 \leq \begin{bmatrix} -X^*T_t & I \end{bmatrix} \begin{bmatrix} A & A^{1/2}X \\ X^*A^{1/2} & D \end{bmatrix} \begin{bmatrix} -T_tX \\ I \end{bmatrix} = D - X^*E_tX \ .$$

So $X^*E_tX \leq D$. Letting t tend to zero yields $X^*X \leq D$ since R_X is contained in $\overline{R_A}$. Conversely, if $D \geq X^*X$,

$$\begin{bmatrix} A & B \\ B^* & D \end{bmatrix} = \begin{bmatrix} A & A^{1/2}X \\ X^*A^{1/2} & X^*X \end{bmatrix} + \begin{bmatrix} 0 & 0 \\ 0 & D-X^*X \end{bmatrix}$$
$$= \begin{bmatrix} A^{1/2} & 0 \\ X^* & 0 \end{bmatrix} \begin{bmatrix} A^{1/2} & X \\ 0 & 0 \end{bmatrix} + \begin{bmatrix} 0 & 0 \\ 0 & D-X^*X \end{bmatrix} \geq 0 \ . \qquad \blacksquare$$

14.14 Corollary. *If* $Q = \begin{bmatrix} A & B \\ B^* & D \end{bmatrix}$ *is positive, then there is an upper triangular operator* $T = \begin{bmatrix} A^{1/2} & X \\ 0 & (D-X^*X)^{1/2} \end{bmatrix}$ *with the properties that* (*i*) $Q = T^*T$ *and* (*ii*) *If for any operator* U, UQ *is upper triangular if and only if* UT^* *is upper triangular.*

Proof. That $T^*T = Q$ is immediate from the previous proof. Write $U = [U_{ij}]_{1 \leq i,j \leq 2}$. Then UT^* is upper triangular exactly when $U_{21}A^{1/2}+U_{22}X^* = 0$. Also UQ is upper triangular exactly when

$$0 = U_{21}A+U_{22}B^* = (U_{21}A^{1/2}+U_{22}X^*)A^{1/2} \ .$$

Since X^* is zero on $R_A^{\perp}$, these two conditions are equivalent. ■

14.15. When $D = X^*X$, $T^* = \begin{bmatrix} A^{1/2} & 0 \\ X^* & 0 \end{bmatrix}$. So if UT^* is upper triangular, it has the form $\begin{bmatrix} * & 0 \\ 0 & 0 \end{bmatrix}$. Thus, the 2-2 entry of UQ is $(U_{21}A^{1/2}+U_{22}X^*)X = 0$. So UQ has the form $\begin{bmatrix} * & * \\ 0 & 0 \end{bmatrix}$. The separate matrix entries of Q belong to the C^*-algebra generated by Q and $C^*(\mathcal{N})$. So as in the proof of the polar decomposition, one obtains that X belongs to the von Neumann algebra generated by Q and the nest. Hence T belongs to this von Neumann algebra as well.

Let $\mathcal{N}$ be a nest and let Q be a positive operator. For each element N of $\mathcal{N}$, decompose Q on $N \oplus N^{\perp}$ as $\begin{bmatrix} A & B \\ B^* & D \end{bmatrix}$. Let X be the operator provided by Lemma 14.13 so that $\begin{bmatrix} A & B \\ B^* & X^*X \end{bmatrix}$ is the minimal positive operator with entries $\begin{bmatrix} A & B \\ B^* & * \end{bmatrix}$. Define $Q[0,N)$ to be this operator. If N and M belong to $\mathcal{N}$ with $N < M$, set

$$Q[N,M) = Q[0,M)-Q[0,N) \ .$$

With respect to the decomposition $N \oplus (M \ominus N) \oplus M^{\perp}$, $Q[N,M)$ has the form

$$\begin{bmatrix} 0 & 0 & 0 \\ 0 & A & B \\ 0 & B^* & X \end{bmatrix} .$$

14.16 Lemma. $\begin{bmatrix} A & B \\ B^* & D \end{bmatrix}$ *is positive if and only if* $D \geq X$.

Proof. First write Q with respect to the given decomposition as $[Q_{ij}]_{1 \leq i,j \leq 3}$. Then

$$Q[0,M) = \begin{bmatrix} Q_{11} & Q_{12} & Q_{13} \\ Q_{21} & Q_{22} & Q_{23} \\ Q_{31} & Q_{32} & Y \end{bmatrix} \text{ and } Q[0,N) = \begin{bmatrix} Q_{11} & Q_{12} & Q_{13} \\ Q_{21} & D_{22} & D_{23} \\ Q_{31} & D_{32} & D_{33} \end{bmatrix} .$$

Furthermore $[D_{ij}]_{2 \leq i,j \leq 3}$ and Y are the minimal operators making $Q[0,M)$ and $Q[0,N)$ positive. In particular, it follows that

$$\begin{bmatrix} Q_{22} & Q_{23} \\ Q_{32} & Y \end{bmatrix} \geq \begin{bmatrix} D_{22} & D_{23} \\ D_{32} & D_{33} \end{bmatrix}$$

and hence $Q[0,M) \geq Q[0,N)$. Thus $Q[N,M)$ is positive. A fortiori, if $D \geq X$, then

$$\begin{bmatrix} A & B \\ B^* & D \end{bmatrix} \geq \begin{bmatrix} A & B \\ B^* & X \end{bmatrix} = \begin{bmatrix} Q_{22} & Q_{23} \\ Q_{32} & Y \end{bmatrix} - \begin{bmatrix} D_{22} & D_{23} \\ D_{32} & D_{33} \end{bmatrix} \geq 0 .$$

On the other hand, if D is such that this matrix is positive, then

$$0 \leq Q[0,N) + \begin{bmatrix} 0 & 0 & 0 \\ 0 & A & B \\ 0 & B^* & D \end{bmatrix} = \begin{bmatrix} Q_{11} & Q_{12} & Q_{13} \\ Q_{21} & Q_{22} & Q_{23} \\ Q_{31} & Q_{32} & D_{33}+D \end{bmatrix} .$$

The minimality of Y yields $D_{33}+D \geq Y$ or $D \geq Y - D_{33} = X$. ∎

14.17 Lemma. *If* $N = N_-$, *then* $Q[0,N) = s - \lim_{M \uparrow N} Q[0,M)$.

Proof. With respect to $N \oplus N^\perp$, $Q[0,N)$ has the form $\begin{bmatrix} A & B \\ B^* & D \end{bmatrix}$. And with respect to $M \oplus M^\perp$, $Q[0,M)$ has the form $\begin{bmatrix} P(M)AP(M) & P(M)B \\ B^*P(M) & D_M \end{bmatrix}$.
The net $Q[0,M)$ is an increasing net of positive operators bounded above by $Q[0,N)$. Hence it converges in the strong operator topology to a limit Q'. From the matrix form, it is clear that Q' has a decomposition with respect to $N \oplus N^\perp$ as $\begin{bmatrix} A & B \\ B^* & D' \end{bmatrix}$. As $0 \leq Q' \leq Q[0,N)$, the minimality of D yields $Q' = Q[0,N)$. ∎

14.18 Lemma. *Let $\mathcal{N}$ be a well ordered nest, and let Q be a positive operator. For each N in $\mathcal{N}$ (except $\mathcal{H}$) , let $Q_N = Q[N,N_+)$. For each finite subnest $\mathcal{F} = \{0:F_0 < F_1 < \ldots < F_n = \mathcal{H}\}$, set $Q_{\mathcal{F}} = \sum_{j=0}^{n-1} Q_{F_j}$. Then $Q_{\mathcal{F}}$ converges to Q in the strong operator topology.*

Proof. Let the order type of $\mathcal{N}$ be $\alpha+1$, and write $\mathcal{N} = \{N_\beta, \beta \leq \alpha\}$. Consider the set S of ordinals γ such that $s-\lim_{\mathcal{F}\subseteq \mathcal{N}_\gamma} Q_{\mathcal{F}} = Q[0,N_\gamma)$ where $\mathcal{N}_\gamma = \{N_\beta : \beta \leq \gamma\}$. This set clearly contains all finite γ. Furthermore, if S contains γ_0, then since $Q[0,N_{\gamma_0+1}) = Q[0,N_{\gamma_0}) + Q[N_{\gamma_0},N_{\gamma_0+1})$, then S contains γ_0+1. Suppose γ is a limit ordinal, $\gamma = \lim \gamma_n$ and γ_n belong to S. Then for every $\mathcal{F} \subseteq N_\gamma$, one has (setting $F_n = N_\gamma$)

$$Q_{\mathcal{F}} = \sum_{j=1}^{n-1} Q[F_j,F_{j+}) \leq \sum_{j=0}^{n-1} Q[F_j,F_{j+1}) = Q[0,N_\gamma) \ .$$

On the other hand

$$s-\lim_{\mathcal{F}\subset \mathcal{N}_\gamma} Q_{\mathcal{F}} \geq s-\lim_{n\to\infty}(s-\lim_{\mathcal{F}\subset \mathcal{N}_{\gamma_n}} Q_{\mathcal{F}}) = s-\lim_{n\to\infty} Q[0,N_{\gamma_n}) = Q[0,N_\gamma)$$

by Lemma 14.17. So equality holds and γ belongs to S. By transfinite induction, α belongs to S. So $s-\lim_{\mathcal{F}} Q_{\mathcal{F}} = Q$. ∎

14.19 Theorem. *Let $\mathcal{N}$ be a well ordered nest. Then every positive operator Q factors as A^*A for some outer operator A in $\mathcal{T}(\mathcal{N})$. Furthermore, A can be chosen in $\{Q,\mathcal{N}\}''$ so that for every operator U, UQ belongs to $\mathcal{T}(\mathcal{N})$ precisely when UA^* does.*

Proof. For each N in $\mathcal{N}$, let $Q_N = Q[N,N_+)$ and let E_N be the orthogonal projection onto $N_+ \ominus N$. By Lemma 14.16 and Corollary 14.14, the positive operator Q_N factors as $A_N^* A_N$ where A_N has the form

$$A_N = \begin{bmatrix} 0 & 0 & 0 \\ 0 & A & X \\ 0 & 0 & 0 \end{bmatrix}$$

with respect to the decomposition $N \oplus E_N\mathcal{H} \oplus N_+^\perp$. Since $A_N = E_N A_N P(N)_-^\perp$, it belongs to $\mathcal{T}(\mathcal{N})$. By Lemma 14.13, R_X is contained in R_A. So $R_{A_N} = R_{A_N E_N}$.

For each finite subnest $\mathcal{F} = \{0 = F_0 < F_1 < \ldots < F_n = \mathcal{H}\}$ of $\mathcal{N}$, let $A_{\mathcal{F}} = \sum_{j=0}^{n-1} A_{F_j}$. Then

$$A_{\mathcal{F}}^{*}A_{\mathcal{F}} = \sum_{j=0}^{n-1} A_{F_j}^{*}A_{F_j} = Q_{\mathcal{F}} \leq Q \ .$$

Thus $\{A_{\mathcal{F}}\}$ is a uniformly bounded family. If x is a vector in the span of a finite set of atoms $\{N_{j+} \ominus N_j, 1 \leq j \leq n\}$, then every finite nest $\mathcal{F}$ containing $\mathcal{F}_0 = \{N_j, 1 \leq j \leq n\}$ satisfies $A_{\mathcal{F}}^{*}x = A_{\mathcal{F}_0}^{*}x$. Thus $\lim_{\mathcal{F}} A_{\mathcal{F}}^{*}x$ exists for a dense set of vectors. The boundedness implies that $s-\lim A_{\mathcal{F}}^{*} = A^{*}$ exists. Indeed, if $E_{\mathcal{F}} = \sum_{F \in \mathcal{F}} E_F$, it follows that $A_{\mathcal{F}}^{*} = E_{\mathcal{F}}A^{*}$. Thus $A_{\mathcal{F}} = AE_{\mathcal{F}}$ and $A = s-\lim A_{\mathcal{F}}$ belongs to $\mathcal{T}(\mathcal{N})$. Moreover, by Lemma 14.18,

$$Q = s-\lim_{\mathcal{F}} Q_{\mathcal{F}} = s-\lim_{\mathcal{F}} A_{\mathcal{F}}^{*}A_{\mathcal{F}} = A^{*}A \ .$$

To show that A is outer, it must be shown that $P(R_A)$ belongs to $\mathcal{N}'$ and $\overline{R_A} \cap N = \overline{AN}$ for all N in $\mathcal{N}$. Since $\overline{R_{A_N}} = \overline{A_N N_+}$ is a subspace of $E_N\mathcal{H}$, it will suffice to show that $\overline{R_A} \cap N = \overline{AN} = \sum_{L<N} \oplus \overline{R_{A_L}}$. For then it is clear that $P(R_A)$ commutes with each E_N and hence is in $\mathcal{N}'$. Let S be the set of N in $\mathcal{N}$ such that this holds. Then S contains 0 trivially. If it contains N, then

$$\begin{aligned}(\overline{R_A} \cap N_+)/\overline{R_A \cap N} &\cong \overline{R_{P(N^{\perp})A}} \cap N_+ \\ &= \overline{R_{A_N+P(N_+)^{\perp}A}} \cap E_N\mathcal{H} = \overline{R_{A_N}}\end{aligned}$$

whereas

$$\overline{AN_+}/(\overline{R_A} \cap N) \cong \overline{P(N)^{\perp}AN_+} = \overline{E_N AN_+} = \overline{A_N N_+} \ .$$

These are equal, and hence

$$\overline{R_A} \cap N_+ = \overline{AN_+} = (\overline{R_A} \cap N) \oplus R_{A_N} = \sum_{L<N_+} \oplus \overline{R_{A_L}} \ .$$

So N_+ belongs to S. Also, if N is the sup of N_k in S, then,

$$\overline{R_A} \cap N \subset \overline{R_{P(N)A}} = \overline{R_{\sum_{L<N} \oplus A_L}} \subset \sum_{L<N} \oplus \overline{R_{A_L}}$$

whereas

$$\overline{AN} = \overline{\bigcup_k AN_k} = \overline{\bigcup \sum_{L<N_k} \oplus \overline{R_{A_L}}} = \sum_{L<N} \oplus \overline{R_{A_L}} \ .$$

So N belongs to S. Since $\mathcal{N}$ is well ordered, $S = \mathcal{N}$ and thus A is outer.

Suppose that U is an operator and consider UQ. This belongs to $\mathcal{T}(\mathcal{N})$ if and only if for every N in $\mathcal{N}$

$$\begin{aligned}0 &= (P(N)^{\perp}UQ)(N) = (P(N)^{\perp}UA^{*}A)(N) \\ &= (P(N)^{\perp}UA^{*}P(N))AN = (P(N)^{\perp}UA^{*}P(N))(\overline{R_A} \cap N) \ .\end{aligned}$$

Clearly, if UA^* belongs to $\mathcal{T}(\mathcal{N})$, then UQ does also. On the other hand, if UQ belongs to $\mathcal{T}(\mathcal{N})$, then since $A^*P(N)$ annihilates $N \ominus (R_A \cap N)$ it follows that $P(N)^\perp UA^*P(N) = 0$. So UA^* belongs to $\mathcal{T}(\mathcal{N})$.

Finally, it follows from the construction of $Q[N,M)$ that it belongs to $\{Q,\mathcal{N}\}''$. So by Remark 14.15, each A_N and hence A belongs to this von Neumann algebra as well. ∎

It is curious that unlike results 14.2, 14.3 and 14.6, the nests for which every T in $\mathcal{T}(\mathcal{N})$ factors as $T = UA$ where U is inner and A is outer is not the same as for factorization of positive operators. The condition turns out to be that every element N of $\mathcal{N}\backslash\{\mathcal{H}\}$ except 0 has an immediate successor.

14.20 Theorem. *Let $\mathcal{N}$ be a nest on a separable Hilbert space.*

i) *If N in $\mathcal{N}\backslash\{0,\mathcal{H}\}$ and $N = N_+$, then there is an operator T in $\mathcal{T}(\mathcal{N})$ such that T does not factor as $T = UA$ where U and A belongs to $\mathcal{T}(\mathcal{N})$, A is outer and U is a partial isometry with $U^*U = P(R_A)$.*

ii) *If every N in $\mathcal{N}\backslash\{0,\mathcal{H}\}$ has an immediate successor, then every T in $\mathcal{T}(\mathcal{N})$ has an inner-outer factorization $T = UA$. Furthermore, the operators U and A may be taken from the von Neumann algebra $W^*(T,\mathcal{N})$.*

Proof. (i) Let x be a unit vector in $N^\perp$ such that $P(N')x \neq 0$ for every $N' > N$. Let y be any unit vetor in N. Then $T = y \otimes x^*$ belongs to $\mathcal{T}(\mathcal{N})$ by Lemma 3.7. If T factors as UA, then $T^*T = x \otimes x^*$ factors as A^*A with A outer. This was showsn to be impossible in Proposition 14.12.

(ii) If $\mathcal{N}$ is well ordered and T belongs to $\mathcal{T}(\mathcal{N})$, then by Theorem 14.19 there is a factorization of T^*T as A^*A where A is outer. Now suppose that 0 has no immediate successor in $\mathcal{N}$, but every N in $\mathcal{N}\backslash\{0,\mathcal{H}\}$ does have one. Let $Q = T^*T$ and define Q_N and A_N as in the proof of Theorem 14.19. For each $N > 0$, one has an outer operator $B_N = \sum_{L \geq N} A_N$ in $\mathcal{T}(\mathcal{N})$ which converges in the strong operator topology such that

$$B_N^*B_N = Q[N,\mathcal{H}) \leq Q \quad .$$

One has $P(N)B_M = B_N$ for $M < N$ in $\mathcal{N}$, and the uniform boundedness ensures that $B = s\text{-}\lim_{N \downarrow 0} B_N$ exists. Furthermore, B belongs to $\mathcal{T}(\mathcal{N})$ and

$$B^*B = s\text{-}\lim_{N\downarrow 0} B_N^*B_N = s\text{-}\lim Q[N,\mathcal{H}) \quad .$$

It must be shown that $Q[N,\mathcal{H})$ converges strongly to Q as N decreases to zero, or that $Q - Q[N,\mathcal{H}) = Q[0,N)$ converges to zero. Now $Q[0,N)$ is the smallest positive operator which has a decomposition with respect to $N \oplus N^\perp$ of the form

$$\begin{bmatrix} P(N)AP(N) & P(N)QP(N)^{\perp} \\ P(N)^{\perp}QP(N) & X \end{bmatrix} .$$

Since $Q = T^*T$ and T belongs to $\mathcal{T}(\mathcal{N})$, it is easy to check that $T^*P(N)T$ has this form. Hence

$$0 \leq s-\lim_{N\downarrow 0} Q\,[0,N) \leq s-\lim_{N\downarrow 0} T^*P(N)T = 0 \ .$$

Furthermore, the proof of Theorem 14.19 shows that each B_N is outer and $\overline{R_{B_N}} = \sum_{L\geq N} \oplus \overline{R_{A_L}}$. Hence $\overline{R_B} = \sum_{N>0} \oplus \overline{R_{A_N}}$ and $P(R_B)$ commutes with $\mathcal{N}$. From this it is easy to verify that B is outer. Thus $T^*T = B^*B$ has an outer factorization.

Let $T = VQ^{1/2}$ and $B = WQ^{1/2}$ be the polar decompositions and let $U = VW^*$. One has $T = UB$. Also,

$$U^*U = WV^*VW^* = WP(R_Q)W^* = WW^* = P(R_A) \ .$$

So U^*U commutes with $\mathcal{N}$. For each N in $\mathcal{N}$, one has

$$UN = U(N \cap \overline{R_B}) = U(\overline{BN}) = \overline{UBN} = \overline{TN} \subseteq N \ .$$

Hence U belongs to $\mathcal{T}(\mathcal{N})$, and is inner. So $T = UB$ is the desired factorization.

From Theorm 14.19, the operators B_N and hence B belong to the von Neumann algebra generated by $\{T^*T, \mathcal{N}\}$. The partial isometries V, W and U belong to the von Neumann algebra generated by $\{T,B\}$. All of these belong to $W^*\{T,\mathcal{N}\}$ as claimed. ∎

14.21 Proposition. *If $\mathcal{N}$ is a nest and T in $\mathcal{T}(\mathcal{N})$ has two inner outer factorizations $T = UA = VB$, then there is a partial isometry W in $\mathcal{D}(\mathcal{N})$ such that $W^*W = P(R_A)$, $WW^* = P(R_B)$, $B = WA$ and $V = UW^*$. Conversely, if W is a partial isometry in $\mathcal{D}(\mathcal{N})$ with $W^*W = P(R_A)$, then $B = WA$ and $V = UW^*$ yields another inner outer factorization VB.*

Proof. Set $W = V^*U$. Since $VV^* = P(R_T) = UU^*$, W is a partial isometry. As in the previous proof, one has $UN = U(N \cap \overline{R_A}) = \overline{TN}$. Hence, $N \cap \overline{R_A} = U^*(\overline{TN})$. Likewise $V^*(\overline{TN}) = N \cap \overline{R_B}$. Thus

$$WN = V^*(\overline{TN}) = N \cap \overline{R_B} \subseteq N \ .$$

So W and likewise W^* belong to $\mathcal{T}(\mathcal{N})$, whence W is in $\mathcal{D}(\mathcal{N})$. The four identities are routine.

Conversely, if W is a partial isometry in $\mathcal{D}(\mathcal{N})$ with initial projection onto $\overline{R_A}$, define $B = WA$ and $V = UW^*$. Then $P(R_B) = WW^*$ belongs to $\mathcal{D}(\mathcal{N}) = \mathcal{N}'$. For N in $\mathcal{N}$,

$$\begin{aligned} N \cap \overline{R_B} &= N \cap W(\overline{R_A}) = (N \cap R_W) \cap W(\overline{R_A}) \\ &= WN \cap W\overline{R_A} = W(N \cap \overline{R_A}) \\ &= W(\overline{AN}) = \overline{BN} \ . \end{aligned}$$

Hence B is outer. Also V is a partial isometry in $\mathcal{T}(\mathcal{N})$ with $V^*V = WW^*$, so V is inner. ∎

Notes and Remarks.

Factorization of this type was first considered by Arveson [7] for nests order isomorphic to subsets of the extended integers. In this context, he obtains Theorem 14.2 (1), Corollary 14.3, Theorem 14.20 (ii) and Proposition 14.21. He applied this to an interpolation problem dealt with in section 15.D. The results, Lemma 14.1 through Corollary 14.6, are due to Larson [4] in his paper on similarity of nests. Lemma 14.4 was proved for continuous nests by Choi and Feintuch [1]. Theorem 14.9 is due to Gohberg and Krein [2], chapter IV. However, they give a different proof of Lemma 14.8. They also obtain necessary and sufficient conditions for $I+K$ with K compact to factor as in Lemma 14.7 in terms of the convergence of certain integrals. (See exercise 14.1.) The notation of inner and outer operators is due to Arveson. The results of 14.12 through 14.21 are due to Power [2, 6, 7, 9]. Lemma 14.13 is due to Lance [2].

Exercises.

14.1 (Gohberg-Krein)

(a) Consider a nest $\mathcal{N}$ and an invertible operator $T = I+K$ for K compact. Show that if T factors as in Theorem 14.9, then $I+P(N)KP(N)$ is invertible for every N in $\mathcal{N}$.

(b) If T above is self-adjoint, then T factors if and only if T is positive.

(c) If K is finite rank, then T factors only if $I+P(N)KP(N)$ is invertible for all N in $\mathcal{N}$. (This is difficult.)

(d) Extend (c) to all K in C_ω.

14.2 Let $L^2 = L^2(T,dm)$ where dm is normalized Lebesgue measure on the unit circle T. This has an orthonormal basis $e_n = e^{in\theta}$, $n \in \mathbb{Z}$. Let $\mathcal{N}$ be the nest given by 0, L^2, and $N_n = span\{e_k, k \geq n\}$. If h is a bounded measurable function (i.e. $h \in L^\infty$), M_h denotes the operator given by $M_h f = hf$. Note that M_h belongs to $\mathcal{T}(\mathcal{N})$ if and only if h belongs to H^∞. A function h in H^∞ is *outer* if $M_h N_0$ is dense in N_0. Show that h is outer if and only if M_h is an outer operator in $\mathcal{T}(\mathcal{N})$.

14.3 Let $\mathcal{N}$ be a countable nest, and let A be an outer operator in $\mathcal{T}(\mathcal{N})$ which is bounded below. Show that there is an invertible element B in $\mathcal{T}(\mathcal{N})^{-1}$ and an isometry U in $\mathcal{D}(\mathcal{N})$ such that $A = UB$.

14.4 Show that every positive operator Q in $\mathcal{B}(\mathcal{H})$ factors as T^*T for some T in $\mathcal{T}(\mathcal{N})$ if and only if either (i) $\mathcal{N}$ is well ordered or (ii) $\mathcal{N}$ has an infinite rank atom A such that if $N = N_+ \in \mathcal{N}$, then N contains A.

14.5 (Power)

(a) Let $\mathcal{N}$ be a nest and let A be an outer operator in $\mathcal{T}(\mathcal{N})$. Suppose X is an operator such that XA belongs to $\mathcal{T}(\mathcal{N})$ and $\ker X$ contains $R_A^{\perp}$. Show that X belongs to $\mathcal{T}(\mathcal{N})$.

(b) Let $\mathcal{N}$ be a well ordered nest. Show that every trace class operator T in $\mathcal{T}(\mathcal{N})$ factors as $T = A_1A_2$ where A_i belong to $\mathcal{T}(\mathcal{N}) \cap \mathcal{C}_2$ and $\|T\|_1 = \|A_1\|_2 = \|A_2\|_2$.

Hint: Write $T = V|T|$ and factor $|T| = A^*A$.

(c) If $\mathcal{N}$ is a nest containing an element $N \neq 0$ or $\mathcal{H}$ such that $N_- = N_+$, find a rank one operator T in $\mathcal{T}(\mathcal{N})$ which cannot be factored as A_1A_2 with A_i in $\mathcal{T}(\mathcal{N}) \cap \mathcal{C}_2$.

(d)* Characterize those nests in which every trace class operator has a "Riesz" factorization as in (b).

14.6* (Larson) Characterize the compact operators K such that $I+K$ is positive, invertible and factors as T^*T for T in $\mathcal{T}(\mathcal{N})^{-1}$ for every nest $\mathcal{N}$.

15. Reflexivity, Ideals and Bimodules

The main result of this chapter shows that the only weak* closed algebra $\mathcal{A}$ containing a maximal abelian von Neumann algebra (masa) such that $Lat\ \mathcal{A} = \mathcal{N}$ is a nest is $\mathcal{T}(\mathcal{N})$. Then we characterize the weak* closed ideals and bimodules of a nest algebra, and show that these are singly generated. (This can be read independently of the first part of this chapter.) From this we deduce that nest algebras are doubly generated as a weak* closed algebra. Then we establish a criterion for a finite set of operators to generate $\mathcal{T}(\mathcal{N})$ as a norm closed left ideal. In the last part ot this chapter, invariant operator ranges for nest algebras are computed.

15.A Utility Grade Tensor Products. First, the notion of a *Hilbert space tensor product* will be needed. Let $\mathcal{H}_1$ and $\mathcal{H}_2$ be Hilbert spaces. The algebraic tensor product of $\mathcal{H}_1$ and $\mathcal{H}_2$ consists of elements of the form $\sum_{i=1}^{n} x_i \otimes y_i$ modulo the identities

$$x \otimes (y_1+y_2) = x \otimes y_1 + x \otimes y_2$$

$$(x_1+x_2) \otimes y = x_1 \otimes y + x_2 \otimes y$$

and

$$(ax) \otimes y = a(x \otimes y) = x \otimes (ay)$$

for all x_i in $\mathcal{H}_1$, y_i in $\mathcal{H}_2$ and a in $\mathbb{C}$.

An inner product is defined on elementary terms by

$$(x_1 \otimes y_1, x_2 \otimes y_2) = (x_1, x_2)(y_1, y_2)$$

and extending by linearity. It is routine to verify that this is well defined. It is also positive definite. To see this, consider a non-zero element $v = \sum_{i=1}^{n} x_i \otimes y_i$. Apply the Gram-Schmidt process to $\{y_i\}$ to obtain an orthonormal set e_j, $1 \leq j \leq m$, so that $y_i = \sum_{j=1}^{i} a_{ij} e_j$. Then

$$v = \sum_{i=1}^{n} x_i \otimes y_i = \sum_{i=1}^{n} \sum_{j=1}^{m} a_{ij} x_i \otimes e_j$$

$$= \sum_{j=1}^{m} \Big(\sum_{i=1}^{n} a_{ij} x_i\Big) \otimes e_j = \sum_{j=1} u_j \otimes e_j \ .$$

As $v \neq 0$, at least one u_j is non-zero. So

$$\|v\|^2 = (v,v) = \sum_{j=1}^{m}\sum_{i=1}^{n}(u_j,u_i)(e_j,e_i) = \sum_{j=1}^{m}\|u_j\|^2 > 0 \quad .$$

Let $\mathcal{H}_1 \otimes \mathcal{H}_2$ denote the Hilbert space completion of this tensor product. Let $\{f_\alpha : \alpha \in A\}$ and $\{e_\beta : \beta \in B\}$ be orthonormal bases for $\mathcal{H}_1$ and $\mathcal{H}_2$. Then $\{f_\alpha \otimes e_\beta : \alpha \in A, \beta \in B\}$ is an orthonormal set in $\mathcal{H}_1 \otimes \mathcal{H}_2$. The closed span of these vectors contains all vectors of the form $x \otimes e_\beta$, and hence all of the form $x \otimes y$. Thus they span all of $\mathcal{H}_1 \otimes \mathcal{H}_2$ and form a basis.

Let S and T be operators in $B(\mathcal{H}_1)$ and $B(\mathcal{H}_2)$. Define $S \otimes T$ in $B(\mathcal{H}_1 \otimes \mathcal{H}_2)$ by setting $S \otimes T(x \otimes y) = Sx \otimes Ty$ and extending by linearity to the algebraic tensor product. It is readily verified that this is well defined. To see that $\|S \otimes T\| = \|S\|\,\|T\|$, consider a vector $v = \sum_{j=1}^{n} x_i \otimes y_i$. Let $M = \operatorname{span}\{y_i, Ty_i : 1 \leq i \leq n\}$, and set $A = P(M)TP(M)$. Then A^*A is positive, and leaves M invariant. So M has an orthonormal basis $\{e_j, 1 \leq j \leq m\}$ consisting of eigenvectors of A^*A, say $A^*Ae_j = \lambda_j e_j$ where $|\lambda_j| \leq \|A^*A\| \leq \|T\|^2$. One can rewrite $y_i = \sum_{j=1}^{m} a_{ij}e_j$, and

$$v = \sum_{i=1}^{n}(x_i \otimes \sum_{j=1}^{m} a_{ij}e_j) = \sum_{j=1}^{m}(\sum_{i=1}^{n} a_{ij}x_i) \otimes e_j = \sum_{j=1}^{m} u_j \otimes e_j \quad .$$

So

$$\begin{aligned}
\|S \otimes Tv\|^2 &= \|\sum_{i=1}^{n} Sx_i \otimes Ty_i\|^2 = \|\sum_{i=1}^{n} Sx_i \otimes Ay_i\|^2 \\
&= \|S \otimes Av\|^2 = (\sum_{j=1}^{m} Su_j \otimes Ae_j, \sum_{j=1}^{m} Su_j \otimes Ae_j) \\
&= \sum_{j=1}^{m}\sum_{k=1}^{m}(Su_j,Su_k)(A^*Ae_j,e_k) = \sum_{j=1}^{m}\lambda_j(Su_j,Su_j) \\
&\leq \|T\|^2\sum_{j=1}^{m}\|Su_j\|^2 \leq \|S\|^2\,\|T\|^2\sum_{j=1}^{m}\|u_j\|^2 \\
&= (\|S\|\,\|T\|\,\|v\|)^2 \quad .
\end{aligned}$$

On the other hand, the supremum of $\|S \otimes T(x \otimes y)\| = \|Sx\|\,\|Ty\|$ over unit vectors x and y is clearly $\|S\|\,\|T\|$. Thus $S \otimes T$ extends by continuity to $\mathcal{H}_1 \otimes \mathcal{H}_2$.

The algebraic tensor product of $B(\mathcal{H}_1)$ and $B(\mathcal{H}_2)$ thus is a subset of $B(\mathcal{H}_1 \otimes \mathcal{H}_2)$. By $B(\mathcal{H}_1) \otimes B(\mathcal{H}_2)$, we mean the weak* closure of this algebraic tensor product. This is readily seen to be all of $B(\mathcal{H}_1 \otimes \mathcal{H}_2)$. To see this, it is enough to get a dense subset of the rank one operators. To avoid confusion, we will write a rank one operator as xy^*. Consider

$u = \sum_{i=1}^{n} x_i \otimes y_i$ and $v = \sum_{j=1}^{m} w_j \otimes z_j$. Then

$$uv^* = \sum_{i=1}^{n}\sum_{j=1}^{m}(x_i \otimes y_i)(w_j \otimes z_j)^*$$
$$= \sum_{i=1}^{n}\sum_{j=1}^{m}(x_i w_j^*) \otimes (y_i z_j^*) \ .$$

So the rank one operator uv^* is the sum of elementary operators $S_{ij} \otimes T_{ij}$. Consequently, the norm closure of $B(\mathcal{H}_1) \otimes B(\mathcal{H}_2)$ contains all finite rank operators, so is weak* dense in $B(\mathcal{H}_1 \otimes \mathcal{H}_2)$.

More generally, if $\mathcal{A}$ and $\mathcal{B}$ are von Neumann algebras, let $\mathcal{A} \otimes \mathcal{B}$ denote the weakly closed algebra generated by the elementary tensors $A \otimes B$ in $B(\mathcal{H}_1) \otimes B(\mathcal{H}_2)$ such that A belongs to $\mathcal{A}$ and B belongs to $\mathcal{B}$. (This is the *spatial tensor product*, and is not the only choice for a tensor product in general.) By abuse of notation, we may write $\mathcal{A} \otimes I$ when we mean $\mathcal{A} \otimes \mathbb{C}I$. We require the following very special case of Tomita's Commutation Theorem.

15.1 Theorem. *Let $\mathcal{A}$ be a von Neumann algebra in $B(\mathcal{H}_1)$. Then commutant of $\mathcal{A} \otimes B(\mathcal{H}_2)$ is $\mathcal{A}' \otimes \mathbb{C}I$, and vice versa.*

Proof. First, suppose that $\mathcal{A} = \mathbb{C}I$. Let T in $B(\mathcal{H}_1 \otimes \mathcal{H}_2)$ commute with $I \otimes B(\mathcal{H}_2)$. For each y in $\mathcal{H}_2$, let P_y denote the projection in $B(\mathcal{H}_2)$ onto $\mathbb{C}y$. Then for each x in $\mathcal{H}_1$, there is a z in $\mathcal{H}_1$ so that

$$T(x \otimes y) = T(I \otimes P_y)x \otimes y = (I \otimes P_y)T(x \otimes y) = z \otimes y \ .$$

Moreover, the linearity of T shows that there is a continuous linear operator B_y on $\mathcal{H}_1$ so that $T(x \otimes y) = (B_y x) \otimes y$. If y_1 and y_2 are linearly independent,

$$B_{y_1+y_2}x \otimes y_1 + B_{y_1+y_2}x \otimes y_2 = B_{y_1+y_2}x \otimes (y_1+y_2)$$
$$= T(x \otimes (y_1+y_2))$$
$$= T(x \otimes y_1)+T(x \otimes y_2)$$
$$= B_{y_1}x \otimes y_1 + B_{y_2}x \otimes y_2 \ .$$

The independence of y_1 and y_2 implies that $B_{y_1} = B_{y_1+y_2} = B_{y_2}$. Hence there is an operator B such that $T(x \otimes y) = (Bx) \otimes y = (B \otimes I)x \otimes y$ for all x and y. Thus $T = B \otimes I$ belongs to $B(\mathcal{H}_1) \otimes I$. On the other hand, it is clear that $B(\mathcal{H}_1) \otimes I$ commutes with $I \otimes B(\mathcal{H}_2)$.

In general, the commutant of $\mathcal{A} \otimes B(\mathcal{H}_2)$ is contained in $(I \otimes B(\mathcal{H}_2))' = B(\mathcal{H}_1) \otimes I$. So, an operator in this set has the form $B \otimes I$ and commutes with $A \otimes I$ for each A in $\mathcal{A}$. Hence B belongs to $\mathcal{A}'$. On the other hand, if B belongs to $\mathcal{A}'$, the $B \otimes I$ commutes with

$A \otimes X$ for every A in $\mathcal{A}$ and X in $\mathcal{B}(\mathcal{H})$. So $\mathcal{A}' \otimes I$ is the commutant of $\mathcal{A} \otimes \mathcal{B}(\mathcal{H})$. By the Double Commutant Theorem, $\mathcal{A} \otimes \mathcal{B}(\mathcal{H})$ is the commutant of $\mathcal{A}' \otimes I$. ∎

Now suppose that $\mathcal{H}_1$ is represented as $L^2(X,\mu)$ where μ is a regular Borel measure on X. Let $L^2(X,\mathcal{H}_2,\mu)$ denote the (equivalence classes of) Borel functions $f : X \to \mathcal{H}_2$ such that

$$\| f \|_2 = (\int \| f(x) \|^2 d\mu)^{1/2}$$

is finite. This is a Hilbert space, (as can be verified in the same way that completeness of $L^2(\mu)$ is proved).

15.2 Lemma. *Let (X,μ) be a regular Borel measure and $\mathcal{H}_1 = L^2(X,\mu)$. Then $\mathcal{H}_1 \otimes \mathcal{H}_2$ is naturally isomorphic to $L^2(X,\mathcal{H}_2,\mu)$ via the unitary operator*

$$U \sum_{i,j} a_{ij}(e_i \otimes y_j) = \sum_{i,j} a_{ij} f_{ij}$$

where $\{e_i\}$ is a basis for $L^2(\mu)$, $\{y_j\}$ is a basis for $\mathcal{H}_2$, and $f_{ij}(x) = e_i(x) y_j$.

Proof. If f belongs to $L^2(\mu)$ and $f = \sum a_i e_i$, then

$$Uf \otimes y_j = U \sum_i a_i e_i \otimes y_i = \sum_i a_i f_{ij}$$

$$= (\sum a_i e_i)(x) y_j = f(x) y_j \ .$$

So U carries $L^2(\mu) \otimes \mathbb{C} y_i$ isometrically onto $L^2(X, \mathbb{C} y_j, \mu)$. It is easy to see that

$$L^2(\mu) \otimes \mathcal{H} = \sum_j \oplus L^2(\mu) \otimes \mathbb{C} y_j$$

and

$$L^2(X,\mathcal{H}_2,\mu) = \sum_j \oplus L^2(X, \mathbb{C} y_j, \mu) \ .$$

So the result follows. ∎

Let $\mathcal{M}$ be a masa in $\mathcal{B}(\mathcal{H})$. By Theorem 7.8, one can realize $\mathcal{M}$ as $L^\infty(X,\mu)$ acting on $L^2(X,\mu)$ for some regular Borel space (X,μ). By Lemma 15.2, we can realize $L^2(X,\mu) \otimes \mathcal{H}$ as $L^2(X,\mathcal{H},\mu)$. There is a von Neumann algebra $L^\infty(X, \mathcal{B}(\mathcal{H}), \mu)$ of μ-essentially bounded Borel functions from X into $\mathcal{B}(\mathcal{H})$ acting on $L^2(X,\mathcal{H},\mu)$ by

$$Tf(x) = T(x) f(x) \ .$$

15.3 Lemma. *The identification between* $L^2(X,\mu) \otimes \mathcal{H}$ *and* $L^2(X,\mathcal{H},\mu)$ *of Lemma* 15.2 *carries* $\mathcal{M} \otimes \mathcal{B}(\mathcal{H})$ *onto* $L^\infty(X,\mathcal{B}(\mathcal{H}),\mu)$.

Proof. Since $(M \otimes I)x \otimes y = (Mx) \otimes y$, it is readily verified that

$$U(M \otimes I)U^* f(x) = M(f(x))$$

for every f in $L^2(X,\mathcal{H},\mu)$. Thus it is clear that $L^\infty(X,\mathcal{B}(\mathcal{H}),\mu)$ is contained in the commutant of $U(\mathcal{M} \otimes I)U^*$. By Theorem 15.1, $L^\infty(X,\mathcal{B}(\mathcal{H}),\mu)$ is contained in $U(\mathcal{M} \otimes \mathcal{B}(\mathcal{H}))U^*$. On the other hand, let M_f belong to $\mathcal{M}$ and A belong to $\mathcal{B}(\mathcal{H})$. For g in $L^2(X,\mu)$ and y in $\mathcal{H}_2$

$$\begin{aligned} U(M_f \otimes T)(g \otimes y) &= U(fg \otimes Ty) = fg(x)Ty \\ &= M_{fT}(gy) = M_{fT}U(g \otimes y) \end{aligned}$$

where fT is the L^∞ function $fT(x) = f(x)T$, and $gy(x) = g(x)y$. So $U(M_f \otimes T)U^* = M_{fT}$ belongs to $L^\infty(X,\mathcal{B}(\mathcal{H}),\mu)$. Thus the weakly closed algebra generated by these tensors is carried into, whence onto, $L^\infty(X,\mathcal{B}(\mathcal{H}),\mu)$. ∎

15.B Reflexivity. The notion of reflexivity for algebras can be generalized to subspaces. A subspace $\mathcal{S}$ of $\mathcal{B}(\mathcal{H}_1,\mathcal{H}_2)$ is *reflexive* if

$$\mathcal{S} = \{T \in \mathcal{B}(\mathcal{H}_1,\mathcal{H}_2) : Tx \in \overline{\mathcal{S}x} \text{ for all } x \in \mathcal{H}_1\} .$$

Reflexive subspaces are always closed in the weak operator topology. The following lemma sets the stage.

15.4 Lemma. *Let* $\mathcal{S}$ *be a weak* closed subspace of* $\mathcal{B}(\mathcal{H}_1)$, *and let* $\mathcal{H}$ *be a separable Hilbert space. Then* $\mathcal{S} \otimes I$ *is a reflexive subspace of* $\mathcal{B}(\mathcal{H}_1,\mathcal{H})$. *For each vector* x *in* $\mathcal{H}_1 \otimes \mathcal{H}$, *let* $M_x = \overline{(\mathcal{S} \otimes I)x}$. *Then for all* T *in* $\mathcal{B}(\mathcal{H}_1)$,

$$dist(T,\mathcal{S}) = \sup_{x \in \mathcal{H}_1 \otimes \mathcal{H}, \|x\|=1} \|P(M_x)^\perp Tx\| .$$

Proof. Let ϕ be a norm one weak* continuous linear functional annihilating $\mathcal{S}$. By Theorem 1.15, there is a sequence $\{s_n, n \geq 1\}$ of positive numbers, and orthonormal sets $\{x_n\}$ and $\{y_n\}$ in $\mathcal{H}_1$ so that $\sum_{n=1}^{\infty} s_n = \|\phi\| = 1$ and

$$\phi(T) = \sum_{n=1}^{\infty} s_n (Tx_n, y_n) .$$

Let $\{e_n, n \geq 1\}$ be an orthonormal set in $\mathcal{H}$. Let $\bar{x} = \sum_{n=1}^{\infty} s_n^{1/2} x_n \otimes e_n$ and $y = \sum_{n=1}^{\infty} s_n^{1/2} y_n \otimes e_n$ be unit vectors in $\mathcal{H}_1 \otimes \mathcal{H}$. Then

$$((T \otimes I)\bar{x},\bar{y}) = \sum_{n=1}^{\infty} s_n(Tx_n,y_n) = \phi(T) \ .$$

Let $M_x = \overline{(S \otimes I)\bar{x}}$. Since $\phi(S) = 0$, it follows that $\bar{y}$ is orthogonal to M_x. Hence

$$\|P(M_x)^{\perp}(T \otimes I)\bar{x}\| \geq |((T \otimes I)\bar{x},\bar{y})| = |\phi(T)| \ .$$

Thus, by the Hahn-Banach Theorem,

$$dist(T,S) = \sup_{\phi \in S_{\perp},\|\phi\|=1} |\phi(T)| \leq \sup_{\|\bar{x}\|=1} \|P(M_x)^{\perp}(T \otimes I)\bar{x}\| \ .$$

On the other hand, $P(M_x)^{\perp}(S \otimes I)\bar{x} = 0$ for S in S. So

$$\begin{aligned} \|P(M_x)^{\perp}(T \otimes I)\bar{x}\| &= \inf_{S \in S} \|P(M_x)^{\perp}((T-S) \otimes I)\bar{x}\| \\ &\leq \inf_{S \in S} \|T-S\| = dist(T,S) \ . \end{aligned}$$

Let A belong to $B(\mathcal{H}_1 \otimes \mathcal{H})$, and suppose that for all v in $\mathcal{H}_1 \otimes \mathcal{H}$, one has

$$Av \in \overline{(S \otimes I)v} \subseteq \overline{B(\mathcal{H}_1) \otimes Iv} \ .$$

Then A belongs to $Alg\,Lat(B(\mathcal{H}_1) \otimes I) = (B(\mathcal{H}_1) \otimes I)'' = B(\mathcal{H}_1) \otimes I$. So A has the form $T \otimes I$. By the previous paragraph, T belongs to S. Thus A belongs to $S \otimes I$ and hence $S \otimes I$ is reflexive. ∎

Note that the previous lemma applies equally well to a subspace S of $B(\mathcal{H}_1,\mathcal{H}_2)$. Now we specialize to subspaces which are *bimodules* for a pair $(\mathcal{M}_2,\mathcal{M}_1)$ of masas, meaning that $\mathcal{M}_2 S \mathcal{M}_1 \subseteq S$. This is the case, for example, if $S = \mathcal{T}(\mathcal{N})$ for a nest $\mathcal{N}$ and $\mathcal{M}_1 = \mathcal{M}_2$ is any masa containing $\mathcal{N}''$. Let $b_1(S)$ denote the unit ball of S.

15.5 Lemma. *Let $\mathcal{M}_i$ be masas in $B(\mathcal{H}_i)$, i=1,2, and let S be a subspace of $B(\mathcal{H}_1,\mathcal{H}_2)$ which is an $(\mathcal{M}_2,\mathcal{M}_1)$ bimodule. Let Q be a projection in $\mathcal{M}_1$, and let $N = \overline{SQ\mathcal{H}_1}$. Then $P(N)$ is the least positive operator A in $\mathcal{M}_2$ such that $A \geq SQS^*$ for all S in $b_1(S)$.*

Proof. Since $SQ = P(N)SQ$ for all S in S, it follows that for S in $b_1(S)$ one has

$$SQS^* = P(N)SQS^*P(N) \leq \|S\|^2 P(N) \leq P(N) \ .$$

Now let A be any positive operator in $\mathcal{M}_2$ such that $A \geq SQS^*$ for all S in $b_1(S)$. Since $\mathcal{M}_2$ is abelian, one has $A \geq P(N)$ if and only if $\|EA\| \geq 1$ for any non-zero projection E in $\mathcal{M}_2$ with $E \leq P(N)$. Let E be such a projection. From the definition of N, there must be a S in S so that $ESQ \neq 0$. Normalize so that $\|ESQ\| = 1$. As S is a bimodule, $T = ESQ$ belongs to $b_1(S)$. Hence

$$EA = EAE \geq ETQT^*E = TT^* \ .$$

So $\|EA\| \geq \|T\|^2 = 1$, and hence $A \geq P(N)$. ■

15.6 Corollary. *Let S be a subspace of $B(\mathcal{H}_1,\mathcal{H}_2)$ which is an $(\mathcal{M}_2,\mathcal{M}_1)$ bimodule. Suppose that $\overline{Sy} = \mathcal{H}_2$ for every $y \neq 0$ in $\mathcal{H}_1$. Let B belong to $(\mathcal{M}_1)_+$. Then the least positive operator A in $\mathcal{M}_2$ such that $A \geq SBS^*$ for all S in $b_1(S)$ is $A = \|B\|I$.*

Proof. For $\epsilon > 0$, let Q_ϵ be the spectral projection $E_B[\|B\|-\epsilon,\|B\|]$. Then $B \geq (\|B\|-\epsilon)Q_\epsilon$, and $\overline{SQ_\epsilon \mathcal{H}_1} = \mathcal{H}_2$ since $Q_\epsilon \neq 0$. So if $A \geq SBS^*$ for all S in $b_1(S)$, then $A \geq (\|B\|-\epsilon)SQ_\epsilon S^*$. By Lemma 15.5, one has $A \geq (\|B\|-\epsilon)I$ for all $\epsilon > 0$. So $A \geq \|B\|I$. ■

15.7 Lemma. *Let $\mathcal{M}$ be a masa in $B(\mathcal{H}_1)$. Let Q be a projection in $\mathcal{M} \otimes B(\mathcal{H})$, and let F be the least projection in $B(\mathcal{H})$ such that $Q \leq I \otimes F$. For each unit vector ξ in $\mathcal{H}$, let E_ξ denote the projection onto $\mathbb{C}\xi$, and set $F(\xi) = \|(I \otimes E_\xi)Q(I \otimes E_\xi)\|$. Then $I \otimes F$ is the least projection P in $\mathcal{M} \otimes B(\mathcal{H})$ satisfying*

$$(I \otimes E_\xi)P(I \otimes E_\xi) \geq F(\xi)I \otimes E_\xi \quad \text{for all } \xi \in \mathcal{H},\ \|\xi\| = 1 \ . \qquad (*)$$

Proof. By Lemma 15.3, we can identify $\mathcal{M} \otimes B(\mathcal{H})$ with $L^\infty(X,B(\mathcal{H}),\mu)$ acting on $L^2(X,\mathcal{H},\mu)$. Thus Q is given by a function $Q(x)$ in $L^\infty(X,B(\mathcal{H}),\mu)$, and since Q is a projection, $Q(x)$ is a projection almost everywhere. So

$$(I \otimes E_\xi)Q(I \otimes E_\xi) = M_{f_\xi} \otimes E_\xi$$

where $f_\xi(x) = (Q(x)\xi,\xi)$, and $F(\xi) = \|f_\xi\|_\infty$. Let $\{\xi_n\}$ be a countable dense subset of $b_1(\mathcal{H})$, and let $f_n = f_{\xi_n}$. Delete a set X_0 of measure zero so that $|f_n(x)| \leq \|f_n\|_\infty$ for x in $X \backslash X_0$, $n \geq 1$ and $Q(x)$ is always a projection. Let $F_0 = \sup_{X\backslash X_0} Q(x)$. Fix x in $X\backslash X_0$ and a unit vector $\xi = Q(x)\xi$. Choose a sequence ξ_{n_i}, converging to ξ. Then

$$F(\xi) = \lim_i F(\xi_{n_i}) \geq \lim_i (Q(x)\xi_{n_i},\xi_{n_i}) = (Q(x)\xi,\xi) = 1 \ .$$

Thus $P \geq I \otimes E_\xi$ and hence $P \geq I \otimes Q(x)$ for each x in $X\backslash X_0$. Consequently, $P \geq I \otimes F_0$.

On the other hand, $Q(x) \leq F_0$ for x in $X\backslash X_0$, so $Q \leq I \otimes F_0$. Thus $F_0 \geq F$. And since $Q \leq I \otimes F$, it readily follows that $I \otimes F$ satisfies (*). By the previous paragraph, $F \geq F_0$. So $F = F_0$ and any P satisfying (*) is greater than $I \otimes F$. ■

15.8 Lemma. *Let $\mathcal{M}_i$ be masas in $\mathcal{B}(\mathcal{H}_i)$ and let $\mathcal{S}$ be a subspace of $\mathcal{B}(\mathcal{H}_1,\mathcal{H}_2)$ which is a $(\mathcal{M}_2,\mathcal{M}_1)$ bimodule. For a vector v in $\mathcal{H}_1 \otimes \mathcal{H}$, let Q be the projection onto $\overline{(\mathcal{M}_1 \otimes I)v}$ and let P be the projection onto $\overline{(\mathcal{S} \otimes I)v}$. For a unit vector ξ in $\mathcal{H}$, let A and B be operators defined by $(I \otimes E_\xi)P(I \otimes E_\xi) = A \otimes E_\xi$ and $(I \otimes E_\xi)Q(I \otimes E_\xi) = B \otimes E_\xi$. Then A and B are positive operators in $\mathcal{M}_2$ and $\mathcal{M}_1$, respectively, and furthermore*

$$A \geq SBS^* \text{ for all } S \text{ in } b_1(\mathcal{S}) \ .$$

Proof. Since $P\mathcal{H}_2 = (\mathcal{M}_2 \otimes I)\overline{(\mathcal{S} \otimes I)v}$ is invariant for $\mathcal{M}_2 \otimes I$, Theorem 15.1 implies that P belongs to $\mathcal{M}_2 \otimes \mathcal{B}(\mathcal{H})$ and A belongs to $(\mathcal{M}_2)_+$. Similarly, Q belongs to $\mathcal{M}_1 \otimes \mathcal{B}(\mathcal{H})$ and B belongs to $(\mathcal{M}_1)_+$. For x in $\mathcal{H}_2$ and S in $b_1(\mathcal{S})$,

$$\begin{aligned}(SBS^*x,x) &= ((S \otimes I)(I \otimes E_\xi)Q(I \otimes E_\xi)(S^* \otimes I)\, x \otimes \xi,\, x \otimes \xi) \\ &= \|Q(S^* \otimes I)x \otimes \xi\|^2 \\ &= \|Q(S^* \otimes I)P(x \otimes \xi)\|^2 \leq \|P(x \otimes \xi)\|^2 \\ &= ((I \otimes E_\xi)P(I \otimes E_\xi)x \otimes \xi, x \otimes \xi) = (Ax,x) \ .\end{aligned}$$

Hence $A \geq SBS^*$. ∎

15.9 Theorem. *Let $\mathcal{M}_i$ be masas in $\mathcal{B}(\mathcal{H}_i)$, and let $\mathcal{S}$ be a subspace of $\mathcal{B}(\mathcal{H}_1,\mathcal{H}_2)$ which is an $(\mathcal{M}_2,\mathcal{M}_1)$ bimodule. Suppose that $\overline{\mathcal{S}y} = \mathcal{H}_2$ for every $y \neq 0$ in $\mathcal{H}_1$. Then $\mathcal{S}$ is weak* dense in $\mathcal{B}(\mathcal{H}_1,\mathcal{H}_2)$.*

Proof. Let v be any unit vector in $\mathcal{H}_1 \otimes \mathcal{H}$, and let ξ be a unit vector in $\mathcal{H}$. Let P, Q, A, B be defined as in Lemma 15.8. Thus $A \geq SBS^*$ for all S in $b_1(\mathcal{S})$. So by Corollary 15.6, $A \geq \|B\|I$. Now

$$\|B\| = \|(I \otimes E_\xi)Q(I \otimes E_\xi)\| = F(\xi)$$

as in Lemma 15.7. So P satisfies (*) of that lemma. Hence $P \geq I \otimes F$ where F is the smallest projection so that $(I \otimes F)v = v$. Hence

$$\mathcal{H}_2 \otimes F\mathcal{H} \subseteq \overline{(\mathcal{S} \otimes I)v} \subseteq \overline{(\mathcal{B}(\mathcal{H}_1,\mathcal{H}_2) \otimes I)v} \subseteq \mathcal{H}_2 \otimes F\mathcal{H} \ .$$

It follows that $\mathcal{S} \otimes I$ has the same closed ranges as $\mathcal{B}(\mathcal{H}_1,\mathcal{H}_2) \otimes I$. By Lemma 15.4, $\mathcal{S}$ is weak* dense in $\mathcal{B}(\mathcal{H}_1,\mathcal{H}_2)$. ∎

15.10 Corollary. *Let $\mathcal{S}$ be a norm closed subspace of the compact operators $\mathcal{K}$ such that $\overline{\mathcal{S}y} = \mathcal{H}$ for every $y \neq 0$. Suppose that $\mathcal{M}_i$ are masas such that $\mathcal{S}$ is an $(\mathcal{M}_2,\mathcal{M}_1)$ bimodule. Then $\mathcal{S} = \mathcal{K}$.*

Proof. By Theorem 15.9, $\mathcal{S}$ is weak* dense in $\mathcal{B}(\mathcal{H})$. If $\mathcal{S}$ were not all of $\mathcal{K}$, there would be a continuous linear functional ϕ in $\mathcal{K}^*$ such that $\phi(\mathcal{S}) = 0$. By Theorem 1.13 and 1.15, $\mathcal{B}(\mathcal{H}) = \mathcal{K}^{**}$. The weak* closure of

S is annihilated by ϕ and thus is not all of $\mathcal{B}(\mathcal{H})$. This contradiction establishes the corollary. ■

In fact, this corollary is quite weak compared to what is now known about transitive algebras containing compact operators. In particular, one can drop the hypotheses about the masas. However, for subalgebras of $\mathcal{B}(\mathcal{H})$, Theorem 15.9 is a good result. This corollary does illustrate, however, the role of duality in dealing with subalgebras of the compact operators. Corollary 15.13 below does not follow from the transitivity results.

15.11 Theorem. *Let $\mathcal{N}$ be a nest, and let $\mathcal{M}$ be a masa in $\mathcal{T}(\mathcal{N})$. Let S be a subspace of $\mathcal{T}(\mathcal{N})$ such that $\mathcal{M}S\mathcal{M} \subseteq S$ and $\overline{Sx} = \overline{\mathcal{T}(\mathcal{N})x}$ for every x in $\mathcal{H}$. Then S is weak* dense in $\mathcal{T}(\mathcal{N})$.*

Proof. Fix $N \neq 0$ in $\mathcal{N}$. Then $P(N)SP(N_-)^\perp$, as a subspace of $\mathcal{B}(N_-^\perp, N)$ satisfies $\overline{P(N)SP(N_-)^\perp y} = \overline{P(N)\mathcal{T}(\mathcal{N})P(N_-)^\perp y} = N$ for all $y \neq 0$ in $N_-^\perp$. By Theorem 15.9, $P(N)SP(N_-)^\perp$ is weak* dense in $\mathcal{B}(N_-^\perp, N)$. By Theorem 3.7, the weak* closure of S contains all finite rank operators in $\mathcal{T}(\mathcal{N})$. By the Erdos Density Theorem 3.11, S is weak* dense in $\mathcal{T}(\mathcal{N})$. ■

15.12 Corollary. *Let $\mathcal{A}$ be a weak* closed algebra containing a masa $\mathcal{M}$ such that Lat $\mathcal{A} = \mathcal{N}$ is a nest. Then $\mathcal{A} = \mathcal{T}(\mathcal{N})$.*

15.13 Corollary. *Let $\mathcal{A}$ be a subalgebra of the compact operators $\mathcal{K}$ such that Lat $\mathcal{A} = \mathcal{N}$. Suppose, furthermore that there is an abelian C^* subalgebra $\mathcal{D}$ of $\mathcal{A}$ such that $\mathcal{D}'$ is abelian. Then $\mathcal{A} = \mathcal{T}(\mathcal{N}) \cap \mathcal{K}$.*

Proof. Let $\overline{\mathcal{A}}$ be the weak* closure of $\mathcal{A}$. Then $\overline{\mathcal{A}}$ contains $\mathcal{D}'' = \mathcal{D}'$ which is a masa. So $\overline{\mathcal{A}} = \mathcal{T}(\mathcal{N})$ by the previous corollary. If $\mathcal{A}$ were less than all of $\mathcal{T}(\mathcal{N}) \cap \mathcal{K}$, the Hahn Banach Theorem would imply that there is a functional ϕ annihilating $\mathcal{A}$ but not all of $\mathcal{T}(\mathcal{N}) \cap \mathcal{K}$. As in Corollary 15.10, this would contradict the weak* density of $\mathcal{A}$ in $\mathcal{T}(\mathcal{N})$. ■

Note that when considering subalgebras, instead of subspaces, one must require $\mathcal{A}$ to contain $\mathcal{M}$, not merely be a bimodule. For example, if $\mathcal{N}$ is a maximal atomic nest and $\mathcal{A} = \{T \in \mathcal{T}(\mathcal{N}) : \Delta(T) = 0\}$, then $\mathcal{M}\mathcal{A}\mathcal{M} = \mathcal{A}$ and $Lat\, \mathcal{A} = \mathcal{N}$, but $\mathcal{A} = \overline{\mathcal{A}}^{w*} \neq \mathcal{T}(\mathcal{N})$. However, one has $\overline{\mathcal{A}N} = N_- \neq \overline{\mathcal{T}(\mathcal{N})N}$ for countably many N in $\mathcal{N}$. So as a subspace, $\mathcal{A}$ is distinguished from $\mathcal{T}(\mathcal{N})$ by its ranges.

15.C Weak* Closed Ideals. Let $\mathcal{N}$ be a nest, and let S be a weak* closed $\mathcal{T}(\mathcal{N})$ bimodule (i.e. $\mathcal{T}(\mathcal{N})S\mathcal{T}(\mathcal{N})=S$). For each N in $\mathcal{N}$, $\widetilde{N} = \overline{SN}$ is invariant for $\mathcal{T}(\mathcal{N})$ and hence belongs to $\mathcal{N}$. Moreover, $\widetilde{0} = 0$, $N \leq N'$ implies $\widetilde{N} \leq \widetilde{N}'$, and if $N = \sup N_k$, then $\widetilde{N} = \sup \widetilde{N}_k$. So the map $\Phi(N) = \widetilde{N}$ is a left continuous, order homomorphism of $\mathcal{N}$ into itself. On the other hand, given such a map Φ, the set

$$S_\Phi = \{T \in \mathcal{B}(\mathcal{H}) : TN \subseteq \Phi(N) \text{ for all } N \text{ in } \mathcal{N}\}$$

is a $\mathcal{T}(\mathcal{N})$ bimodule. It is a routine exercise to verify that $\overline{S_\Phi N} = \Phi(N)$.

15.14 Theorem. *Every weak* closed $\mathcal{T}(\mathcal{N})$ bimodule S has the form S_Φ, where Φ is the left continuous order homomorphism of $\mathcal{N}$ into itself with $\Phi(0) = 0$ given by $\Phi(N) = \overline{SN}$.*

Proof. Clearly, S is contained in S_Φ; so it suffices to prove that S_Φ is contained in S. Fix T in S_Φ. By the Erdos Density Theorem 3.10, there is a net F_α of finite rank contractions in $\mathcal{T}(\mathcal{N})$ converging to I in the * strong topology. So TF_α belong to S_Φ and converge to T in the * strong topology. By Proposition 3.8, F_α is the sum of rank one operators in $\mathcal{T}(\mathcal{N})$. So TF_α is the sum of operators of the form TR where R is a rank one operator in $\mathcal{T}(\mathcal{N})$. It suffices, then, to show that all such operators belong to S.

Let $x \otimes y^*$ be a rank one operator in $\mathcal{T}(\mathcal{N})$. Let N be the element of $\mathcal{N}$ provided by Lemma 3.7 such that x belongs to N and y belongs to $(N_-)^\perp$. Then Tx belongs to $\Phi(N) = \overline{SN}$. So there are operators S_n in S such that $\lim S_n x = Tx$. Hence $\lim S_n x \otimes y^* = Tx \otimes y^*$ belongs to S. Thus $S = S_\Phi$ by the argument of the previous paragraph. ∎

Now we consider the question of generators. Notice that if you are only interested in S as a $\mathcal{T}(\mathcal{N})$ bimodule, you do not need the previous section on reflexivity.

15.15 Theorem. *Let $\mathcal{N}$ be a nest on a separable space, and let $\mathcal{M}$ be a masa containing $\mathcal{N}''$. Then every $\mathcal{T}(\mathcal{N})$ bimodule is singly generated as an $\mathcal{M}$ bimodule.*

Proof. Let S be a $\mathcal{T}(\mathcal{N})$ bimodule, and set $\Phi(N) = \overline{SN} \equiv \widetilde{N}$ for N in $\mathcal{N}$. By Theorem 15.14, $S = S_\Phi$. The weak* closed $\mathcal{M}$ bimodule generated by an operator S is $S_0 = \overline{\mathcal{M}S\mathcal{M}}^{w*}$. For S_0 to equal S, it is necessary that for every vector x,

$$\overline{Sx} = \overline{S_0x} = \overline{\mathcal{M}S\mathcal{M}x} \ .$$

By Theorem 15.14, this is also sufficient to show that S generates S as a $\mathcal{T}(\mathcal{N})$ bimodule. To see that this suffices for S_0 to equal S, fix N in $\mathcal{N}$ and consider $P(\widetilde{N})S_0P(N_-)^\perp$ as a subspace of $\mathcal{B}(N_-^\perp, \widetilde{N})$. For every x in $N_-^\perp$,

$$\begin{aligned} P(\widetilde{N})\overline{S_0 P(N_-)^\perp x} &= P(\widetilde{N})\overline{Sx} = P(\widetilde{N})\overline{S\,T(\mathcal{N})x} \\ &= P(\widetilde{N})\overline{SN} = P(\widetilde{N})\mathcal{H} \ . \end{aligned}$$

By Theorem 15.9, $P(\widetilde{N})S_0P(N_-)^\perp$ equals $\mathcal{B}(N_-^\perp,\widetilde{N})$. In particular, S_0 contains every rank one operator in $\mathcal{S}$. So by the proof of Theorem 15.14, $S_0 = \mathcal{S}$.

Now we construct an operator S with this property. If E is a projection $\mathcal{M}$, one can always find a vector x with $\mathrm{supp}(x) = E$ (in $\mathcal{M}$) in the sense that $Ex = x$ but $Fx \neq x$ for any projection $F < E$ in $\mathcal{M}$. By Proposition 2.13, $\mathcal{N}$ is compact and metrizable in the order topology and has countably many atoms. So there is a sequence $\{N_k, k \geq 0\}$ of distinct elements of $\mathcal{N}$ which is dense in the order topology, $N_0 = \{0\}$, $N_1 = \mathcal{H}$, and which contains the endpoints of every atom. For $k \geq 2$, there are unique integers $i = i(k)$ and $j = j(k)$ such that $0 \leq i,\ j \leq k-1$, $N_i < N_k < N_j$, and if $0 \leq \ell \leq n-1$, then $N_\ell \leq N_i$ or $N_\ell \geq N_j$.

Now we define S. If $\mathcal{H}_- \neq \mathcal{H}$, choose unit vectors e_1 and f_1 so that $\mathrm{supp}(e_1) = P(\widetilde{\mathcal{H}})$ and $\mathrm{supp}(f_1) = P(\mathcal{H}_-)^\perp$. For $k \geq 2$, choose unit vectors e_k and f_k so that

$$\mathrm{supp}(e_k) = P(\widetilde{N}_k)P(\widetilde{N}_{i(k)})^\perp \quad \text{and} \quad \mathrm{supp}(f_k) = P(N_{k-})^\perp P(N_{j(k)-}) \ .$$

Set $S_k = 2^{-k} e_k \otimes f_k$ and $S = \sum_{k=1}^{\infty} S_k$. Fix k, $i = i(k)$ and $j = j(k)$ and consider

$$P(\widetilde{N}_i)^\perp P(\widetilde{N}_k) S P(N_{k-})^\perp P(N_{j-}) \ .$$

For $\ell < k$ and $N_\ell \leq N_i$, $P(\widetilde{N}_i)^\perp S_\ell = 0$; and $\ell < k$ and $N_\ell \geq N_j$ implies $S_\ell P(N_{j-}) = 0$. For $\ell > k$ and $N_\ell < N_k$, $N_{j(\ell)} \leq N_k$ so $S_\ell P(N_{k-})^\perp = 0$; and if $\ell > k$ and $N_\ell > N_k$, then $N_{i(\ell)} \geq N_k$ so $P(\widetilde{N}_k)S_\ell = 0$. Consequently,

$$P(\widetilde{N}_i)^\perp P(\widetilde{N}_k) S P(N_{k-})^\perp P(N_{j-}) = S_k \ .$$

Hence $S_0 = \overline{\mathcal{M} S \mathcal{M}}^{w*}$ contains each S_k for $k \geq 1$.

Let x be a non-zero vector and let E be the projection onto $\overline{\mathcal{M}x}$. Let N be the least element of $\mathcal{N}$ such that $E \leq P(N)$. Then $\overline{Sx} = \widetilde{N}$ and $\overline{S_0 x} = \overline{S_0 E \mathcal{H}}$. If $N \neq N_-$, then $N = N_k$ for some k and $P(N_-)^\perp E \neq 0$. Thus $Ee_k \neq 0$, so $S_0 E\mathcal{H}$ contains $S_k E e_k$ which is a non-zero multiple of f_k with support $P(\widetilde{N}_{i(k)})^\perp P(\widetilde{N}_k)$. Likewise, $Ee_{i(k)} \neq 0$ and $S_0 E \mathcal{H}$ contains a non-zero multiple of $f_{i(k)}$ with support $P(\widetilde{N}_{i(i(k))})^\perp P(\widetilde{N}_{i(k)})$. In at most k steps, one obtains vectors in $S_0 E \mathcal{H}$ with joint support equal to $P(\widetilde{N})$.

Now suppose $N = N_-$. Recursively define a sequence k_n by setting $k_0 = 0$ and k_{n+1} to be the least integer so that $k_{n+1} > k_n$ and $N_{k_n} < N_{k_{n+1}} < N$. So $i(k_{n+1}) = k_n$ and $N_{j(k_n)} \geq N$ for every $n \geq 1$. Since $\{N_n\}$ is dense in $\mathcal{N}$, one has $\sup N_{k_n} = N$. From the definition of N, one has $EP(N_{k_n})^\perp P(N) \neq 0$ and hence $Ee_{k_n} \neq 0$. As above, we obtain that $S_0 E \mathcal{H}$ contains f_{k_n}, which has support

$$P(\widetilde{N}_{k_n})P(\widetilde{N}_{i(k_n)})^{\perp} = P(\widetilde{N}_{k_n})P(\widetilde{N}_{k_{n-1}})^{\perp} .$$

Jointly, these vectors have support $\sup P(\widetilde{N}_{k_n}) = P(\widetilde{N})$. As $\overline{S_0 E \mathcal{H}}$ is $\mathcal{M}$ invariant, it must be all of $\widetilde{N} = \overline{Sx}$.

Thus it has been established that S generates $\mathcal{S}$ as an $\mathcal{M}$ bimodule. ■

15.16 Corollary. *Every weak* closed ideal in $\mathcal{T}(\mathcal{N})$ is principal.*

Proof. Immediate. ■

15.17 Corollary. *$\mathcal{T}(\mathcal{N})$ is doubly generated as a weak* closed algebra.*

Proof. Let $\mathcal{M}$ be a masa containing $\mathcal{N}''$. By Remark 7.5, there is a positive operator A such that $\mathcal{M}$ is the weak* closed algebra generated by A. Since $\mathcal{T}(\mathcal{N})$ is a bimodule over itself, it is singly generated as an $\mathcal{M}$ bimodule by some operator T. The weak* closed algebra generated by A and T contains $\mathcal{M}$, so is an $\mathcal{M}$ bimodule contained in $\mathcal{T}(\mathcal{N})$ containing T. So it equals $\mathcal{T}(\mathcal{N})$. ■

15.D Interpolation. A well known problem in Banach algebras is determining when a finite set $A_1,...,A_n$ generate the algebra as a left ideal. This is equivalent to the existence of elements $B_1,...,B_n$ such that $\sum_{i=1}^{n} B_i A_i = I$. In a C^*-algebra, it is easy to verify that a necessary and sufficient condition is that $\sum_{i=1}^{n} A_i^* A_i$ be invertible, or equivalently, that there is an $\epsilon > 0$ so that

$$\sum_{i=1}^{n} \|A_i x\|^2 \geq \epsilon \|x\|^2 \text{ for all } x \text{ in } \mathcal{H} .$$

In a nest algebra, this is necessary but not sufficient. For example, let $\mathcal{N} = \{\{0\}, \mathbb{C}e_1, \mathbb{C}^2\}$ be the standard nest on $\mathbb{C}^2 = span\ \{e_1, e_2\}$. Let $A_1 = \begin{bmatrix} 1 & 0 \\ 0 & 0 \end{bmatrix}$ and $A_2 = \begin{bmatrix} 0 & 1 \\ 0 & 0 \end{bmatrix}$. Now $A_1^* A_1 + A_2^* A_2 = I$, but if B_1 and B_2 are upper triangular, then $(B_1 A_1 + B_2 A_2)^* e_2 = 0$.

Further necessary conditions are obtained as follows: Suppose $\sum_{i=1}^{n} B_i A_i = I$ and let N belong to $\mathcal{N}$. Since compression to $N^{\perp}$ is a homomorphism (Proposition 2.15),

$$\sum_{i=1}^{n} P(N)^{\perp} B_i P(N)^{\perp} A_i P(N)^{\perp} = P(N)^{\perp} .$$

Applying this to a vector x yields

$$\begin{aligned}\|P(N)^{\perp}x\| &\leq \sum_{i=1}^{n}\|B_i\|\,\|P(N)^{\perp}A_i x\| \\ &\leq (\sum_{i=1}^{n}\|B_i\|^2)^{1/2}(\sum_{i=1}^{n}\|P(N)^{\perp}A_i x\|^2)^{1/2} \\ &= (\sum_{i=1}^{n}\|B_i\|^2)^{1/2}(\sum_{i=1}^{n}A_i^*P(N)^{\perp}A_i x, x)\ .\end{aligned}$$

Thus one obtains $\sum_{i=1}^{n}A_i^*P(N)^{\perp}A_i \geq \epsilon P(N)^{\perp}$, or equivalently

$$\sum_{i=1}^{n}\|P(N)^{\perp}A_i x\|^2 \geq \epsilon^2\|P(N)^{\perp}x\|^2$$

for all x in $\mathcal{H}$ and N in $\mathcal{N}$, where $\epsilon = (\sum_{i=1}^{n}\|B_i\|^2)^{-1/2} > 0$. It will be shown that this is sufficient.

15.18 Lemma. *Let A be a partial isometry in $\mathcal{T}(\mathcal{N})$ such that $D_A = A^*A$ belongs to $\mathcal{N}'$ (an inner operator* 14.11). Suppose $\epsilon > 0$ and $\|P(N)^{\perp}Ax\| \geq \epsilon\|P(N)^{\perp}x\|$ for all N in $\mathcal{N}$ and x in $D_A\mathcal{H}$, the initial space of A. Then there is an operator B in $\mathcal{T}(\mathcal{N})$ so that $BA = D_A$. Moreover, $\|B\| \leq 4\epsilon^{-2}$.

Proof. For N in $\mathcal{N}$ and x in $D_A\mathcal{H}$,

$$\begin{aligned}\|P(N)AP(N)^{\perp}x\|^2 &= \|AP(N)^{\perp}x\|^2 - \|P(N)^{\perp}AP(N)^{\perp}x\|^2 \\ &\leq (1-\epsilon^2)\|P(N)^{\perp}x\|^2 \leq (1-\epsilon^2)\|x\|^2\ .\end{aligned}$$

On the other hand, if x is in $D_A^{\perp}\mathcal{H}$, $P(N)AP(N)^{\perp}x = 0$. Thus by Arveson's Distance Formula (Theorem 9.5),

$$dist\,(A^*,\mathcal{T}(\mathcal{N})) = \sup_{N\in\mathcal{N}}\|P(N)^{\perp}A^*P(N)\| \leq (1-\epsilon^2)^{1/2}\ .$$

Choose C in $\mathcal{T}(\mathcal{N})$ so that $\|A^*-C\| \leq (1-\epsilon^2)^{1/2}$. Without loss of generality, we may suppose that $C = D_AC$. Let $T = I-D_A+CA$. Then

$$\|I-T\| = \|A^*A-CA\| \leq \|A^*-C\| \leq (1-\epsilon^2)^{1/2} < 1\ .$$

So T^{-1} belongs to $\mathcal{T}(\mathcal{N})$ and $\|T^{-1}\| \leq (1-(1-\epsilon^2)^{1/2})^{-1}$. Let $B = T^{-1}C$. Then

$$\begin{aligned}BA &= T^{-1}CA = (I-D_A+CA)^{-1}(I-D_A+CA-I+D_A) \\ &= I-(I-D_A+D_ACAD_A)^{-1}(I-D_A) = D_A\ .\end{aligned}$$

Finally, $\|B\| \leq (1-(1-\epsilon^2)^{1/2})^{-1}(1+(1-\epsilon^2)^{1/2}) \leq (2\epsilon^{-2})2 = 4\epsilon^{-2}$. ∎

15.19 Theorem. *Let $\mathcal{N}$ be a nest, let $\epsilon > 0$, and let $A_1,\dots,A_n$ be operators in $\mathcal{T}(\mathcal{N})$ satisfying $\sum_{i=1}^{n} \|P(N)^{\perp}A_i x\|^2 \geq \epsilon^2 \|P(N)^{\perp}x\|^2$ for all x in $\mathcal{H}$ and N in $\mathcal{N}$. Then there are operators $B_1,\dots,B_N$ in $\mathcal{T}(\mathcal{N})$ such that $\sum_{i=1}^{n} B_i A_i = I$. Moreover, if $\|A_i\| \leq 1$ for all i, one may take $\|B_i\| \leq 4n\epsilon^{-3}$ for $1 \leq i \leq n$.*

Proof. Scale each A_i so that $\|A_i\| \leq 1$, and adjust ϵ if necessary. First suppose that $\mathcal{N}$ is a finite nest. Since $\sum_{i=1}^{n} A_i^* A_i \geq \epsilon^2 I$, Theorem 14.2 provides an invertible element C of $\mathcal{T}(\mathcal{N})$ so that $C^*C = \sum_{i=1}^{n} A_i^* A_i$. Hence $A_i' = A_i C^{-1}$ satisfy $\sum_{i=1}^{n} A_i'^* A_i' = I$. Now $\|C\|^2 \leq \sum_{i=1}^{n} \|A_i\|^2 \leq n$, so

$$\begin{aligned} \|P(N)^{\perp}x\| &= \|P(N)^{\perp}CP(N)^{\perp}C^{-1}P(N)^{\perp}x\| \\ &\leq \|C\| \, \|P(N)^{\perp}C^{-1}P(N)^{\perp}x\| \\ &\leq n^{1/2} \|P(N)^{\perp}C^{-1}P(N)^{\perp}x\| . \end{aligned}$$

Hence

$$\sum_{i=1}^{n} \|P(N)^{\perp}A_i C^{-1}x\|^2 \geq \epsilon^2 \|P(N)^{\perp}C^{-1}x\|^2 \geq \frac{\epsilon^2}{n} \|P(N)^{\perp}x\|^2 .$$

So A_i' satisfy the same hypotheses with $\epsilon' = \epsilon n^{-1/2}$.

Consider the Hilbert space $\mathcal{H}^{(n)} = \mathcal{H} \otimes \mathbb{C}^n$. Then $\mathcal{B}(\mathcal{H}^{(n)}) = \mathcal{B}(\mathcal{H}) \otimes \mathcal{M}_n$ is the algebra of $n \times n$ matrices with entries in $\mathcal{B}(\mathcal{H})$. The nest $\mathcal{N}^{(n)} = \{N \otimes \mathbb{C}^n : N \in \mathcal{N}\}$ gives rise to the nest algebra $\mathcal{T}(\mathcal{N}^{(n)}) = \mathcal{T}(\mathcal{N}) \otimes \mathcal{M}_n$. This algebra contains the operator

$$A = \begin{bmatrix} A_1' & 0 & \dots & 0 \\ A_2' & 0 & \dots & 0 \\ & & & \\ A_n' & 0 & \dots & 0 \end{bmatrix} .$$

Now $A^*A = \begin{bmatrix} I & 0 & & 0 \\ 0 & 0 & & 0 \\ & & & \\ 0 & 0 & & 0 \end{bmatrix}$ is a projection which commutes with $\mathcal{N}^{(n)}$. So A is inner. Moreover,

$$A^*P(N^{(n)})^{\perp}A = \begin{bmatrix} \sum_{i=1}^{n} A_i'^*P(N)^{\perp}A_i' & 0 & 0 \\ 0 & 0 & 0 \\ 0 & 0 & 0 \end{bmatrix} \geq (\epsilon')^2 D_A P(N^{(n)})^{\perp} \ .$$

So by Lemma 15.18, there is an operator B in $\mathcal{T}(\mathcal{N}^{(n)})$ so that $BA = D_A$. The first row of the $n \times n$ matrix is $\begin{bmatrix} B_1' & \dots & B_n' \end{bmatrix}$ which yields $\sum_{i=1}^{n} B_i'A_i' = I$. The norm condition $\|B\| \leq 4(\epsilon')^{-2} = 4n\epsilon^{-2}$ implies the same condition for each B_i'. Let $B_i = C^{-1}B_i'$, so $\|B_i\| \leq 4n\epsilon^{-3}$. Then

$$\sum_{i=1}^{n} B_iA_i = C^{-1}(\sum_{i=1}^{n} B_i'A_i')C = I \ .$$

Now turn to a general nest $\mathcal{N}$. For each finite subnest $\mathcal{F}$, there are operators $B_i^{\mathcal{F}}$ in $\mathcal{T}(\mathcal{F})$ with $\|B_i^{\mathcal{F}}\| \leq 4n\epsilon^{-3}$ and $\sum_{i=1}^{n} B_i^{\mathcal{F}}A_i = I$. Thus the set

$$B_{\mathcal{F}} = \{(B_1,\dots,B_n) \in \mathcal{T}(\mathcal{F})^{(n)} : \sum_{i=1}^{n} B_iA_i = I, \ \|B_i\| \leq 4n\epsilon^{-3}\}$$

is a non-empty set which is compact in the weak* topology (because multiplication by fixed operators is weak* continuous). Since $B_{\mathcal{F}_1} \cap \dots \cap B_{\mathcal{F}_n} = B_{\mathcal{F}}$ where $\mathcal{F}$ is the nest $\bigcup_{i=1}^{n} \mathcal{F}_i$, these sets have the finite intersection property. Thus by compactness, the intersection over all finite subnests is non-empty. Let $(B_1,\dots,B_n)$ be an element of this intersection. Then $\sum_{i=1}^{n} B_iA_i = I$, $\|B_i\| \leq 4n\epsilon^{-3}$, and each B_i belongs to $\mathcal{T}\{0,N,\mathcal{H}\}$ for every N in $\mathcal{N}$, and thus belongs to $\mathcal{T}(\mathcal{N})$. ∎

15.E Invariant Operator Ranges. An *operator range* is the (non-closed) range of a bounded operator. The operator ranges invariant for an algebra forms a lattice under sums and intersection which is usually bigger than the invariant subspace lattice, and hence contains more information. In this section, the invariant operator ranges of a nest algebra will be computed.

The operator ranges in $\mathcal{H}$ are of the form $A\mathcal{H}'$ where A belongs to $\mathcal{B}(\mathcal{H}',\mathcal{H})$. But clearly there is no loss of generality in taking $\mathcal{H}' = \mathcal{H}$. Also the range $\mathcal{R}(A) = \mathcal{R}((AA^*)^{1/2})$, so one may always take A to be positive. On the other hand, it is often convenient to restrict A to $(\ker A)^{\perp}$, so that it has the same range but is also one to one.

15.20 Lemma. *The operator ranges of* $\mathcal{H}$ *form a lattice. Moreover*

$$\mathcal{R}(A)+\mathcal{R}(B) = \mathcal{R}((AA^*+BB^*)^{1/2}) = \mathcal{R}([A \;\; B])$$

$$\mathcal{R}(A) \cap \mathcal{R}(B) = \mathcal{R}((AXA^*)^{1/2})$$

where $X = P(\mathcal{H})P\,|\,\mathcal{H}$ *and* P *is the projection in* $\mathcal{B}(\mathcal{H} \oplus \mathcal{H})$ *onto* $\ker[A \;\; B]$.

Proof. Let $T = [A \;\; B]$ It is clear that $\mathcal{R}(T) = (\mathcal{R}(A)+\mathcal{R}(B))$. This equals $\mathcal{R}((TT^*)^{1/2}) = \mathcal{R}((AA^*+BB^*)^{1/2}$. Let $P = \begin{bmatrix} X & Z \\ Z^* & Y \end{bmatrix}$ be the projection onto $\ker T$. then $0 = TP$ implies $AX = -BZ^*$ and $AZ = -BY$. So

$$\mathcal{R}(AX)+\mathcal{R}(AZ) = \mathcal{R}(BZ^*)+\mathcal{R}(BY) \subseteq \mathcal{R}(A) \cap \mathcal{R}(B) \;.$$

On the other hand, if $w = Au = Bv$, then $\begin{pmatrix} u \\ -v \end{pmatrix}$ belongs to $\ker T$. So

$$\begin{pmatrix} u \\ -v \end{pmatrix} = P\begin{pmatrix} u \\ -v \end{pmatrix} = \begin{pmatrix} Xu-Zv \\ Z^*u-Yv \end{pmatrix} \;.$$

Thus $w = A(Xu-Zv) = AXu-AZv$ belongs to $\mathcal{R}(AX)+\mathcal{R}(AZ)$. Hence

$$\mathcal{R}(A) \cap \mathcal{R}(B) = \mathcal{R}(AX)+\mathcal{R}(AZ) = \mathcal{R}((A(XX^*+ZZ^*)A^*)^{1/2})$$
$$= \mathcal{R}((AXA^*)^{1/2})$$

since $P = P^2$ implies $XX^*+ZZ^* = X$. ■

The first main result is to find the algebras with no proper invariant operator ranges. For technical reasons, certain unbounded operators will be considered. Let $\mathcal{D}(X)$ denote the domain of a linear map, and write $X \subset Y$ to mean $\mathcal{D}(X) \subset \mathcal{D}(Y)$ and $Xx = Yx$ for all x in $\mathcal{D}(X)$.

15.21 Lemma. *Let* $\mathcal{S}$ *be a set of operators with no proper invariant operator ranges. Suppose that* X *is a linear map such that the graph* $\mathcal{G}(X) = \{x \oplus Xx : x \in \mathcal{D}(X)\}$ *is an operator range and such that* $SX \subset XS$ *for all* S *in* $\mathcal{S}$. *Then* X *is a scalar multiple of the identity.*

Proof. Let A be an injective operator in $\mathcal{B}(\mathcal{H},\mathcal{H} \oplus \mathcal{H})$ with range $\mathcal{G}(X)$, and let $P_1(P_2)$ be the projection of $\mathcal{H} \oplus \mathcal{H}$ onto its first (second) coordinate. Then $\mathcal{D}(X) = \mathcal{R}(P_1A)$ is an operator range which is invariant for $\mathcal{S}$. Thus $\mathcal{D}(X) = \mathcal{H}$. So P_1A is a bounded bijection and hence has a bounded inverse by Banach's Open Mapping Theorem. Thus $X = P_2A(P_1A)^{-1}$ is bounded. The identity $SX = XS$ for S in $\mathcal{S}$ implies that $\mathcal{R}(X)$ is an invariant range for $\mathcal{S}$, so equals 0 or $\mathcal{H}$. Choose λ so that $X-\lambda I$ is not invertible and hence $\mathcal{R}(X-\lambda I) \neq \mathcal{H}$. As $\mathcal{R}(X-\lambda I)$ is also invariant for $\mathcal{S}$, it must equal $\{0\}$; so $X = \lambda I$. ■

The next lemma is the weak operator topology analogue of Lemma 15.4.

15.22 Lemma. *Let S be a unital subalgebra of $B(\mathcal{H})$. An operator A belongs to the weak operator closure of S if and only if $A^{(n)}$ belongs to $Alg\,Lat\,S^{(n)}$ for all $n \geq 1$.*

Proof. Let $\phi(T) = \sum_{j=1}^{n}(Tx_j,y_j)$ be a weakly continuous function annihilating S. Let $x = (x_1,...,x_n)$ and $y = (y_1,...,y_n)$ be vectors in $\mathcal{H}^{(n)}$. Clearly $(S^{(n)}x,y) = 0$ for all S in S. So $M = \overline{S^{(n)}x}$ is an invariant subspace of $S^{(n)}$ containing x and orthogonal to y. If $A^{(n)}$ belongs to $Alg\,Lat\,S^{(n)}$ for all $n \geq 1$, then

$$(A^{(n)}x,y) = (P(M)^{\perp}A^{(n)}P(M)x,y) = 0 \quad .$$

So by the Hahn-Banach Theorem, A belongs to $\overline{S}^{WOT}$. The converse is clear. ■

15.23 Lemma. *Let S be a unital algebra with no invariant operator ranges. Let L be an invariant operator range for $S^{(n)}$ which contains a vector $x_1 \oplus ... \oplus x_n$ with linearly independent components. Then $L = \mathcal{H}^{(n)}$.*

Proof. The case $n = 1$ is given by hypothesis. Assume it is valid for all $m < n$. If L contains a vector $0 \oplus ... \oplus 0 \oplus x$ with $x \neq 0$, then

$$L_n = \{y \in \mathcal{H}: 0 \oplus ... \oplus 0 \oplus y \in L \cap 0 \oplus ... \oplus 0 \oplus \mathcal{H}\}$$

is a non-zero operator range for S. Thus $L_n = \mathcal{H}$. So $L = L' \oplus L_n$ where $L' = \{y \in \mathcal{H}^{(n-1)}: y \oplus 0 \in L\}$ is an invariant operator range for $S^{(n-1)}$. Now L' contains $x_1 \oplus ... \oplus x_{n-1}$, so by the induction hypothesis, $L' = \mathcal{H}^{(n-1)}$ whence $L = \mathcal{H}^{(n)}$.

Suppose on the other hand that $L_n = \{0\}$. Let L' be the projection of L onto the first $n-1$ coordinates. Then there is a linear map X of L' into $\mathcal{H}$ so that $L = \{y \oplus Xy : y \in L'\}$. Hence

$$S^{(n)}(y \oplus Xy) = S^{(n-1)}y \oplus SXy = S^{(n-1)}y \oplus XS^{(n-1)}y \quad .$$

In particular, L' is an invariant operator range for $S^{(n-1)}$ containing $x_1 \oplus ... \oplus x_{n-1}$. By induction, $L' = \mathcal{H}^{(n-1)}$. Therefore there are linear maps X_i of $\mathcal{H}$ into $\mathcal{H}$ so $X(y_1,...,y_{n-1}) = \sum_{i=1}^{n-1} X_i y_i$. It follows that $M_i = \{y \oplus X_i y : y \in \mathcal{D}(\mathcal{H})\}$ is an invariant operator range for $S^{(2)}$ (unitarily equivalent to the intersection of L with $0 \oplus ... \oplus \mathcal{H} \oplus 0 ... \oplus \mathcal{H}$). By Lemma 15.21, $X_i = \lambda_i I$. Thus $x_n = \sum_{i=1}^{n-1} \lambda_i x_i$ for every $(x_1,...,x_n)$ in L. This is contrary to hypothesis, so this case does not occur. ■

15.24 Theorem. *Let S be a WOT closed, unital algebra with no proper invariant operator ranges. Then $S = B(\mathcal{H})$.*

Proof. Let A belong to $B(\mathcal{H})$. Let $x = x_1 \oplus \ldots \oplus x_n$ be a vector in $\mathcal{H}^{(n)}$, and let $M = \overline{S^{(n)}x}$ be the smallest invariant subspace containing x. Without loss of generality, one may assume that $x_1,\ldots,x_k$ are linearly independent and $x_j = \sum_{i=1}^{k} c_{ji} x_i$ for $k+1 \le j \le n$. Let L be the projection of M onto $\mathcal{H}^{(k)}$. This is an invariant operator range of $S^{(k)}$. So Lemma 15.23 implies $L = \mathcal{H}^{(k)}$. Hence M equals

$$\{y_1 \oplus \ldots y_n : y_i \in \mathcal{H},\ 1 \le i \le k,\ y_j = \sum_{i=1}^{k} c_{ji} y_i,\ k+1 \le j \le n\} .$$

Clearly, this is invariant for $A^{(m)}$. By Lemma 15.22, A belongs to S. But A was arbitrary, so $S = B(\mathcal{H})$. ∎

Now consider a norm closed unital algebra that does have invariant operator ranges.

15.25 Lemma. *Let S be a closed unital algebra, and let A be an injective operator in $B(\mathcal{H}',\mathcal{H})$ so that $A\mathcal{H}'$ is invariant under S. Then there is a continuous representation π of S in $B(\mathcal{H}')$ so that $SA = A\pi(S)$ for all S in S.*

Proof. For each x in $\mathcal{H}'$ and S in S, SAx belongs to $A\mathcal{H}'$ and hence there is a unique vector $y = y(S,x)$ in $\mathcal{H}'$ so that $Ay = SAx$. The uniqueness forces $y(S,x)$ to be linear in both variables. Thus for each S, there is a linear map $\pi(S)x = y(S,x)$ which is everywhere defined. It will be shown that $\pi(S)$ is closed, and hence is bounded by the Closed Graph Theorem. So suppose that $\lim x_n = 0$ and $\lim \pi(S)x_n = y_0$. Then

$$Ay_0 = \lim_{n\to\infty} A\pi(S)x_n = \lim_{n\to\infty} SAx_n = 0 .$$

As A is injective, $y_0 = 0$ and $\pi(S)$ is closed.

Thus π is a linear map of S into $B(\mathcal{H}')$. Apply the Closed Graph Theorem again. Suppose $\lim S_n = 0$ and $\lim \pi(S_n) = X$ in $B(\mathcal{H}')$. Then

$$AX = \lim_{n\to\infty} A\pi(S_n) = \lim_{n\to\infty} S_n A = 0 .$$

Again the injectivity of A implies $X = 0$. So π is continuous. ∎

15.26 Theorem. *Let $\mathcal{A}$ be an abelian C^* algebra, and let L be an invariant operator range for $\mathcal{A}$. Then there is a positive operator A in $\mathcal{A}'$ such that $L = A\mathcal{H}$.*

Proof. Let $L = B\mathcal{H}'$ for some injective operator B. By Lemma 15.25, there is a representation π of $\mathcal{A}$ in $\mathcal{B}(\mathcal{H}')$ so that $SB = B\pi(S)$ for all S in $\mathcal{A}$. Let $\mathcal{U}$ be the unitary group of $\mathcal{A}$, so $\pi(\mathcal{U})$ is a bounded abelian group in $\mathcal{B}(\mathcal{H}')$. By Corollary 17.2, there is an operator T so that $T\pi(U)T^{-1}$ is unitary for all U in $\mathcal{U}$. Let $B_0 = BT^{-1}$ and $\pi_0(S) = T\pi(S)T^{-1}$. Then $\mathcal{R}(B_0) = \mathcal{R}(B) = L$ and π_0 is a $*$-representation satisfying

$$SB_0 = SBT^{-1} = B\pi(S)T^{-1} = B_0\pi_0(S) \quad .$$

The polar decomposition of B_0^* yields the factorization $B_0 = AU$ where $A = (B_0B_0^*)^{1/2}$ is a positive operator in $\mathcal{B}(\mathcal{H})$ and U is an isometry (since B_0 is injective) of $\mathcal{H}'$ onto $\overline{L}$. Let $\pi_1(S) = U\pi_0(S)U^*$ be a $*$-representation of $\mathcal{A}$ into $\mathcal{B}(\mathcal{H})$. Then for all S in $\mathcal{A}$,

$$SAU = AU\pi_0(S) = A\pi_1(S)U \quad .$$

Since U is an isometry onto $\overline{L} = (\ker A)^{\perp} \supseteq (\ker \pi_1(S))^{\perp}$, $SA = A\pi_1(S)$. Now $S^*A = A\pi_1(S^*) = A\pi_1(S)^*$, so

$$SA^2 = A\pi_1(S)A = A(A\pi_1(S)^*)^* = A(S^*A)^* = A^2S \quad .$$

So A^2, and hence A, belongs to $\mathcal{A}'$. ■

Now we are in a position to compute the invariant operator ranges of algebras with commutative subspace lattices (CSL's). (See Chapter 22 for detailed information about this class.)

15.27 Lemma. *Let $\mathcal{H} = \sum_{n\geq 1} \oplus \mathcal{H}_n$ and let T_{ij} be operators in $\mathcal{B}(\mathcal{H}_j,\mathcal{H}_i)$. Suppose $B = (b_{ij})$ is a bounded operator on ℓ_2 such that $\|T_{ij}\| \leq b_{ij}$ for every $i,j \geq 1$. Then the matrix $T = (T_{ij})$ represents a bounded operator from $\mathcal{H}$ to itself with $\|T\| \leq \|B\|$.*

Proof. Let $x = (x_k)$ be a unit vector in $\mathcal{H}$. Then $u_i = \|x_i\|$ determines a unit vector $u = (u_1,u_2,\ldots)$ in ℓ_2. So

$$\|Tx\|^2 = \sum_{i=1}^{\infty}\|\sum_{j=1}^{\infty}T_{ij}x_j\|^2 \leq \sum_{i=1}^{\infty}(\sum_{j=1}^{\infty}\|T_{ij}\|\ \|x_j\|)^2$$

$$\leq \sum_{i=1}^{\infty}(\sum_{j=1}^{\infty}b_{ij}u_j)^2 = \|Bu\|^2 \leq \|B\|^2 \quad .$$ ■

15.28 Lemma. *Let $\mathcal{T}$ be a norm closed algebra and let $\mathcal{M}$ be a masa such that $\mathcal{M}\mathcal{T}\mathcal{M} = \mathcal{T}$. Suppose A is a positive contraction in $\mathcal{M}$ and $L = \mathcal{R}(A)$ is invariant for $\mathcal{T}$. Set $E_n = E_A(2^{-n},1]$ and let $L_n = \overline{\mathcal{T}E_n\mathcal{H}}$. There is an integer k so that $P(L_n) \leq E_{n+k}$ for all $n \geq 1$.*

Proof. If the lemma is false, it is due to one of three possibilities:

Case 1. There exists an n_0 such that $E_A\{0\}L_{n_0} \neq 0$.

Case 2. There exists an n_0 so that $E_n^{\perp}L_{n_0} \neq 0$ for all $n \geq 0$.

Case 3. There is a sequence $n_1 < n_2 < \ldots$ such that $E_{n_k+3k}^{\perp}L_{n_k} \neq 0$ for $k \geq 1$.

Exclude from Cases 2 and 3 any situation handled by the previous case.

Case 1. Since $\mathcal{T}E_{n_0}\mathcal{H}$ is dense in L_{n_0}, there is an operator T in $\mathcal{T}$ so that $E_A\{0\}TE_{n_0} \neq 0$. But $E_A\{0\}A = 0$ while $E_{n_0}\mathcal{H}$ is contained in $\mathcal{R}(A)$. Thus $\mathcal{R}(A)$ is not invariant for T, contrary to hypothesis.

Case 2. As case 1 is excluded, $\mathcal{T}E_{n_0} = E_A\{0\}^{\perp}\mathcal{T}E_{n_0}$. Since $E_A\{0\}^{\perp} = \sup E_n$, one can extract a subsequence $n_0 < n_1 < n_2 < \ldots$ so that $n_k \geq 4k$ and $P_k = E_{n_{k+1}} - E_{n_k}$ satisfies $P_k\mathcal{T}E_{n_0} \neq 0$. Choose $T_k = P_kT_kE_{n_0}$ in $\mathcal{T}$ of norm one (possible since $\mathcal{M}\mathcal{T}\mathcal{M} = \mathcal{T}$), and set $T = \sum_{k=1}^{\infty} 2^{-k}T_k$. Choose unit vectors y_k in $E_{n_0}\mathcal{H}$ so that $\|T_ky_k\| > .9$. Inductively define $\epsilon_k = 0$ or 1, and $w_k = \sum_{j=1}^{k} \epsilon_j 4^{-j}y_j$ as follows: if $\|T_kw_{k-1}\| < 4^{-k}/2$, set $\epsilon_k = 1$; otherwise set $\epsilon_k = 0$. Let $y = \sum_{k=1}^{\infty} \epsilon_k 4^{-k}y_k$. Then

$$\|P_kTy\| = \|2^{-k}T_ky\| \geq 2^{-k}\Big|\|T_kw_{k-1}\| - \epsilon_k 4^{-k}\|T_ky_k\|\Big| - 2^{-k}\sum_{j>k} 4^{-j}$$
$$\geq 2^{-k}\cdot 4^{-k}(.9 - .5) - 2^{-k}\cdot 4^{-k}/3 = 8^{-k}/15 \ .$$

Now $y = E_{n_0}y$ belongs to $\mathcal{R}(A)$. However $\|A\,|P_k\mathcal{H}\| \leq 2^{-n_k} \leq 2^{-4k}$, so $\|A^{-1}P_kTy\| \geq 2^k/15$. It follows that Ty is not in $\mathcal{R}(A)$. This contradicts the invariance of L, and so rules out case 2.

Case 3. As case 2 is ruled out, for each n_k there is an integer m_k so that $E_{m_k}^{\perp}\mathcal{T}E_{n_k} = 0$. Drop to a subsequence (also labelled n_k for convenience) so that $n_{k+1} > m_k$. Let $P_k = E_{n_k} - E_{n_{k-1}}$ and $Q_k = E_{m_k} - E_{n_k+3k}$. Then $Q_k\mathcal{T}P_k \neq 0$ but $Q_k\mathcal{T}E_{n_{k-1}} = 0$. Choose operators $T_k = Q_kT_kP_k$ of norm one in $\mathcal{T}$, and choose unit vectors $y_k = P_ky_k$ so that $\|T_ky_k\| > ½$. Set $T = \sum_{k=1}^{\infty} 2^{-k}T_k$. Now $AP_k \geq 2^{-n_k}P_k$ so there is a (unique) vector

$z_k = P_k z_k$ such that $Az_k = y_k$; and $\|z_k\| \le 2^{n_k}$. Set $z = \sum_{k=1}^{\infty} 2^{-n_k-k} z_k$. Then $y \equiv Az = \sum_{k=1}^{\infty} 2^{-n_k-k} y_k$ belongs to $\mathcal{R}(A)$. But

$$Ty = \sum_{k=1}^{\infty} 2^{-n_k-k} 2^{-k} T_k y_k = \sum_{k=1}^{\infty} 2^{-n_k-2k} Q_k T_k y_k \ .$$

Now $\|AQ_k\| \le 2^{-n_k-3k}$, so $\|A^{-1}Q_k Ty\| \ge 2^k \|T_k y_k\| > 2^{k-1}$. Consequently, Ty is not in $\mathcal{R}(A)$. This again is a contradiction, so the lemma is established. ■

15.29 Theorem. *Let $\mathcal{T}$ be a norm closed algebra such that $(\mathcal{T} \cap \mathcal{T}^*)''$ contains a masa $\mathcal{M}$ and $\mathcal{M}\mathcal{T}\mathcal{M} = \mathcal{T}$. Let $\mathcal{L} = Lat\,\mathcal{T}$. Then the following are equivalent:*

(1) *L is an invariant operator range for $\mathcal{T}$.*

(2) *L is an invariant operator range for $Alg\,\mathcal{L}$.*

(3) *$L = A\mathcal{H}$ where $A = \sum_{n\ge 1} 2^{-n} P(L_n)$ where $\{L_n, n \ge 1\}$ is an increasing sequence of elements of $\mathcal{L}$.*

(3′) *$L = A\mathcal{H}$ where A is on the norm closed convex hull of $\{P(L): L \in \mathcal{L}\}$.*

(4) *$L = A\mathcal{H}$ where A is in the weak operator closed convex hull of $\{P(L): L \in \mathcal{L}\}$.*

(4′) *$L = A\mathcal{H}$ where A belongs to $\mathcal{M}_+$ and the ranges of the spectral projections $E_A(\epsilon,\infty)$ belong to $\mathcal{L}$ for all $\epsilon \ge 0$.*

Proof. Clearly, (3) ⇒ (3′) ⇒ (4) ⇒ (4′). If (4′) holds, normalize so that $\|A\| = 1$. Let $P_0 = 0$ and $P_n = E_A(2^{-n},\infty)$. By hypothesis, there is an increasing sequence L_n in $\mathcal{L}$ so that $P_n = P(L_n)$. Let

$$A_0 = \sum_{n=1}^{\infty} 2^{-n} P_n = \sum_{n=1}^{\infty} 2^{1-n}(P_n - P_{n-1}) = \sum_{n=1}^{\infty} 2^{1-n} E_A(2^{-n}, 2^{1-n}] \ .$$

Then $\frac{1}{2}A_0 \le A \le A_0$, hence there is an invertible operator M in $\mathcal{M}$ so that $A_0 = AM$. Thus $\mathcal{R}(A_0) = \mathcal{R}(A) = L$, whence (4′) ⇒ (3).

If (3) holds, set $Q_n = P(L_n) - P(L_{n-1})$ for $n \ge 1$ and $L_0 = 0$. Fix T in $Alg\,\mathcal{L}$, and let $T_{nm} = Q_n T Q_m | Q_m \mathcal{H}$. Since $TQ_m = P(L_m)TQ_m$, $T_{nm} = 0$ for $n > m$. Now $AQ_m = 2^{1-m} Q_m$, thus "$A^{-1}T_{nm}A$" $= 2^{n-m} T_{nm}$. Set $b_{nm} = 0$ for $n > m$ and $b_{nm} = 2^{n-m}\|T\|$ if $n \le m$. Then $B = (b_{nm})$ is a bounded operator on ℓ_2 with $\|B\| \le \sum_{k\ge 0} 2^{-k}\|T\| = 2\|T\|$. By Lemma 15.27, the operator S on $\mathcal{H} = \sum_{n=1}^{\infty} \oplus\, Q_n \mathcal{H}$ with matrix $(2^{n-m} T_{nm})$ is bounded. Moreover A has the diagonal matrix form $diag(2^{1-n} Q_n)$, so AS and TA both have the matrix form $(2^{1-n} T_{mn})$. Thus $TA = AS$ and consequently $\mathcal{R}(A)$ is invariant for T. So (3) implies (2). Clearly (2) implies (1).

Suppose (1) holds. Then L is invariant for $\mathcal{T} \cap \mathcal{T}^*$. Hence by Theorem 15.26, there is a positive operator A in $(\mathcal{T} \cap \mathcal{T}^*)'$ so that $L = A\mathcal{H}$. Since $(\mathcal{T} \cap \mathcal{T}^*)''$ contains $\mathcal{M}$, $(\mathcal{T} \cap \mathcal{T}^*)' = (\mathcal{T} \cap \mathcal{T}^*)'''$ is contained in $\mathcal{M}$. Normalize A so that $\|A\| = 1$, and let $E_n = E_A(2^{-n},1]$. Let $L_n = \overline{\mathcal{T}E_n\mathcal{H}}$ be the smallest subspace in $Lat\,\mathcal{T}$ containing $\mathcal{R}(E_n)$. By Lemma 15.28, there is an integer k so that $P(L_n) \leq E_{n+k}$ for all $n \geq 1$. Once this is established, set

$$A_0 = \sum_{n=1}^{\infty} 2^{-n} P(L_n) \leq \sum_{n=1}^{\infty} 2^{-n} E_{n+k} \leq 2^k A \ .$$

Then $2^{-k}A_0 \leq A \leq A_0$. Hence there is an invertible operator M in $\mathcal{M}$ so that $A_0 = AM$. Thus $L = \mathcal{R}(A) = \mathcal{R}(A_0)$ and (1) implies (3). ∎

15.30 Corollary. *Let $\mathcal{N}$ be a nest such that $\mathcal{N}^{\perp}$ is well ordered. Then the only invariant operator ranges of $\mathcal{T}(\mathcal{N})$ are the elements of $\mathcal{N}$ itself.*

Proof. Let L be an invariant operator range of $\mathcal{T}(\mathcal{N})$. By Theorem 15.29, there is an increasing sequence N_k in $\mathcal{N}$ so that $L = A\mathcal{H}$ where $A = \sum_{k \geq 1} 2^{-k} P(N_k)$. But the sequence N_k is eventually constant by the well ordering hypothesis, say $N_k = N_{k_0}$ for $k \geq k_0$. Then $2^{-k_0} P(N_{k_0}) \leq A \leq P(N_{k_0})$ and thus $L = N_{k_0}$. ∎

15.31 Remark. The condition $(\mathcal{T} \cap \mathcal{T}^*)''$ contains $\mathcal{M}$ is essential. Let $\mathcal{T}_0$ be the algebra of strictly upper triangular operators with respect to a basis $\{e_n, n \geq 1\}$. This algebra has invariant operator ranges which are not the range of a diagonal operator. The proof of Theorem 15.29 shows however that every range of a diagonal operator invariant for $\mathcal{T}_0$ is also invariant for $\mathcal{T} = Alg\,Lat\,\mathcal{T}_0$.

Notes and Remarks.

A deeper treatment of tensor products can be found in C^*-algebra texts such as Kadison-Ringrose, [3] vol. 2 or Takesaki [2]. The notion of reflexivity for subspaces is due to Loginov and Sul'man [1]. Even further generalization of this notion is found in Erdos [12]. The duality argument, Lemma 15.4, and Theorems 15.9 and 15.11 and their corollaries are due to Arveson [6]. The treatment given here is new. Radjavi and Rosenthal [1] proved Corollary 15.12 for the weak operator topology. Theorem 15.14 is due to Erdos and Power [1]. Theorem 15.15 is somewhat new, but the ideas are mostly contained in Power [9] who proves Corollary 15.16. Longstaff [1] proved Corollary 15.17. Theorem 15.19 was proved for a special nest by Arveson [7]. The proof in full generality is basically the same. Operator ranges were introduced by Foias [1] who proved most of the results of the last section. The proof of Lemma 15.20 is taken from Fillmore and Williams [1]. Lemma 15.27 is due to Nordgren, Radjabalipour, Radjavi and Rosenthal [1]. Lemma 15.19 and Theorem 15.29 are due to Davidson [3], but the main ideas are already in Foias's paper. Davidson [3] contains the example referred to in Remark 15.31 (See Exercise 15.11). Ong [1, 2] has some related results.

Exercises.

15.1 Show that there is a constant C so that

$$dist\,(T, \mathcal{A} \otimes I) \leq C \sup_{P \in Lat\, \mathcal{A} \otimes I} \|P^{\perp} T P\|$$

for every T in $\mathcal{B}\,(\mathcal{H}_1 \otimes \mathcal{H})$ and every weak* closed subalgebra of $\mathcal{B}\,(\mathcal{H}_1)$.

15.2* Prove Lemma 15.7 without measure theory.

15.3 One has to go to certain extra trouble in Lemmas 15.5 and 15.7 due to the fact that the positive operators do not form a lattice. The following exercises are to emphasize this fact.

(a) Show that a 2X2 Hermitian matrix A is positive if and only if $tr\, A \geq 0$ and $\det A \geq 0$.

(b) Let $A = \begin{bmatrix} 2 & 0 \\ 0 & 0 \end{bmatrix}$ and $B = \begin{bmatrix} 1 & 0 \\ 0 & 1 \end{bmatrix}$. Find two (or all) positive operators P which are greater than both A and B and are minimal with this property.

(c) Show that the self-adjoint elements in an abelian von Neumann algebra do form a complete lattice.

(d) Let $X = \begin{bmatrix} 1 & 1 \\ 1 & 1 \end{bmatrix}$. Find two (or all) positive, diagonal operators $D \geq X$ which are minimal with this property.

15.4 (Rosenthal) Show that the algebra $\mathcal{A}$ consisting of all operators on $\mathcal{H} \oplus \mathcal{H}$ of the form $\begin{bmatrix} \alpha I & T \\ 0 & \beta I \end{bmatrix}$ where T belongs to $\mathcal{B}\,(\mathcal{H})$ and α, β are scalars is a reflexive algebra which is not finitely generated.

15.5 (Erdos-Power) Let $\mathcal{S} = \mathcal{S}_\Phi$ be a weak* closed $\mathcal{T}\,(\mathcal{N})$ bimodule.

(a) Show that $\mathcal{S}$ is an ideal if and only if $\Phi(N) \leq N$ for all N in $\mathcal{N}$.

(b) Show that $\mathcal{S}$ is an algebra if and only if $\Phi(\Phi(N)) \leq \Phi(N)$ for all N in $\mathcal{N}$.

15.6 (Erdos-Power) Show that $\mathcal{S}_\Phi$ is the second dual of $\mathcal{S}_\Phi \cap \mathcal{K}$.

15.7 (Erdos-Power) Prove that $S_\Phi + K$ is norm closed.

15.8* (Power) Does every nest have the property: if A_i belong to $\mathcal{T}(\mathcal{N})$ and $\sum_{i=1}^{n} A_i^* A_i = P$ is invertible, then there is an operator C in $\mathcal{T}(\mathcal{N})^{-1}$ such that $C^*C = P$.

15.9 (Arveson)

(a) Use Theorem 8.9 to construct an expectation of $B(\mathcal{H}^2)$ onto the set of Toeplitz operators, and takes operators in $\mathcal{T}(\mathcal{P})$ to co-analytic Toeplitz operators where

$$\mathcal{P} = \{P_n = span\{1, z, \ldots, z^n\}, n \geq 0\} .$$

(b) If $f_1, \ldots, f_n$ belong to H^∞ and $\sum_{i=1}^{n} \| T_{f_i}^* x \|^2 \geq \epsilon^2 \| x \|^2$ for all x in H^2, prove that there exist $g_1, \ldots, g_n$ in H^∞ so that $\sum_{i=1}^{n} g_i f_i = 1$.

15.10 Let $\mathcal{T}_0^*$ denote the strictly lower triangular operators with respect to a basis $\{e_n, n \geq 1\}$. Find all invariant operator ranges of $\mathcal{T}_0^*$.

15.11 (Davidson) Let $\mathcal{T}_0$ be the strictly upper triangular operators with respect to a basis $\{e_n, n \geq 1\}$. Let $\{f_n, g_n, n \geq 1\}$ be a basis for a Hilbert space $\mathcal{H}'$, and set $x_n = f_n + 2^{-k^2/2} k g_k$ for $2^{k-1} \leq n < 2^k$, $k \geq 1$. Let $\mathcal{X} = span\{x_n, n \geq 1\}$. Define A in $B(\mathcal{X}, \mathcal{H})$ by $Ax_n = 2^{-n^2} e_n$ and extend linearly. For each T in $\mathcal{T}_0$ with matrix (t_{ij}), define $\pi(T)$ in $B(X)$ by $\pi(T)x_n = \sum_{i=1}^{n-1} 2^{i^2 - n^2} t_{in} x_i$.

(a) Prove that π is continuous and $TA = A\pi(T)$ for all T in $\mathcal{T}_0$. Hence $\mathcal{R}(A)$ is invariant for $\mathcal{T}_0$.

(b) Prove that π does not extend to $\mathcal{T}$, and hence $\mathcal{R}(A)$ is not invariant for $\mathcal{T}$.

16. Duality

In this chapter, we show that any nest algebra is the second dual of $\mathcal{T}(\mathcal{N}) \cap \mathcal{K}$. Knowledge of the extreme points of the pre-annihilator of $\mathcal{T}(\mathcal{N})$ will yield new proofs of the Arveson Distance Formula and the Erdos Density Theorem. Because we wish to obtain these results as corollaries, we must avoid their use in the development used here. The reader will probably recognize a number of short cuts based on the material developed back in Chapter 3. (We will use the easy Lemma 3.7.) In the last half of this chapter, we investigate the decomposition of trace class operators in $\mathcal{T}(\mathcal{N})$ into sums of rank one operators.

Let

$$\mathcal{A} = \mathcal{T}(\mathcal{N})_{\perp} = \{A \in \mathcal{C}_1 : tr(TA)=0 \text{ for all } T \text{ in } \mathcal{T}(\mathcal{N})\},$$

and let

$$\begin{aligned}\mathcal{A}_0 &= \{A \in \mathcal{C}_1 \cap \mathcal{T}(\mathcal{N}) : \Delta(A) = 0\} \\ &= \{A \in \mathcal{C}_1 : P(N_-)^{\perp} A P(N) = 0 \text{ for all } N \text{ in } \mathcal{N}\} \ .\end{aligned}$$

Also let $\mathcal{R} = \mathcal{T}(\mathcal{N}) \cap \mathcal{K}$ and let $\mathcal{R}_0$ be the norm closed span of the rank one operators in $\mathcal{R}$. (By Corollary 3.12, we know that $\mathcal{R}_0 = \mathcal{R}$.) The first step will be to compute the extreme points of $\mathcal{A}_0$. To do this, we need a decomposition argument based on the 2×2 matrix results Lemma 14.13 and Corollary 14.14.

16.1 Lemma. *Let* $T = \begin{bmatrix} T_{11} & T_{12} \\ 0 & T_{22} \end{bmatrix}$ *represent a trace class operator from* $\mathcal{H}_1 \oplus \mathcal{H}_2$ *to* $\mathcal{K}_1 \oplus \mathcal{K}_2$. *Then there is a decomposition* $T = T_1 + T_2$ *such that* $T_1 = \begin{bmatrix} T_{11} & Y_1 \\ 0 & 0 \end{bmatrix}$, $T_2 = \begin{bmatrix} 0 & Y_2 \\ 0 & T_{22} \end{bmatrix}$, *and* $\|T\|_1 = \|T_1\|_1 + \|T_2\|_1$.

Proof. Let $T = UQ$ be the polar decomposition of T. Write $Q = \begin{bmatrix} A & B \\ B^* & D \end{bmatrix}$ and apply Lemma 14.13. There is a bounded operator X from $\mathcal{H}_2$ to $\mathcal{H}_1$ so that $B = A^{1/2}X$. Also, $Q_1 = \begin{bmatrix} A & B \\ B^* & X^*X \end{bmatrix}$ is positive and less than Q; where $Q_2 = \begin{bmatrix} 0 & 0 \\ 0 & D - X^*X \end{bmatrix}$ is positive. Since Q is trace class, so are Q_1 and Q_2 and

$$\|Q\|_1 = trQ = trQ_1 + trQ_2 = \|Q_1\|_1 + \|Q_2\|_1 \ .$$

Let $U = [U_{ij}]$, and define $T_1 = UQ_1$ and $T_2 = UQ_2$. Clearly, T_2 has the form $\begin{bmatrix} 0 & * \\ 0 & * \end{bmatrix}$, so T_1 has the form $\begin{bmatrix} T_{11} & * \\ 0 & * \end{bmatrix}$. Since $UQ = T$ is triangular, Corollary 14.14 yields the identity $U_{21}A^{½} + U_{22}X^* = 0$. So

$$T_1 = UQ_1 = \begin{bmatrix} T_{11} & Y_1 \\ 0 & U_{21}B + U_{22}X^*X \end{bmatrix}$$

$$= \begin{bmatrix} T_{11} & Y_1 \\ 0 & (U_{21}A^{½} + U_{22}X^*)X \end{bmatrix} = \begin{bmatrix} T_{11} & Y_1 \\ 0 & 0 \end{bmatrix} .$$

Thus $T_2 = T - T_1 = \begin{bmatrix} 0 & Y_2 \\ 0 & T_{22} \end{bmatrix}$ where $Y_2 = T_{12} - Y_1$. Finally,

$$\|T\|_1 = \|Q\|_1 = \|Q_1\|_1 + \|Q_2\|_1 \geq \|T_1\|_1 + \|T_2\|_1 \geq \|T\|_1 \ .$$

So equality follows. ■

16.2 Lemma. *$\mathcal{A}_0$ is the annihilator of $\mathcal{R}_0$. The extreme points of $b_1(\mathcal{A}_0)$ are the rank one operators $x \otimes y^*$ where $\|x\| = \|y\| = 1$, and for some N in $\mathcal{N}$, $x \in N$ and $y \in N^\perp$.*

Proof. By Lemma 3.7, $\mathcal{R}_0$ is spanned by $u \otimes v^*$ where $u \in N$ and $v \in N_-^\perp$ for some N in $\mathcal{N}$. Thus the annihilator of $\mathcal{R}_0$ consists of all trace class operators A such that

$$0 = tr(Au \otimes v^*) = (Au, v)$$

for all such u and v. Hence AN is orthogonal to $(N_-)^\perp$. That is, $P(N_-)^\perp AP(N) = 0$ for all N in $\mathcal{N}$. Conversely, if this holds, A annihilates each rank one operator in $\mathcal{R}_0$. This condition characterizes $\mathcal{A}_0$, so $\mathcal{A}_0 = \mathcal{R}_0^\perp$.

Let A be an extreme point of $b_1(\mathcal{A}_0)$. Let N belong to $\mathcal{N}$. Since $\Delta(A) = 0$, A maps N into N_-. So as an operator from $\mathcal{H} = N \oplus N^\perp$ to $\mathcal{H} = N_- \oplus N_-^\perp$, A has the form $A = \begin{bmatrix} A_{11} & A_{12} \\ 0 & A_{22} \end{bmatrix}$. If A_{11} and A_{22} are both non-zero, Lemma 16.1 decomposes $A = A_1 + A_2$ so that $\|A\|_1 = \|A_1\|_1 + \|A_2\|_1$ and $A_1 = \begin{bmatrix} A_{11} & Y_1 \\ 0 & 0 \end{bmatrix}$ and $A_2 = \begin{bmatrix} 0 & Y_2 \\ 0 & A_{22} \end{bmatrix}$. Both A_i are non-zero. Since A_1 and A_2 belong to $\mathcal{T}(\mathcal{N})$ and $\Delta(A) = \Delta(A_1) \oplus \Delta(A_2) = 0$, it follows that both A_1 and A_2 belong to $\mathcal{A}_0$. Hence

$$A = \|A_1\|_1(\|A_1\|_1^{-1}A_1) + \|A_2\|_1(\|A_2\|_1^{-1}A_2)$$

is not an extreme point. Consequently, for each N in $\mathcal{N}$, either

$A = P(N_-)A$ or $A = AP(N)^\perp$.

Let N be the greatest element of $\mathcal{N}$ such that $A = AP(N)^\perp$. For $N' > N$, $A = P(N'_-)A$. Take the infimum over all $N' > N$, and since $(N_+)_- \leq N$, one obtains $A = P(N)AP(N)^\perp$. Now $P(N)\mathcal{B}(\mathcal{H})P(N)^\perp$ is contained in $\mathcal{A}_0$. By section 1.2, $A = \sum_{n\geq 1} s_n e_n \otimes f_n^*$ where s_n are the s-numbers of A and $\|e_n\| = \|f_n\| = 1$. Since $\sum_{n\geq 1} s_n = \|A\|_1 = 1$, $e_n \in N$, and $f_n \in N^\perp$, A is the convex combination of elements $e_n \otimes f_n^*$ in $b_1(\mathcal{A}_0)$. As A is extreme, $s_1 = 1$ and $s_n = 0$ for $n \geq 2$. That is, A is rank one of the desired form. ■

16.3 Lemma. *Let* $T = \begin{pmatrix} T_1 & T_2 \\ T_3 & T_4 \end{pmatrix}$ *be a trace class operator from one space to another. Suppose* $\|T\|_1 \leq 1$ *and* $\|T_1\|_1 \geq 1-\epsilon$ *for* $0 \leq \epsilon \leq 1$. *Then* $\left\| \begin{bmatrix} 0 & T_2 \\ T_3 & T_4 \end{bmatrix} \right\|_1 \leq 4\epsilon^{1/2}$.

Proof. Let A_i, $1 \leq i \leq 4$, be norm one operators such that $tr(T_iA_i) = \|T_i\|_1$. Then

$$1 \geq tr(T\begin{bmatrix} A_1 & 0 \\ 0 & A_4 \end{bmatrix}) = tr(T_1A_1)+tr(T_4A_4)$$

$$= \|T_1\|_1+\|T_4\|_1 \ .$$

Hence $\|T_4\| \leq \epsilon$. Also if c and s are positive reals with $c^2+s^2 = 1$,

$$1 \geq tr(T\begin{bmatrix} cA_1 & 0 \\ sA_2 & 0 \end{bmatrix}) = c\ tr(T_1A_1)+s\ tr(T_2A_2)$$

$$= c\|T_1\|_1+s\|T_2\|_1 \ .$$

Maximizing over all choices of c yields $\|T_1\|_1^2+\|T_2\|_1^2 \leq 1$, so $\|T_2\|_1 \leq (2\epsilon)^{1/2}$. Likewise, $\|T_3\|_1 \leq (2\epsilon)^{1/2}$. So

$$\left\| \begin{bmatrix} 0 & T_2 \\ T_3 & T_4 \end{bmatrix} \right\|_1 \leq \|T_2\|_1+\|T_3\|_1+\|T_4\|_1$$

$$\leq (2\sqrt{2}+1)\epsilon^{1/2} \leq 4\epsilon^{1/2} \ .$$

■

16.4 Corollary. *Suppose that* $\|T_n\|_1 \leq 1$ *for* $n \geq 1$, $\|T\|_1 = 1$ *and* T_n *converge weak* to* T *(i.e. in the* $\tau(C_1,K)$ *topology). Then* $T_n \to T$ *in norm.*

Proof. Let $T = \sum_{i\geq 1} s_ie_i \otimes f_i^*$ and let P_k and Q_k be the projections onto $span\{e_1,\ldots,e_k\}$ and $span\{f_1,\ldots,f_k\}$ respectively. Then P_kTQ_k converges to

T in the trace norm, and $\|T - P_kTQ_k\|_1 = 1 - \|P_kTQ_k\|_1$. Given $\epsilon > 0$, choose k so large that $\|P_kTQ_k\| > 1-\epsilon$. Now $P_kT_nQ_k$ converges weak* to P_kTQ_k (for fixed k). Since $P_kC_1Q_k$ is finite dimensional, $P_kT_nQ_k$ converges to $P_kT_nQ_k$ in norm. Choose n_0 so large that for all $n \geq n_0$, $\|P_k(T-T_n)Q_k\|_1 < \epsilon$ and $\|P_kT_nQ_k\|_1 > 1-\epsilon$. By Lemma 16.3, $\|T_n - P_kT_nQ_k\|_1 < 4\epsilon^{1/2}$. So

$$\|T-T_n\|_1 \leq \|P_k(T-T_n)Q_k\| + \|T-P_kTQ_k\| + \|T_n - P_kT_nQ_k\|$$
$$< \epsilon + \epsilon + 4\epsilon^{1/2} .$$

Hence T_n converges to T in norm. ∎

16.5 Corollary. *The ball $b_1(\mathcal{A}_0)$ is the norm closed convex hull of its extreme points.*

Proof. By Lemma 16.2, $\mathcal{A}_0$ is the annihilator of $\mathcal{R}_0$, and thus is a dual space (of $\mathcal{K}/\mathcal{R}_0$). So by the Krein-Milman Theorem, $b_1(\mathcal{A}_0)$ is the weak* closed convex hull of its extreme points. By Corollary 16.4, the boundary of $b_1(\mathcal{A}_0)$ is in the norm closure of the convex hull. The result follows. ∎

16.6 Theorem. *$\mathcal{A}_0 = \{A \in \mathcal{C}_1 \cap \mathcal{T}(\mathcal{N}) : \Delta(A) = 0\}$ is the pre-annihilator of $\mathcal{T}(\mathcal{N})$ and the annihilator of $\mathcal{T}(\mathcal{N}) \cap \mathcal{K}$. Thus $\mathcal{T}(\mathcal{N})$ is the second dual of $\mathcal{T}(\mathcal{N}) \cap \mathcal{K}$.*

Proof. By Corollary 16.5, an operator T annihilates $\mathcal{A}_0$ if and only if it annihilates each extreme point. By Lemma 16.2, this holds exactly when $0 = (Tx,y)$ for every $x \in N$, $y \in N^{\perp}$, $N \in \mathcal{N}$. That is, precisely when T belongs to $\mathcal{T}(\mathcal{N})$. By the Hahn-Banach Theorem, $\mathcal{A}_0$ is the pre-annihilator of $\mathcal{T}(\mathcal{N})$. The pre-annihilator of $\mathcal{A}_0$ is $\mathcal{R}_0$, and this necessarily includes $\mathcal{T}(\mathcal{N}) \cap \mathcal{K}$. Since the reverse inclusion is trivial, $\mathcal{R}_0 = \mathcal{T}(\mathcal{N}) \cap \mathcal{K}$. The dual of $\mathcal{T}(\mathcal{N}) \cap \mathcal{K}$ is $\mathcal{C}_1/\mathcal{A}_0$ and the dual of $\mathcal{C}_1/\mathcal{A}_0$ is $\mathcal{A}_0^{\perp} = \mathcal{T}(\mathcal{N})$. ∎

16.7 Corollary. (Erdos Density Theorem) *The finite rank operators in $b_1(\mathcal{T}(\mathcal{N}))$ are norm dense in $b_1(\mathcal{T}(\mathcal{N}) \cap \mathcal{K})$ and weak* dense in $b_1(\mathcal{T}(\mathcal{N}))$.*

Proof. $\mathcal{T}(\mathcal{N}) \cap \mathcal{K} = \mathcal{R}_0$ is the closed span of the rank one operators. By Goldstines' Theorem, the unit ball of any Banach space is weak* dense in the ball of its second dual. ∎

16.8 Corollary. (Arveson's Distance Formula) *If $\mathcal{N}$ is a nest and A is an operator, then $dist(A, \mathcal{T}(\mathcal{N})) = \sup_{\mathcal{N}} \|P(N)^{\perp}AP(N)\|$.*

Proof. As $\mathcal{T}(\mathcal{N})$ is weak* closed, a standard separation theorem yields

$$dist(A,\mathcal{T}(\mathcal{N})) = \sup_{T\in\mathcal{T}(\mathcal{N})_\perp, \|T\|_1=1} |tr\,AT|$$
$$= \sup\{\,|(Ax,y)| : N \in \mathcal{N}, x \in N,\ y \in N^\perp, \|x\|=\|y\|=1\}$$
$$= \sup_{\mathcal{N}} \|P(N)^\perp AP(N)\| \ . \qquad \blacksquare$$

16.9 Corollary. (Lidskii's Theorem) *Let T be a trace class operator. Then $tr\,T$ equals the sum of its eigenvalues counting algebraic multiplicity.*

Proof. First, suppose T is quasinilpotent. By Corollary 3.2, there is a maximal nest $\mathcal{N}$ such that T belongs to $\mathcal{T}(\mathcal{N})$. Let $E = P(N)-P(N_-)$ be an atom of $\mathcal{N}$, and let $ETE = \alpha(E)E$. For $\lambda \neq 0$, $\lambda I-T$ is invertible in $\mathcal{T}(\mathcal{N})$ since $(\lambda I-T)^{-1} = \lambda^{-1}\sum_{n\geq 0}(\lambda^{-1}T)^n$. So

$$E = E(\lambda I-T)(\lambda I-T)^{-1}E$$
$$= E(\lambda I-T)E(\lambda I-T)^{-1}E$$
$$= (\lambda-\alpha(E))E(\lambda I-T)^{-1}E$$

Thus $\alpha(E) = 0$ for every E. So T belongs to $\mathcal{T}(\mathcal{N})_\perp$ and since $I \in \mathcal{T}(\mathcal{N})$,

$$tr\,T = tr(TI) = 0 \ .$$

In general, the Riesz theory shows that the non-zero spectral values of T are eigenvalues λ_i of finite multiplicity n_i. The eigenvalues correspond to eigenspaces $E_T(\lambda_i) = ker(T-\lambda_i)^{n_i}$. One can choose a basis for $E_T(\lambda_i)$ so that $T\,|E_T(\lambda_i)$ is upper triangular with λ_i on the diagonal. Applying the Gram-Schmidt process to the union yields a basis for $\mathcal{H}_0 = E_T(\mathbb{C}\setminus\{0\})$ such that T is triangular, and each λ_i occurs on the diagonal exactly n_i times. The compression T_0 of T to $\mathcal{H}_0^\perp$ is quasinilpotent. So

$$tr\,T = tr\,P(\mathcal{H}_0)TP(\mathcal{H}_0)+tr\,P(\mathcal{H}_0)^\perp TP(\mathcal{H}_0)^\perp$$
$$= \sum_{k\geq 1} n_i\lambda_i + trT_0 = \sum_{i\geq 1} n_i\lambda_i \ . \qquad \blacksquare$$

16.10 Corollary. *Let T belong to $\mathcal{T}(\mathcal{N})_\perp$. Then given $\epsilon > 0$, there are rank one operators R_n in $\mathcal{T}(\mathcal{N})_\perp$ such that $T = \sum_{n\geq 1} R_n$ and $\sum_{n\geq 1}\|R_n\|_1 \leq \|T\|_1+\epsilon$.*

Corollary 16.10 is a simple result, and it is left as an exercise. However, it raises the question of when an exact decomposition into rank one operators is possible. That is, when can one write $T = \sum_{n\geq 1} R_n$ as a sum of rank one operators in $\mathcal{T}(\mathcal{N})_\perp$ such that

$$\|T\|_1 = \sum_{n\geq 1} \|R_n\|_1 \,?$$

It happens that this *exact decomposition* is always possible precisely when the nest is countable. In general, an integral version is needed. Our analysis is based on a careful examination of the decomposition used in Lemma 16.1.

First, recall from Lemma 16.2 that rank one operators in $\mathcal{T}(\mathcal{N})_\perp$ are of the form $R = P(N)RP(N)^\perp$. If a trace class operator has the form $T = P(N)TP(N)^\perp$, then T belongs to $\mathcal{T}(\mathcal{N})_\perp$. One can (see Chapter 1) write $T = \sum_{n\geq 1} s_n R_n$ where $s_n \geq 0$, R_n are norm one, rank one operators and $\sum_{n\geq 1} s_n = \|T\|_1$. Clearly,

$$T = \sum_{n\geq 1} s_n P(N)R_n P(N)^\perp \quad ,$$

so T has an exact decomposition into rank one operators in $\mathcal{T}(\mathcal{N})_\perp$. So actually, it suffices to decompose an operator into a sum of operators of this type.

16.11 Proposition. *Let T be a rank n operator in $\mathcal{T}(\mathcal{N})_\perp$. Then there are n rank one operators $R_1,\ldots,R_n$ in $\mathcal{T}(\mathcal{N})_\perp$ such that $T = \sum_{i=1}^n R_i$ and $\|T\|_1 = \sum_{i=1}^n \|R_i\|_1$.*

Proof. First we show that the decomposition obtained in Lemma 16.1 has the addition property

$$rank\,T = rank\,T_1 + rank\,T_2 \quad .$$

Using the notation of that lemma, one has

$$Q_1 = \begin{bmatrix} I \\ X^* \end{bmatrix} A \begin{bmatrix} I & X \end{bmatrix} \quad .$$

So $rank\,Q_1 = rank\,A$, and $Ran(Q_1) = \begin{bmatrix} I \\ X^* \end{bmatrix} Ran(A)$ has trivial intersection with $\mathcal{H}_2$. But $Q_2 = \begin{bmatrix} 0 & 0 \\ 0 & D - X^*X \end{bmatrix}$ has range contained in $\mathcal{H}_2$. Thus

$$rank\,T = rank\,Q = rank\,Q_1 + rank\,Q_2 = rank\,T_1 + rank\,T_2 \quad .$$

Now, proceed by induction on n. For $n = 1$, it is trivial. Suppose there is a element N in $\mathcal{N}$ so that both $T_{11} = TP(N)$ and $T_{22} = P(N)^\perp T$ are non-zero. Then the decomposition of Lemma 16.1 with respect to $\mathcal{H} = N \oplus N^\perp$ and $\mathcal{K} = N_- \oplus N_-^\perp$ yields $T = T_1 + T_2$ where

$T_1 = \begin{bmatrix} T_{11} & * \\ 0 & 0 \end{bmatrix}$ and $T_2 = \begin{bmatrix} 0 & * \\ 0 & T_{22} \end{bmatrix}$ are both non-zero, and hence have rank at least one. By the previous paragraph, the rank of T_1 and T_2 is strictly less than n. So by the induction hypothesis, each can be written as an exact sum of *rank* T_i rank-one operators. Adding yields the decomposition for T. Otherwise, let N be the sup of all N' in $\mathcal{N}$ such that $TP(N') = 0$. Clearly, $TP(N) = 0$. Also, if $N' > N$, then $TP(N') \neq 0$ so $P(N'_-)^{\perp}T = 0$. Consequently, $P(N)^{\perp}T = 0$ (Verify!). So $T = P(N)^{\perp}TP(N)$ decomposes exactly by the method given in the paragraph preceding this proposition. By induction, this decomposition is valid for all finite rank operators in $\mathcal{T}(\mathcal{N})_{\perp}$. ■

The next theorem will immediately be subsumed by its sequel. However, it shows somewhat more simply exactly how the decomposition works.

16.12 Theorem. *Let $\mathcal{N}$ be a well ordered nest. Then every trace class operator T in $\mathcal{T}(\mathcal{N})$ is exactly decomposable, i.e. $T = \sum_{n\geq 1} R_n$, $\|T\|_1 = \sum_{n\geq 1} \|R_n\|_1$, and R_n belong to $\mathcal{T}(\mathcal{N})$.*

Proof. We use the results of section 14.13-14.18. Let $Q = (T^*T)^{1/2}$, and let $T = UQ$ be the polar decomposition of T. By Lemma 14.18, one associates to each N in $\mathcal{N}$ a positive operator Q_N such that $Q = \sum_{N\in\mathcal{N}} Q_N$ converges in the strong operator topology. From section 14.15 and Lemma 14.16, one finds that with respect to the decomposition $N \oplus (N_+/N) \oplus N_+^{\perp}$, Q_N has the form

$$Q_N = \begin{bmatrix} 0 & 0 & 0 \\ 0 & A & B \\ 0 & B^* & X \end{bmatrix}$$

where X is the least positive operator such that Q_N is positive. Or in other words, using Lemma 14.13, Q is decomposed into $Q[0,N)+Q[N,\mathcal{H})$ and then $Q[N,\mathcal{H})$ is likewise decomposed as $Q_N+Q[N_+,\mathcal{H})$. From the proof of Lemma 16.1, one obtains that operators $T_N = UQ_N$ belong to $\mathcal{T}(\mathcal{N})$. Clearly, $T = \sum T_N$. Moreover, $T_NP(N) = 0$ and $P(N_+)^{\perp}T_N = 0$.

So T_N belongs to $P(N_+)\mathcal{B}(\mathcal{H})P(N)^{\perp}$, which is a subspace of $\mathcal{T}(\mathcal{N})$ by Lemma 2.8. As in the paragraph preceding Proposition 16.11, T_N can be exactly decomposed as the sum of rank one operators in this subspace. Finally,

$$\|T\|_1 = \|Q\|_1 = trQ = \sum trQ_N = \sum \|Q_N\|_1 = \sum \|T_N\|_1 \ .$$

So the desired decomposition is obtained. ∎

16.13 Theorem. *Let $\mathcal{N}$ be a countable nest. Then every trace class operator T in $\mathcal{T}(\mathcal{N})$ is exactly decomposable.*

Proof. It is necessary to extend the definition of Q_N of the previous proof to all N in $\mathcal{N}$, even when $N = N_+$, in such a way that

$$Q = \sum_{N \in \mathcal{N}} Q_N$$

is still valid. Set $Q_N = s-\lim_{N' \downarrow N} Q[N,N')$, which exists since $Q[N,N')$ is an increasing function of N'. For each finite subnest

$$\mathcal{F} = \{0 = F_0 < F_1 < \dots < F_n = \mathcal{H}\}$$

of $\mathcal{N}$, let $Q_{\mathcal{F}} = \sum_{j=0}^{n-1} Q_{F_j}$. It will be shown that $s-\lim_{\mathcal{F}} Q_{\mathcal{F}} = Q$ in the strong operator topology.

First, $Q_{F_i} \le Q[F_i, F_{i+1})$, so

$$Q_{\mathcal{F}} \le \sum_{i=0}^{n-1} Q[F_i, F_{i+1}) = Q[0,\mathcal{H}) = Q \ .$$

Also, if $\mathcal{F}_1$ is contained in $\mathcal{F}_2$, then clearly $Q_{\mathcal{F}_1} \le Q_{\mathcal{F}_2}$. So this is an increasing nest, bounded above by Q. Thus it has a limit. The same can be said for the restriction of this sum to those F such that $N \le F < M$ say. So let $\mathcal{E}$ consist of all intervals $[N,M)$ such that the equality

$$Q[N,M) = \sum_{N \le F < M} Q_F$$

holds. The collection $\mathcal{E}$ is non-empty since it contains all atoms $[N,N_+)$ and $\mathcal{N}$ is atomic. It is also hereditary in the sense that if $[N',M') \subseteq [N,M)$ and $[N,M)$ belongs to $\mathcal{E}$, so does $[N',M')$. Also, if $[N,M)$ and $[M,L)$ belong to $\mathcal{E}$, so does $[N,L)$ since $Q[N,L) = Q[N,M) + Q[M,L)$ by definition of $Q[\cdot,\cdot)$. Most importantly, suppose $[N_k,M_k)$ belong to $\mathcal{E}$ and $N_{k+1} \le N_k < M_k \le M_{k+1}$. Then $[N_0,M_0)$ belongs to $\mathcal{E}$ where $N_0 = \inf N_k$ and $M_0 = \sup M_k$. To see this, first use Lemma 14.17 to obtain

$$\begin{aligned} Q[N_1,M_0) &= s-\lim_{k \to \infty} Q[N_1,M_k) \\ &= s-\lim_{k \to \infty} \sum_{N_1 \le F < M_k} Q_F = \sum_{N_1 \le F < M_0} Q_F \ . \end{aligned}$$

If $N_0 = N_k$ for some k, the result follows. Otherwise from the definition of Q_{N_0}, one has

$$Q_{N_0} = s-\lim_{k\to\infty} Q[N_0,M_0) - Q[N_k,M_0)$$

whence

$$Q[N_0,M_0) = Q_{N_0} + s-\lim_{k\to\infty} Q[N_k,N_1) + Q[N_1,M_0)$$

$$= Q_{N_0} + s-\lim_{k\to\infty} \sum_{N_k \le F < N_0} Q_F = \sum_{N_0 \le F < M_0} Q_F \ .$$

Now $\mathcal{N}$ is atomic and for each atom, one can use Zorn's lemma to obtain a maximal interval in $\mathcal{E}$ containing that atom. No two maximal intervals can intersect, or even abut like $[N,M)$ and $[M,L)$. Thus if there are two or more intervals, then between each two intervals, there must be another. So these maximal intervals form a dense order type. This implies that $\mathcal{N}$ is uncountable. As $\mathcal{N}$ is in fact countable, the maximal interval must be all of $[0,\mathcal{H})$. So $Q = \sum_{N\in\mathcal{N}} Q_N$ as desired.

The proof is completed as in Theorem 16.12. ■

16. 14 Lemma. *Let $\mathcal{N}$ be a nest. Suppose T in $\mathcal{T}(\mathcal{N})_\perp$ has an exact decomposition as a sum of rank one operators in $\mathcal{T}(\mathcal{N})$. Then it has an exact decomposition as a sum of rank one operators in $\mathcal{T}(\mathcal{N})_\perp$.*

Proof. Let $T = \sum_{n\ge1} R_n$ where $\|T\|_1 = \sum_{n\ge1} \|R_n\|_1$ and R_n are rank one in $\mathcal{T}(\mathcal{N})$. Each R has the form $R = P(N)RP(N_-)^\perp$ for some N in $\mathcal{N}$. But if $E = P(N)-P(N_-) \ne 0$, it may occur that $ER_nE \ne 0$ for some n. Let T_E be the sum of all R_n such that $ER_nE \ne 0$. Then $T_E = P(N)T_EP(N_-)^\perp$, and since $ETE = 0$, it follows that $ET_EE = 0$. So with respect to the decomposition $N_- \oplus E\mathcal{H} \oplus N^\perp$, T_E has the form

$$T_E = \begin{bmatrix} 0 & * & * \\ 0 & 0 & * \\ 0 & 0 & 0 \end{bmatrix} .$$

Apply Lemma 16.1 to T_E as an operator from $N \oplus N^\perp$ to $N_- \oplus N_-^\perp$. This decomposes T_E as an exact sum $T_E = T_1 + T_2$ where $T_1 = P(N_-)T_1P(N_-)^\perp$ and $T_2 = P(N)T_2P(N)^\perp$. By the remark preceding Proposition 16.11, both T_1 and T_2 have exact decompositions into rank one operators in $\mathcal{T}(\mathcal{N})_\perp$. So T_E has such a decomposition. This procedure may be carried out for each of the (countably many) atoms for which $T_E \ne 0$. The result is an exact decomposition in $\mathcal{T}(\mathcal{N})_\perp$. ■

16.15 Corollary. *Let $\mathcal{N}$ be a countable nest, and let T belong to $\mathcal{T}(\mathcal{N})_\perp$. Then T decomposes as $T = \sum_{n\ge1} R_n$ where R_n are rank one operators in $\mathcal{T}(\mathcal{N})_\perp$, and $\|T\|_1 = \sum_{n\ge1} \|R_n\|_1$.*

Next, we develop the appropriate analogue of Theorem 16.13 for arbitrary nests.

16.16 Lemma. *Let $\mathcal{N}$ be a nest on a separable space. Let Q be a positive operator, and define $Q[N,M)$ as in 14.15. Then there is an extension $Q(\Delta)$ defined on all Borel subsets of $\mathcal{N}$ with values in the positive operators which is countably additive in the strong operator topology.*

Proof. By definition, $Q[N,M)+Q[M,L) = Q[N,L)$, so Q is finitely additive in this restricted sense. Moreover, if $[N,M)$ is the countable disjoint union of $[N_k,M_k)$, then each endpoint M_k equals N_j for some j except when $M_k = M$. Thus the nest generated by $\{N_k, k \geq 1\}$ is precisely $\{0, N_k, k \geq 1, M, \mathcal{H}\}$ which is countable and in fact is well ordered. By Lemma 14.17, $Q[N,M) = \sum_{k \geq 1} Q[N_k,M_k)$.

Since $B(\mathcal{H})$ is the dual of C_1 (Theorem 1.15), for each T in C_1 one can define

$$Q_T[N,M) = tr(Q[N,M)T) \ .$$

This is a positive valued function on half open intervals which is countably additive when the union is a half open interval. By a theorem of Cartheodory (Royden [1], p. 257), there is a regular Borel measure μ_T on $\mathcal{N}$ so that $\mu_T[N,M) = Q_T[N,M)$ for all N, M in $\mathcal{N}$.

Clearly, $\|\mu_T\| \leq \|Q\| \, \|T\|_1$. Also since

$$\lambda_1 \mu_{T_1}[N,M)+\lambda_2 \mu_{T_2}[N,M) = \mu_{\lambda_1 T_1+\lambda_2 T_2}[N,M)$$

for all N and M in $\mathcal{N}$, it follows that $\mu(T) = \mu_T$ is a continuous linear map of C_1 into the space of bounded regular Borel measures on $\mathcal{N}$.

For each fixed Borel set Δ, the map $f_\Delta(T) = \mu_T(\Delta)$ is hence a positive linear functional on C_1. So there is a positive operator $Q(\Delta)$ so that

$$\mu_T(\Delta) = tr\, Q(\Delta)T$$

for every T in C_1. Moreover, if Δ is the disjoint union of Borel sets Δ_i

$$\begin{aligned} tr\, Q(\Delta)T &= \mu_T(\Delta) = \textstyle\sum \mu_T(\Delta_i) \\ &= \textstyle\sum tr\, Q(\Delta_i)T = tr(\sum Q(\Delta_i)T) \ . \end{aligned}$$

(The sum $\sum Q(\Delta_i)$ is a monotone sequence bounded above, and hence converges strongly.) This identity and the Hahn-Banach Theorem forces $Q(\Delta) = \sum Q(\Delta_i)$, so $Q(\cdot)$ is countably additive. ∎

16.17 Corollary. *Suppose that Q is trace class in Lemma 16.16. Then $Q(\cdot)$ is a trace class valued measure which is countably additive in the norm topology. So $\tau(\Delta) = tr\, Q(\Delta)$ is a scalar valued measure mutually absolutely continuous to $Q(\cdot)$.*

Proof. Since $Q(\Delta) \le Q$ for every Δ, it must be trace class. (Of course, the theorem could have been proven using the duality $C_1 = \mathcal{K}^*$ instead.) Let Δ be the disjoint union of Borel sets Δ_i. Then $Q_n = \sum_{i=1}^{n} Q(\Delta_i) \le Q(\Delta)$ for each n, and hence $\|Q_n\|_1 \le \|Q(\Delta)\|_1$. By Lemma 16.16, $Q(\Delta) = s\text{-}\lim Q_n$. By Corollary 16.4, Q_n converges to $Q(\Delta)$ in norm. Thus $\tau(\Delta) = tr\, Q(\Delta)$ is a countably additive Borel measure on $\mathcal{N}$. Clearly, $\tau(\Delta) = 0$ if and only if $0 = tr\, Q(\Delta) = \|Q(\Delta)\|_1$. So $\tau(\cdot)$ and $Q(\cdot)$ are mutually absolutely continuous. ■

Now we need a Radon-Nikodym theorem.

16.18 Theorem. *Let $Q(\cdot)$ be a countably additive, trace class valued Borel measure on a compact metric space X, and let $\tau(\cdot) = tr\, Q(\cdot)$. There is an integrable Borel function $D(x)$ with values in $b_1(C_1)_+$ such that*

$$Q(\Delta) = \int_\Delta D\, d\tau \ .$$

Proof. For each K in $\mathcal{K}$, the measure $\mu_K(\Delta) = tr\, Q(\Delta)K$ is absolutely continuous with respect to τ. By the Radon-Nikodym Theorem, there is an essentially unique positive Borel function $D_K(x)$ so that

$$\mu_K(\Delta) = \int_\Delta D_K\, d\tau$$

for all Borel sets Δ. Since

$$\left|\int_\Delta D_K d\tau\right| = |tr\, Q(\Delta)K| \le \|Q(\Delta)\|_1\, \|K\| = \tau(\Delta)\|K\| \ ,$$

$|D_K(x)| \le \|K\|$ a.e.τ. If K_1 and K_2 are both compact, then

$$\begin{aligned}\mu_{\lambda_1 K_1 + \lambda_2 K_2}(\Delta) &= \lambda_1 \mu_{K_1}(\Delta) + \lambda_2 \mu_{K_2}(\Delta)\\ &= \int_\Delta \lambda_1 D_{K_1} + \lambda_2 D_{K_2} d\tau = \int_\Delta D_{\lambda_1 K_1 + \lambda_2 K_2} d\tau \ .\end{aligned}$$

So $D_{\lambda_1 K_1 + \lambda_2 K_2} = \lambda_1 D_{K_1} + \lambda_2 D_{K_2}$ almost everywhere $d\tau$.

Fix a basis $e_1, e_2, \ldots$ and let $\mathcal{K}_0$ be finite linear combinations of $e_i \otimes e_j^*$ with coefficients in $\mathbb{Q} + i\mathbb{Q}$. As this set is countable, one may delete a set N of τ measure 0 so that the map taking K to D_K is linear on $\mathcal{K}_0$. By the Monotone Convergence Theorem,

$$\begin{aligned}\int_X \sum_{n\ge 1} D_{e_n \otimes e_n^*} d\tau &= \sum_{n \ge 1} \int_X D_{e_n \otimes e_n^*} d\tau\\ &= \sum_{n\ge 1} (Q(X)e_n, e_n) = tr\, Q(X) \ .\end{aligned}$$

Hence $\sum_{n \ge 1} D_{e_n \otimes e_n^*}(x)$ is finite almost everywhere. So we will assume that this is finite on $X \backslash N$, and that $|D_K(x)| \le \|K\|$ for x in $X \backslash N$ for all K

in $\mathcal{K}_0$. For each x in $X \backslash N$, the map on $\mathcal{K}_0$ given by $\phi_x(K) = D_K(x)$ is a linear map such that $|\phi_x(K)| \leq \|K\|$. Thus ϕ_x extends to a norm one positive linear functional on $\mathcal{K}$. So there is a positive trace class operator $D(x)$ of norm at most one so that $\phi_x(K) = tr\, D(x)K$. Hence

$$tr\, Q(\Delta)K = \mu_K(\Delta) = \int_\Delta tr(D(x)K)d\tau = tr \int_\Delta D\; d\tau K$$

for all K in $\mathcal{K}_0$. By the Hahn-Banach Theorem, $Q(\Delta) = \int_\Delta D\; d\tau$. Finally

$$\|Q(X)\|_1 \leq \int \|D\|_1 d\tau \leq \int d\tau = \|Q(X)\|_1 \ .$$

So $\|D(x)\|_1 = 1$ a.e.τ. ∎

16.19 Theorem. *Let $\mathcal{N}$ be a nest on a separable space, and let T be a trace class operator in $\mathcal{T}(\mathcal{N})$. Then there is a regular Borel measure τ on $\mathcal{N}$ with $\|\tau\| = \|T\|_1$, and a Borel function $T(N)$ with values in $b_1(\mathcal{C}_1 \cap \mathcal{T}(\mathcal{N}))$ such that $T(N) = P(N_+)T_N P(N)^\perp$, $T = \int T(N)d\tau(N)$, and $\|T\|_1 = \int \|T(N)\|_1 d\tau(N)$.*

Proof. Let $T = UQ$ be the polar decomposition of T. By Theorem 16.16 and Corollary 16.17, there is a positive trace class valued measure $Q(\cdot)$ such that $Q[N,M)$ agrees with definition 14.15. From the proof of Lemma 16.1, $UQ[N,M)$ belongs to $\mathcal{T}(\mathcal{N}) \cap P(M)\mathcal{B}(\mathcal{H})P(N)^\perp$ for every pair $N < M$ in $\mathcal{N}$. By Theorem 16.18, there is a Borel function $D(N)$ of positive, norm one trace class operators such that $Q(\Delta) = \int_\Delta D\, d\tau$ where $\tau(\Delta) = tr\, Q(\Delta)$. Let $T(N) = UD(N)$. Then

$$T = UQ = \int UD\, d\tau = \int T(N) d\tau(N)$$

and

$$\|T\|_1 \leq \int \|T(N)\|_1 d\tau \leq \int \|D(N)\|_1 d\tau = \|Q\|_1 = \|T\|_1 \ .$$

Now if a pair $N < M$ belong to N, then for every interval $[N',M')$ contained in $[N,M)$ one has

$$UQ[N',M') = P(M')T[N',M')P(N')^\perp = P(M)T[N',M')P(N)^\perp \ .$$

Thus $UQ(\Delta) = P(M)UQ(\Delta)P(N)^\perp$ for every Borel subset Δ of $[N,M)$. So

$$\int_\Delta P(M)T(L)P(N)^\perp d\tau(L) = \int_\Delta T(L)d\tau(L)$$

for all such Δ. Hence $T(L) = P(M)T(L)P(N)^\perp$ a.e.τ for L in $[N,M)$. Let $\{[N_k,M_k), k \geq 1\}$ be a countable dense set of pairs in the set of intervals of $\mathcal{N}$. Then one can delete a set Z of τ-measure zero so that off this set one has $T(L) = P(M_k)T(L)P(N_k)^\perp$ whenever $N_k \leq L < M_k$ and L belongs to $\mathcal{N} \backslash Z$. Since these pairs are dense, one obtains $T(L) = P(L_+)T(L)P(L)^\perp$. ∎

In spite of this positive result, there are trace class operators in uncountable nests which are not exactly decomposable.

16.20 Lemma. *Let Q be a positive trace class operator. If $Q = Q_1+Q_2$ and $\|Q\|_1 = \|Q_1\|_1+\|Q_2\|_1$, then Q_1 and Q_2 are both positive.*

Proof. $\|Q\|_1 = tr\,Q = tr\,Q_1+tr\,Q_2 \le \|Q_1\|_1+\|Q_2\|_1 = \|Q\|$. Hence $\|Q_i\| = tr\,Q_i$. Now if $T = \sum_{n=1}^{\infty} s_n e_n \otimes f_n^*$ is trace class with s-numbers s_n, one has $\|T\|_1 = \sum_{n=1}^{\infty} s_n$. If $tr\,T = \|T\|_1$, then

$$\|T\|_1 = tr\,T = \sum_{n=1}^{\infty} s_n(e_n,f_n) \le \sum_{n=1}^{\infty} s_n\|e_n\|\,\|f_n\| = \|T\|_1 \ .$$

This forces $e_n = f_n$, so T is positive. Thus both Q_1 and Q_2 are positive. ∎

16.21 Lemma. *Let $T = \begin{bmatrix} A & B \\ 0 & 0 \end{bmatrix}$ be a trace class operator on $\mathcal{H}_1 \oplus \mathcal{H}_2$. Suppose that $Ran\,A$ is dense in $\mathcal{H}_1$. Suppose $T = T_1+T_2$ where $T_1 = \begin{bmatrix} A & C \\ 0 & 0 \end{bmatrix}$ and $T_2 = \begin{bmatrix} 0 & B-C \\ 0 & 0 \end{bmatrix}$, and $\|T\|_1 = \|T_1\|_1+\|T_2\|_1$. Then $T_2 = 0$.*

Proof. Let $T = UQ$ be the polar decomposition of T, and let $Q_i = U^*T_i$. Then $Q = Q_1+Q_2$ and $\|Q\|_1 \ge \|Q_1\|_1+\|Q_2\|_1 \ge \|Q\|_1$. So by Lemma 16.20, both Q_i are positive. Since $\|Q_i\|_1 = \|T_i\|_1$, it follows that $T_i = UQ_i$. Now Q_2 has the form $U^*\begin{pmatrix} 0 & * \\ 0 & * \end{pmatrix} = \begin{pmatrix} 0 & * \\ 0 & * \end{pmatrix}$, and thus the (1,2) entry is also 0. Hence Q_1 agrees with Q except possibly in the (2,2) entry. One will have $Q_1 = Q$ if it can be shown that Q is the smallest positive operator of the form $\begin{pmatrix} Q_{11} & Q_{12} \\ Q_{21} & * \end{pmatrix}$.

Now $Q^2 = T^*T = \begin{bmatrix} A^* \\ B^* \end{bmatrix} \begin{bmatrix} A & B \end{bmatrix}$ Let $X = (AA^*+BB^*)^{1/4}$. Since $AA^* \le X^4$ and $BB^* \le X^4$, Lemma 10.1 implies that there are norm one operators C and D so that $A = XC$ and $B = XD$. So

$$X^4 = AA^*+BB^* = X(CC^*+DD^*)X \ .$$

By hypothesis, $Ran(X)$ is dense in $\mathcal{H}_1$ so X is one to one. Thus $CC^*+DD^* = X^2$. We claim that $Q = \begin{bmatrix} C^* \\ D^* \end{bmatrix}\begin{bmatrix} C & D \end{bmatrix} = \begin{bmatrix} C^*C & C^*D \\ D^*C & D^*D \end{bmatrix}$. To see this, compute

$$\begin{bmatrix} C^*C & C^*D \\ D^*C & D^*D \end{bmatrix}^2 = \begin{bmatrix} C^* \\ D^* \end{bmatrix} \begin{bmatrix} C & D \end{bmatrix} \begin{bmatrix} C^* \\ D^* \end{bmatrix} \begin{bmatrix} C & D \end{bmatrix} = \begin{bmatrix} C^* \\ D^* \end{bmatrix} X^2 \begin{bmatrix} C & D \end{bmatrix}$$

$$= \begin{bmatrix} C^*X \\ D^*X \end{bmatrix} \begin{bmatrix} XC & XD \end{bmatrix} = \begin{bmatrix} A^* \\ B^* \end{bmatrix} \begin{bmatrix} A & B \end{bmatrix} = Q^2 .$$

Now $\overline{Ran\, C} = \overline{Ran\, A} = \mathcal{H}_1$, so by the second half of Lemma 14.13, Q is the smallest positive operator of the form $\begin{bmatrix} C^*C & C^*D \\ D^*C & * \end{bmatrix}$.

Hence $Q_1 = Q$, and $Q_2 = 0 = T_2$. ■

16.22 Corollary. *Let* $T = \begin{bmatrix} T_{11} & T_{12} \\ 0 & T_{22} \end{bmatrix}$ *be a trace class operator on* $\mathcal{H}_1 \oplus \mathcal{H}_2$ *such that either* $\overline{Ran\, T_{11}} = \mathcal{H}_1$ *or* $ker\, T_{22} = \{0\}$. *Then the exact decomposition* $T = T_1 + T_2$ *with* $\|T\| = \|T_1\|_1 + \|T_2\|_1$, $T_1 = \begin{bmatrix} T_{11} & Y_1 \\ 0 & 0 \end{bmatrix}$, *and* $T_2 = \begin{bmatrix} 0 & Y_2 \\ 0 & T_{22} \end{bmatrix}$ *is unique. So if both* T_{11} *and* T_{22} *are non-zero, one cannot write* $T = R + S$ *with* $\|T\|_1 = \|R\|_1 + \|S\|_1$ *and* $R = \begin{bmatrix} 0 & * \\ 0 & 0 \end{bmatrix} \neq 0$.

Proof. First assume that $ker\, T_{22} = \{0\}$, and suppose $T = T_1 + T_2$ as described. Let $T = UQ$ be the polar decomposition, and set $Q_i = U^* T_i$, $i = 1,2$. As in Lemma 16.21, it follows that Q_1 and Q_2 are positive and Q_2 has the form $\begin{bmatrix} 0 & 0 \\ 0 & * \end{bmatrix}$. Thus the decomposition produced in Lemma 16.1 has the minimal possible choice of Q_1. Let us fix Q_1 and Q_2 as in Lemma 16.1. Non-uniqueness can occur only if there is a positive operator $0 \neq Q' \leq Q_2$ such that UQ' has the form $\begin{bmatrix} 0 & * \\ 0 & 0 \end{bmatrix}$. Equivalently, this occurs only if $T_2 = \begin{bmatrix} 0 & Y_2 \\ 0 & T_{22} \end{bmatrix}$ has a non-trivial exact decomposition of the form $\begin{bmatrix} 0 & * \\ 0 & * \end{bmatrix} + \begin{bmatrix} 0 & * \\ 0 & 0 \end{bmatrix}$. Taking adjoints puts this into the form of Lemma 16.21, and $\overline{Ran\, T_{22}^*} = (ker\, T_{22})^{\perp} = \mathcal{H}_2$. So Lemma 16.21 shows that this decomposition is unique.

If instead, $Ran\, T_{11}$ is dense in $\mathcal{H}_1$, note that T^* is unitarily equivalent to $\begin{bmatrix} T_{22}^* & T_{12}^* \\ 0 & T_{11}^* \end{bmatrix}$ and $ker\, T_{11}^* = 0$. So the first case shows that the decomposition is unique.

If T decomposed as $R+S$ with $R = \begin{bmatrix} 0 & * \\ 0 & 0 \end{bmatrix} \neq 0$, decompose S by Lemma 16.1 as S_1+S_2. Then $T = (S_1+R)+S_2 = S_1+(R+S_2)$ gives two different exact decompositions, contrary to fact. ■

16.23 Theorem. *Let $\mathcal{N}$ be an uncountable nest on a separable space. Then $\mathcal{T}(\mathcal{N})_\perp$ contains trace class operators which are not exactly decomposable into a sum of rank one operators in $\mathcal{T}(\mathcal{N})$.*

Proof. First, suppose that $\mathcal{N}$ is not purely atomic. Let x be a vector supported on the continuous part, and let P be the projection onto $\overline{\mathcal{N}''x}$. Then P commutes with $\mathcal{N}$, and $\mathcal{N}_0 = P\mathcal{N}|P\mathcal{H}$ is a continuous nest with cyclic vector. By Proposition 7.17, $\mathcal{N}_0$ is unitarily equivalent to the Volterra nest $\mathcal{M}$ on $L^2(0,1)$. If T belongs to $\mathcal{T}(\mathcal{N}_0)_\perp$, then PTP belongs to $\mathcal{T}(\mathcal{N})_\perp$. An exact decomposition $PTP = \sum_{n\geq 1} R_n$ in $\mathcal{T}(\mathcal{N})_\perp$ leads to the exact decomposition $T = \sum_{n\geq 1} PR_n|P\mathcal{H}$ in $\mathcal{T}(\mathcal{N}_0)_\perp$. We show that certain operators in $\mathcal{T}(\mathcal{M})$ are not exactly decomposable.

Let V be the usual Volterra operator (see Chapter 5). Now V is a Hilbert-Schmidt operator in $\mathcal{T}(\mathcal{M})$, so V^2 belongs to $\mathcal{T}(\mathcal{M}) \cap C_1$ which equals $\mathcal{T}(\mathcal{M})_\perp$ since $\mathcal{M}$ has no atoms. For each M_t in $\mathcal{M}$, $V_t = P(M_t)V|M_t$ has no kernel so $V_t^2 = P(M_t)V^2|M_t$ has no kernel. Thus by Corollary 16.22, V^2 cannot be written as $V^2 = R+S$ where $R = P(M_t)RP(M_t)^\perp \neq 0$ and $\|R\|_1+\|S\|_1 = \|V^2\|_1$. This holds for every t in $(0,1)$. As every rank one operator has this form, it follows that if R is rank one in $\mathcal{T}(\mathcal{N})$ then

$$\|V^2-R\|_1+\|R\|_1 > \|V^2\| \ .$$

So V^2 does not have an exact decomposition.

If $\mathcal{N}$ is atomic but uncountable, there is a projection P in $\mathcal{N}'$ so that $\mathcal{N}_0 = R\mathcal{N}|P\mathcal{H}$ has a dense order on its atoms, each of which is one dimensional. (Such a nest is unitarily equivalent to the Cantor nest of Example 2.6.) Let the atoms be $\{E_n, n \geq 1\}$ and let the order on the atoms be $\ll$. Choose a sequence a_n of positive real numbers such that $\sum_{n=1}^{\infty} a_n \leq 1$. Let $\{E_{nm}\}$ be a set of matrix units, and define a trace class operator in $\mathcal{T}(\mathcal{N}_0)$ by $T = \sum_{E_m \ll E_n} a_n a_m E_{mn}$. Let e_n be a unit vector in E_n and set $x = \sum_{n\geq 1} a_n e_n$.

Let N belong to $\mathcal{N}_0$. Because the order is dense, either $N = N_-$ or $N = N_+$. If $N = N_-$, it follows that the range of $T|N$ equals $span\{P(M)x : 0 \leq M \leq N\}$. This is all of N. Similarly, if $N = N_+$, the range of $T^*|N^\perp$ is $span\{P(M)^\perp x : N \leq M \leq \mathcal{H}\}$ which is all of $N^\perp$. As before, Corollary 16.22 implies that there are no rank one operators R in $\mathcal{T}(\mathcal{N}_0)_\perp$ such that $\|T\|_1 = \|T-R\|_1+\|R\|_1$. By Lemma 16.14, it does not even have an exact decomposition in $\mathcal{T}(\mathcal{N}_0)$. Again, as in the first paragraph, one deduces that PTP does not have an exact decomposition in $\mathcal{T}(\mathcal{N})$. ∎

Notes and Remarks.

Lemma 16.1 is due to Lance [2] who used it to give the proof 16.8 of the distance formula. Theorem 16.6 is due to Fall-Arveson-Muhly [1]. The treatment leading to Theorem 16.6 independent of Erdos' Density Theorem or Ringrose's characterization of the radical is due to Power [9], as are the proofs of 16.7 and 16.9. The decomposition results Corollary 16.10 through Theorem 16.19 are due to Power [2]. The failure of exact decomposition, Theorem 16.28, is new.

Exercises.

16.1 Prove that the unit ball of $\mathcal{T}(\mathcal{N}) \cap \mathcal{C}_1$ is the norm closed convex hull of its rank one operators.

16.2 Prove Corollary 16.10. Hint: Use successive approximations.

16.3 Let $\mathcal{N}$ be a countable nest. Describe the measure $Q(\cdot)$ and the integral of Theorem 16.18 in terms of the notation of Theorem 16.13.

16.4 Derive the analogue of Theorem 16.18 for $\mathcal{T}(\mathcal{N})_\perp$.

17. Isomorphisms

In this chapter, we classify the isomorphisms between two nest algebras and between two quasitriangular algebras. We make use of two important features: a nest algebra contains a maximal abelian von Neumann algebra (necessarily intermediate to $\mathcal{N}''$ and $\mathcal{D}(\mathcal{N})$) and an adequate supply of rank one operators. To make use of rank one operators, it is shown that they are characterized algebraically within $\mathcal{T}(\mathcal{N})$. It will be shown that every isomorphism α of one nest algebra onto another is spatial, meaning that there is an invertible operator S so than $\alpha = Ad\,S$ where $Ad\,S(X) = SXS^{-1}$ for every X. In particular, these isomorphisms are always continuous in both the norm and weak* topologies. Then the structure of the outer automorphism groups of $\mathcal{T}(\mathcal{N})$ and $Q\mathcal{T}(\mathcal{N})$ will be computed.

17.1 Theorem. *Let G be an amenable group. Then every representation of G on a Hilbert space bounded by K is similar to a unitary representation by a positive operator S satisfying $K^{-1}I \le S \le KI$.*

Proof. Let m be an invariant mean on $\ell^\infty(G)$, and let V_g be a bounded representation of G on $\mathcal{H}$. Let $K = \sup\{\|V_g\| : g \in G\}$. Define an inner product on $\mathcal{H}$ by defining

$$F_{x,y}(g) = (V_g^* x, V_g^* y)$$

and setting $<x,y> = m(F_{x,y})$. It is routine to verify that this is a sesquilinear form. Moreover,

$$|<x,y>| \le \sup_G |(V_g^* x, V_g^* y)| \le K^2 \|x\|\,\|y\|$$

and

$$|<x,x>| \ge \inf_G |(V_g^* x, V_g^* x)| \ge K^{-2} \|x\|^2 \ .$$

Thus there is a positive operator S with $K^{-1}I \le S \le KI$ such that

$$<x,y> = (S^2 x, y) = (Sx, Sy) \ .$$

The invariance of m implies that for h in G,

$$<V_h^* x, V_h^* y> = m({}_hF_{x,y}) = m(F_{x,y}) = <x,y> \ .$$

Thus

$$(SV_h^*S^{-1}x, SV_h^*S^{-1}y) = \langle V_h^*S^{-1}x, V_h^*S^{-1}y\rangle$$
$$= \langle S^{-1}x, S^{-1}y\rangle = (x,y) \ .$$

Hence $SV_h^*S^{-1}$ is unitary, and consequently $S^{-1}V_hS$ is unitary for every h in G. ∎

17.2 Corollary. *Every bounded abelian group of operators on Hilbert space is similar to a group of unitary operators.*

Proof. By Theorem 8.7, every abelian group is amenable. ∎

A *Boolean algebra* of idempotents on Hilbert space is a commuting family $\mathcal{B}$ of idempotents containing 0 and I which is complemented ($P \in \mathcal{B}$ implies $I-P \in \mathcal{B}$) and is closed under lattice operations $P \wedge Q = PQ$ and $P \vee Q = P+Q-PQ$. This latter equation is redundant since

$$I-(P \vee Q) = (I-P) \wedge (I-Q) \ .$$

Notice that $Ran\,(P \wedge Q)$ is the intersection of $Ran\,P$ and $Ran\,Q$, and $Ran\,(P \vee Q)$ is the span of those two subspaces.

17.3 Corollary. *Let $\mathcal{B}$ be a Boolean algebra of idempotents on Hilbert space bounded by M. Then there is a positive, invertible operator S with $(2M)^{-1}I \le S \le (2M)I$ such that $S\mathcal{B}S^{-1}$ consists of orthogonal projections.*

Proof. The set $\{2P-I : P \in \mathcal{B}\}$ is an abelian group since

$$(2P-I)(2Q-I) = 2(2PQ-P-Q+I)-I$$
$$= 2\{(P \wedge Q) \vee [(I-P) \wedge (I-Q)]\}-I \ ,$$

and in particular, $(2P-I)^2 = I$. Also,

$$\|2P-I\| \le \|P\|+\|I-P\| \le 2M \ .$$

Let S be obtained by applying Theorem 17.1. For P in $\mathcal{B}$, $P' = SPS^{-1}$ is idempotent and normal, and hence is an orthogonal projection. ∎

17.4 Theorem. *Let $\mathcal{A}$ and $\mathcal{B}$ be two maximal abelian von Neumann algebras, and let α be an algebra isomorphism of $\mathcal{A}$ onto $\mathcal{B}$. Then there is a unitary operator U such that $\alpha = Ad\,U$.*

Proof. The isomorphism preserves spectrum and hence spectral radius. As $\mathcal{A}$ and $\mathcal{B}$ are abelian, every element is normal so has norm equal to the spectral radius. Hence α is isometric. If P is a projection in $\mathcal{A}$, then $\alpha(P)$ is an idempotent in $\mathcal{B}$ and thus also a projection. Let $A = \sum_{i=1}^{n} a_iP_i$ be a linear combination of projections in $\mathcal{A}$. Then

$$\alpha(A^*) = \sum_{i=1}^{n} \overline{a_i}\alpha(P_i) = (\sum_{i=1}^{n} a_i\alpha(P_i))^* = \alpha(A)^* \ .$$

The elements A of this form are dense in $\mathcal{A}$, and hence α is a $*$-isomorphism.

Now let $\{A_i, i \in I\}$ be a bounded set of positive elements of $\mathcal{A}$ with supremum A. It will be shown that $\alpha(A)$ is the supremum of $S = \{\alpha(A_i), i \in I\}$. First, α preserves spectrum, hence it preserves positivity and order. So $A_i \leq A$ implies $\alpha(A_i) \leq \alpha(A)$, and hence $\alpha(A)$ is an upper bound for S. Suppose that B is any upper bound for S. Then $A_i \leq \alpha^{-1}(B)$ for every i. Hence $A \leq \alpha^{-1}(B)$, whence $\alpha(A) \leq B$. So $\alpha(A)$ is the supremum of S. In particular, if P_n is a countable family of pairwise orthogonal projections in $\mathcal{A}$, one obtains countable additivity:

$$\alpha(s-\sum_{n=1}^{\infty} P_n) = s-\sum_{n=1}^{\infty} \alpha(P_n) \ .$$

It is clear that $\mathcal{A}$ and $\mathcal{B}$ may be replaced by unitarily equivalent algebras without loss. By Theorem 7.8, it can then be assumed that there is a finite regular Borel space $(\mathcal{X},\mu)$ so that $\mathcal{A}$ is given by the canonical representation of $L^\infty(\mu)$ acting on $L^2(\mu)$. Likewise, we can suppose there is a finite Borel space $(\mathcal{Y},\nu)$ so that $\mathcal{B}$ acts as $L^\infty(\nu)$ on $L^2(\nu)$. Define a measure $\tilde{\mu}$ on the Borel sets of $\mathcal{Y}$ by

$$\tilde{\mu}(Y) = \|\alpha^{-1}(\chi_Y)\|_2^2 \ .$$

That is, the characteristic function χ_Y is sent by α^{-1} to an idempotent, hence characteristic function, χ_X in $L^\infty(\mu)$ and $\tilde{\mu}(Y) = \mu(X)$. The countable additivity of $\tilde{\mu}$ follows from the previous paragraph. Furthermore, if $\nu(Y) = 0$, then $\alpha^{-1}(\chi_Y) = 0$ so $\tilde{\mu}(Y) = 0$; and if $\nu(Y) \neq 0$, then $\alpha^{-1}(\chi_Y) \neq 0$ so $\tilde{\mu}(Y) \neq 0$. Consequently, ν and $\tilde{\mu}$ are mutually absolutely continuous.

By the Radon-Nikodym Theorem, there is a strictly positive, integrable function h in $L^1(\nu)$ such that $\tilde{\mu} = hd\nu$. Define a map U from simple functions on X into $L^2(\nu)$ by

$$Uf = h^{1/2}\alpha(f) \ .$$

If χ_X is a characteristic function, then $\alpha(\chi_X) = \chi_Y$ in $\mathcal{B}$ and

$$\begin{aligned} \|\chi_X\|_2^2 &= \|\alpha^{-1}(\chi_Y)\|_2^2 = \tilde{\mu}(Y) \\ &= \int \chi_Y h d\nu = \int |U\chi_X|^2 d\nu = \|U\chi_X\|_2^2 \ . \end{aligned}$$

Thus if X_i are pairwise disjoint, one can choose disjoint Y_i so that $\alpha(\chi_{X_i}) = \chi_{Y_i}$. Hence if $f = \sum_{i=1}^{n} a_i\chi_{X_i}$

$$\|Uf\|_2^2 = \|\sum_{i=1}^{n} a_i h^{1/2}\chi_{Y_i}\|_2^2 = \sum |a_i|^2 \|h^{1/2}\chi_{Y_i}\|_2^2$$
$$= \sum_{i=1}^{n} |a_i|^2 \|\chi_{X_i}\|_2^2 = \|f\|_2^2 \ .$$

It follows easily that U is an isometric map from the simple functions of $L^2(\mu)$ onto a dense subset of $L^2(\nu)$. Hence it extends to a unitary operator of $L^2(\mu)$ onto $L^2(\nu)$ which we also denote by U. In particular, if f is in $L^\infty(\mu)$, one still has $Uf = h^{1/2}\alpha(f)$.

Let f and g belong to $L^\infty(\mu)$. Then

$$(UM_gU^*)Uf = UM_gf = Ugf = h^{1/2}\alpha(g)\alpha(f) = M_{\alpha(g)}Uf \ .$$

The vectors of the form Uf for f in $L^\infty(\mu)$ are dense in $L^2(\nu)$, and hence $UM_gU^* = \alpha(M_g)$ for every g in $L^\infty(\mu)$. That is, $\alpha = Ad\,U$. ■

Now we turn our attention to nonselfadjoint algebras. In particular, consider algebras $\mathcal{A}$ which are *reflexive* and contain a maximal abelian von Neumann algebra $\mathcal{M}$ (masa). The invariant subspaces lattice $\mathcal{L} = Lat\,\mathcal{A}$ is a sub-lattice of $Lat\,\mathcal{M} = Proj\,\mathcal{M}$, the projection lattice of $\mathcal{M}$. Hence the orthogonal projections onto elements of $\mathcal{L}$ all commute. These lattices are called *commutative subspace lattices* (CSL). These lattices will be investigated more thoroughly in Chapter 22. For the moment, our interest lies only in the fact that this class contains all nest algebras.

17.5 Theorem. *Let $\mathcal{L}_1$ and $\mathcal{L}_2$ be commutative subspace lattices and let α be an isomorphism of $Alg\,\mathcal{L}_1$ onto $Alg\,\mathcal{L}_2$. Fix a masa $\mathcal{M}$ in $Alg\,\mathcal{L}_1$. Then there is an invertible operator S and an automorphism β of $Alg\,\mathcal{L}_1$ such that $\alpha = Ad\,S \circ \beta$ and $\beta(M) = M$ for all M in $\mathcal{M}$.*

Proof. Since $\mathcal{M}$ is a maximal abelian subalgebra of $\mathcal{B}(\mathcal{H})$ and a fortiori of $Alg\,\mathcal{L}_1$, it follows that $\alpha(\mathcal{M})$ is maximal abelian in $Alg\,\mathcal{L}_2$. The norm closure (even weak operator closure) of $\alpha(\mathcal{M})$ is abelian and contains $\alpha(\mathcal{M})$, so $\alpha(\mathcal{M})$ is norm closed. Let ϕ be the restriction isomorphism of $\mathcal{M}$ onto $\alpha(\mathcal{M})$. For every M in $\mathcal{M}$,

$$\sigma(M) = \sigma_{\mathcal{M}}(M) = \sigma_{\alpha(\mathcal{M})}(\phi(M)) \ .$$

The norm in $\mathcal{M}$ is equal to the spectral radius, and the spectral radius is a lower bound for the norm in $\alpha(\mathcal{M})$. Hence $\|\phi(M)\| \geq \|M\|$ for every M in $\mathcal{M}$. Consequently, ϕ^{-1} is contractive and thus a continuous bijection of one Banach space onto another. By Banach's Isomorphism Theorem, ϕ is continuous.

Let $\mathcal{U}$ be the unitary group of $\mathcal{M}$. Then $\phi(\mathcal{U})$ is a bounded abelian group of operators. By Corollary 17.2, there is an invertible operator T so that $T\phi(\mathcal{U})T^{-1}$ is a group of unitaries. Let $\mathcal{B} = T\phi(\mathcal{M})T^{-1}$. Since $\mathcal{M}$ is spanned by $\mathcal{U}$, it follows that $\mathcal{B}$ is spanned by the abelian unitary group

$T\phi(\mathcal{U})T^{-1}$. Hence it is an abelian von Neumann algebra. Moreover, it is maximal abelian in the algebra $\mathcal{A} = T(Alg\,\mathcal{L}_2)T^{-1} = Alg(T\mathcal{L}_2)$.

Let P be a projection onto an element L of $\mathcal{L}_2$. Then P belongs to $Alg\,\mathcal{L}_2$ and $Q = \alpha^{-1}(P)$ is an idempotent in $Alg\,\mathcal{L}_1$. Furthermore, if A belongs to $Alg\,\mathcal{L}_1$,

$$\alpha((I-Q)AQ) = P^{\perp}\alpha(A)P = 0 \ .$$

Hence, the range of Q is invariant for $Alg\,\mathcal{L}_1$ and consequently for $\mathcal{M}$. So there is a projection R in $\mathcal{M}$ with the same range, characterized by the identities $RQ = Q$ and $QR = R$. Hence $R' = T\alpha(R)T^{-1}$ is an idempotent (hence projection) in $\mathcal{B}$. Let $Q' = TPT^{-1}$, which is an idempotent with range TL. Since $R'Q' = Q'$ and $Q'R' = R'$, it follows that R' is the orthogonal projection onto TL.

Now we show that $\mathcal{B}$ is maximal abelian. Let X belong to $\mathcal{B}'$. Then X commutes with the projection R' onto each element in $T\mathcal{L}_2$. So X belongs to $\mathcal{A}$ and thus belongs to $\mathcal{B}$. Consequently, $Ad\,T\circ\phi$ is an algebra isomorphism between two masas. By Theorem 17.4, it has the form $Ad\,U$ where U is unitary. Let $S = T^{-1}U$, and $\beta = Ad\,S^{-1}\circ\alpha$. Clearly, $\alpha = Ad\,S\circ\beta$. Moreover, the restriction of β to $\mathcal{M}$ equals $Ad\,U^{*}\circ Ad\,T\circ\phi$ which is the identity. It remains to show that β takes $Alg\,\mathcal{L}_1$ onto itself.

If P is the orthogonal projection onto an element L of $Lat\,(Alg(\mathcal{L}_1))$, then P belongs to $\mathcal{M}$. Hence for any A in $Alg\,\mathcal{L}_1$,

$$0 = \beta(P^{\perp}AP) = P^{\perp}\beta(A)P \ .$$

Thus $\beta(A)$ belongs to $Alg\,\mathcal{L}_1$. On the other hand, if P is a projection onto an invariant subspace of $\beta(Alg\,\mathcal{L}_1)$, then it is in particular invariant for $\mathcal{M}$ and thus P belongs to $\mathcal{M}$. So for A in $Alg\,\mathcal{L}_1$,

$$0 = P^{\perp}\beta(A)P = \beta(P^{\perp}AP) \ .$$

And thus the range of P belongs to $Lat\,(Alg\,\mathcal{L}_1)$. (It is true but not needed here that $Lat\,(Alg\,\mathcal{L}_1) = \mathcal{L}_1$. See Chapter 22.) In other words, $Alg\,\mathcal{L}_1$ and $\beta(Alg\,\mathcal{L}_1)$ have the same invariant subspace lattice. Since

$$\beta(Alg\,\mathcal{L}_1) = S^{-1}\alpha(Alg\,\mathcal{L}_1)S = S^{-1}(Alg\,\mathcal{L}_2)S = Alg(S^{-1}\mathcal{L}_2) \ ,$$

one knows that $\beta(Alg\,\mathcal{L}_1)$ is reflexive. Therefore, $\beta(Alg\,\mathcal{L}_1) = Alg\,\mathcal{L}_1$; so β is an automorphism. ∎

Now, we consider the usefulness of finite rank operators. The following lemma is very easy and the proof is omitted. It should be compared with the much more difficult proof in the nest algebra case (Lemma 17.9).

17.6 Lemma. *An operator T has rank one if and only if for all operators A and B, $ATB = 0$ implies either $AT = 0$ or $TB = 0$.*

17.7 Theorem. *Let $\mathcal{A}$ and $\mathcal{B}$ be norm closed algebras containing the finite rank operators. Let α be an isomorphism of $\mathcal{A}$ onto $\mathcal{B}$. Then there is an invertible operator S such that $\alpha = Ad\, S$.*

Proof. Lemma 17.6 characterizes rank one operators algebraically. So the isomorphism α carries the set of rank one operators onto itself.

Fix a non-zero vector y in $\mathcal{H}$. Since $y \otimes y^*$ is rank one, there are vectors u_y and v_y so that $\alpha(y \otimes y^*) = u_y \otimes v_y^*$. If x is any vector, $\alpha(x \otimes y^*)$ is rank one of the form $a \otimes b^*$. Hence

$$\begin{aligned}\alpha(x \otimes y^*) &= \alpha((x \otimes y^*)\|y\|^{-2}(y \otimes y^*)) = \|y\|^{-2}(a \otimes b^*)(u_y \otimes v_y^*) \\ &= \|y\|^{-2}(u_y,b)a \otimes v^* = u_x \otimes v_y^* \ .\end{aligned}$$

Furthermore, the linearity of α shows that the map S_y taking x to u_x is linear. And S_y is one to one since α is.

Likewise, for each non-zero vector x, there is a vector u_x and a linear one to one map T_x such that $\alpha(x \otimes y^*) = u_x \otimes (T_x y)^*$. Hence $S_y x$ is a multiple of u_x for every y, and $T_x y$ is a multiple v_y for every x.

Pick unit vectors x_0 and y_0. For each y, replace v_y and S_y by scalar multiples so that $S_y x_0 = S_{y_0} x_0$ (which is non-zero since S_{y_0} is one to one). We show that $S_y = S_{y_0}$. For if x is not a multiple of x_0, there are constants c and d so that $S_y x = cS_{y_0} x$ and $S_y(x+x_0) = dS_{y_0}(x+x_0)$. Compute

$$dS_{y_0}x+dS_{y_0}x_0 = S_y x+S_y x_0 = cS_{y_0}x+S_{y_0}x_0 \ .$$

So

$$(d-c)S_{y_0}x = (1-d)S_{y_0}x_0 \ .$$

But $S_{y_0}x$ and $S_{y_0}x_0$ are linearly independent, whence $c = d = 1$. Let S be the operator $S_{y_0} = S_y$.

Now scale each T_x so that

$$\alpha(x \otimes y^*) = Sx \otimes (T_x y)^* \ .$$

As above, it follows that $T_x = T_{x_0}$ for every x. We will denote this by T. So, $\phi(x \otimes y^*) = Sx \otimes (Ty)^*$ for every x and y in $\mathcal{H}$. As α takes the rank one operators *onto* the rank one operators, S and T are surjective.

Now, apply the multiplicative nature of α to

$$(x,y)x_0 \otimes y_0^* = (x_0 \otimes y^*)(x \otimes y_0^*)$$

to obtain

$$(x,y)Sx_0 \otimes (Ty_0)^* = (Sx_0 \otimes (Ty)^*)(Sx \otimes (Ty_0)^*)$$
$$= (Sx,Ty)Sx_0 \otimes (Ty_0)^* \ .$$

Thus

$$(x,y) = (Sx,Ty) = (T^*Sx,y)$$

for every x and y in $\mathcal{H}$. So $T = (S^*)^{-1}$. So

$$\alpha(x \otimes y^*) = Sx \otimes (S^*)^{-1}y = S(x \otimes y^*)S^{-1} \ .$$

Let A be any element of $\mathcal{A}$, and let x belong to $\mathcal{H}$. One obtains

$$S(Ax \otimes y_0)S^{-1} = \alpha(Ax \otimes y_0) = \alpha(A)\alpha(x \otimes y_0)$$
$$= \alpha(A)S(x \otimes y_0^*)S^{-1} \ .$$

Therefore $SAx = \alpha(A)Sx$ for all x in $\mathcal{H}$, so $\alpha(A) = SAS^{-1}$.

It remains to show that S is continuous. For A in $\mathcal{A}$ and x, y in $\mathcal{H}$ one has

$$|(\alpha(A)x,y)| = |(AS^{-1}x,S^*y)| \leq \|A\| \ \|S^{-1}x\| \ \|S^*y\| \ .$$

Hence

$$\sup_{\|A\| \leq 1} |(\alpha(A)x,y)| \leq \|S^{-1}x\| \ \|S^*y\| \ .$$

So, the set $\{\alpha(A)x : \|A\| \leq 1\}$ for fixed x as a set of functionals on $\mathcal{H}$ are bounded pointwise. By the uniform boundedness principle,

$$\sup_{\|A\| \leq 1} \|\alpha(A)x\| = C_x < \infty \ .$$

A second application yields

$$\sup_{\|A\| \leq 1} \|\alpha(A)\| < \infty \ .$$

Thus α is norm continuous. Let e be any unit vector. Then

$$\|Sx\| = \|Sx \otimes e^*\| = \|\alpha(x \otimes (S^*e)^*)\| \leq \|\alpha\| \ \|S^*e\| \ \|x\| \ .$$

So S is continuous, and likewise S^{-1} is continuous. ■

17.8 Corollary. *Let $\mathcal{N}$ and $\mathcal{M}$ be nests. Then $\mathcal{QT}(\mathcal{N})$ and $\mathcal{QT}(\mathcal{M})$ are isomorphic if and only if they are unitarily equivalent. Every such isomorphism is spatially implemented, and factors as $Ad\,U \cdot Ad\,A$ where U is unitary and A is invertible in $\mathcal{QT}(\mathcal{N})$.*

Proof. Since these algebras contain the finite rank operators, Theorem 17.7 shows that every isomorphism is spatial. If $\alpha = Ad\,S$ is an isomorphism, then $\mathcal{QT}(\mathcal{M}) = Ad\,S(\mathcal{QT}(\mathcal{N})) = \mathcal{QT}(S\mathcal{N})$. By the Similarity Theorem 13.20, there is a unitary operator U and a compact operator K with $\|K\| < ½$ such that $SN = (U+K)N$ for every N in $\mathcal{N}$. Thus $U^*SN = (I+U^*K)N$ for every N in $\mathcal{N}$. Since K is compact,

$Q\,\mathcal{T}((I+U^*K)\mathcal{N}) = Q\,\mathcal{T}(\mathcal{N})$. Thus

$$UQ\,\mathcal{T}(\mathcal{N})U^* = UQ\,\mathcal{T}((I+U^*K)\mathcal{N})U^* = Q\,\mathcal{T}((U+K)N) = Q\,\mathcal{T}(\mathcal{M})\ .$$

Moreover, $T = (I+U^*K)^{-1}U^*S$ is invertible in $\mathcal{T}(\mathcal{N})$, hence $A = (1+U^*K)T$ is invertible in $Q\,\mathcal{T}(\mathcal{N})$. So $Ad\,S = Ad\,U \cdot Ad\,T$. ∎

Now, we turn to nest algebras. The first step is an algebraic description of the rank one elements as in Lemma 17.6.

17.9 Lemma. *Let $\mathcal{N}$ be a nest, and let T belong to $\mathcal{T}(\mathcal{N})$. Then T has rank one if and only if whenever A and B belong to $\mathcal{T}(\mathcal{N})$ and $ATB = 0$ then either $AT = 0$ or $TB = 0$.*

Proof. If $T = x \otimes y^*$ is rank one, then $ATB = Ax \otimes (B^*y)^*$. So $ATB = 0$ if and only if $Ax = 0$ or $B^*y = 0$. Since $AT = Ax \otimes y^*$ and $TB = x \otimes (B^*y)^*$, it is clear that the latter condition is equivalent to $AT = 0$ or $TB = 0$.

Suppose T in $\mathcal{T}(\mathcal{N})$ has rank at least two. Let

$$N' = \bigvee\{N \in \mathcal{N}: rank\,TP(N) \leq 1\}\ .$$

Since the rank one operators are strongly closed, $TP(N')$ has rank at most one and so $N' < \mathcal{H}$. Choose any $N_0 > N'$ for which $(N_0)_- \neq \mathcal{H}$. Since $TP(N_0)$ has rank at least two, one can repeat this process for its adjoint to obtain N_1 in $\mathcal{N}$ so that $(N_1)_+ \neq \{0\}$ and $T' = P(N_1)^\perp TP(N_0)$ has rank at least two. Let S be the operator in $\mathcal{B}(N_0, N_1^\perp)$ agreeing with T'.

Pick any unit vector y in N_0 so that $Sy \neq 0$. Since $(ker\,S^*)^\perp$ has dimension at least two, it has non trivial intersection with $\{Sy\}^\perp$. Choose a unit vector x in this intersection. Then x belongs to $N_1^\perp$ and satisfies $T^*x \neq 0$ but $(Ty,x) = 0$. Choose unit vectors z_0 in $(N_0)_-^\perp$ and z_1 in $(N_1)_+$. By Lemma 2.8, the operators $A = z_1 \otimes x^*$ and $B = y \otimes z_2^*$ belong to $\mathcal{T}(\mathcal{N})$. Furthermore, $AT = z_1 \otimes (T^*x)^* \neq 0$ and $TB = Ty \otimes z_0^* \neq 0$, yet $ATB = (Ty,x)z_1 \otimes z_0^* = 0$. ∎

17.10 Lemma. *Let $\mathcal{A}$ be a maximal abelian von Neumann algebra. Let x_n, $n \geq 1$ be vectors in $\mathcal{H}$ such that $\sum_{n=1}^{\infty} \|x_n\| < \infty$. Then there is a vector z and elements A_n in the unit ball of $\mathcal{A}$ such that $A_n z = x_n$ for $n \geq 1$.*

Proof. By Theorem 7.8, $\mathcal{A}$ is unitarily equivalent to the canonical representation of $L^\infty(\mu)$ on $L^2(\mu)$ for a finite regular Borel measure μ. So x_n correspond to functions f_n in $L^2(\mu)$. Let $g = \sum_{n=1}^{\infty} |f_n|$, which belongs to $L^2(\mu)$ and hence converges almost everywhere. Let $h_n = g^{-1}f_n$ where h_n is defined to be zero on $f_n^{-1}\{0\}$. By construction, $\|h_n\|_{L^\infty(\mu)} \leq 1$ and $h_n g = f_n$. So choose z to correspond to g, and A_n corresponding to M_{h_n} ∎

17.11 Corollary. *Let T be a linear map defined on all of $\mathcal{H}$. If T commutes with every A in a masa $\mathcal{A}$, then T belongs to $\mathcal{A}$.*

Proof. It suffices to prove that T is continuous. If it is not continuous, one can choose vectors x_n such that $\|x_n\| \leq 2^{-n}$ and $\|Tx_n\| \geq n$. Let z and A_n be provided by Lemma 17.10. Then for all n,

$$n \leq \|Tx_n\| = \|TA_n z\| = \|A_n Tz\| \leq \|Tz\| \ .$$

This is impossible. So T is bounded. ∎

17.12 Theorem. *Let $\mathcal{N}$ be a nest, and let $\mathcal{A}$ be a masa contained in $\mathcal{T}(\mathcal{N})$. Suppose that α is an automorphism of $\mathcal{T}(\mathcal{N})$ such that $\alpha(A) = A$ for all A in $\mathcal{A}$. Then there is an invertible element S in $\mathcal{A}$ such that $\alpha = Ad\, S$.*

Proof. Let $\mathcal{N}_1$ denote the set of all N in $\mathcal{N}$ such that $N \neq 0$ and $N_- \neq \mathcal{H}$. This consists of all $N \neq 0$ or $\mathcal{H}$, together with $\mathcal{H}$ if $\mathcal{H}$ has an immediate predecessor. Fix N in $\mathcal{N}_1$. Let x belong to N and let y belong to $(N_-)^{\perp}$. So $x \otimes y^*$ belongs to $\mathcal{T}(\mathcal{N})$ by Lemma 3.7. By Lemma 17.6, $\alpha(x \otimes y^*)$ is rank one, and thus can be written as $u \otimes v^*$. Moreover,

$$u \otimes v^* = \alpha(P(N)(x \otimes y^*)P(N_-)^{\perp}) = P(N)u \otimes (P(N_-)^{\perp}v)^* \ .$$

So u belongs to N and v belongs to $(N_-)^{\perp}$. The vectors u and v are unique up to a scalar λ since

$$u \otimes v = (\lambda u) \otimes (\overline{\lambda}^{-1}v)^* \ .$$

Fix two unit vectors x_0 in N and y_0 in $(N_-)^{\perp}$. Choose u_0 and v_0 so that $\alpha(x_0 \otimes y_0^*) = u_0 \otimes v_0^*$. If x_1, x_2 belong to N, use Lemma 17.9 to obtain a vector x in N and A_i in $\mathcal{A}$ so that $A_i x = x_i$, $0 \leq i \leq 2$. Let $\alpha(x \otimes y_0^*) = u \otimes v^*$, normalized so that $(v, v_0) = \|v_0\|^2$. If λ_0, λ_1, λ_2 are scalars,

$$\alpha(\sum_{i=0}^{2}\lambda_i x_i \otimes y_0^*) = \alpha(\sum_{i=0}^{2}\lambda_i A_i x \otimes y_0^*) \tag{17.10.1}$$

$$= (\sum_{i=0}^{2}\lambda_i A_i)u \otimes v^* = \sum_{i=0}^{2}\lambda_i (A_i u) \otimes v^* \ .$$

Taking $\lambda_0 = 1$, $\lambda_1 = \lambda_2 = 0$ yields $u_0 \otimes v_0^* = A_0 u \otimes v^*$. The normalization guarantees that $v = v_0$ and $A_0 u = u_0$. Let $u_i = A_i u$ for $i = 1,2$. Then $\alpha(x_i \otimes y_0^*) = u_i \otimes v_0^*$. Consequently, $u_i = U(x_i)$ is independent of the choice of x. By (17.10.1), U is a linear map.

Similarly, there is a linear map V on $(N_-)^{\perp}$ such that $\alpha(x_0 \otimes y^*) = u_0 \otimes (Vy)^*$. Let y_1 belong to $(N_-)^{\perp}$. Let y belong to $(N_-)^{\perp}$ and let B_0, B_1 belong to $\mathcal{A}$ such that $B_i y = y_i$ for $i = 0,1$ (Lemma 17.9). Then as above, write $\alpha(x \otimes y) = u \otimes v^*$ and compute

$$\alpha((\lambda_0x_0+\lambda_1x_1)\otimes(\mu_0y_0+\mu_1y_1)^*) = (\lambda_0A_0+\lambda_1A_1)\alpha(x\otimes y^*)(\mu_0B_0+\mu_1B_1)^*$$
$$= (\lambda_0A_0u+\lambda_1A_1u)\otimes(\mu_0B_0v+\mu_1B_1v)^* \ .$$

With $\lambda_0 = \mu_0 = 1$ and $\lambda_1 = \mu_1 = 0$, one sees that u and v can be normalized so that $A_0u = u_0$ and $B_0v = v_0$. With this done, $\lambda_0 = \mu_1 = 0$ and $\lambda_1 = \mu_0 = 1$ yields $A_1u = Ux_1$. Likewise, $\lambda_0 = \mu_1 = 1$ and $\lambda_1 = \mu_0 = 0$ yields $B_1v = Vy_1$. So plugging in $\lambda_0 = \mu_0 = 0$ and $\lambda_1 = \mu_1 = 1$ gives

$$\alpha(x_1\otimes y_1) = Ux_1\otimes(Vy_1)^* \ .$$

Next, notice that if u belongs to N and v belongs to $(N_-)^\perp$, then $\alpha^{-1}(u\otimes v^*) = x\otimes y$ is rank one and as before, x belongs to N and y belongs to $(N_1)^\perp$. So Ux and Vy are multiples of u and v respectively. Hence U and V are surjective, as well as one to one since α is one to one. If A belongs to $\mathcal{A}$ and x is in N,

$$UAx\otimes v_0^* = \alpha(Ax\otimes y_0^*) = A\alpha(x\otimes y_0^*) = AUx\otimes v_0^* \ .$$

So U (as an operator on N) commutes with the restriction of $\mathcal{A}$ to N which is a masa. By Corollary 17.11, U belongs to $\mathcal{A}$. Likewise V belongs to $\mathcal{A}$. In particular, U is invertible in $\mathcal{B}(N)$ and V is invertible in $\mathcal{B}(N_-)^\perp$.

So far, N has been fixed. For each N, there are operators U_N in $\mathcal{A}\,|N$ and V_N in $\mathcal{A}\,|(N_-)^\perp$ such that

$$\alpha(x\otimes y^*) = U_Nx\otimes(V_Ny)^*$$

for all x in N and y in $(N_-)^\perp$. Suppose $N_1 < N_2$ belong to $\mathcal{N}_1$. Then for each x in N_1 and y in $(N_2)_-^\perp$, one obtains

$$U_{N_1}x\otimes(V_{N_1}y)^* = U_{N_2}x\otimes(V_{N_2}y)^* \ .$$

Thus there is a non-zero scalar λ such that

$$U_{N_2}|N_1 = \lambda U_{N_1} \quad V_{N_1}|(N_2)_-^\perp = \bar{\lambda}V_{N_2} \ .$$

Pick an N_0 and replace U_N and V_N by scalar multiples so that they agree with U_{N_0} and V_{N_0} on their common domains. It follows that there is an operator U defined on $\bigcup\{N\in\mathcal{N}: N_- < \mathcal{H}\}$ such that $U\,|N = U_N$ for each N. Likewise, there is an operator V defined on $\bigcup\{N_-^\perp : N \neq 0\}$ such that $V\,|N_-^\perp = V_N$. Furthermore,

$$\alpha(x\otimes y^*) = Ux\otimes(Vy)^*$$

for every rank one operator in $\mathcal{T}(\mathcal{N})$.

Let $x_1\otimes y_1^*$ and $x_2\otimes y_2^*$ be two non-zero rank one operators in $\mathcal{T}(\mathcal{N})$. Then

$$\begin{aligned}(x_2,y_1)Ux_1 \otimes (Vy_2)^* &= \alpha((x_1 \otimes y_1^*)(x_2 \otimes y_2^*)) \\ &= \alpha(x_1 \otimes y_1^*)\alpha(x_2 \otimes y_2^*) \\ &= (Ux_2,Vy_1)Ux_1 \otimes (Vy_2)^* \ .\end{aligned}$$

Then

$$(x_2,y_1) = (Ux_2,Vy_1) = (V^*Ux_2,y_1) \ .$$

This holds for every x_2 in N_2 and y_1 in $(N_-)^\perp$ and every N_1, N_2 in $\mathcal{N}_1$. Hence

$$V\,|N_2 \ominus (N_1)_- = (U\,|N_2 \ominus (N_1)_-)^{*-1} \ .$$

With N_2 fixed, one has $U\,|N_2$ is invertible so V is uniformly bounded on $\bigcup\{N_2 \ominus (N_1)_- : N_1 > 0\}$, and thus V extends to be continuous on N_2. As it is bounded on $(N_2)_-$, V is bounded on $\mathcal{H}$. Likewise, U extends to be bounded on $\mathcal{H}$. Moreover, $V\,|N_2 = (U\,|N_2)^{*-1}$ and likewise $V\,|(N_2)_-^\perp = (U\,|(N_2)_-^{\perp *})^{-1}$ so $V = U^{*-1}$. In other words, U is an invertible operator in $\mathcal{A}$ such that

$$\alpha(x \otimes y^*) = Ux \otimes (U^{*-1}y)^* = U(x \otimes y^*)U^{-1}$$

for every rank one operator in $\mathcal{T}(\mathcal{N})$.

Let T belong to $\mathcal{T}(\mathcal{N})$. For every $x \otimes y^*$ in $\mathcal{T}(\mathcal{N})$,

$$\begin{aligned}U(Tx \otimes y^*)U^{-1} &= \alpha(T(x \otimes y^*)) = \alpha(T)\alpha(x \otimes y^*) \\ &= \alpha(T)U(x \otimes y^*)U^{-1} \ .\end{aligned}$$

Therefore, with $u = Ux$, one has

$$(UTU^{-1} - \alpha(T))u = 0 \ .$$

This holds for every u in $\bigcup\{N : N_- < \mathcal{H}\}$. This is dense in $\mathcal{H}$, so $\alpha(T) = UTU^{-1}$ for every T in $\mathcal{T}(\mathcal{N})$. ∎

Combining Theorems 17.5 and 17.12, one immediately obtains

17.13 Corollary. *Every isomorphism between nest algebras is spatial.*

This is not the end of the story. Let $\mathcal{N}$ be a nest and let $\alpha = Ad\,S$ be an automorphism of $\mathcal{T}(\mathcal{N})$. Then

$$\mathcal{T}(\mathcal{N}) = S\,\mathcal{T}(\mathcal{N})S^{-1} = \mathcal{T}(S\mathcal{N}) \ .$$

So $\theta_S(N) = SN$ is a dimension preserving order isomorphism of $\mathcal{N}$. If S is an invertible element of $\mathcal{T}(\mathcal{N})$, then $\theta_S = id$. Conversely, if $\theta_S = id$, then $SN = N$ for every N and thus S and S^{-1} belong to $\mathcal{T}(\mathcal{N})$. The automorphisms $Ad\,S$ for S in $\mathcal{T}(\mathcal{N})^{-1}$ are the *inner automorphisms*, denoted $Inn(\mathcal{T}(\mathcal{N}))$. This is a normal subgroup of $Aut\,\mathcal{T}(\mathcal{N})$ because $\alpha Ad\,S\alpha^{-1} = Ad\,\alpha(S)$. The quotient group of the automorphism group

$Aut(\mathcal{T}(\mathcal{N}))$ by $Inn(\mathcal{T}(\mathcal{N}))$ is denoted $Out(\mathcal{T}(\mathcal{N}))$ for *outer* automorphism group. Let $Aut(\mathcal{N})$ denote the group of dimension preserving order isomorphisms of $\mathcal{N}$.

17.14 Corollary. *The map $\Theta : Ad\, S \rightarrow \theta_S$ of $Aut(\mathcal{T}(\mathcal{N}))$ into $Aut(\mathcal{N})$ is surjective, and provides a natural isomorphism between $Out\, \mathcal{T}(\mathcal{N})$ and $Aut(\mathcal{N})$.*

Proof. The discussion prior to the statement shows that the kernel of Θ is $Inn(\mathcal{T}(\mathcal{N}))$, and it is routine to verify that Θ is a homomorphism. If θ belongs to $Aut(\mathcal{N})$, then by the Similarity Theorem 13.20, there is a invertible operator S such that $\theta_S = \theta$. So Θ is surjective. Hence $Out\, \mathcal{T}(\mathcal{N})$ is naturally isomorphic to $Aut(\mathcal{N})$. ∎

17.15 Corollary. *If α belongs to $Aut(\mathcal{T}(\mathcal{N}))$ and $\|\alpha - id\| < 1$, then α is inner. Hence $Inn(\mathcal{T}(\mathcal{N}))$ is open and closed, and $Out\, \mathcal{T}(\mathcal{N})$ has the discrete topology.*

Proof. If $\alpha = Ad\, S$ is not inner, then $\theta_S \neq id$. So there is some N in $\mathcal{N}$ with $SN \neq N$. But then $\alpha(P(N)) = SP(N)S^{-1}$ is an idempotent with range SN in $\mathcal{N}$. Thus $\|\alpha(P(N)) - P(N)\| \geq 1$. (For example, if $SN > N$, choose a unit vector x in $SN \ominus N$. Then $\alpha(P(N)) - P(N)x = x$.) So id is in the interior of $Inn(\mathcal{T}(\mathcal{N}))$. Hence $Inn\, \mathcal{T}(\mathcal{N})$ is open. The complement of $Inn\, \mathcal{T}(\mathcal{N})$ is the union of cosets, so is also open. ∎

Now we analyze the outer automorphisms of $Q\mathcal{T}(\mathcal{N})$. The main tool is Theorem 12.8. The first step, Lemma 17.19, is to make the perturbations involved more restricted.

17.16 Lemma. *Let $\mathcal{N}$ be a nest and let Δ be the unique expectation onto $\mathcal{D}_a(\mathcal{N})$, the atomic part of the diagonal. Suppose S is invertible, and let Ψ be the corresponding expectation onto $\mathcal{D}_a(S\mathcal{N})$. Then $\Psi \circ Ad\, S\, |\mathcal{D}_a(\mathcal{N})$ and $\Delta \circ Ad\, S^{-1}\, |\mathcal{D}_a(S\mathcal{N})$ are isomorphisms, reciprocal to one another.*

Proof. Let $E = P(N_+) - P(N)$ be an atom of $\mathcal{N}$ and let $\hat{E} = P(SN_+) - P(SN)$ be the corresponding atom of $S\mathcal{N}$. Let $A = \hat{E}SE$ as an element of $\mathcal{B}(E\mathcal{H}, \hat{E}\mathcal{H})$. As an operator from $N \oplus E\mathcal{H} \oplus N_+^{\perp}$ to $SN \oplus \hat{E}\mathcal{H} \oplus (SN_+)^{\perp}$, S is upper triangular with A as its 2–2 entry. Likewise, S^{-1} is upper triangular from $SN \oplus \hat{E}\mathcal{H} \oplus (SN_+)^{\perp}$ to $N \oplus E\mathcal{H} \oplus N_+^{\perp}$. So a computation shows that its 2–2 entry must be A^{-1}. If X belongs to $\mathcal{B}(E)$,

$$\Psi \circ Ad\, S(EXE) = \Psi \begin{bmatrix} * & * & * \\ 0 & A & * \\ 0 & 0 & * \end{bmatrix} \begin{bmatrix} 0 & 0 & 0 \\ 0 & X & 0 \\ 0 & 0 & 0 \end{bmatrix} \begin{bmatrix} * & * & * \\ 0 & A^{-1} & * \\ 0 & 0 & * \end{bmatrix}$$

$$= \Psi \begin{bmatrix} 0 & * & * \\ 0 & AXA^{-1} & * \\ 0 & 0 & 0 \end{bmatrix} = \begin{bmatrix} 0 & 0 & 0 \\ 0 & AXA^{-1} & 0 \\ 0 & 0 & 0 \end{bmatrix} .$$

Similarly, if Y belongs to $\mathcal{B}(\hat{E})$

$$\Delta \circ Ad\, S^{-1}(\hat{E}Y\hat{E}) = E(A^{-1}YA)E \quad .$$

Summing over all atoms yields the lemma. ■

17.17 Corollary. *With the same notation, $\Psi \circ Ad\, S \mid \mathcal{D}_a(\mathcal{N}) \cap \mathcal{N}''$ is a $*$-isomorphism onto $\mathcal{D}_a(S\mathcal{N}) \cap (S\mathcal{N})''$ with inverse $\Delta \circ Ad\, S^{-1} \mid \mathcal{D}_a(S\mathcal{N}) \cap (S\mathcal{N})''$.*

Proof. $\mathcal{D}_a(\mathcal{N}) \cap \mathcal{N}''$ is the centre of $\mathcal{D}_a(\mathcal{N})$ and hence is carried onto the centre of $\mathcal{D}_a(S\mathcal{N})$, namely $\mathcal{D}_a(S\mathcal{N}) \cap (S\mathcal{N})''$. These two algebras are abelian von Neumann algebras. Projections are thus taken to projections. Hence on the span of the projections, one sees that adjoints are preserved. By continuity, these maps are $*$-isomorphisms. ■

17.18 Lemma. *Let $\mathcal{N}$ be a nest, S an invertible operator, and let F be a finite dimensional subspace such that $P(F)$ commutes with $S\mathcal{N}$. Then there is a subspace F_1 such that $P(F_1)$ belongs to $\mathcal{N}'$ and an invertible compact perturbation S_1 of S so that $S_1(\mathcal{N}^{F_1}) = (S\mathcal{N})^F$.*

Proof. Let Δ and Ψ be as in the previous lemma. Since $P(F)$ is finite rank in $(S\mathcal{N})'$, there are finitely many atoms $\hat{E}_1, \ldots, \hat{E}_n$ so that $P(F) = \sum_{i=1}^{n} \hat{E}_i P(F) \hat{E}_i$. Hence

$$R_1 \equiv \Delta Ad\, S^{-1}(P(F)) = \sum_{i=1}^{n} E_i S^{-1} P(F) S E_i \quad .$$

Let F_1 be the range of R_1. As F_1 is the direct sum of subspaces of atoms, $P(F_1)$ belongs to $\mathcal{N}'$. Let

$$S_1 = P(F)SP(F_1) + P(F)^{\perp}SP(F_1)^{\perp} \quad .$$

This is a compact perturbation of S. Moreover, $R_1 P(F_1) = P(F_1)$ and $P(F_1)R_1 = R_1$. Let $R = \Psi Ad\, S(P(F))$. By Lemma 17.16, $\Psi Ad\, S(R) = P(F)$ whence $P(F)R = R$ and $RP(F) = P(F)$. So R is an idempotent with range F. Thus $S_1 F_1 = SF_1 = F$. As operators from $F_1 \oplus F_1^{\perp}$ to $F \oplus F^{\perp}$ and back, S and S^{-1} are triangular. So S_1 is invertible. If N belongs to $\mathcal{N}$, $S_1(F_1 \vee N) = F \vee SN$ belongs to $(S\mathcal{N})^F$. Hence S_1 carries $\mathcal{N}^{F_1}$ onto $(S\mathcal{N})^F$ as desired. ■

It is desirable to arrange it so that only perturbations of a fixed nest $\mathcal{N}$ need be considered. Moreover, it would help if these perturbations always commute. To this end, one arranges that the perturbations consist of sum of finite atoms and sub projections of infinite atoms in a fixed masa.

17.19 Lemma. *Let $\mathcal{N}$ be a nest. Extend $\mathcal{N}''$ to a von Neumann algebra $\mathcal{M}$ whose restriction to each infinite atom is an atomic masa, but agrees with $\mathcal{N}''$ on the complement of the infinite atoms. Let S be an invertible operator such that $Q\mathcal{T}(S\mathcal{N}) = Q\mathcal{T}(\mathcal{N})$. Then there is an invertible, compact perturbation S_0 of S and two finite dimensional subspaces F and G with $P(F)$ and $P(G)$ in $\mathcal{M}$ such that $S_0\mathcal{N}^F = \mathcal{N}^G$.*

Proof. By Theorem 12.8, there are finite dimensional subspaces F_1 and G_1 with $P(F_1)$ in $(S\mathcal{N})'$ and $P(G_1)$ in $\mathcal{N}'$ and a compact operator K so that $(I+K)(S\mathcal{N})^{F_1} = \mathcal{N}^{G_1}$. By the previous lemma, there is a subspace F_2 with $P(F_2)$ in $\mathcal{N}'$ and a compact perturbation S_1 of S so that $S_1\mathcal{N}^{F_2} = (S\mathcal{N})^{F_1}$. Thus $(I+K)S_1\mathcal{N}^{F_2} = \mathcal{N}^{G_1}$. Let $S_2 = (I+K)S_1$. It remains to modify F_2, G_1 and S_2 so that the subspaces have projections in $\mathcal{M}$ instead of just $\mathcal{N}'$.

Let $H_1,\dots,H_n$ be the atoms of $\mathcal{N}$ such that $H_jP(G_1) \neq 0$ and let $H = \sum_{j=1}^{n} H_j$. Let $E_1,\dots,E_m$ be the atoms E of $\mathcal{N}$ such that either $P(F_2)E \neq 0$ or $\theta_{S_2}(E) \leq H$. Let F be the sum of the finite dimensional spaces $E_i\mathcal{H}$, $1 \leq i \leq n'$, together with subspace M_i of the infinite atoms $E_i\mathcal{H}$, $n' < i \leq n$, so that $P(M_i)$ belongs to $\mathcal{M}$ and $\dim M_i = \dim(F_2 \wedge E_i\mathcal{H})$. Let U be a unitary of the form $I + \sum_{i=n'+1}^{n} E_iK_1E_i$ where K_1 is finite rank, and $UM_i = F_2 \wedge E_i\mathcal{H}$ for each infinite atom. Similarly, if H_j is an infinite atom, choose a subspace L_j so that $P(L_j)$ belongs to $\mathcal{M}$ and $\dim L_j = \dim(G_1 \wedge H_j\mathcal{H})$. Let V be a unitary of the form $I + \sum_{j=1}^{m} H_jK_2H_j$ where K_2 is finite rank and $VL_j = G_1 \wedge H_j\mathcal{H}$. Let $F_3 = U^*F_2$, $G_2 = V^*G_1$, and $S_3 = V^*S_2U$. Note that S_3 takes $\mathcal{N}^{F_3}$ onto $\mathcal{N}^{G_2}$. Now $P(F)$ is the sum of atoms of $\mathcal{N}^{F_3}$. Thus $\theta_{S_3}(P(F)) = P(G)$ is the sum of atoms of $\mathcal{N}^{G_2}$ which includes all finite rank atoms H_j, as well as $\theta_{S_3}(E_iP(F_2)^\perp)$ when E_i is finite. So G is the sum of atoms of $\mathcal{N}$ together with the L_j for infinite atoms. Hence F and G have the desired form.

Set $S_0 = P(G)S_3P(F)+P(G)^\perp S_3P(F)^\perp$. It must be shown that S_0 is invertible. The atoms comprising F in $\mathcal{N}^{F_3}$ split $\mathcal{N}^{F_3}$ into intervals $A_1,\dots,A_k$ where the odd intervals belong to F, and the even intervals to $F^\perp$. Likewise, $\mathcal{N}^{G_2}$ is split into intervals $B_1,\dots,B_k$. The operators S_3 and S_3^{-1} written as operator matrix from $\sum \oplus A_i$ to $\sum \oplus B_i$ and back are both upper triangular. Hence their diagonal entries are invertible. From this it follows that $P(G)S_3P(F)$ and $P(G)^{\perp}S_3P(F)^{\perp}$ are invertible (and upper triangular) as operators in $B(F,G)$ and $B(F^{\perp},G^{\perp})$ respectively. So S_0 is invertible. A review of the definition of S_0 shows that $S-S_0$ is compact. It is clear that $S_0\mathcal{N}^F = \mathcal{N}^G$. ■

At this stage, one can prove the analogue of Corollary 17.15.

17.20 Theorem. *Let α be an automorphism of $\mathcal{Q}\mathcal{T}(\mathcal{N})$ such that $\|\alpha-id\| < 1$. Then α is inner via S in $\mathcal{Q}\mathcal{T}(\mathcal{N})^{-1}$ such that $\|S-I\| \le 2\|\alpha-id\|$.*

Proof. By Theorem 17.7, $\alpha = Ad\,S$ for some invertible operator S. Normalize S so that $\|S\| = 1$. Let $\epsilon = \|\alpha-id\|$. For each K in $\mathcal{K}$,

$$\|SK-KS\| = \|(\alpha(K)-K)S\| \le \|\alpha-id\|\ \|K\|\ \|S\| = \epsilon\|K\| \quad .$$

Thus by the weak* continuity of δ_S, $\|\delta_S\| \le \epsilon$. By Theorem 9.6, there is a scalar λ so that $\|S-\lambda I\| \le \epsilon$. Since $\|S\| = 1$, $|1-\lambda| \le \|S-\lambda I\|$. Hence $\|S-I\| \le 2\epsilon$. It remains to show that S belongs to $\mathcal{Q}\mathcal{T}(\mathcal{N})^{-1}$.

Now $\mathcal{Q}\mathcal{T}(\mathcal{N}) = \alpha(\mathcal{Q}\mathcal{T}(\mathcal{N})) = \mathcal{Q}\mathcal{T}(S\mathcal{N})$. So by Lemma 17.19, there is an invertible, compact perturbation S_0 of S so that $S_0\mathcal{N}^F = \mathcal{N}^G$, and $P(F)$ and $P(G)$ are finite rank projections in $\mathcal{M}$ (as described in that lemma). Now $S_0S^{-1}-I = K$ is compact, so $\alpha_0 = Ad\,S_0$ factors as $\alpha_0 = Ad\,(I+K)\cdot\alpha$. Let $\theta = \theta_{S_0}$ be the order isomorphism of $\mathcal{N}^F$ onto $\mathcal{N}^G$ induced by S_0, and consider it extended to all intervals. By Theorem 8.3, there is an expectation Φ of $\mathcal{B}(\mathcal{H})$ onto $\mathcal{M}$. For each interval E, $\alpha_0(E)$ is an idempotent with range $\theta(E)$. So $\theta(E) = \Phi\alpha_0(E)$. Since Φ is contractive and $\Phi(E) = E$,

$$\|\theta(E)-E\| \le \|\alpha_0(E)-E\| \le \|(\alpha_0-\alpha)(E)\|+\|\alpha-id\| \quad .$$

Let $\{E_n\}$ be any family of pairwise orthogonal intervals. Then since K is compact, $\|(\alpha_0-\alpha)(E_n)\| = \|Ad\,(I+K)\alpha(E_n)\| < 1-\epsilon$ for all except finitely many n. Now $\theta(E_n)$ and E_n belong to $\mathcal{M}$ and thus commute; as $\|\theta(E_n)-E_n\| < 1$, then $\theta(E_n) = E_n$ except finitely often. Also,

$$\|\theta(E)-E\|_e \le \|\alpha-id\| < 1$$

so $\theta(E)-E$ is compact and hence finite rank. In particular, $\theta(A) = A$ for all but a finite set of atoms $\{A_1,\dots,A_n\}$. Let $Q = \bigvee_{i=1}^{n}(A_i-\theta(A_i))^2$, the least projection in $\mathcal{M}$ dominating $|A_i-\theta(A_i)|$ for all i. Then Q is finite rank. Let E be an interval of $\mathcal{N}^F$, and write $\theta(E)-E = B_1-B_2$ where B_i are pairwise orthogonal projections on $\mathcal{M}$. As B_i are finite rank, they are contained in the atomic part of $\mathcal{N}^F$ or $\mathcal{N}^G$. If A is an atom of $\mathcal{N}^F$ such that $AB_2 \ne 0$, then $\theta(A) \le \theta(E)$, whence $AB_2 = (A-\theta(A))B_2 \le QB_2$. Similarly, if A is an atom such that $\theta(A)B_1 \ne 0$, then $A \le E$ so $AB_1 = 0$.

Thus $\theta(A)B_2 = (\theta(A)-A)B_2 \le QB_2$. It follows that $Q^{\perp}\theta(E) = Q^{\perp}E$ for every interval. In particular, if N belongs to $\mathcal{N}^F$,

$$Q^{\perp}N = Q^{\perp}\theta(P(N))\mathcal{H} = Q^{\perp}S_0N \ .$$

So $Q^{\perp}S_0$ belongs to $\mathcal{T}(\mathcal{N}^F)$. Likewise $Q^{\perp}S_0^{-1}$ belongs to $\mathcal{T}(\mathcal{N}^G)$. As S and S^{-1} are compact perturbations of these operators, they both belong to $Q\,\mathcal{T}(\mathcal{N})$. ■

17.21. Let S, F, G, S_0, $\mathcal{N}^F$, and $\mathcal{N}^G$ be as in Lemma 17.19. Let F_0 and G_0 be the sums of finite atoms dominated by F and G respectively. Now S_0 induces an automorphism θ_{S_0} of $\mathcal{N}^F$ onto $\mathcal{N}^G$ which extends naturally to intervals. In particular, each infinite atom of $\mathcal{N}^F$ is taken to an infinite atom of $\mathcal{N}^G$. These infinite atoms are finite rank perturbations of atoms of $\mathcal{N}$. Let $\mathcal{N}_{F_0}$ (and $\mathcal{N}_{G_0}$) denote the restriction of $\mathcal{N}$ to $F_0^{\perp}\mathcal{H}$. Then θ_{S_0} induces an element $\tilde{\theta} = \tilde{\theta}_{S_0}$ in $ISO(\mathcal{N}_{F_0},\mathcal{N}_{G_0})$ by identifying each infinite atom in $\mathcal{N}$ with the corresponding infinite atom of $\mathcal{N}_{F_0}$ (and $\mathcal{N}_{G_0}$). This isomorphism $\tilde{\theta}$ is not uniquely determined by S. For example, if $E = \sum_{i=1}^{k} E_i$ is a sum of finite atoms of $\mathcal{N}_{F_0}$, let $H = \tilde{\theta}(E) \equiv \sum_{i=1}^{k}\tilde{\theta}(E_i)$ be the corresponding sum of atoms in $\mathcal{N}_{G_0}$. Set $S_1 = ES_0H + E^{\perp}S_0H^{\perp}$. One argues as in the last paragraph in the proof of Lemma 17.19 that S_1 is invertible and S_1 takes $\mathcal{N}^{F+E}$ to $\mathcal{N}^{G+H}$. Moreover, $\tilde{\theta}_{S_1}$ in $ISO(\mathcal{N}_{F_0+E},\mathcal{N}_{G_0+H})$ is just the restriction of $\tilde{\theta}_{S_0}$ to $\mathcal{N}_{F_0+E}$.

This discussion leads to the following group of "almost automorphisms" of $\mathcal{N}$. Let $\mathcal{P}(\mathcal{N})$ denote the set of pairs (F,G) of sums of atoms of $\mathcal{N}$, with the further stipulation in the case that $\mathcal{N}$ has no infinite rank atoms that $rank\,F = rank\,G$. Let $\mathcal{N}_F = \mathcal{N}|F^{\perp}$ and let $ISO(\mathcal{N}_F,\mathcal{N}_G)$ denote the dimension preserving isomorphisms of $\mathcal{N}_F$ onto $\mathcal{N}_G$. If θ is such a map and $E = \sum_{i=1}^{k} E_i$ is the sum of finite rank atoms of $\mathcal{N}_F$, let $H = \sum_{i=1}^{k}\theta(E_i)$ and define θ_E in $ISO(\mathcal{N}_{F+E},\mathcal{N}_{G+H})$ by restriction. Let $\mathcal{G}$ denote the union $\bigcup\{ISO(\mathcal{N}_F,\mathcal{N}_G):(F,G)\in\mathcal{P}(\mathcal{N})\}$. Put an equivalence relation on $\mathcal{G}$ by $\theta \sim \eta$ if there are sums of atoms E and F so that $\theta_E = \eta_F$. Let $[\theta]$ denote the equivalence class of θ in $\mathcal{G}/\sim$.

Define an associative multiplication on $\mathcal{G}/\sim$ as follows: Given θ in $ISO(\mathcal{N}_F,\mathcal{N}_G)$ and η in $ISO(\mathcal{N}_E,\mathcal{N}_H)$, let $A = \theta^{-1}(G^{\perp}E)$ and $B = E^{\perp}G$. Then θ_A belongs to $ISO(\mathcal{N}_{F+A},\mathcal{N}_{G\vee E})$ and η_B belongs to $ISO(\mathcal{N}_{G\vee E},\mathcal{N}_{H+\eta(B)})$. So $\eta_B\theta_A$ is defined in $ISO(\mathcal{N}_{F+A},\mathcal{N}_{H+\eta(B)})$. Let $[\eta]\,[\theta] = [\eta_B\theta_A]$. It is left to the reader to verify that this makes $\mathcal{G}/\sim$ a group with identity element $[id]$, and the relation $[\theta]\,[\theta^{-1}] = [id]$. This group will be denoted by $a\text{--}Aut(\mathcal{N})$.

Define a map from $Aut\, Q\,\mathcal{T}(\mathcal{N})$ into $a{-}Aut\,\mathcal{N}$ as follows. For each automorphism $\alpha = Ad\,S$, use Lemma 17.19 to obtain a compact perturbation S_0 such that θ_{S_0} belongs to $ISO(\mathcal{N}^F,\mathcal{N}^G)$. As above, form $\tilde\theta_{S_0}$. Define $\Theta(\alpha) = [\tilde\theta_{S_0}]$.

17.22 Theorem. *The map Θ is a well defined homomorphism of $Aut\, Q\,\mathcal{T}(\mathcal{N})$ onto $a{-}Aut\,\mathcal{N}$ with kernel $Inn\, Q\,\mathcal{T}(\mathcal{N})$. Hence $Out\, Q\,\mathcal{T}(\mathcal{N})$ is naturally isomorphic to $a{-}Aut\,\mathcal{N}$.*

Proof. First show that Θ is well defined. Let $\alpha = Ad\,S$ belong to $Aut\, Q\,\mathcal{T}(\mathcal{N})$ and let S_0 and S_1 be two invertible compact perturbations of S which implement isomorphisms θ_i of $\mathcal{N}^{F_i}$ onto $\mathcal{N}^{G_i}$, $i = 0,1$. Let $F_i^0(G_i^0)$ denote the sum of finite atoms dominated by F_i (G_i) and let $\tilde\theta_i$ be the induced map in $ISO(\mathcal{N}_{F_i^0},\mathcal{N}_{G_i^0})$. Let $\alpha_i = Ad\,S_i$ and $K = S_1S_0^{-1}-I$. Then $\alpha_1 = (Ad\,I+K)\alpha_0$. For any operator A in $Q\,\mathcal{T}(\mathcal{N})$, $\alpha_1(A) - \alpha_0(A)$ is compact and hence $\Phi\alpha_1(A)-\Phi\alpha_0(A)$ is compact (where Φ is the expectation onto $\mathcal{M}$). In particular, each infinite atom of $\mathcal{N}$ is taken to the same infinite atom by both $\tilde\theta_i$, $i = 0,1$. Also, if E is an interval of $\mathcal{N}$ orthogonal to F_0+F_1 then

$$\tilde\theta_1(E)-\tilde\theta_2(E) = \Phi\alpha_1(E)-\Phi\alpha_0(E) = \Phi(Ad\,(I+K)-id)\alpha_0(E)$$

is compact; and thus is of the form A_1-A_0 where A_1 and A_0 are finite sums of finite rank atoms of $\mathcal{N}$. In particular, this difference has norm zero or one. The compactness of K ensures that if $\{E_n\}$ are pairwise orthogonal intervals, then $\tilde\theta_1(E_n) = \tilde\theta_0(E_n)$ except finitely often. Let $\{E_1,\dots,E_n\}$ be the atoms such that $\tilde\theta_1(E_k) \neq \tilde\theta_0(E_k)$, and one may suppose that the first m are finite rank. Let $H_i = F_i^{0\perp}(F_0^0 \vee F_1^0+\sum_{i=1}^{m} E_i)$. It will be shown that $\tilde\theta_{0H_0} = \tilde\theta_{1H_1}$. Let N belong to $\mathcal{N}$ and consider $H^\perp P(N)$ where $H = H_0 \vee H_1$. Now $\tilde\theta_1(H^\perp P(N))-\tilde\theta_0(H^\perp P(N)) = A_1-A_0$ as above. But as in the proof of Theorem 17.20, it follows that $A_0 = A_1 = 0$. For if E is an atom of $\mathcal{N}$ and $E \leq A_0$, then $E \leq \tilde\theta_0(H^\perp P(N))$ so $E' = \tilde\theta_0^{-1}(E) \leq H^\perp P(N)$; but $E\tilde\theta_1(H^\perp P(N)) = 0$ so $\tilde\theta_1(E)$ is orthogonal to $H^\perp P(N)$. This means $E' \leq H$, which is absurd. Hence it has been shows that $[\tilde\theta_0] = [\tilde\theta_1]$, and Θ is well defined.

It is readily apparent that Θ is a homomorphism. For let $\alpha_i = Ad\,S_i$, $i=1,2$, be automorphisms. Let S'_i be compact perturbations implementing $\tilde\theta_i$ in $ISO(\mathcal{N}_{F_i},\mathcal{N}_{G_i})$. Choose sums of atoms H_1 and H_2 so that $\tilde\theta = \tilde\theta_{1H_1}\tilde\theta_{2H_2}$ is defined. By Remark 17.21, there is a compact perturbation S''_i $(= H_iS'_iH_i+H_i^\perp S'_iH_i^\perp)$ which implements $\tilde\theta_{iH_i}$. So $S''_1S''_2$ is an invertible compact perturbation of S_1S_2. It is clear that $S''_1S''_2$ implements $\tilde\theta$. So $\Theta(\alpha_1)\Theta(\alpha_2) = \Theta(\alpha_1\alpha_2)$.

Suppose $\Theta(\alpha) = [id]$. Then $\alpha = Ad\,S$ and S has a compact perturbation S_0 mapping $\mathcal{N}^F$ onto $\mathcal{N}^G$ such that $S_0(N \vee F) = N \vee G$ for every N in $\mathcal{N}$. So S and S^{-1} belong to $Q\,\mathcal{T}(\mathcal{N})$, and thus α is inner.

Conversely, suppose S belongs to $Q\,\mathcal{T}(\mathcal{N})^{-1}$, and S_0 is an invertible compact perturbation mapping $\mathcal{N}^F$ onto $\mathcal{N}^G$. Now S_0 is bounded below by some $\epsilon > 0$, so for each N in $\mathcal{N}^F$, $S_0 b_1(N)$ contains $b_\epsilon(\theta(N))$. Write $S_0 = T+K$ where T belongs to $\mathcal{T}(\mathcal{N}^F)$ and K is compact. Then $P(N)^\perp K b_1(N) = P(N)^\perp S_0 b_1(N)$ contains $b_\epsilon(P(N)^\perp \theta(N))$, and thus this is finite rank. By considering S_0^{-1} as well, we conclude that $P(\theta(N)) - P(N)$ is finite rank. Hence $\theta(E) - E$ is finite rank for every interval, and hence is of the form $A_1 - A_2$ where A_i are orthogonal projections in $\mathcal{M}$ which are sums of finite atoms of $\mathcal{N}$, and subprojections of G and F respectively. Now if $\{E_n, n \geq 1\}$ are the finite rank atoms of $\mathcal{N}$ orthogonal to F, then $\theta(E_n) > E_n$ in the order of $\mathcal{N}$ implies that $\theta(E_n) K E_n$ is bounded below by ϵ on E_n. So this occurs only finitely often. So $\theta(E) = E$ except for a finite set $E_1, \ldots, E_n$. If E is an infinite atom, $\theta(E)$ is an infinite atom in $\mathcal{N}^G$ and $E - \theta(E)$ is compact; so they correspond to the same atom of $\mathcal{N}$. Furthermore, $E = \theta(E)$ except those finitely many which meet F or G. Finally, we argue as before to show that if F is an interval orthogonal to the finitely many atoms for which $\theta(E) \neq E$, then $\theta(F) = F$. Hence $[\tilde{\theta}] = [id]$. Thus $ker\,\Theta = Inn\,Q\,\mathcal{T}(\mathcal{N})$.

Now we show that Θ is surjective. Let θ in $ISO(\mathcal{N}_F, \mathcal{N}_G)$ representative of $[\theta]$. By the Similarity Theorem 13.20, there is an invertible operator S in $\mathcal{B}(F^\perp, G^\perp)$ which implements θ. If $rank\,F = rank\,G$, S can be extended to an invertible operator in $\mathcal{B}(\mathcal{H})$ of the form $S' = G^\perp S F^\perp + GUF$ where U is unitary. Clearly, $Ad\,S$ takes $\mathcal{T}(\mathcal{N}^F)$ onto $\mathcal{T}(\mathcal{N}^G)$, hence $Ad\,S$ is an automorphism of $Q\,\mathcal{T}(\mathcal{N})$. Also, it is apparent that $\Theta(Ad\,S) = [\theta]$. If $rank F \neq rank\,G$, then $\mathcal{N}$ has an infinite atom A. Suppose that $rank\,F < rank\,G$. Let E be a projection in $\mathcal{M}$ with $E \leq A$ and $rank\,E + rank\,F = rank\,G$. Let U be a unitary operator on $\mathcal{H}$ which carries $(E+F)\mathcal{H}$ onto $G\mathcal{H}$, takes $AE^\perp\mathcal{H}$ onto $A\mathcal{H}$, and is the identity on $(A+F)^\perp\mathcal{H}$. Let $S' = U(F + G^\perp S F^\perp)$. This is invertible by construction, and $S'(F+E)\mathcal{H} = G\mathcal{H}$. If N belongs to $\mathcal{N}_F$, then

$$S'(N \vee F+E) = US'N \vee G = U\theta(N) \vee G = \theta(N) \vee G \quad .$$

So $\Theta(Ad\,S') = [\theta]$. If $rank\,F > rank\,G$, obtain S' as above so that $\Theta(Ad\,S') = [\theta^{-1}]$. Then $\Theta(Ad\,S'^{-1}) = [\theta]$.

The map Θ is "natural" in the sense made clear in the previous paragraph. Namely, $\alpha + Inn\,Q\,\mathcal{T}(\mathcal{N})$ contains a representative $Ad\,S$ which implements a representative θ in $\Theta(\alpha)$. ∎

17.23 Corollary. *Let $\mathcal{N}$ be a continuous nest. Then $Out\, \mathcal{Q}\,\mathcal{T}(\mathcal{N})$ is isomorphic to $Homeo_0[0,1]$, the order preserving homeomorphisms of $[0,1]$.*

Proof. As $\mathcal{N}$ has no atoms, $a{-}Aut\ \mathcal{N} = Aut\ \mathcal{N}$. As $\mathcal{N}$ is order isomorphic to $[0,1]$, the result follows. ∎

17.24 Corollary. *Let $\mathcal{N}$ be a nest of order type $\omega = \mathbb{N} \cup \{\infty\}$, and let the atoms by $\{A_k, k \geq 1\}$. Then $Out\, \mathcal{Q}\,\mathcal{T}(\mathcal{N}) = 0$ or $\mathbb{Z}$. The latter occurs if and only if there is a finite, non-zero number of infinite rank atoms and $rank\, A_k$ is eventually periodic.*

Proof. Let E (and F) be a sum of e (and f) finite atoms of $\mathcal{N}$. Then $\mathcal{N}_E$ and $\mathcal{N}_F$ are nests of order type ω. So $ISO(\mathcal{N}_E, \mathcal{N}_F)$ is either empty, or has exactly one element θ taking the n-th atom of $\mathcal{N}_E$ to the n-th atom of $\mathcal{N}_F$. In the latter case, this implies that $rank(A_{n+e}) = rank(A_{n+f})$ for n sufficiently large. If $e = f$, this is equivalent to $[id]$. But if $k = e-f \neq 0$, then there is an integer n_0 so that $rank\, A_{n+k} = rank\, A_n$ for all $n \geq n_0$. In this case the span of the first n_0+k atoms is sent onto the first n_0 atoms. This can only occur if this span is infinite dimensional. So there are infinite atoms. The nests $\mathcal{N}_E$ and $\mathcal{N}_F$ have exactly the same infinite atoms in exactly the same order. So they must be mapped identically onto themselves ($\theta(A) = A$ for every infinite atom). Thus if $k \neq 0$, there are only finitely many infinite atoms. Conversely suppose $\mathcal{N}$ has a finite non-zero number of infinite atoms, including A_j say, and that there are positive integers n_0 and k_0 so that $rank\, A_{n+k_0} = rank\, A_n$ for all $n \geq n_0$. Also suppose that k_0 is minimal. Let E be the sum of all finite atoms A_i with $i < n_0+k_0$, and let F be the sum of the finite atoms A_i with $i < n_0$. Define a map θ_0 taking each infinite atom to itself and $\theta_0(A_{n+k_0}) = A_n$ for $n \geq n_0$. This is an element of $ISO(\mathcal{N}_E, \mathcal{N}_F)$. If $[\theta]$ is any element of $a{-}Aut\ \mathcal{N}$, then the integer $k = e-f$ is a multiple $k_0 p$ of k_0. It follows that $[\theta] = [\theta_0]^p$. So $a{-}Aut\ \mathcal{N}$ is isomorphic to $\mathbb{Z}$. ∎

Notes and Remarks.

Theorem 17.1 on unitarizing groups is due to Dixmier [2]. Corollary 17.3 was proven independently by Wermer [1] who used it to show that commuting spectral operators are similar to commuting normal operators. (See Dunford and Schwartz [1], volume 3). Theorem 17.4 is from Dixmier [4] Chapter III, Part 3 § 2. Theorem 17.12 and Corollary 17.13 are due to Ringrose [4]. Theorem 17.5 is a generalization of this work due to Gilfeather and Moore [1]. Theorem 17.7 is essentially due to Rickart [1], 2.5.9. The applications of the Similarity Theorem, Corollaries 17.14 and 17.15, are due to Davidson and Wagner [1]. Lemmas 17.16 and 17.17 are taken from the ideas in Apostol-Gilfeather [1] and Apostol-Davidson [1]. Theorem 17.20 is due to Wagner [2], and Theorem 17.22 is due to

Davidson-Wagner [1]. The approach taken here to these results is a modified version of the latter paper. Partial results on this subject are also contained in Wagner's thesis [1].

One can also study isomorphisms between the quotient algebras $Q\mathcal{T}(\mathcal{N})/\mathcal{K}$. In this case, there are no finite rank operators, so the analysis is more subtle. It is shown in Apostol-Gilfeather [1] and Apostol-Davidson [1] that any two such algebras which are isomorphic are unitarily equivalent (in the Calkin algebra). For example, suppose $\mathcal{N}$ and $\mathcal{M}$ have order type $\omega = \mathbb{N} \cup \{\infty\}$ with atoms of $rank\,\{d_n, n \geq 1\}$ and $\{e_n, n \geq 1\}$ respectively. Then $Q\mathcal{T}(\mathcal{N})/\mathcal{K}$ and $Q\mathcal{T}(\mathcal{M})/\mathcal{K}$ are isomorphic if and only if there are integers n_0 and m_0 so that $d_{n_0+k} = d_{m_0+k}$ for $k \geq 1$. However, it is not known even in this special case whether all such isomorphisms are spatially implemented.

Exercises.

17.1 Let $\mathcal{A}$ be a C^*-algebra which is generated as a C^*-algebra by an **amenable subgroup $\mathcal{G}$ of the unitary group.** Show that every bounded representation of $\mathcal{A}$ is similar to a unitary representation.

17.2 (a) If $\mathcal{N}$ is a nest of order type $\omega = \mathbb{N} \cup \{\infty\}$, show that every automorphism of $\mathcal{T}(\mathcal{N})$ is inner.

(b) If $\mathcal{N}$ is a nest of order type $\omega^*+\omega = \{-\infty\} \cup \mathbb{Z} \cup \{\infty\}$, show that $Out\,\mathcal{T}(\mathcal{N})$ is 0 or $\mathbb{Z}$. When is it $\mathbb{Z}$?

17.3 (Wagner) Classify $Out\,Q\mathcal{T}(\mathcal{N})$ when $\mathcal{N}$ has order type $\omega^*+\omega$. There are five possibilities for the infinite rank atoms: (ϕ) none; (F) finite, non-zero; (N) countably many, order type ω; (N^*) countably many, order type ω^*; and (Z) countably many, order type $\omega^*+\omega$. There are four possibilities for partial periodicity of $d_n = rank\,A_n$: (P_+), there are positive integers n_0 and k_0 so that $d_{n+k_0} = d_n$ for $n \geq n_0$; (P_-), there are negative integers m_0 and ℓ_0 so that $d_{m+\ell_0} = d_m$ for $m \leq m_0$; (DP) = $P_+ \cup P_-$; and (NP) neither P_+ nor P_-. Show that $Out\,Q\mathcal{T}(\mathcal{N})$ is given by the following chart:

	ϕ	F	N	N^*	Z
NP	0	0	0	0	0
P_+	0	$\mathbb{Z}$	0	$\mathbb{Z}$	0
P_-	0	$\mathbb{Z}$	$\mathbb{Z}$	0	0
DP	$\mathbb{Z}$	$\mathbb{Z} \oplus \mathbb{Z}$	$\mathbb{Z}$	$\mathbb{Z}$	$\mathbb{Z}$

17.4 (a) Show that every automorphism of $\mathcal{T}(\mathcal{N})$ extends uniquely to an automorphism $Q\mathcal{T}(\mathcal{N})$. Thus there is an injection $i: Aut\,\mathcal{T}(\mathcal{N}) \rightarrow Aut\,Q\mathcal{T}(\mathcal{N})$.

(b) Show that $i^{-1}(Inn\, Q\,\mathcal{T}(\mathcal{N})) = Inn\,\mathcal{T}(\mathcal{N})$. Hence $Out\,\mathcal{T}(\mathcal{N})$ imbeds as a subgroup of $Aut\, Q\,\mathcal{T}(\mathcal{N})$.

(c) Show that the map $j{:}Aut(\mathcal{N}) \to a{-}Aut\,\mathcal{N}$ given by $j(\theta) = [\theta]$ is an injection, and that this diagram commutes:

$$\begin{array}{ccc} Aut\,\mathcal{T}(\mathcal{N}) & \xrightarrow{i} & Aut\,Q\,\mathcal{T}(\mathcal{N}) \\ \downarrow \pi & & \downarrow \pi \\ Aut\,\mathcal{N} & \xrightarrow{j} & a{-}Aut\,\mathcal{N} \end{array}$$

17.5 Let $\mathcal{N}$ be a nest and let A be an invertible element of $Q\,\mathcal{T}(\mathcal{N})$. Show that A factors as $A = (I+K)VT$ where K is compact, T is an invertible element of $\mathcal{T}(\mathcal{N})$, $E_1,..,E_n$ are infinite rank atoms with $E = \sum_{i=1}^{n} E_i$, and V is a unitary operator of the form $V = E^{\perp}+EVE$ such that VE_iV^* belongs to $\mathcal{N}'$ and VE_i-E_iV is compact, $1 \le i \le n$. The converse is easy.

17.6 (Gilfeather-Moore)

(a) Let $\mathcal{A}_4$ consist of all operators of the form

$$\begin{bmatrix} a & b & 0 & h \\ 0 & c & 0 & 0 \\ 0 & d & e & f \\ 0 & 0 & 0 & g \end{bmatrix}$$

Show that this is a CSL algebra. Define a map ρ by leaving all coordinates the same except the (1.4) entry, which is replaced by $-h$. Show that ρ is an automorphism which does not preserve rank. Hence ρ is not spatial.

(b) Let $\mathcal{A}_6$ be the 6×6 analogue consisting of matrices

$$\begin{bmatrix} * & * & & & & * \\ & * & & & & \\ & * & * & * & & \\ & & & * & & \\ & & & * & * & * \\ & & & & & * \end{bmatrix}.$$

Again, define ρ by switching the sign of the (1,6) entry. Show that ρ is an automorphism. Show that ρ preserves rank one operators but not rank two operators.

(c) Let $\mathcal{A}_\infty$ denote the algebra of operators of the form

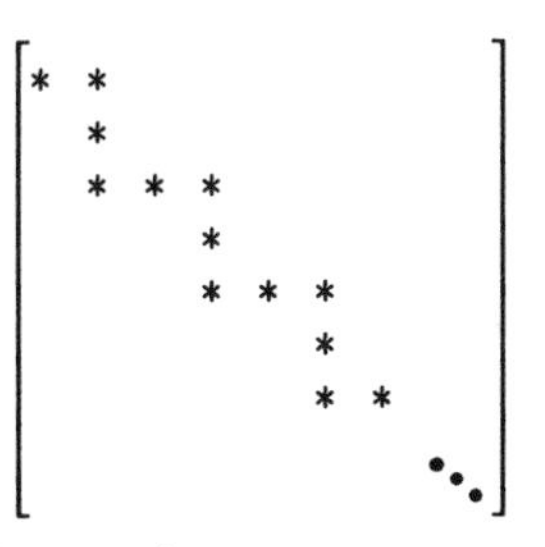

Let $T = diag(1,2,3,...)$ and define ρ by $\rho(A) = TAT^{-1}$. Verify that ρ is a continuous automorphism which is not spatial.

17.7 (Gilfeather-Moore) Let $\mathcal{L}$ be a CSL and let β be an automorphism of $Alg\,\mathcal{L}$ such that $\beta\,|\,\mathcal{M} = id$ for some masa $\mathcal{M}$.

(a) For each projection E in $\mathcal{M}$, let $\beta_E(T) = \beta(T)E$. Let $\mathcal{E}$ be the set of projections such that β_E is continuous. Show that $\mathcal{E}$ is closed under span.

(b) Use the Uniform Boundedness Principle to prove that $\{\|\beta_E\| : E \in \mathcal{E}\}$ is bounded. Hence show that $\mathcal{E}$ is closed under countable unions. Hence deduce that $\mathcal{E}$ has a greatest element E_0.

(c) Show that $E_0^{\perp}$ is finite dimensional. Hint: If $E_0^{\perp}$ can be split into a set $\{E_n, n \geq 1\}$ of pairwise orthogonal projections in $\mathcal{M}$, build an operator $A = \sum A_n E_n$ such that $\beta(A)$ is not bounded.

(d) Similarly, find a maximal projection E_1 in $\mathcal{M}$ so that $\beta E_1(T) = E_1\beta(T)$ is continuous.

(e) Show that the map $\alpha(T) = E_1^{\perp}\beta(T)E_0^{\perp}$ is continuous. Hence deduce that β is continuous.

(f) Show that every algebraic isomorphism between CSL algebras is continuous.

18. Perturbations of Operator Algebras

Perturbation results strive to show that two subalgebras of a larger algebra with "close" unit balls are spatially related (unitary equivalence, similarity, etc.). Such results exist for nest algebras and for certain classes of von Neumann algebras. Here we will deal only with the abelian and abelian commutant case. There is a close connection between perturbation results and results on cohomology and derivations. These will be dealt with in the next chapter.

The distance between two subspaces M and N of a Banach space will be given by the *Hausdorff distance* between their unit balls $b_1(M)$ and $b_1(N)$:

$$d(M,N) = d_H(b_1(M),b_1(N))$$
$$= \max\{ \sup_{m \in b_1(M)} \inf_{n \in b_1(N)} \|n-m\|, \sup_{n \in b_1(N)} \inf_{m \in b_1(M)} \|n-m\|\} .$$

The first theorem deals with more general reflexive algebras. Then we specialize to nest algebras.

18.1 Lemma. *Let B be a unital Banach algebra. For each B in $b_1(B)$, there is a idempotent F satisfying $\|B-F\| \le 8\|B-B^2\|$. When $\|B-B^2\| \le 1/8$, one can achieve $\|B-F\| \le 5\|B-B^2\|$.*

Proof. If $\|B-B^2\| > 1/8$, take $F = 0$. Suppose $\|B-B^2\| = \epsilon \le 1/8$. define

$$T = [I-4(B-B^2)]^{-\frac{1}{2}} = I+\sum_{n=1}^{\infty} \binom{2n}{n}(B-B^2)^n .$$

This series converges absolutely, and

$$\|T-I\| \le \sum_{n=1}^{\infty} \binom{2n}{n}\epsilon^n = (1-4\epsilon)^{-\frac{1}{2}}-1$$
$$= \frac{4\epsilon}{1-4\epsilon+\sqrt{1-4\epsilon}} \le \frac{4\epsilon}{1/2+1/\sqrt{2}} < \frac{10}{3}\epsilon .$$

Furthermore, T commutes with B and $(I-2B)^2T^2 = I$. Define F by the identity $I-2F = (1-2B)T$. Squaring yields $F = F^2$. Moreover,

$$\|F-B\| = \frac{1}{2}\|(I-2B)-(I-2F)\|$$
$$= \frac{1}{2}\|(I-2B)(I-T)\| < \frac{1}{2}\cdot 3\cdot \frac{10}{3}\epsilon = 5\epsilon \quad . \qquad \blacksquare$$

18.2 Theorem. *Let $\mathcal{A} = Alg\,\mathcal{L}$ be a reflexive algebra with a commutative subspace lattice $\mathcal{L}$. Let $\mathcal{B}$ be a norm closed subalgebra of $\mathcal{B}(\mathcal{H})$ such that $d(\mathcal{A},\mathcal{B}) = \delta < 1/20$. Then there is a complete lattice isomorphism θ of $\mathcal{M} = Lat\,\mathcal{B}$ onto $\mathcal{L}$ such that $\|\theta - id\| \le 2\delta$.*

Proof. Fix M in $\mathcal{M}$ and let $P = P(M)$. The algebra $\mathcal{A}$ contains the von Neumann algebra $\mathcal{A} \cap \mathcal{A}^* = \mathcal{L}'$ which has abelian commutant $\mathcal{L}''$. For each A in $b_1(\mathcal{L}')$, pick B and C in $b_1(\mathcal{B})$ such that $\|A-B\| \le \delta$ and $\|A^*-C\| \le \delta$. Then

$$\|PA-AP\| = \|PAP^{\perp}-P^{\perp}AP\|$$
$$= \max\{\|P(A-C^*)P^{\perp}\|,\|P^{\perp}(A-B)P\|\} \le \delta \quad .$$

Since $\mathcal{L}''$ is abelian, Theorem 9.6 implies that $d(P,\mathcal{L}'') \le \delta$. Choose a self-adjoint element R in $\mathcal{L}''$ such that $\|P-R\| \le \delta$. Then the spectrum $\sigma(R)$ is contained in $[-\delta,\delta] \cup [1-\delta,1+\delta]$. By the functional calculus, there is a projection Q in $\mathcal{L}''$ such that $\|Q-R\| \le \delta$, hence $\|P-Q\| \le 2\delta$.

The projection Q belongs to $\mathcal{L}$. For otherwise, $\mathcal{A}$ contains an operator $A = Q^{\perp}AQ$ of norm one. Choose B in $b_1(\mathcal{B})$ with $\|A-B\| \le \delta$. Then

$$1 = \|A\| = \|Q^{\perp}AQ-P^{\perp}BP\|$$
$$\le \|(P-Q)AQ\|+\|P^{\perp}(A-B)Q\|+\|P^{\perp}B(Q-P)\|$$
$$\le 2\delta+\delta+2\delta = 5\delta < 1 \quad .$$

This contradiction establishes our claim. The projection Q is uniquely determined in $\mathcal{L}''$ since $\|Q-Q'\| = 1$ for distinct projections. Let L in $\mathcal{L}$ be the subspace such that $Q = P(L)$. Define $\theta(M) = L$.

Conversely, if $Q = P(L)$ for some L in $\mathcal{L}$, choose B in $\mathcal{B}$ so that $\|(2Q-I)-(2B-I)\| \le \delta$ and $\|2B-I\| \le 1$. Then $\|B\| \le 1$ and $\|Q-B\| \le \delta/2$. Next, obtain an idempotent in $\mathcal{B}$ close to B. Compute:

$$\|B-B^2\| = \|Q^{\perp}B+(Q-B)B\|$$
$$\le \|Q^{\perp}(B-Q)\|+\|Q-B\|\,\|B\| \le \delta \quad .$$

By Lemma 18.1, there is an idempotent F in $\mathcal{B}$ with $\|B-F\| < 5\delta$. The range $M = F\mathcal{H}$ must belong to $\mathcal{M}$, for otherwise there is an operator $C = (I-F)CF$ in $\mathcal{B}$ of norm one. Choose A in $b_1(\mathcal{A})$ with $\|A-C\| \le \delta$. Then since $\|Q-F\| < 6\delta$, one obtains

$$1 = \|C\| = \|(1-F)CF - Q^{\perp}AQ\|$$
$$\leq \|(Q-F)CF\| + \|Q^{\perp}C(F-Q)\| + \|Q^{\perp}(C-A)Q\|$$
$$< 6\delta + 6\delta + \delta = 13\delta < 1 \ .$$

This contradiction shows that M indeed belongs to $\mathcal{M}$. And since $P(M)F = F$ and $FP(M) = P(M)$,

$$\|Q - P(M)\| = \|P(M)^{\perp}(Q - P(M)) + [(Q - P(M))P(M)]^{*}\|$$
$$= \|P(M)^{\perp}(Q-F) + [(Q-F)P(M)]^{*}\| < 12\delta \ .$$

Therefore, $\|Q - P(\theta(M))\| < 14\delta < 1$, so $\theta(M) = L$. Thus θ is surjective.

Next, we show that θ is injective. Let M belong to $\mathcal{M}$, and let $L = \theta(M)$. Set $Q = P(L)$ and let F be the idempotent in $\mathcal{B}$ found in the previous paragraph. To show that θ is injective, it suffices to show that $F\mathcal{H} = M$. Decompose $\mathcal{H}$ as $M \oplus M^{\perp}$, and write F as a matrix $F = \begin{bmatrix} F_{11} & F_{12} \\ 0 & F_{22} \end{bmatrix}$ with respect to this decomposition. Now

$$\|F - P(M)\| \leq \|F - Q\| + \|Q - P(M)\| < 6\delta + 2\delta = 8\epsilon < 1 \ .$$

As F is idempotent, both F_{11} and F_{22} are idempotent and $\|I - F_{11}\| < 1$ and $\|F_{22}\| < 1$. Thus $F_{11} = I$, $F_{22} = 0$ and $F\mathcal{H} = M$. So θ is a bijection.

To see that θ is a lattice isomorphism, suppose that M and N belong to $\mathcal{M}$. Let Q_M and Q_N be the orthogonal projections onto $\theta(M)$ and $\theta(N)$ respectively, and let F_M and F_N be idempotents in $\mathcal{B}$ with

$$\|Q_M - F_M\| < 6\delta \quad \text{and} \quad \|Q_N - F_N\| < 6\delta \ .$$

The operator $F_M F_N$ is an idempotent in $\mathcal{B}$ with range $M \wedge N$ since

$$(F_M F_N)^2 = F_M(F_N F_M F_N) = F_M(F_M F_N) = F_M F_N \ .$$

Similarly, $F_M + F_N - F_M F_N$ is an idempotent in $\mathcal{B}$ with range $M \vee N$. A computation yields

$$\|Q_M Q_N - F_M F_N\| \leq \|Q_M(Q_M - F_N)\| + \|(Q_M - F_M)F_N\|$$
$$< 6\delta + 6\delta(1 + 6\delta) < 14\delta \ .$$

Let R be the element of $\mathcal{M}$ such that $Q_M Q_N = P(\theta(R))$. Then $\|P(R) - F_M F_N\| < 16\delta < 1$. So as in the previous paragraph, R equals the range of $F_M F_N$, namely $M \wedge N$. So

$$\theta(M \wedge N) = \theta(M) \wedge \theta(N) \ .$$

A similar computation yields $\theta(M \vee N) = \theta(M) \vee \theta(N)$ as well.

Finally, consider arbitrary intersections in $\mathcal{M}$. Let $\{M_\lambda, \lambda \in \Lambda\}$ belong to $\mathcal{M}$ and let $L_\lambda = \theta(M_\lambda)$. Set $M = \bigwedge M_\lambda$, $L = \bigwedge L_\lambda$. Since $M \leq M_\lambda$, $\theta(M) \leq \theta(M_\lambda) = L_\lambda$ and hence $\theta(M) \leq L$. On the other hand, $\theta^{-1}(L) \leq \theta^{-1}(L_\lambda) = M_\lambda$ so $\theta^{-1}(L) \leq M$ whence $L \leq \theta(M)$. So $\theta(M) = L$.

In a like manner, one shows that θ also preserves arbitrary sups. So θ is a complete lattice isomorphism. ■

Now we specialize to nest algebras.

18.3 Lemma. *Let $\mathcal{N}$ and $\mathcal{M}$ be nests on a separable Hilbert space. Suppose that θ is an order isomorphism of $\mathcal{N}$ onto $\mathcal{M}$ such that $\|\theta - id\| = \epsilon < 1/2$. Then there is an invertible operator S with $\|S-I\| < 2\epsilon$ which implements θ.*

Proof. Notice that θ preserves dimension. For if $N_1 < N_2$ in $\mathcal{N}$ and $M_i = \theta(N_i)$ for $i=1,2$, then

$$d(N_2 \ominus N_1, M_2 \ominus M_1) \leq d(N_2,M_2) + d(N_1,M_1) \leq 2\|\theta - id\| < 1 \ .$$

So $dim\, N_2 \ominus N_1 = dim\, M_2 \ominus M_1$. By the Similarity Theorem 13.20, there is an invertible operator T which implements θ. Furthermore, it can be arranged that there is a unitary operator U a such that $\|T-U\| < \delta$ where δ is a small positive number satisfying $\delta < \epsilon(1-2\epsilon)/2$.

Apply Arveson's Distance Formula (Theorem 9.5) to T:

$$\begin{aligned} d(T, \mathcal{T}(\mathcal{N})) &= \sup_{N \in \mathcal{N}} \|P(N)^{\perp} T P(N)\| \\ &= \sup_{N \in \mathcal{N}} \|P(N)^{\perp} P(\theta(N)) T P(N)\| \\ &\leq \|\theta - id\| \, \|T\| < \epsilon(1+\delta) < \epsilon + \delta \ . \end{aligned}$$

Thus there is an element V in $\mathcal{T}(\mathcal{N})$ with $\|T-V\| < \epsilon+\delta$. Hence $\|U-V\| < \epsilon + 2\delta < 1$, so V is invertible.

It will be shown that V^{-1} belongs to $\mathcal{T}(\mathcal{N})$. This is equivalent to showing that $VN = N$ for all N in $\mathcal{N}$. If it were otherwise, there is some N in $\mathcal{N}$ such that VN is a proper subspace of N. Pick any unit vector x in $N \ominus VN$ and choose y in the unit ball of $M = \theta(N)$ with $\|x-y\| \leq \epsilon$. Since $T^{-1}M = N$, the vector $z = T^{-1}y$ belongs to N. And,

$$\begin{aligned} 1 = \|x - Vz\| &\leq \|x-y\| + \|(T-V)T^{-1}y\| \\ &< \epsilon + (\epsilon+\delta)(1-\delta)^{-1} < (2\epsilon+\delta)(1-\delta)^{-1} < 1 \ . \end{aligned}$$

Hence V^{-1} must belong to $\mathcal{T}(\mathcal{N})$.

Therefore $S = TV^{-1}$ implements θ, and

$$\|S-I\| \leq \|T-V\| \, \|V^{-1}\| < \frac{\epsilon+\delta}{1-\epsilon-\delta} < 2\epsilon \ .$$ ■

18.4 Theorem. *Let $\mathcal{N}$ be a nest, and let B be a norm closed algebra such that $d(\mathcal{T}(\mathcal{N}), B) = \epsilon < 1/20$. Then B is a nest algebra which is similar to $\mathcal{T}(\mathcal{N})$ via an invertible operator S satisfying $\|S-I\| < 4\epsilon$.*

Proof. By Theorem 18.2, $\mathcal{M} = Lat\, B$ is a nest and there is an order isomorphism θ of $\mathcal{N}$ onto $\mathcal{M}$ with $\|\theta - id\| \leq 2\epsilon$. By Lemma 18.3, θ is implemented by an invertible operator S with $\|S-I\| < 4\epsilon$. Now $S\mathcal{T}(\mathcal{N})S^{-1} = \mathcal{T}(\mathcal{M})$. So compute $d(\mathcal{T}(\mathcal{N}),\mathcal{T}(\mathcal{M}))$. If A belongs to $\mathcal{T}(\mathcal{N})$, the distance formula yields

$$\begin{aligned} d(A,\mathcal{T}(\mathcal{M})) &= \sup_{N \in \mathcal{N}} \|P(\theta N)^{\perp} A P(\theta N)\| \\ &= \sup_{N \in \mathcal{N}} \|P(\theta N)^{\perp} A P(\theta N) - P(N)^{\perp} A P(N)\| \\ &\leq \sup_{N \in \mathcal{N}} \|(P(N)-P(\theta N))AP(\theta N) + P(N)^{\perp} A (P(\theta N)-P(N))\| \\ &\leq 4\epsilon \ . \end{aligned}$$

Hence $d(A,b_1(\mathcal{T}(\mathcal{M}))) \leq 8\epsilon$. Likewise, if B belongs to $b_1(\mathcal{T}(\mathcal{M}))$, $d(B,b_1(\mathcal{T}(\mathcal{N}))) \leq 8\epsilon$. Now B is contained in $\mathcal{T}(\mathcal{M})$, and

$$\begin{aligned} d(B,\mathcal{T}(\mathcal{M})) &\leq d(B,\mathcal{T}(\mathcal{N})) + d(\mathcal{T}(\mathcal{N}),\mathcal{T}(\mathcal{M})) \\ &\leq \epsilon + 8\epsilon = 9\epsilon < 1 \ . \end{aligned}$$

Hence $B = \mathcal{T}(\mathcal{M})$ and is similar to $\mathcal{T}(\mathcal{N})$ as desired. ∎

The following Corollary is immediate from the proofs.

18.5 Corollary. *Let $\mathcal{N}$ and $\mathcal{M}$ be nests on a separable Hilbert space. The following are equivalent*

1) *$\mathcal{N}$ and $\mathcal{M}$ are close,*
2) *$\mathcal{T}(\mathcal{N})$ and $\mathcal{T}(\mathcal{M})$ are close,*
3) *$\mathcal{N}$ and $\mathcal{M}$ are similar via an invertible operator close to I.*

Now we turn to the von Neumann algebra case. If $\mathcal{A}$ is a von Neumann algebra, let $\mathcal{A}_p$ denote the set of projections in $\mathcal{A}$.

18.6 Lemma. *Suppose $\mathcal{A}$ and $\mathcal{B}$ are von Neumann algebras such that $d(\mathcal{A},\mathcal{B}) = \delta < 1$. Then $d_H(\mathcal{A}_p,\mathcal{B}_p) \leq \delta$.*

Proof. Fix P in $\mathcal{A}_p$. Then $2P-I$ belongs to $b_1(\mathcal{A})$, hence there is a B in $b_1(\mathcal{B})$ with $\|(2P-I)-B\| \leq \delta$. Replace B by $(B+B^*)/2$ so that it is self-adjoint and $\|B\| \leq 1$. Then for every vector x in $\mathcal{H}$,

$$\|Bx\| \geq \|(2P-I)x\| - \delta\|x\| = (1-\delta)\|x\| \ .$$

Thus the spectrum $\sigma(B)$ is contained in $[-1,-1+\delta] \cup [1-\delta,1]$. Let Q be the spectral projection $E_B[1-\delta,1]$. Then

$$\|B-(2Q-I)\| \leq \delta \ .$$

Hence,

$$\|P-Q\| = \frac{1}{2}\|(2P-I)-B+B-(2Q-I)\| \leq \delta \ .$$

Repeat the argument with $\mathcal{B}$ and $\mathcal{A}$ interchanged. ■

18.7 Lemma. *Suppose that $\mathcal{A}$ and $\mathcal{B}$ are von Neumann algebras such that $d(\mathcal{A},\mathcal{B}) = \delta < 1/3$, and let $Z(\mathcal{A})$ and $Z(\mathcal{B})$ be their centres. Then $d_H(Z(\mathcal{A})_p, Z(\mathcal{B})_p) \leq \delta$.*

Proof. Let P be a central projection in $\mathcal{A}$. By Lemma 18.6, there is a projection Q in $\mathcal{B}$ such that $\|P-Q\| \leq \delta$. It suffices to show that Q belongs to $Z(\mathcal{B})$. Otherwise, there is an operator B in $\mathcal{B}$ of the form $B = QBQ^{\perp} \neq 0$. Normalize B so that $\|B\| = 1$, and choose A in $b_1(\mathcal{A})$ so that $\|A-B\| \leq \delta$. Then

$$\begin{aligned} 1 = \|B\| &= \|QBQ^{\perp}-PAP^{\perp}\| \\ &\leq \|(Q-P)BQ^{\perp}\|+\|P(B-A)Q^{\perp}\|+\|PA(P-Q)\| \\ &\leq 3\delta < 1 \ . \end{aligned}$$

This is absurd, and the lemma follows. ■

18.8 Corollary. *Suppose that $\mathcal{A}$ and $\mathcal{B}$ are von Neumann algebras such that $d(\mathcal{A},\mathcal{B}) < 1/3$. If $\mathcal{A}$ is abelian, then $\mathcal{B}$ is also abelian.*

Proof. Let Q be any projection in $\mathcal{B}$. Lemma 18.6 provides a projection P in $\mathcal{A}$ such that $\|P-Q\| \leq \delta$. Now P is central since $\mathcal{A}$ is abelian. Thus by the proof of Lemma 18.7, Q is also central. As every projection in $\mathcal{B}$ is central, $\mathcal{B}$ is abelian. ■

18.9 Lemma. *Let $\mathcal{A}$ and $\mathcal{B}$ be abelian von Neumann algebras such that $d_H(\mathcal{A}_p,\mathcal{B}_p) = \delta < ½$. Then there exists a unique $*$-isomorphism Φ of $\mathcal{A}$ onto $\mathcal{B}$ such that $\sup_{P \in \mathcal{A}_p} \|\Phi(P)-P\| \leq \delta$.*

Proof. For each projection P in $\mathcal{A}$, there is a projection Q in $\mathcal{B}$ such that $\|P-Q\| \leq \delta$. If Q' were another projection in $\mathcal{B}$ with $\|P-Q'\| \leq \delta$, then $\|Q-Q'\| \leq 2\delta < 1$. As $\mathcal{B}$ is abelian, this forces $Q' = Q$. So Q is unique and thus one can define $\Phi(P) = Q$. The same argument shows that Φ is one to one. Interchanging the role of $\mathcal{A}$ and $\mathcal{B}$ yields a map Ψ from $\mathcal{B}_p$ to $\mathcal{A}_p$ with the same properties. Moreover, $\|\Phi\Psi(Q)-Q\| < 1$ whence Φ is a bijection and $\Psi = \Phi^{-1}$.

Suppose P_1 and P_2 are projections in $\mathcal{A}$ with $P_1 \leq P_2$. Let $Q_i = \Phi(P_i)$, $i=1,2$. Then

$$\begin{aligned} \|Q_1-Q_2Q_1\| &= \|Q_1-P_1+P_2(P_1-Q_1)+(P_2-Q_2)Q_1\| \\ &\leq \|P_2^{\perp}(Q_1-P_1)\|+\|(P_2-Q_2)Q_1\| \leq 2\delta < 1 \ . \end{aligned}$$

Thus $Q_1 = Q_2Q_1$, whence $Q_2 \leq Q_2$. So Φ is monotone. The same goes for Φ^{-1}.

Now let P_1 and P_2 be mutually orthogonal projections in $\mathcal{A}$ and let $Q_i = \Phi(P_i)$, i=1,2. Then

$$\|Q_1Q_2\| = \|Q_1Q_2 - P_1P_2\|$$
$$\leq \|(Q_1-P_1)Q_2 + P_1(Q_2-P_2)\| \leq 2\delta < 1 \ .$$

Since Q_1 and Q_2 commute, $Q_1Q_2 = 0$, so these projections are orthogonal. Thus Q_1+Q_2 is a projection. By monotonicity

$$\Phi(P_1+P_2) \geq Q_1+Q_2 = \Phi(\Psi(Q_1+Q_2))$$
$$\geq \Phi(\Psi(Q_1)+\Psi(Q_2)) = \Phi(P_1+P_2) \ .$$

Thus $\Phi(P_1+P_2) = Q_1+Q_2 = \Phi(P_1)+\Phi(P_2)$.

Now let P_1 and P_2 be arbitrary projections in $\mathcal{A}$. Then $P_1P_2 \leq P_i$ for $i = 1$ and 2, hence $\Phi(P_1P_2) \leq \Phi(P_i)$ for $i = 1$ and 2 and so $\Phi(P_1P_2) \leq \Phi(P_1)\Phi(P_2)$. This holds for Φ^{-1} as well, so

$$P_1P_2 \leq \Phi^{-1}(\Phi(P_1)\Phi(P_2)) \leq \Phi^{-1}(\Phi(P_1))\Phi^{-1}(\Phi(P_2)) = P_1P_2 \ .$$

Consequently, $\Phi(P_1P_2) = \Phi(P_1)\Phi(P_2)$. Also, consider

$$P_1 \vee P_2 = P_1+P_2-P_1P_2$$

which is the projection onto the span of $P_1\mathcal{H}$ and $P_2\mathcal{H}$. One has

$$\Phi(P_1 \vee P_2) = \Phi(I-P_1^{\perp}P_2^{\perp}) = I-\Phi(P_1)^{\perp}\Phi(P_2)^{\perp} = \Phi(P_1) \vee \Phi(P_2) \ .$$

So Φ is a lattice isomorphism of $\mathcal{A}_p$ onto $\mathcal{B}_p$.

Extend Φ to all finite linear combinations $\mathcal{F}$ of $\mathcal{A}_p$ by linearity. Let $P_1,\ldots,P_n$ be pairwise orthogonal projections in $\mathcal{A}$ with sum I. Consider a self-adjoint element $A = \sum_{i=1}^{n}\lambda_iP_i$ with $|\lambda_i| \leq 1$ for $1 \leq i \leq n$. Then A is in the convex hull of

$$\{\sum_{i=1}^{n}\epsilon_iP_i : \epsilon_i = \pm 1\} = \{2P-I : P = \sum_{i \in S} P_i, S \subseteq \{1,\ldots,n\}\} \ .$$

Thus

$$\|\Phi(A)-A\| \leq \sup_{P \in \mathcal{A}_p} \|\Phi(2P-I)-(2P-I)\|$$
$$= 2 \sup_{P \in \mathcal{A}_p} \|\Phi(P)-P\| \ . \tag{18.9.1}$$

Applying this to the real and imaginary parts of an element X in $\mathcal{F}$ yields $\|\Phi(X)-X\| \leq 4\delta\|X\|$.

Thus Φ is norm continuous. Since $\mathcal{F}$ is norm dense in $\mathcal{A}$, one can extend Φ to a map of $\mathcal{A}$ into $\mathcal{B}$. The range of Φ contains the dense subspace of all finite linear combinations of $\mathcal{B}_p$. Also Φ is bounded below by $1-2\delta$ on the self-adjoint part of $\mathcal{A}$. Hence Φ is surjective. It is easy to see that Φ is multiplicative in $\mathcal{F}$, and so the extension is multiplicative. So Φ is a $*$-isomorphism of $\mathcal{A}$ onto $\mathcal{B}$. ■

18.10 Lemma. *Let Φ be a linear map of a real subspace $\mathcal{D}$ in $\mathcal{B}(\mathcal{H})_{s.a.}$ into $\mathcal{B}(\mathcal{H})_{s.a.}$ such that $\|\Phi(A)-A\| \le \delta\|A\|$. Then for any pairwise orthogonal family of projections $\{P_\lambda : \lambda \in \Lambda\}$ in $\mathcal{D}$,*

$$\sum_\lambda (\Phi(P_\lambda)-P_\lambda)^2 \le \delta^2 I \ .$$

Proof. It suffices to prove this for any finite family, say $P_1,\dots,P_n$. For each ϵ in $\{-1,1\}^n$, let $P_\epsilon = \sum_{j=1}^n \epsilon_j P_j$. Then $\|P_\epsilon\| = 1$, so

$$\delta^2 I \ge (\Phi(P_\epsilon)-P_\epsilon)^2 = \sum_{i=1}^n\sum_{j=1}^n \epsilon_i\epsilon_j(\Phi(P_i)-P_i)(\Phi(P_j)-P_j) \ .$$

If one sums $\epsilon_i\epsilon_j$ over all possible choices of ϵ for fixed i, j, one obtains $2^n\delta_{ij}$. Thus averaging over all 2^n values of ϵ yields

$$\delta^2 I \ge \sum_{j=1}^n (\Phi(P_j)-P_j)^2 \ .$$

■

18.11 Lemma. *Let $X = UR$ be the polar decomposition of an operator X. Suppose that $\|I-X\| = \delta < 1$. Then $\|I-U\| \le \sqrt{2}\delta(1+(1-\delta^2)^{1/2})^{-1/2}$.*

Proof. First, X is invertible so U is unitary. Moreover,

$$\begin{aligned}\|Im\ U\| &= \|Im(U-R)\| \le \|U-R\| \\ &= \|U^*-R\| = \|U(U^*-R)\| = \|I-X\| = \delta \ .\end{aligned}$$

So the spectrum $\sigma(U)$ belongs to

$$\{\lambda \in \mathbb{C} : |\lambda| = 1, |Im\ \lambda| \le \delta, dist(\lambda,[0,1]) \le \delta\} \ .$$

For such λ,

$$|\lambda-1|^2 \le |\sqrt{1-\delta^2}+i\delta-1|^2 = \frac{2\delta^2}{1+\sqrt{1-\delta^2}} \ .$$

The norm estimate is now immediate since $\|U-I\| = spr(U-I)$. ■

18.12 Theorem. *Let $\mathcal{A}$ be an abelian von Neumann algebra, and let $\mathcal{B}$ be a von Neumann algebra such that $d(\mathcal{A},\mathcal{B}) = \delta < 1/3$. Then there is a unitary operator U in $(\mathcal{A} \cup \mathcal{B})''$ such that $\mathcal{B} = U^*\mathcal{A}U$, and $\|U-I\| \le 2\sqrt{2}\delta(1+(1-4\delta^2)^{1/2})^{-1/2} < \frac{15}{7}\delta$.*

Proof. By Corollary 18.8, $\mathcal{B}$ is abelian. By Lemmas 18.6 and 18.9, there is a $*$-isomorphism Φ of $\mathcal{A}$ onto $\mathcal{B}$ such that $\sup_{P \in \mathcal{A}_p} \|\Phi(P)-P\| \leq \delta$. So by (18.9.1), $\|\Phi(A)-A\| \leq 2\delta\|A\|$ for every self-adjoint element of $\mathcal{A}$. By Lemma 18.10, if $P_1,\dots,P_n$ are pairwise orthogonal projections in $\mathcal{A}$,

$$\sum_{j=1}^{n}(\Phi(P_j)-P_j)^2 \leq 4\delta^2 I \quad .$$

Consider the collection $\mathcal{E}$ of finite partitions of the identity into pairwise orthogonal projections in $\mathcal{A}$. Put an order on $\mathcal{E}$ given by $(P_1,\dots,P_n) \ll (Q_1,\dots,Q_m)$ if each P_j is the sum of certain Q_i's. For each α in $\mathcal{E}$, let

$$X_\alpha = \sum_{j=1}^{n}\Phi(P_j)P_j \quad .$$

Note that $X_\alpha P_j = \Phi(P_j)X_\alpha$ for $1 \leq j \leq n$. Hence if $(P,P^\perp) \ll \alpha$, $X_\alpha P = \Phi(P)X_\alpha$.

Each X_α is a norm one element in $(\mathcal{A} \cup \mathcal{B})''$. Let X be a weak* limit of a cofinal subnet $\{X_\beta : \beta \in \mathcal{E}'\}$ of $\{X_\alpha\}$. Then for each projection P in $\mathcal{A}$, there is a $\beta \gg (P,P^\perp)$ in $\mathcal{E}'$. Hence $XP = \Phi(P)X$ for all P in $\mathcal{A}_p$. By linearity,

$$XA = \Phi(A)X$$

for all A in $\mathcal{A}$. Now, for each β

$$\begin{aligned} 0 \leq I-X_\beta^* X_\beta &= \sum_{j=1}^{n}P_j-\sum_{j=1}^{n}P_j\Phi(P_j)P_j \\ &= \sum_{j=1}^{n}P_j(P_j-\Phi(P_j))P_j \leq \delta I \quad . \end{aligned}$$

Hence X is invertible. Moreover, for each β

$$\begin{aligned} (I-X_\beta^*)(I-X_\beta) &= (I-X_\beta)+(I-X_\beta^*)-(I-X_\beta^* X_\beta) \leq I-X_\beta+I-X_\beta^* \\ &= \sum_{j=1}^{n}P_j-\Phi(P_j)P_j+\Phi(P_j)-P_j\Phi(P_j) \\ &= \sum_{j=1}^{n}(\Phi(P_j)-P_j)^2 \leq 4\delta^2 I \quad . \end{aligned}$$

Hence $\|I-X_\beta\| \leq 2\delta$, whence $\|I-X\| \leq 2\delta$.

Let $X = U(X^*X)^{1/2}$ be the polar decomposition of X. Since X is invertible, U is unitary and it belongs to $(\mathcal{A} \cup \mathcal{B})''$. For A in $\mathcal{A}$,

$$X^*XA = X^*\Phi(A)X = (\Phi(A^*)X)^*X = AX^*X \quad .$$

Thus $(X^*X)^{1/2}$ commutes with $\mathcal{A}$. So

$$(UA-\Phi(A)U)(X^*X)^{1/2} = U(X^*X)^{1/2}A-\Phi(A)U(X^*X)^{1/2}$$
$$= XA-\Phi(A)X = 0 \ .$$

But $(X^*X)^{1/2}$ is invertible. So $A = U^*\Phi(A)U$ for all A in $\mathcal{A}$. So $U^*\mathcal{B}U = \mathcal{A}$. By Lemma 18.11,

$$\|U-I\| \leq 2\sqrt{2}\delta(1+(1-4\delta^2)^{1/2})^{-1/2}$$
$$< \delta(2\sqrt{2})(1+(1-4/9)^{1/2})^{-1/2} < \frac{15}{7}\delta \ .$$

■

18.13 Corollary. *Let $\mathcal{A}$ be a von Neumann algebra with abelian commutant, and let $\mathcal{B}$ be a von Neumann algebra such that $d(\mathcal{A},\mathcal{B}) = \delta < 1/6$. Then there is a unitary operator U in $(\mathcal{A} \cup \mathcal{B})''$ such that $\mathcal{B} = U^*\mathcal{A}U$ and*

$$\|U-I\| \leq 2\sqrt{2}\delta(1+1-4\delta^2)^{1/2})^{-1/2} < (2.03)\delta \ .$$

Proof. By Lemma 18.7 and Theorem 18.12, there is a unitary operator U satisfying the norm condition in $(Z(A) \cup Z(\mathcal{B}))'' \subseteq (\mathcal{A} \cup \mathcal{B})''$ such that

$$Z(U\mathcal{B}U^*) = UZ(\mathcal{B})U^* = Z(A) = \mathcal{A}' \ .$$

Thus $U\mathcal{B}U^*$ is contained in $Z(U\mathcal{B}U^*)' = \mathcal{A}$. Compute $d(\mathcal{B},U\mathcal{B}U^*)$. Every B in $b_1(\mathcal{B})$ satisfies

$$d(B,Ub_1(\mathcal{B})U^*) \leq \|B-UBU^*\| = \|BU-UB\| \leq 2\|U-I\| < 5\delta \ .$$

The same holds for B' in $Ub_1(\mathcal{B})U^*$, so $d(\mathcal{B},U\mathcal{B}U^*) < 5\delta$. Hence

$$d(\mathcal{A},U\mathcal{B}U^*) < \delta+5\delta < 1 \ .$$

If $U\mathcal{B}U^*$ were a proper subspace of $\mathcal{A}$, it would be at a distance of one. Thus $U\mathcal{B}U^* = \mathcal{A}$. ■

Notes and Remarks.

Perturbations of von Neumann algebras were first considered by Kadison and Kastler [1]. They proved that close von Neumann algebras can be decomposed into central summands corresponding to type I_n, II_1, II_∞, and III each of which is close to the corresponding summand of the other algebras. The material in this chapter on type I von Neumann algebras is due to Christensen [1]. His general result is for all type I algebras, and the ideas are sketched in the exercises. J. Phillips [1] got these results also except for the estimate on $\|U-I\|$. There is the related interesting question for separable C^*-algebras. Non-separable C^*-algebras which are close need not be unitarily equivalent (Choi-Christensen [1]). The interested reader should also consult Phillips [2], Johnson [4], and Christensen [6].

Perturbations of nest algebras were considered by Lance [2] who proved Lemma 18.3 and Corollary 18.5 using cohomological methods. The approach taken here using the Similarity Theorem and Theorem 18.2 is due to Davidson [7] and Choi-Davidson [1].

Exercises

18.1 (Choi-Davidson) Let $\mathcal{L}$ be a finite commutative subspace lattice. Show that there is an $\epsilon > 0$ and $C < \infty$ so that if $\mathcal{A}$ is a norm closed algebra such that $dist(Alg\,\mathcal{L},\mathcal{A}) < \epsilon$, then $\mathcal{A}$ is similar to $Alg\,\mathcal{L}$ via an invertible operator S with $\|S-I\| \leq C\epsilon$.

18.2 (Choi-Davidson)

(a) For t in $\mathbb{C}$, let

$$\mathcal{A}_t = \left\{ \begin{bmatrix} \lambda & a \\ 0 & \lambda+at \end{bmatrix} : \lambda, a \in \mathbb{C} \right\} .$$

Show that $dist(\mathcal{A}_0,\mathcal{A}_t) \leq |t|$, but $\mathcal{A}_0$ is not algebraically isomorphic to $\mathcal{A}_t$ for any $t \neq 0$.

(b) For t in $\mathbb{C}$, let $S_t \subseteq M_2$ and $\mathcal{A}_t \subseteq M_4$ be given by

$$S_t = \left\{ \begin{bmatrix} a & b \\ c & at \end{bmatrix} : a,b,c \in \mathbb{C} \right\}, \ \mathcal{A}_t = \left\{ \begin{bmatrix} \lambda I & S \\ 0 & \lambda I \end{bmatrix} : \lambda \in \mathbb{C}, S \in S_t \right\} .$$

Define $\phi_t : \mathcal{A}_0 \rightarrow \mathcal{A}_t$ by

$$\phi_t \begin{bmatrix} \lambda & 0 & a & b \\ 0 & \lambda & c & 0 \\ 0 & 0 & \lambda & 0 \\ 0 & 0 & 0 & \lambda \end{bmatrix} = \begin{bmatrix} \lambda & 0 & a & b \\ 0 & \lambda & c & at \\ 0 & 0 & \lambda & 0 \\ 0 & 0 & 0 & \lambda \end{bmatrix} .$$

Show that ϕ_t is an isomorphism and $\|\phi_t - \phi_0\| = |t|$. Show that $\mathcal{A}_t$ is not similar to $\mathcal{A}_0$ if $t \neq 0$.

18.3 (Christensen) Let $\mathcal{A}$ be a type I von Neumann algebra, and let $\mathcal{B}$ be a von Neumann algebra such that $d(\mathcal{A},\mathcal{B}) = \delta < \frac{1}{200}$. Let P be a maximal abelian projection in $\mathcal{A}$, and let Q be a projection in $\mathcal{B}$ such that $\|P-Q\| \leq \delta$.

(a) Show that $\{Z(A) \cup P\}''$ and $\{Z(B) \cup Q\}''$ are close, and apply Theorem 18.7 to replace $\mathcal{B}$ by $\widetilde{\mathcal{B}}$ with $Z(\mathcal{A}) = Z(\widetilde{\mathcal{B}})$ and $P = \tilde{Q}$.

(b) Prove that $\tilde{Q}$ is an abelian projection in $\widetilde{\mathcal{B}}$ and hence deduce that $\mathcal{B}$ is type I. Show that $P\mathcal{A}P$ and $P\widetilde{\mathcal{B}}P$ are close, and apply Theorem 18.7 again to replace $\widetilde{\mathcal{B}}$ by $\mathcal{C}$ in which $Z(\mathcal{A}) = Z(\mathcal{C})$, $P \in \mathcal{C}$, and $P\mathcal{A}P = P\mathcal{C}P$, and $d(\mathcal{A},\mathcal{C})$ is small.

(c) Let Φ_A and Φ_C be the restriction maps of $\mathcal{A}'$ and $\mathcal{C}'$ to $P\mathcal{H}$ respectively, and note that $\Phi = \Phi_C^{-1}\Phi_A$ is a $*$-isomorphism. Show that $d(\mathcal{A}',\mathcal{C}') \le 4d(\mathcal{A},\mathcal{C})$ and hence deduce $\|\Phi - id\| < 8\delta$.

(d) Let $1 \le m \le \infty$ and let P_m be the central summand of $\mathcal{A}$ corresponding to the type I_m part of $\mathcal{A}'$. Let E_{ij} be a set of matrix units for $P_m\mathcal{A}'$. Set $X_m = \sum_{j=1}^{m} \Phi(E_{j1})E_{1j}$. Show that $X_m A = \Phi(A)X_m$ for all A in $P_m\mathcal{A}'$. Show that X_m is invertible, and hence deduce that $\mathcal{A}'$ and $\mathcal{C}'$ are unitarily equivalent.

(e) Estimate $\|I - X_m\|$ and hence find a unitary U such that $U\mathcal{B}U^* = \mathcal{A}$ and $\|U - I\| \le C\delta$ for a universal constant C. (Caveat: this is very hard.)

19. Derivations of Nest Algebras

In this chapter, derivations of nest algebras will be computed. Derivations are defined in Chapter 10. A brief consideration of higher cohomology is contained in the notes. Also, there is an appendix on the K_0 group of a nest algebra. The first result is automatic continuity, and it parallels Theorem 10.3.

19.1 Theorem. *Let $\mathcal{A}$ be a weak* closed subalgebra of $B(\mathcal{H})$. Let M be a dual normal $\mathcal{A}$ module, and let $\delta : \mathcal{A} \to M$ be a derivation. Then there is a central projection P in $\mathcal{A} \cap \mathcal{A}^*$ such that $D_P(A) = \delta(AP)$ is continuous and $(\mathcal{A} \cap \mathcal{A}^*)(I-P)$ is finite dimensional.*

Proof. As in Theorem 10.3, let $\mathcal{J}$ be the set of elements J in $\mathcal{A} \cap \mathcal{A}^*$ such that $D_J(A) = \delta(AJ)$ is a continuous map. Let $\mathcal{R}_X$ denote multiplication on the right by X. If B belongs to $\mathcal{A} \cap \mathcal{A}^*$, then

$$D_{BJ}(A) = \delta(ABJ) = D_J \mathcal{R}_B(A)$$

and

$$D_{JB}(A) = \delta(AJ)B + AJ\delta(B) = (\mathcal{R}_B D_J + \mathcal{R}_{J\delta(B)})(A) \ .$$

Consequently, $\mathcal{J}$ is a two sided ideal in $\mathcal{A} \cap \mathcal{A}^*$.

By Theorem 10.3, the restriction of δ to $\mathcal{A} \cap \mathcal{A}^*$ is bounded, say by K. Suppose that J_α is a net of elements in the unit ball of $\mathcal{J}$ converging weak* to J. Then

$$\begin{aligned} D_J(A) &= \delta(A)J + A\delta(J) \\ &= w^*\lim \delta(A)J_\alpha + A\delta(J) = w^*\lim D_{J_\alpha}(A) + A\delta(J - J_\alpha) \ . \end{aligned}$$

So D_J is the pointwise limit of the continuous maps $T_\alpha = D_{J_\alpha} + \mathcal{R}_{\delta(J-J_\alpha)}$. Moreover, from the second line,

$$\|T_\alpha(A)\| \le \|\delta(A)\| + \|A\|K$$

for all α. By the Uniform Boundedness Principle, $\sup\|T_\alpha\| < \infty$. Hence D_J is continuous. By the Krein-Smulyan Theorem, it follows that $\mathcal{J}$ is weak* closed.

It now follows that $\mathcal{J}$ contains a maximum idempotent P which is central in $\mathcal{A} \cap \mathcal{A}^*$ such that $\mathcal{J} = (\mathcal{A} \cap \mathcal{A}^*)P$. Suppose that $(\mathcal{A} \cap \mathcal{A}^*)P^\perp$ is not finite dimensional. Then there are pairwise orthogonal projections E_n in $(\mathcal{A} \cap \mathcal{A}^*)P^\perp$. By the definition of $\mathcal{J}$, D_{E_n} is unbounded. Choose A_n in $\mathcal{A}$

with $\|A_n\| < 2^{-n}$ and $\|\delta(A_nE_n)\| > 2^n$. Let $A = \sum_{n=1}^{\infty} A_nE_n$. Then

$$\begin{aligned} 2^n &< \|\delta(A_nE_n)\| = \|\delta(AE_n)\| \\ &= \|\delta(A)E_n + A\delta(E_n)\| \le \|\delta(A)\| + K\|A\| \ . \end{aligned}$$

This contradiction shows that $(\mathcal{A} \cap \mathcal{A}^*)P^{\perp}$ is finite dimensional. ■

19.2 Corollary. *Let δ be a derivation on a CSL algebra $\mathcal{A}$. Then δ is continuous.*

Proof. By Theorem 19.1, there is a central projection P in $\mathcal{A} \cap \mathcal{A}^* = \mathcal{L}'$ (where $\mathcal{L} = Lat\,\mathcal{A}$) such that $(\mathcal{A} \cap \mathcal{A}^*)P^{\perp}$ is finite dimensional and D_P is continuous. The projection $P^{\perp}$ is the sum $\sum_{i=1}^{k} P_i$ of minimal projections in $\mathcal{L}''$, each of which is finite rank since $(\mathcal{A} \cap \mathcal{A}^*)P^{\perp} = \sum_{i=1}^{k} \oplus\, \mathcal{B}(P_i\mathcal{H})$ is finite dimensional. Applying Theorem 19.1 to $\delta^*(X) = \delta(X^*)^*$ on $\mathcal{A}^*$, one obtains a central projection Q in $\mathcal{A} \cap \mathcal{A}^*$ such that $Q^{\perp}(\mathcal{A} \cap \mathcal{A}^*)$ is finite dimensional and ${}_QD(A) = \delta(QA)$ is continuous. As above, $Q^{\perp} = \sum_{j=1}^{\ell} Q_j$ is the sum of minimal finite rank projections in $\mathcal{L}''$. Hence

$$\delta(A) = \delta(AP) + \delta(QAP^{\perp}) + \delta(Q^{\perp}AP^{\perp}) \ .$$

So $\delta = D_P + {}_QD\,\mathcal{R}_{P^{\perp}} + \delta\,\mathcal{L}_{Q^{\perp}}\mathcal{R}_{P^{\perp}}$ is the sum of two continuous maps and a linear map acting on the finite dimensional space $Q^{\perp}\mathcal{A}P^{\perp}$. Thus δ is continuous. ■

19.3 Lemma. *Let δ be a bounded derivation on a Banach algebra $\mathcal{A}$. Then $\alpha_t = e^{t\delta}$, $t \in \mathbb{R}$, is a continuous one parameter group of automorphisms of $\mathcal{A}$, and $\delta = \lim_{t\to 0} t^{-1}(\alpha_t - id)$.*

Proof. If δ is a derivation, then $\alpha = e^{\delta} = \sum_{n=0}^{\infty} \delta^n/n!$ is a bounded linear map. One readily verifies by induction the formula

$$\delta^n(xy) = \sum_{k=0}^{n} \binom{n}{k} \delta^k(x)\delta^{n-k}(y) \ .$$

So

$$\begin{aligned} \alpha(xy) &= \sum_{n=0}^{\infty} \frac{\delta^n(xy)}{n!} = \sum_{n=0}^{\infty} \sum_{k=0}^{n} \frac{\delta^k(x)}{k!} \, \frac{\delta^{n-k}(y)}{n-k!} \\ &= \sum_{k=0}^{\infty} \frac{\delta^k(x)}{k!} \left(\sum_{j=0}^{\infty} \frac{\delta^j(y)}{j!} \right) = \alpha(x)\alpha(y) \ . \end{aligned}$$

Now $t\delta$ is a derivation for every δ, thus α_t is a homomorphism for every t. Moreover, $\alpha_s\alpha_t = e^{s\delta}e^{t\delta} = e^{(s+t)\delta} = \alpha_{s+t}$ so this forms a one parameter group of automorphisms. Finally,

$$t^{-1}(\alpha_t - id) = \delta + t\sum_{n=2}^{\infty} t^{n-2}\frac{\delta^n}{n!} \quad .$$

For $|t| \leq 1$, $\| \sum_{n\geq 2} t^{n-2}\delta^n/n!\| < e^{\|\delta\|}$. So $\lim_{t\to 0} t^{-1}(\alpha_t - id) = \delta$. ■

19.4 Lemma. *Let $\mathcal{N}$ be a nest. For every operator A in $\mathcal{B}(\mathcal{H})$,*

$$dist(A,\mathbb{C}I) \leq 2\|\delta_A | \mathcal{T}(\mathcal{N})\| \quad .$$

Proof. By Theorem 8.3 and Corollary 8.5, there is an expectation Φ of $\mathcal{B}(\mathcal{H})$ onto $\mathcal{N}''$ so that $B = \Phi(A)$ belongs to the weak* closed convex hull of $\{UAU^*: U \in \mathcal{U}(\mathcal{N}')\}$. Hence

$$\begin{aligned}\|B-A\| &\leq \sup\{\|UAU^*-A\|: U \in \mathcal{U}(\mathcal{N}')\} \\ &= \sup\{\|\delta_A(U^*)\|: U \in \mathcal{U}(\mathcal{N}')\} \leq \|\delta_A | \mathcal{T}(\mathcal{N})\| \quad .\end{aligned}$$

Also, for T in $\mathcal{T}(\mathcal{N})$,

$$\begin{aligned}\|\delta_B(T)\| &\leq \sup\{\|[UAU^*,T]\|: U \in \mathcal{U}(\mathcal{N}')\} \\ &= \sup\{\|[A,U^*TU]\|: U \in \mathcal{U}(\mathcal{N}')\} \leq \|\delta_A | \mathcal{T}(\mathcal{N})\| \, \|T\| \quad .\end{aligned}$$

Fix N in $\mathcal{N}$, and let $x \in N$ and $y \in N^\perp$ be unit vectors. Then

$$|(\delta_B(x \otimes y^*)y,x)| = |(Bx,x)-(By,y)| \leq \|\delta_B | \mathcal{T}(\mathcal{N})\| \quad .$$

Let λ and μ be points in $\sigma(B)$ so that $|\lambda-\mu|$ is maximal. Since B commutes with $\mathcal{N}$, it is possible to find an N in $\mathcal{N}$ so that one point, say λ, belongs to $\sigma(B|N)$ and the other, μ, belongs to $\sigma(B|N^\perp)$. As B is normal, it is easy to approximate λ by (Bx,x) where x is a unit vector in N and approximate μ by (By,y) where y is a unit vector in $N^\perp$. Thus

$$\|\delta_B | \mathcal{T}(\mathcal{N})\| \geq diam(\sigma(B)) \geq dist(B,\mathbb{C}I) \quad .$$

Hence

$$dist(A,\mathbb{C}I) \leq \|A-B\| + dist(B,\mathbb{C}I) \leq 2\|\delta_A | \mathcal{T}(\mathcal{N})\| \quad .$$ ■

19.5 Corollary. *The commutant of $\mathcal{T}(\mathcal{N})$ is trivial.*

Proof. If $\delta_A | \mathcal{T}(\mathcal{N}) = 0$, then A is scalar by Lemma 19.4. ■

19.6 Corollary. *Let α be an automorphism of a nest algebra $\mathcal{T}(\mathcal{N})$. Then there is an invertible operator A with $\|A-I\| \leq 4\|\alpha - id\|$ such that $\alpha = Ad\,A$.*

Proof. By Theorem 17.12, α is spatial, say $\alpha = Ad\,A$. By Corollary 19.5, A is unique up to a scalar factor. Normalize A so that $\|A\| = 1$. For T in $\mathcal{T}(\mathcal{N})$,

$$\begin{aligned}\|\delta_A(T)\| &= \|AT-TA\| = \|(\alpha(T)-T)A\| \\ &\leq \|\alpha-id\|\ \|A\|\ \|T\| \ .\end{aligned}$$

By Lemma 19.4, there is a scalar λ so that $\|A-\lambda I\| \leq 2\|\alpha-id\|$. As $\|A\| = 1$, it follows that $|1-|\lambda|\,| \leq \|A-\lambda I\|$, so

$$\|A-|\lambda|^{-1}\lambda I\| \leq 4\|\alpha-id\| \ .$$

Replace A by $|\lambda|^{-1}\bar{\lambda}A$. ∎

19.7 Theorem. *Every derivation δ of a nest algebra $\mathcal{T}(\mathcal{N})$ is inner.*

Proof. By Corollary 19.2, δ is bounded. Let $\alpha_t = e^{t\delta}$, $t \in \mathbb{R}$, be the continuous group of automorphism generated by δ. There is an $\epsilon > 0$ so that $|t| < \epsilon$ implies $\|\alpha_t-id\| < 1$. By Corollary 17.15, α_t is inner. Let A_t be an invertible element in $\mathcal{T}(\mathcal{N})$ so that $\alpha_t = Ad(A_t)$. By Corollary 19.6, A_t can be normalized so that $\|A_t-I\| \leq 4\|\alpha_t-id\|$. By Lemma 19.3, $t^{-1}\|\alpha_t-id\|$ is bounded. So $D_n = n(A_{1/n}-I)$ is a bounded net. So there is a weak* convergent subnet D_{n_γ} with limit D in $\mathcal{T}(\mathcal{N})$. Again by Lemma 19.3, one obtains

$$\begin{aligned}\delta_D(T) &= \lim D_{n_\gamma}T-TD_{n_\gamma} = \lim n_\gamma(A_{1/n_\gamma}T-TA_{1/n_\gamma}) \\ &= \lim_\gamma n_\gamma(\alpha_{1/n_\gamma}-id)(T)A_{1/n_\gamma} = \delta(T) \ .\end{aligned}$$

So $\delta = \delta_D$ is inner. ∎

There is the analogous theorem for $Q\mathcal{T}(\mathcal{N})$, based on Theorem 17.20.

19.8 Lemma. *Let $\mathcal{A}$ be an algebra of operators containing the compact operators $\mathcal{K}$. Then every derivation δ on $\mathcal{A}$ is spatial.*

Proof. Every compact operator K factors as $K = K_1K_2$ where both K_1 and K_2 are compact. So $\delta(K) = K_1\delta(K_2)+\delta(K_1)K_2$ is compact. By Theorem 10.6, $\delta\,|\mathcal{K}$ is weak* continuous so extends to a derivation δ' of $\mathcal{B}(\mathcal{H})$. By Theorem 10.8, $\delta' = \delta_X$ for some X in $\mathcal{B}(\mathcal{H})$. For any operators A in $\mathcal{A}$ and K in $\mathcal{K}$,

$$\begin{aligned}\delta(A)K &= \delta(AK) - A\delta(K) \\ &= \delta_X(AK) - A\delta_X(K) = \delta_X(A)K \ .\end{aligned}$$

This holds for all K, whence $\delta = \delta_X$. ■

19.9 Lemma. $e^{\delta_X} = Ad\, e^X$.

Proof. Let A be an operator. Then

$$\begin{aligned}Ad\, e^X(A) &= \left(\sum_{k=0}^{\infty} \frac{X^k}{k!}\right) A \left(\sum_{j=0}^{\infty} \frac{(-X)^j}{j!}\right) \\ &= \sum_{n=0}^{\infty} \sum_{j=0}^{n} \frac{X^{n-j}}{n-j!} A \frac{(-X)^j}{j!} \\ &= \sum_{n=0}^{\infty} \frac{1}{n!} \sum_{j=0}^{n} \binom{n}{j} (-1)^j X^{n-j} A X^j\end{aligned}$$

and

$$e^{\delta_X}(A) = \sum_{n=0}^{\infty} \frac{1}{n!} \delta_X^n(A) \ .$$

Now $\delta_X^0(A) = A$ and $\delta_X(A) = XA - AX$. One readily establishes by induction that $\delta_X^n(A) = \sum_{j=0}^{n} \binom{n}{j} (-1)^j X^{n-j} A X^j$. So equality is evident. ■

19.10 Theorem. *Every derivation δ of a quasitriangular algebra $\mathcal{QT}(\mathcal{N})$ is inner.*

Proof. By Lemma 19.8, $\delta = \delta_X$ for some X in $\mathcal{B}(\mathcal{H})$. By Lemma 19.3, one defines the continuous automorphism group $\alpha_t = e^{t\delta} = Ad\, e^{tX}$ by Lemma 19.9. For $|t|$ sufficiently small, $\|\alpha_t - id\| < 1$. So by Theorem 17.20, α_t is inner so $\alpha_t = Ad\, S_t$ and S_t belongs to $\mathcal{QT}(\mathcal{N})^{-1}$. But S_t is determined up to a scalar factor since $\mathcal{K}' = \mathbb{C}I$. Thus e^{tX} belongs to $\mathcal{QT}(\mathcal{N})$. So

$$X = \lim_{t \to 0} \frac{e^{tX} - I}{t}$$

belongs to $\mathcal{QT}(\mathcal{N})$. That is, $\delta = \delta_X$ is inner. ■

Next, we consider derivations into the compact operators. But first we require a more general result.

19.11 Proposition. *Let $\mathcal{N}$ be a nest, and let B be a weak* closed algebra containing $\mathcal{T}(\mathcal{N})$. Then every derivation δ of $\mathcal{T}(\mathcal{N})$ into B is inner.*

Proof. Let δ' be the restriction of δ to $\mathcal{N}'$, and note that B is a dual normal $\mathcal{N}'$ module. By Theorem 10.8, $\delta' = \delta_E$ for some E in B. Let $\delta_1 = \delta - \delta_E$. Then $\delta_1(P) = 0$ for every P in $\mathcal{N}'$. For every T in $\mathcal{T}(\mathcal{N})$ and N in $\mathcal{N}$,

$$P(N)^{\perp}\delta_1(T)P(N) = \delta_1(P(N)^{\perp}TP(N)) = 0 \ .$$

So δ_1 is a derivation of $\mathcal{T}(\mathcal{N})$ into itself. By Theorem 19.7, $\delta_1 = \delta_A$ for some A in $\mathcal{T}(\mathcal{N})$. Thus $\delta = \delta_{A+E}$ is inner. ∎

19.12 Theorem. *Every derivation δ of a nest algebra $\mathcal{T}(\mathcal{N})$ into the compact operators $\mathcal{K}$ is inner.*

Proof. First, one has that δ maps $\mathcal{T}(\mathcal{N})$ into $\mathcal{B}(\mathcal{H})$. So by Proposition 19.11, $\delta = \delta_X$ for some X in $\mathcal{B}(\mathcal{H})$. The restriction of δ to $\mathcal{N}'$ is compact valued. Thus by the Johnson-Parrott Theorem 10.12, X belongs to $\mathcal{N}''+\mathcal{K}$. So there is a compact operator K and an operator T in $\mathcal{N}''$ so that $\delta_T = \delta - \delta_K$ is zero on $\mathcal{N}'$, and is compact valued on $\mathcal{T}(\mathcal{N})$. Now consider four cases, depending on the structure of $\mathcal{N}$:

Case 1. $\mathcal{N}$ contains an element N such that N and $N^{\perp}$ are infinite dimensional. Let U be any isometry of $N^{\perp}$ onto N. Let A belong to $P(N)\mathcal{B}(\mathcal{H})P(N)$. Then AU belongs to $\mathcal{T}(\mathcal{N})$. Let $T_1 = TP(N)$ and $T_2 = TP(N)^{\perp}$. Then

$$\delta_{T_1}(A) = \delta_T(A)P(N) = \delta_T(A)UU^* = (\delta_T(AU) - A\delta_T(U))U^*$$

is compact. Thus δ_{T_1} is a derivation of $\mathcal{B}(N)$ into $\mathcal{K}$. So $T_1 = \lambda_1 P(N) + K_1$ where K_1 is a compact operator on N. Similarly, $T_2 = \lambda_2 P(N)^{\perp} + K_2$. Finally,

$$\delta_T(U) = (\lambda_1 - \lambda_2)U + K_1 U - UK_2 \ .$$

So $\lambda_1 = \lambda_2$, and T is scalar plus compact. Thus $\delta_T = \delta_{K_1+K_2}$, whence $\delta = \delta_{K+K_1+K_2}$ is inner.

Case 2. $\mathcal{N}$ contains an element N_∞ of finite codimension, and $\{N \in \mathcal{N}: N < N_\infty\} = \{0 = N_0 < N_1 < \ldots < N_k < \ldots\}$ is a sequence of finite dimensional subspaces with $\bigvee_{k \geq 1} N_k = N_\infty$. Let $A_k = P(N_k) - P(N_{k-1})$ for $k \geq 1$ be the atoms of $\mathcal{N}$. Then $T = \sum_{k \geq 1} \lambda_k A_k + T_\infty$ where $T_\infty = P(N_\infty)^{\perp} T_\infty P(N_\infty)^{\perp}$ is compact. It will be shown that $\lim \lambda_k = \lambda_\infty$ exists. For otherwise, there is an $\epsilon > 0$ and a sequence $k_1 < k_2 < \ldots$ such that $|\lambda_{k_i} - \lambda_{k_{i+1}}| \geq \epsilon$. Let e_i be a unit vector in $A_{k_i}\mathcal{H}$. Then $S = \sum_{i \geq 1} e_i \otimes e_{i+1}^*$ belongs to $\mathcal{T}(\mathcal{N})$. So

$$\delta_T(S) = \sum_{i\geq 1} (\lambda_{k_i} - \lambda_{k_{i+1}}) e_i \otimes e^*_{i+1} \quad .$$

This is not compact, contrary to hypothesis. So $\lim\lambda_k$ must exist. Hence $K_1 = T - \lambda_\infty I$ is compact, and $\delta = \delta_{K+K_1}$ is inner.

Case 3. $\mathcal{N}$ contains an N_∞ of finite dimension, and

$$\{N \in \mathcal{N}: N > N_\infty\} = \{\ldots N_k < \ldots < N_1 < N_0 = \mathcal{H}\}$$

is a decreasing sequence of subspaces of finite codimension with $\bigwedge_{k\geq 1} N_k = N_\infty$. Then δ_{T^*} is a derivation of $\mathcal{T}(\mathcal{N}^\perp)$ into $\mathcal{K}$, so T^* is scalar plus compact by case 2. Hence so is T, and thus δ is inner.

Case 4. $\mathcal{N}$ is a finite nest consisting only of subspaces of finite dimension or codimension. Thus $\mathcal{N}$ has one atom A such that $A^\perp$ is finite rank. So $T = \lambda A + A^\perp T A^\perp$, and hence $T - \lambda I$ is compact. Again, δ is inner.

It is readily verified that these four cases are exhaustive. ■

19.13 Corollary. *The essential commutant of* $\mathcal{Q}\mathcal{T}(\mathcal{N})$ *is* $\mathbb{C}I + \mathcal{K}$.

A linear map T between Banach spaces X and Y is called *weakly compact* if $\overline{Tb_1(X)}$ is weakly compact. This notion coincides with boundedness when X is reflexive. Now we classify those derivations of $\mathcal{T}(\mathcal{N})$ which are compact or weakly compact.

19.14 Theorem. *Let* $\delta = \delta_X$ *be a derivation of* $\mathcal{B}(\mathcal{H})$ *and let* $\mathcal{N}$ *be a nest. The following are equivalent:*

i) δ *is weakly compact,*

ii) $\delta\,|\mathcal{T}(\mathcal{N})$ *is weakly compact,*

iii) *Ran* (δ) *is contained in* $\mathcal{K}$,

iv) *Ran* $(\delta\,|\mathcal{T}(\mathcal{N}))$ *is contained in* $\mathcal{K}$,

v) $\delta = \delta_K$ for some K in $\mathcal{K}$.

Proof. Clearly i) implies ii). Now δ maps $\mathcal{K}$ into itself. Let $\delta_0 = \delta\,|\mathcal{K}\cap\mathcal{T}(\mathcal{N})$ belong to $\mathcal{B}(\mathcal{K}\cap\mathcal{T}(\mathcal{N}),\mathcal{K})$. By Theorem 16.7, $(\mathcal{K}\cap\mathcal{T}(\mathcal{N}))^{**} = \mathcal{T}(\mathcal{N})$ and by Theorem 1.15, $\mathcal{K}^{**} = \mathcal{B}(\mathcal{H})$. A simple computation using the weak* continuity of δ shows that $\delta\,|\mathcal{T}(\mathcal{N}) = \delta_0^{**}$. The image of the unit ball $B_0 = \overline{\delta_0 b_1(\mathcal{K}\cap\mathcal{T}(\mathcal{N}))}$ is closed and convex, hence $\mathcal{K}^*$-weakly closed in $\mathcal{K}$. Now B_0 is a subset of $B = \overline{\delta b_1(\mathcal{T}(\mathcal{N}))}$ which is $\mathcal{B}(\mathcal{H})^*$-weakly compact. The $\mathcal{K}^*$-topology (or weak* topology) on B is a weaker Hausdorff topology, and thus must agree with the $\mathcal{B}(\mathcal{H})^*$ topology in B. So B is $\mathcal{K}^*$ compact and thus so is B_0. But the unit ball of any Banach space is weak* dense in the ball of its second dual (Goldstine's Theorem). Hence B_0 is $\mathcal{K}^*$ dense in B, whence $B = B_0$ is contained in $\mathcal{K}$.

This proves ii) implies iv). By Theorem 19.12, iv) implies v). Clearly, v) implies iii), and iii) implies iv).

Suppose $\delta = \delta_F$ and F is finite rank; say $F = PFP$ where P is a finite rank projection. Then $\overline{\delta b_1(B(\mathcal{H}))}$ is contained in

$$\|F\|(b_1(PB(\mathcal{H}))+b_1(B(\mathcal{H})P)) \ .$$

Now $B(\mathcal{H})P$ is contained in $\mathcal{K}$, so the $B(\mathcal{H})^*$ and $\mathcal{K}^*$ topologies coincide. Clearly $b_1(B(\mathcal{H})P)$ is weak* compact. Likewise $b_1(PB(\mathcal{H}))$ is weakly compact, so δ_F is weakly compact. Now if K is a compact operator, choose finite rank operators F_n converging to K. Then δ_K is the limit of δ_{F_n}, and hence is weakly compact. This established v) implies i). ■

19.15 Theorem. *Let $\mathcal{N}$ be a nest. Let N_f (and N_c) be the sup (inf) of all finite (cofinite) dimensional elements of $\mathcal{N}$. Let $\delta = \delta_X$ be a derivation of $\mathcal{T}(\mathcal{N})$ onto $B(\mathcal{H})$. The following are equivalent:*

i) *δ is compact,*

ii) *$\delta = \delta_K$ where $K = P(N_f)KP(N_c)^\perp$ is compact,*

iii) *δ is the limit of finite rank derivations of $\mathcal{T}(\mathcal{N})$ into $\mathcal{T}(\mathcal{N}) \cap \mathcal{K}$.*

Proof. Assume i) holds. Then δ is weakly compact by Theorem 19.14, and hence $\delta = \delta_K$ for some compact operator K. Let $N = N_f^+$. This is the least infinite dimensional element in $\mathcal{N}$, and $N_f = N_-$. Suppose that $P(N_f)^\perp K \neq 0$. Choose a unit vector y in $N_f^\perp$ such that $K^*y = z \neq 0$. For every unit vector x in N, $x \otimes y^*$ belongs to $b_1(\mathcal{T}(\mathcal{N}))$ by Proposition 2.8. Now

$$\delta_K(x \otimes y^*) = Kx \otimes y^* - x \otimes z^* \ .$$

If z is not a multiple of y, fix a vector v so that $(v,y) = 0$ and $(v,z) = 1$. The the set $\{\delta_K(x \otimes y^*) : x \in b_1(N)\}$ of rank 2 operators takes v onto $b_1(N)$. Hence it is not a precompact set, contrary to hypothesis. On the other hand, if $z = \lambda y \neq 0$, then $\delta_K(x \otimes y^*)y = (K-\overline{\lambda}I)x$. Since $K-\lambda I$ is Fredholm, its range is bounded below on a subspace of finite codimension in $b_1(N)$ and a similar contradiction is reached. So $P(N_f)^\perp K = 0$.

Similarly, $KP(N_c) = 0$. Hence $K = P(N_f)KP(N_c)^\perp$. This shows that i) implies ii).

There are finite dimensional elements N_k in $\mathcal{N}$ such that $P(N_k)$ converges strong* to $P(N_f)$. Similarly, there are cofinite elements M_k with $P(M_k)$ converging strong* to $P(N_c)$. By Proposition 1.18, $F_k = P(N_k)KP(M_k)^\perp$ converges in norm to K. For T in $\mathcal{T}(\mathcal{N})$,

$$\begin{aligned}\delta_{F_k}(T) &= P(N_k)KP(M_k)^\perp T - TP(N_k)KP(M_k)^\perp \\ &= P(N_k)\delta_{F_k}(T)P(M_k)^\perp \ .\end{aligned}$$

This has range in the finite dimensional space $P(N_k)B(\mathcal{H})P(M_k)^{\perp}$, which is contained in $\mathcal{T}(\mathcal{N}) \cap K$. As δ_{F_k} converge to δ_K, ii) implies iii). Finally, iii) implies i) trivially. ■

Next, we will show that a sequence of derivations converging pointwise in fact converges in the norm. This result is akin to the well-known result that a sequence in ℓ_1 which converges weakly must converge in norm. Indeed, our first lemma is a strengthening of this fact. (A proof using the ℓ_1 result is sketched in Exercise 19.3).

19.16 Phillip's Lemma. *Let ϕ_n be a sequence in ℓ_∞^* which converges to 0 in the weak* topology. Then $\|\phi_n|c_o\| = \sum_{k=1}^{\infty} |\phi_n(e_k)|$ converges to zero. (e_k is the standard basis for c_0).*

Proof. Suppose that $\limsup \|\phi_n|c_0\| > \epsilon > 0$. Now $\lim_{n\to\infty} \phi_n(e_k) = 0$ for every $k \geq 1$. Iteratively choose increasing sequences of integers $0 < N_1 < N_2 < \ldots$ and $0 < n_1 < n_2 < \ldots$ such that $\|\phi_{n_k}|c_0\| > \epsilon$ and

$$\sum_{i=1}^{N_{k-1}} |\phi_{n_k}(e_i)| + \sum_{i>N_k} |\phi_{n_k}(e_i)| < \epsilon/4 \ . \qquad (*)$$

Once $N_1,\ldots,N_k$ and $n_1,\ldots,n_k$ are defined, choose n_{k+1} so large that $\sum_{i=1}^{N_k} |\phi_{n_{k+1}}(e_i)| < \epsilon/4$ and $\|\phi_{n_{k+1}}|c_0\| > \epsilon$. Now choose N_{k+1} large enough that (*) holds. Let B_k denote the interval $\{i : N_{k-1} < i \leq N_k\}$.

If Σ is a subset of $\mathbb{N}$, let

$$\|\phi|\Sigma\| = \sup\{|\phi(x)| : x \in \ell_\infty, \|x\| \leq 1, \mathrm{supp}(x) \leq \Sigma\} \ .$$

The next step is to produce a subsequence $k_1, k_2, \ldots$ so that

$$\|\phi_{n_{k_i}}|\bigcup_{j\neq i} B_{k_j}\| < \epsilon/4 \ .$$

Take $k_1 = 1$. Split $\mathbb{N}\backslash\{1\}$ into disjoint infinite sets $\Delta_1, \Delta_2, \ldots$ and let $\Sigma_i = \bigcup_{j\in\Delta_i} B_j$. At most $M = [4\|\phi_{n_1}\|\epsilon^{-1}]$ of these have $\|\phi_{n_1}|\Sigma_i\| \geq \epsilon/4$. To see this, choose x_i with support Σ_i, $\|x_i\| = 1$, and $\phi_{n_1}(x_i) \geq \epsilon/4$ for i in a set I of cardinality at least $M+1$. Let $x = \sum_{i\in I} x_i$. Then $\|x\| = 1$ and

$$\phi_{n_1}(x) = \Sigma\phi_{n_1}(x_i) > (M+1)\epsilon/4 > \|\phi_{n_1}\| \ .$$

So one obtains infinite subsets Δ and $\Sigma = \bigcup_{j\in\Delta} B_j$ so that $\|\phi_{n_1}|\Sigma\| < \epsilon/4$. Consider the subsequence $k_1 \cup \Delta$. Pick k_2 in Δ and proceed as above to obtain an infinite subset Δ_2 of Δ and $\Sigma_2 = \bigcup_{j\in\Delta_2} B_j$ so that

$\|\phi_{n_{k_2}} | B_{k_1} \cup \Sigma_2\| < \epsilon/4$. Proceeding iteratively, one obtains the desired subsequence. For convenience, let us relabel this subsequence as $\phi_1, \phi_2, \ldots$ and corresponding sets $B_1, B_2, \ldots$.

From $\|\phi_i | c_0\| > \epsilon$ and (*), one has $\|\phi_i | B_i\| > 3\epsilon/4$. Choose x_i on ℓ_∞ with $\mathrm{supp}(x_i)$ contained in B_i, $\|x_i\| = 1$, and $\phi_i(x_i) > 3\epsilon/4$. Let $x = \sum_{i \geq 1} x_i$. Then $\mathrm{supp}(\sum_{j \neq i} x_j)$ is contained in $\Sigma_i = \bigcup_{j \neq i} B_j$ and hence

$$\phi_i(x) > \phi_i(x_i) - |\phi_i(\sum_{j \neq i} x_j)|$$

$$> 3\epsilon/4 - \|\phi_i | \Sigma_i\| > \epsilon/2 \ .$$

This contradicts the fact that $\lim \phi_i(x) = 0$. So the lemma must be valid. ∎

19.17 Lemma. *Let δ_n be a sequence of derivations from a von Neumann algebra $\mathcal{A}$ into $B(\mathcal{H})$ such that $\lim_{n \to \infty} \delta_n(A) = 0$ for all A in $\mathcal{A}$. If P and Q are mutually orthogonal projections in the centre $Z(\mathcal{A})$, then*

$$\limsup \|\delta_n | (P+Q)\mathcal{A}\| = \max\{\limsup \|\delta_n | P\mathcal{A}\|, \limsup \|\delta_n | Q\mathcal{A}\|\} \ .$$

Proof. For all A in $\mathcal{A}$,

$$\delta_n(PA) = \delta_n(PPAP) = \delta_n(P)PA + P\delta_n(PA)P + PA\delta_n(P) \ .$$

Since $\lim \|\delta_n(P)\| = 0$, one obtains $\lim \|\delta_n(PA) - P\delta_n(PA)P\| = 0$ uniformly for $\|A\| \leq 1$. Similarly, this holds for Q. Since

$$\|PAP + QBQ\| = \max\{\|PAP\|, \|QBQ\|\} \ ,$$

one obtains for $\|A\| \leq 1$,

$$\begin{aligned}\limsup \|\delta_n((P+Q)A)\| &= \limsup \|\delta_n(PA) + \delta_n(QA)\| \\ &= \limsup \|P\delta_n(PA)P + Q\delta_n(QA)Q\| \\ &= \limsup \max\{\|P\delta_n(PA)P\|, \|Q\delta_n(QA)Q\|\} \\ &= \limsup \max\{\|\delta_n(PA)\|, \|\delta_n(QA)\|\} \\ &\leq \max\{\limsup \|\delta_n | P\mathcal{A}\|, \limsup \|\delta_n | Q\mathcal{A}\|\} \ .\end{aligned}$$

The reverse inequality is immediate. ∎

19.18 Theorem. *Let $\mathcal{A}$ be an abelian von Neumann algebra, and suppose δ_n is a sequence of derivations converging pointwise to zero. Then $\lim_{n \to \infty} \|\delta_n\| = 0$.*

Proof. By the Uniform Boundedness Principle, $\sup \|\delta_n\| < \infty$. So we will suppose that $\|\delta_n\| \leq 1$. Assume that $\limsup \|\delta_n\| > \epsilon > 0$. . We will choose a sequence $n_1 < n_2 < \ldots$ and pairwise orthogonal projections

$P_1,P_2,\ldots$ in $\mathcal{A}$ so that $\|\delta_{n_k}(P_k)\| > \epsilon/8$. Assume $P_1,\ldots,P_k$ have been chosen, and let $Q_k = (\sum_{i=1}^{k} P_i)^{\perp}$. Furthermore, we also assume that $\limsup_{n\to\infty} \|\delta_n |Q_k\mathcal{A}\| > \epsilon$. Then choose $n_{k+1} > n_k$ large enough so that $\|\delta_{n_{k+1}} |Q_k\mathcal{A}\| > \epsilon$, and $\|\delta_{n_{k+1}}(Q_k)\| < \epsilon/8$. Since the convex combinations of projections in $Q_k\mathcal{A}$ are norm dense in the positive cone, there is a projection $P < Q_k$ so that $\|\delta_{n_{k+1}}(P)\| > \epsilon/4$. Thus $\|\delta_{n_{k+1}}(Q_k-P)\| > \epsilon/8$. By Lemma 19.17, either

$$\limsup \|\delta_n |PA\| > \epsilon \quad or \quad \limsup \|\delta_n |(Q_k-P)\mathcal{A}\| > \epsilon \ .$$

Take P_{k+1} to be the other projection, (Q_k-P) or P respectively. Then

$$\|\delta_{n_{k+1}}(P_{k+1})\| > \epsilon/8 \ \text{ and } \ \limsup_{n\to\infty} \|\delta_n |Q_{k+1}\mathcal{A}\| > \epsilon \ .$$

The sequences n_k and P_k are now produced inductively.

Now choose norm one linear functionals f_k on $B(\mathcal{H})$ so that $f_k(\delta_{n_k}(P_k)) > \epsilon/8$. Define ϕ_k in ℓ_∞^* by

$$\phi_k(\{x_n\}) = f_k(\delta_{n_k}(\sum_{n=1}^{\infty} x_n P_n)) \ .$$

Then $\|\phi_k\| \le \|f_k\| \, \|\delta_{n_k}\| \le 1$. For each x in ℓ_∞ and $A = \sum_{n=1}^{\infty} x_n P_n$,

$$\lim_{n\to\infty} \phi_k(x) = \lim_{n\to\infty} f_k(\delta_{n_k}(A)) = 0 \ .$$

So ϕ_k converge weak* to 0. By Phillip's Lemma 19.16, $\|\phi_k |c_0\|$ tends to 0. But $\phi_k(e_k) = f_k(\delta_{n_k}(P_k)) > \epsilon/8$. This contraction establishes the theorem. ∎

19.19 Lemma. *Let δ_n be a sequence of derivations on $B(\mathcal{H})$, where $\mathcal{H}$ is a Hilbert space of finite or infinite dimension, which converges pointwise to zero. Then* $\lim_{n\to\infty} \|\delta_n\| = 0$.

Proof. By Theorem 10.8, $\delta_n = \delta_{X_n}$ for X_n in $B(\mathcal{H})$. Fix a unit vector e_0 and add a scalar multiple of the identity to X_n so that $(X_n e_0, e_0) = 0$. For every vector x in $\mathcal{H}$,

$$[X_n, x \otimes e_0^*]e_0 = X_n x(e_0,e_0) - x(Xe_0,e_0) = X_n x \ .$$

This converges to zero. Likewise, $e_0^*[X_n, e_0 \otimes x^*] = -(X^*x)^*$ converges to zero. Thus X_n converges to 0 in the *strong topology. If $\mathcal{H}$ is finite dimensional, X_n converges to 0 in norm. So we will suppose that $\mathcal{H}$ is infinite dimensional.

Suppose that $\limsup \|X_n\| = \epsilon > 0$. One can choose a subsequence $n_1 < n_2 < \ldots$ and an orthonormal set $e_0, e_1, e_2, \ldots$ so that $\|X_{n_k} e_k\| > \epsilon/2$ and $\|X_{n_k} e_{k-1}\| < \epsilon/4$. To see this, suppose $e_0, \ldots, e_k$, $n_1, \ldots, n_k$ are chosen. Take N so large that

$$\|X_n \,| span\{e_0, \ldots, e_k\}\| < \epsilon/4 \text{ for } n \geq N \ .$$

Then choose $n_k \geq N$ so $\|N_{n_k}\| > 3\epsilon/4$ and note that $\|X_{n_k} \,| span\{e_0, \ldots, e_k\}^{\perp}\| > \epsilon/2$. Choose e_{k+1} appropriately, and proceed iteratively. Now set $A = \sum_{k=1}^{\infty} e_k \otimes e_{k-1}^*$. Then

$$\begin{aligned}\|\delta_{n_k}(A) e_{k-1}\| &= \|X_{n_k} A e_{k-1} - A X_{n_k} e_{k-1}\| \\ &\geq \|X_{n_k} e_k\| - \|X_{n_k} e_{k-1}\| > \epsilon/4 \ .\end{aligned}$$

This contradicts $\lim \|\delta_n(A)\| = 0$. So $\lim \|X_n\| = 0$ follows. ∎

19.20 Theorem. *Let $\mathcal{A}$ be a von Neumann algebra with abelian commutant. Let δ_n be a sequence of derivations into $B(\mathcal{H})$ converging pointwise to zero. Then* $\lim_{n \to \infty} \|\delta_n\| = 0$.

Proof. By Theorem 19.8, the restriction δ'_n of δ_n to $\mathcal{A}'$ converges to zero in norm. By Theorem 10.8, $\delta'_n = \delta_{X_n}$ where by Theorem 9.6, $dist(X_n, \mathcal{A}') \leq \|\delta'_n\|$. So it may be supposed that $\|X_n\|$ tends to zero. Replace δ_n by $\overline{\delta}_n = \delta_n - \delta_{X_n}$, and now one has a sequence of derivation converging pointwise to zero such that $\overline{\delta}_n \,| \mathcal{A}' = 0$. So if A belongs to $\mathcal{A}$ and B to $\mathcal{A}'$, then

$$B\overline{\delta}_n(A) = \overline{\delta}_n(BA) = \overline{\delta}_n(AB) = \overline{\delta}_n(A)B \ .$$

Thus $\overline{\delta}_n$ maps $\mathcal{A}$ into itself. If P and Q are mutually orthogonal projections in $\mathcal{A}'$ and $A = AP$ and $B = BQ$ are norm one elments of $\mathcal{A}$, then

$$\delta_n(A+B) = P\delta_n(A)P + Q\delta_n(B)Q \ .$$

So $\|\delta_n \,|(P+Q)\mathcal{A}\| = \max\{\|\delta_n \,| P\mathcal{A}\|, \|\delta_n \,| Q\mathcal{A}\|\}$.

If $\limsup \|\overline{\delta}_n\| > 0$, drop to a subsequence and normalize so that the sequence, say δ_n, satisfies $\|\delta_n\| = 1$. It still converges pointwise to 0, and vanishes on $\mathcal{A}'$. Let $\mathcal{E}$ be a maximal *chain* of projections in $\mathcal{A}'$ such that $\limsup \|\delta_n \,| E\mathcal{A}\| = 1$, and let $E_0 = \inf\{E : E \in \mathcal{E}\}$. If E_0 belongs to $\mathcal{E}$, then E_0 is not an atom of $\mathcal{A}'$ for then $E_0\mathcal{A} = B(E_0)$ contradicting Lemma 19.19. So $E_0 = E_1 + E_2$ where E_1 and E_2 are non-zero projections in $\mathcal{A}$. But then

$$\begin{aligned}1 &= \limsup_{n \to \infty} \|\delta_n \,| E_0\mathcal{A}\| \\ &= \limsup_{n \to \infty} \max_{i=1,2}\{\|\delta_n \,| E_i\mathcal{A}\|\} = \max_{i=1,2}\{\limsup_{n \to \infty} \|\delta_n \,| E_i\mathcal{A}\|\} \ .\end{aligned}$$

So either E_1 or E_2 belongs to $\mathcal{E}$, contradicting maximality. Thus E_0 is not in $\mathcal{E}$. For E in $\mathcal{E}$, $\|\delta_n\, |E\mathcal{A}\| = \max\{\|\delta_n\, |EE_0^\perp\mathcal{A}\|, \|\delta_n\, |E_0\mathcal{A}\|\}$. Since $\limsup\|\delta_n\, |E_0\mathcal{A}\| < 1$, it follows that $\limsup\|\delta_n\, |EE_0^\perp\mathcal{A}\| = 1$. Let $\mathcal{E}_0 = \{EE_0^\perp : E \in \mathcal{E}\}$ and note that $\inf\{E : E \in \mathcal{E}_0\} = 0$.

By Theorem 10.3, δ_n is weak* continuous. So if F belongs to $\mathcal{E}_0$, then $\sup_{E\in\mathcal{E}_0}\|\delta_n\, |(F-E)\mathcal{A}\| = \|\delta_n\, |F\mathcal{A}\|$. Choose a subsequence $n_1 < n_2 < \ldots$ and projections $E_0 > E_1 > \ldots$ in $\mathcal{E}_0$ so that $\|\delta_{n_i}\, |(E_{i-1}-E_i)\mathcal{A}\| > ½$. To achieve this, suppose $n_1,\ldots,n_k$ and $E_1,\ldots,E_k$ are defined. Choose $n_{k+1} > n_k$ so that $\|\delta_{n_{k+1}}\, |E_k\mathcal{A}\| > ½$. Then by the sentence above, one can choose E_{k+1} in $\mathcal{E}_0$ so that $\|\delta_{n_{k+1}}\, |(E_k-E_{k+1})\mathcal{A}\| > ½$. Let $A_k = A_k(E_{k-1}-E_k)$ be norm one elements of $\mathcal{A}$ so that $\|\delta_{n_k}(A_k)\| > ½$. Let $A = \sum_{k\geq 1} A_k$. Then A belongs to $\mathcal{A}$, and

$$\|\delta_{n_k}(A)\| \geq \|\delta_{n_k}(A)(E_{k-1}-E_k)\| = \|\delta_{n_k}(A_k)\| > ½$$

contradicting pointwise convergence. Hence $\lim_{n\to\infty}\|\delta_n\| = 0$ follows. ∎

19.21 Theorem. *Let $\mathcal{N}$ be a nest, and let δ_n be a sequence of derivations of $\mathcal{T}(\mathcal{N})$ into $\mathcal{B}(\mathcal{H})$ which converge pointwise to 0. Then* $\lim\|\delta_n\| = 0$.

Proof. The restriction δ'_n of δ_n to $\mathcal{N}'$ converge to 0 in norm by Theorem 19.20. By Theorems 10.8 and 9.6, $\delta'_n = \delta_{X_n}$ with $\|X_n\| \leq \|\delta'_n\|$. So $\overline{\delta}_n = \delta_n - \delta_{X_n}$ also converges to 0 pointwise, and $\overline{\delta}_n\, |\mathcal{N}' = 0$. By Proposition 19.11, $\overline{\delta}_n = \overline{\delta}_{Y_n}$ and thus Y_n belongs to $\mathcal{N}''$. If $\limsup\|\overline{\delta}_n\| > \epsilon > 0$, drop to a subsequence labelled $\delta_n = \delta_{Y_n}$ so that $\|\delta_n\| > \epsilon$ for all $n \geq 1$. So $dist\,(Y_n, \mathbb{C}I) > \epsilon/2$.

If $\mathcal{N}$ has atoms, fix one atom $A_0 = P(N_1)-P(N_0)$. Without loss of generality, it may be assumed that $Y_nA_0 = 0$. If e is a unit vector in N_1 and $e_0 = A_0e_0$, then $T = e \otimes e_0^*$ belongs to $\mathcal{T}(\mathcal{N})$. So

$$0 = \lim_{n\to\infty}\|\delta_n(T)e_n\| = \lim_{n\to\infty}\|Y_ne\| \ .$$

Similarly, one obtains $\lim_{n\to\infty}\|Y_ne\| = \lim_{n\to\infty}\|Y_n^*e\| = 0$ for e in $N_0^\perp$. So Y_n tends strongly to 0, and in particular $\lim_{n\to\infty}\|Y_nA\| = 0$ for every atom A. Let E be the sum of all atoms and suppose that $\limsup\|Y_nE\| > \eta > 0$. As in Lemma 19.19, find a subsequence $n_1 < n_2 < \ldots$ and distinct atoms A_k so that $\|Y_{n_k}A_k\| > \eta$ and $\|Y_{n_k}A_{k-1}\| < \eta/2$. Pick unit vectors e_k in $A_k\mathcal{H}$, and let $T_k = e_{2k-1} \otimes e_{2k}^*$ if $A_{2k-1} \ll A_{2k}$ or $T_k = e_{2k} \otimes e_{2k-1}^*$ if $A_{2k} \ll A_{2k-1}$. Set $T = \sum_{k=1}^{\infty} T_k$. A computation shows that

$$\|\delta_{n_{2k}}(T)\| \ge \|(A_{2k-1}+A_{2k})\delta_{n_{2k}}(T)\| = \|\delta_{n_{2k}}(T_k)\| > \eta/2 \ .$$

This contradiction establishes $\lim_{n\to\infty}\|Y_nE\| = 0$.

If $\mathcal{N}$ is not purely atomic yet has an atom, it follows that $Y_nE^\perp$ tends strongly to 0 but $\|Y_nE^\perp\| > \epsilon/2$. So it can and will be assumed that $Y_nE^\perp$ is not bounded below by $\epsilon/4$. On the other hand, if $\mathcal{N}$ is nonatomic, Y_n can be assumed to be non-invertible and $\|Y_nE^\perp\| > \epsilon/2$. For each n, choose non-zero projections E_n and F_n in $\mathcal{N}'E^\perp$ so that $\|Y_nE_n\| < \epsilon/4$ and Y_nF_n is bounded below by $\epsilon/2$ on F_n. One can choose N_n in $\mathcal{N}$ so that either $E_nP(N_n) \ne 0$ and $F_nP(N_n)^\perp \ne 0$ or $E_nP(N_n)^\perp \ne 0$ and $F_nP(N_n) \ne 0$. Without loss of generality, the first case occurs infinitely often and by dropping to a subsequence, it occurs always. Let μ be a continuous scalar valued measure absolutely continuous to $\mathcal{N}E^\perp$, and let $\mu(S)$ denote the μ-measure of the corresponding Borel set S. Choose non-zero projections E'_n and F'_n in $\mathcal{N}''$ so that (i) $E'_n \le E_nP(N_n)$, $F'_n \le F_nP(N_n)^\perp$, (ii) $\mu(E'_n) = \alpha_n$, $\mu(F'_n) = \beta_n$, and (iii) $\alpha_{n+1} < \alpha_n/3$ and $\beta_{n+1} < \beta_n/3$. Then $E''_n = E'_n(\sum_{k>n} E'_k)^\perp$ and $F''_n = F'_n(\sum_{k>n} F'_n)^\perp$ are non-zero, and $\{E''_n\}$ and $\{F''_n\}$ are pairwise orthogonal projections. Pick unit vectors $e_n = E''_ne_n$ and $f_n = F''_nf_n$. Then $\sum_{n=1}^{\infty} e_n \otimes f_n^*$ belongs to $\mathcal{T}(\mathcal{N})$. So

$$\begin{aligned}\|\delta_n(T)\| &\ge \|E''_n\delta_n(T)F''_n\| = \|Y_ne_n \otimes f_n^* - e_n \otimes (Y_n^*f_n)^*\| \\ &\ge \|Y_n^*f_n\| - \|Y_ne_n\| = \|Y_nf_n\| - \|Y_ne_n\| \\ &> \epsilon/2 - \epsilon/4 = \epsilon/4 \ .\end{aligned}$$

This contradiction implies that $\lim_{n\to\infty}\|Y_nE^\perp\| = 0$, and hence $\lim\|Y_n\| = 0$. ∎

Appendix: K_0 of a Nest Algebra.

K theory is a generalized homology theory for Banach algebras which extends the corresponding notions from algebraic topology. Its main significance has been its use as an isomorphism invariant to distinguish certain C^*-algebras. (E.g. AF C^*-algebras and irrational rotation algebras.) The interested reader can consult Taylor [1] for a nice introduction to K theory of Banach algebras, and Blackadar [1] for a full treatment of K theory for C^*-algebras and many references. The computation of K_0 of a nest algebra, which we present here, is due to Pitts [2].

Let $\mathcal{A}$ be a unital Banach algebra, and let $\mathcal{M}_n(\mathcal{A})$ denote the algebra of $n \times n$ matrices with coefficients in $\mathcal{A}$. Let $P_n(\mathcal{A})$ denote the set of all idempotents in $\mathcal{M}_n(\mathcal{A})$. For $n < m$, imbed $P_n(\mathcal{A})$ into $P_m(\mathcal{A})$ by sending an idempotent E to $E \oplus O_{m-n}$ where O_{m-n} is the zero matrix in $\mathcal{M}_{m-n}(\mathcal{A})$. An equivalence relation $\sim$ is put on $\bigcup_{n\ge 1} P_n(\mathcal{A})$ by setting

$E \sim F$ if there are integers r, s and n such that $E \oplus O_r$ and $F \oplus O_s$ are similar in $M_n(\mathcal{A})$. Let $S(\mathcal{A})$ denote the set of equivalence classes. The equivalence class of E will be denoted by $[E]$. An addition is defined on $S(\mathcal{A})$ by

$$[E]+[F] = [E \oplus F] \ .$$

This is easily seen to be well defined, and makes $S(\mathcal{A})$ into an abelian monoid with $[0]$ as a neutral element.

$K_0(\mathcal{A})$ is the abelian group obtained from $S(\mathcal{A})$ by introducing formal inverses and the cancellation law. Formally, this is done by considering all pairs (a,b), (c,d), etc. in $S(\mathcal{A}) \times S(\mathcal{A})$. Define $(a,b)+(c,d) = (a+c,b+d)$. Then introduce an equivalence relation $(a,b) \approx (c,d)$ if there is an element e in $S(\mathcal{A})$ such that $a+d+e = b+c+e$. It is routine to verify that the set of equivalence classes $\{a,b\}$ forms an abelian group, called $K_0(\mathcal{A})$. For example, $\{b,a\}$ is the inverse of $\{a,b\}$. If $S(\mathcal{A})$ satisfies cancellation, meaning that $a+e = b+e$ implies $a = b$, then the map taking a to $\{a,0\}$ is an injective homomorphism of $S(\mathcal{A})$ into $K_0(\mathcal{A})$. Otherwise, this does not hold.

If ϕ is a continuous homomorphism of a unital Banach algebra $\mathcal{A}$ into another unital algebra $\mathcal{B}$, then ϕ induces a natural homomorphism $\phi^{(n)}$ of $M_n(\mathcal{A})$ into $M_n(\mathcal{B})$. This determines a map of $\bigcup_{n \geq 1} P_n(\mathcal{A})$ into $\bigcup_{n \geq 1} P_n(\mathcal{B})$. If E and F are similar in $M_n(\mathcal{A})$ via S, then $\phi^{(n)}(E)$ is similar to $\phi^{(n)}(F)$ via $\phi^{(n)}(S)$. Therefore, ϕ induces a homomorphism of $S(\mathcal{A})$ into $S(\mathcal{B})$. This uniquely determines a homomorphism ϕ_* of $K_0(\mathcal{A})$ into $K_0(\mathcal{B})$. Suppose ψ is a homomorphism of $\mathcal{B}$ into $\mathcal{C}$. It is routine to verify that $\psi_*\phi_* = (\psi\phi)_*$. In particular, if ϕ is an isomorphism then $(\phi^{-1})_* = (\phi_*)^{-1}$; so ϕ_* is a isomorphism.

The first lemma shows that algebras which are "infinite" have trivial K-theory because of cancellation.

19.22 Lemma. *Let $\mathcal{A}$ be a unital, weak* closed subalgebra of $\mathcal{B}(\mathcal{H})$. Then $K_0(\mathcal{A} \otimes \mathcal{B}(\mathcal{H})) = 0$.*

Proof. First note that $M_n(\mathcal{A} \otimes \mathcal{B}(\mathcal{H})) \simeq \mathcal{A} \otimes \mathcal{B}(\mathcal{H})$. So it suffices to show that every idempotent E in $\mathcal{A} \otimes \mathcal{B}(\mathcal{H})$ has $[E] = [0]$. Now $\mathcal{A} \otimes \mathcal{B}(\mathcal{H}) \simeq \mathcal{A} \otimes \mathcal{B}(\mathcal{H}) \otimes \mathcal{B}(\mathcal{H})$, and the latter algebra can be thought of as infinite matrices with coefficients in $\mathcal{A} \otimes \mathcal{B}(\mathcal{H})$. In particular, the diagonal matrix $E^{(\infty)}$ belongs to this algebra. We exhibit it in $\mathcal{A} \otimes \mathcal{B}(\mathcal{H})$ explicitly as follows. Let W_n, $n \geq 1$, be a sequence of isometries in $\mathcal{B}(\mathcal{H})$ with pairwise orthogonal ranges such that $\sum_{n \geq 1} \tilde{W}_n \tilde{W}_n^* = I$ in the strong operator topology. Then $W_n = I \otimes \tilde{W}_n$ are isometries in $\mathcal{A} \otimes \mathcal{B}(\mathcal{H})$ with the same property. Let

$$F = \sum_{n \geq 1} W_n E W_n^* \ .$$

This converges SOT and thus is an idempotent in $\mathcal{A} \otimes \mathcal{B}(\mathcal{H})$ unitarily equivalent to $E^{(\infty)}$.

Let S denote the shift operator taking the n-th "copy" of E to the $(n+1)$st for all $n \geq 1$. Precisely, $S = \sum_{n \geq 1} W_{n+1} W_n^*$. Note that

$$F - SFS^* = \sum_{n \geq 1} W_n E W_n^* - \sum_{n \geq 2} W_n E W_n^* = W_1 E W_1^* \ .$$

Also let $\tilde{V}_1$ and $\tilde{V}_2$ be isometries of $\mathcal{B}(\mathcal{H})$ with $\tilde{V}_1\tilde{V}_1^* + \tilde{V}_2\tilde{V}_2^* = I$. Set $V_i = I \otimes \tilde{V}_i$, which are isometries in $\mathcal{A} \otimes \mathcal{B}(\mathcal{H})$ with the same property. Note that $\begin{bmatrix} 0 & 0 & V_1^* \\ W_1 & S & 0 \\ 0 & 0 & V_2^* \end{bmatrix}$ is a unitary operator in $M_3(\mathcal{A} \otimes \mathcal{B}(\mathcal{H}))$. Moreover,

$$\begin{bmatrix} 0 & 0 & V_1^* \\ W_1 & S & 0 \\ 0 & 0 & V_2^* \end{bmatrix} \begin{bmatrix} E & 0 & 0 \\ 0 & F & 0 \\ 0 & 0 & 0 \end{bmatrix} \begin{bmatrix} 0 & W_1^* & 0 \\ 0 & S^* & 0 \\ V_1 & 0 & V_2 \end{bmatrix} = \begin{bmatrix} 0 & 0 & 0 \\ 0 & W_1EW_1^* + SFS^* & 0 \\ 0 & 0 & 0 \end{bmatrix}$$

$$= \begin{bmatrix} 0 & 0 & 0 \\ 0 & F & 0 \\ 0 & 0 & 0 \end{bmatrix} .$$

It follows that $[E]+[F]+[0] = [F]+[0]$. Hence $[E] = [0]$. But E is arbitrary, so $K_0(\mathcal{A} \otimes \mathcal{B}(\mathcal{H})) = 0$. ■

19.23 Corollary. *Let $\mathcal{N}$ be a continuous nest. Then $K_0(\mathcal{T}(\mathcal{N})) = 0$.*

Proof. By Theorem 13.10, every continuous nest is similar to a nest of uniform infinite multiplicity (see Example 13.11). Let $\mathcal{N}_0 = \mathcal{N} \otimes I$ be the infinite ampliation of $\mathcal{N}$, and let S be an invertible operator implementing a similarity. Then $\mathcal{T}(\mathcal{N})$ is similar to $\mathcal{T}(\mathcal{N}_0) \simeq \mathcal{T}(\mathcal{N}) \otimes \mathcal{B}(\mathcal{H})$. Thus by Lemma 19.22, $K_0(\mathcal{T}(\mathcal{N})) = 0$. ■

Let $\mathcal{N}_0$ be a nest, and let Δ be the unique expectation onto the atomic part $\mathcal{D}_a(\mathcal{N}_0)$ of the diagonal. Let S be any invertible operator, let $\mathcal{N}_1 = S\mathcal{N}_0$, and let Δ_1 be the corresponding expectation onto $\mathcal{D}_a(\mathcal{N}_1)$. Let $P_0 = \Delta(I)$ and $P_1 = \Delta_1(I)$. Since θ_S preserves dimension, one can choose a partial isometry W which carries each atom A onto $\theta_S(A)$ such that $W^*W = P_0$ and $WW^* = P_1$.

19.24 Lemma. *Let $\mathcal{N}_0$, $\mathcal{N}_1$, S and W be given above. Let E be any idempotent in $\mathcal{T}(\mathcal{N})$ such that $E = EP_0 = P_0E$. Then $[E] = [S^{-1}WEW^*S]$.*

Proof. Note that W^*S and $S^{-1}W$ belong to $\mathcal{T}(\mathcal{N})$. A simple computation shows that

$$\begin{bmatrix} S^{-1}W & S^{-1}(I-P_1)S \\ P_0-I & W^*S \end{bmatrix} \quad \text{and} \quad \begin{bmatrix} W^*S & P_0-I \\ S^{-1}(I-P_1)S & S^{-1}W \end{bmatrix}$$

are inverses in $\mathcal{M}_2(\mathcal{T}(\mathcal{N}))$. Furthermore,

$$\begin{bmatrix} S^{-1}W & S^{-1}(I-P_1)S \\ P_0-I & W^*S \end{bmatrix} \begin{bmatrix} E & 0 \\ 0 & 0 \end{bmatrix} \begin{bmatrix} W^*S & P_0-I \\ S^{-1}(I-P_1)S & S^{-1}W \end{bmatrix} = \begin{bmatrix} S^{-1}WEW^*S & 0 \\ 0 & 0 \end{bmatrix} .$$

Thus $[E] = [S^{-1}WEW^*S]$ as claimed. ∎

19.25 Theorem. *Let $\mathcal{N}$ be a nest on a separable Hilbert space. Let $\mathcal{D}_f(\mathcal{N})$ denote the atomic part of the diagonal algebra corresponding to finite rank atoms; and let Δ_f be the unique expectation onto $\mathcal{D}_f(\mathcal{N})$. Then Δ_{f*} is an isomorphism of $K_0(\mathcal{T}(\mathcal{N}))$ onto $K_0(\mathcal{D}_f(\mathcal{N}))$.*

Proof. Let i be the injection of $\mathcal{D}_f(\mathcal{N})$. Then $\Delta_f \circ i = id$, and hence

$$\Delta_{f*} \circ i_* = id_{K_0(\mathcal{D}_f(\mathcal{N}))} \ .$$

It will be shown that $i_* \circ \Delta_{f*} = id_{K_0(\mathcal{T}(\mathcal{N}))}$. To do this, it suffices to show that $[E] = [\Delta_f(E)]$ in $K_0(\mathcal{T}(\mathcal{N}))$ for all idempotents E in $\mathcal{M}_n(\mathcal{T}(\mathcal{N}))$. Since $\mathcal{M}_n(\mathcal{T}(\mathcal{N})) \simeq \mathcal{T}(\mathcal{N}^{(n)})$ and $\Delta_f^{(n)}$ is the diagonal expectation in $\mathcal{T}(\mathcal{N}^{(n)})$, it suffices to verify this for $n = 1$.

Let E be any idempotent in $\mathcal{T}(\mathcal{N})$. Let X be any invertible operator such that $SES^{-1} = F = F^*$ is a projection. For example,

$$S = (E^*E+(I-E)^*(I-E))^{1/2}$$

will suffice. Let $\mathcal{N}_1 = S\mathcal{N}$, and let Δ and Δ_1 be the expectations onto the atomic part of the diagonal (including infinite rank atoms in $\mathcal{N}$ and $\mathcal{N}_1$ respectively). Since $Ad\,S$ is an isomorphism of $\mathcal{T}(\mathcal{N})$ onto $\mathcal{T}(\mathcal{N}_1)$, $\sigma_S = (Ad\,S)_*$ is an isomorphism on the K_0 groups. The projection F is self-adjoint and thus $F = \Delta_1(F)+(F-\Delta_1(F))$ splits into its atomic and nonatomic part. The restriction $\mathcal{N}_1^c$ of $\mathcal{N}$ to $P_1^\perp\mathcal{H}$ is a continuous nest, and $F-\Delta_1(F) = P_1^\perp F$ identified with the restriction F_c of F to $P_1^\perp\mathcal{H}$. By Corollary 19.23, $K_0(\mathcal{T}(\mathcal{N}_1^c)) = 0$ and thus $[F_c] = 0$ in $K_0(\mathcal{T}(\mathcal{N}_1^c))$. Let j be the injection of $\mathcal{T}(\mathcal{N}_1^c)$ onto $\mathcal{T}(\mathcal{N}_1)$. Then it is easy to see that $0 = j_*[F_c] = [P_1^\perp F]$ in $K_0(\mathcal{T}(\mathcal{N}_1))$. Hence

$$\sigma_S([E]) = [F] = [\Delta_1(F)] \ .$$

Let W be a partial isometry as in Lemma 19.23 which carries the atoms of $\mathcal{N}$ onto the atoms of $\mathcal{N}_1$. By that lemma,

$$\sigma_S[\Delta(E)] = \sigma_S[S^{-1}W\Delta(E)W^*S] = [W\Delta(E)W^*] \ .$$

The idempotent $W\Delta(E)W^*$ belongs to $\mathcal{D}_a(\mathcal{N}_1)$. Moreover, it is similar in $\mathcal{D}_a(\mathcal{N}_1)$ to F. To see this, let $T = \Delta_1(SW^*)$ and note that $\Delta_1(WS^{-1})$ is its inverse in $\mathcal{D}_a(\mathcal{N}_1)$. Then

$$\begin{aligned} TW\Delta(E)W^*T^{-1} &= \Delta_1(SW^*)\Delta_1(W\Delta(E)W^*)\Delta_1(WS^{-1}) \\ &= \Delta_1(S\Delta(E)S^{-1}) = \Delta_1(SES^{-1}) = \Delta_1(F) \ . \end{aligned}$$

To verify the last line, check it on each atom. The computation is identical to the proof of Lemma 17.16.

Consequently, $\sigma_S([\Delta(E)]) = \sigma_S([E])$. Since σ_S is an isomorphism, $[\Delta(E)] = [E]$. The part of $\mathcal{D}_a$ corresponding to infinite atoms has trivial K_0 by Lemma 19.22. So it follows that $[E] = [\Delta_f(E)]$. This completes the proof. ■

19.26 Corollary. *Let $\mathcal{N}$ be a nest on a separable space. Let $\mathbf{A}_f$ denote the set of finite rank atoms of $\mathcal{N}$. Then $K_0(\mathcal{T}(\mathcal{N}))$ is isomorphic to*

$$\{g:\mathbf{A}_f \to \mathbb{Z} : \sup_{A \in \mathbf{A}_f} \frac{|g(A)|}{dim\, A} < \infty\}$$

if $\mathbf{A}_f$ is nonempty, and 0 otherwise.

The proof is left as Exercise 19.7. Note that if there are n finite atoms, then $K_0(\mathcal{T}(\mathcal{N})) \simeq \mathbb{Z}^n$. Also if there are countably many atoms and the dimensions of the atoms are bounded above, then $K_0(\mathcal{T}(\mathcal{N}))$ is all bounded sequences of integers. In general, the sequence is bounded by $(C\, dim\, A : A \in \mathbf{A}_f)$ for some constant C.

Notes and Remarks.

Theorem 19.1 and Corollary 19.2 are due to Christensen [2], as is Lemma 19.4 and Theorem 19.7. Theorem 19.10 is due to Wagner [2]. Proposition 19.11 is again due to Christensen. Theorem 19.12 is in Christensen and Peligrad [1]. Theorem 19.14 is a combination of Peligrad [1] and a special case of a theorem of Akemann and Wright [1]. Theorem 19.15 is also due to Peligrad. Lemma 19.16 is due to R. Phillips [1]. Theorems 19.18 and 19.20 are due to Akemann and Johnson [1]. Theorem 19.21 is due to Johansen [1].

We now present a brief outline of cohomology. There is an extensive literature about cohomology of Banach algebras and C^*-algebras. In particular, see Johnson [2] and Ringrose [7].

Let $\mathcal{A}$ be a Banach algebra and let $\mathcal{M}$ be a Banach $\mathcal{A}$ bimodule. The space $C^n(\mathcal{A},\mathcal{M})$ of *n-cochains* is the space of all bounded n-linear maps of $\mathcal{A}^{(n)}$ into $\mathcal{M}$. By convention, $C^0(\mathcal{A},\mathcal{M}) = \mathcal{M}$. A *coboundary operator* ∂_n is defined from $C^n(\mathcal{A},\mathcal{M})$ into $C^{n+1}(\mathcal{A},\mathcal{M})$ by

$$\begin{aligned}(\partial_n\phi)(A_0,\dots,A_n) = {} & A_0\phi(A_1,\dots,A_n) \\ & + \sum_{j=1}^{n}(-1)^j\phi(A_0,\dots,A_{j-2},A_{j-1}A_j,A_{j+1},\dots,A_n) \\ & + (-1)^{n+1}\phi(A_0,\dots,A_{n-1})A_n \quad .\end{aligned}$$

The ∂_0 map on $C^0(\mathcal{A},\mathcal{M})$ is given for M in $\mathcal{M}$ by the inner derivation $\partial M(A) = AM-MA$. By convention, we use ∂ instead of ∂_n since n is usually clear in context. A routine but tedious computation shows that $\partial^2 = 0$. For $n \geq 1$, the space $Z^n(\mathcal{A},\mathcal{M})$ of *n-cocyles* is the kernel of ∂_n. The space $B^n(\mathcal{A},\mathcal{M})$ of *n-coboundaries* is the image of ∂_{n-1}. Since $\partial^2 = 0$, it follows that $B^n(\mathcal{A},\mathcal{M})$ is a subspace of $Z^n(\mathcal{A},\mathcal{M})$. The *nth cohomology space of $\mathcal{A}$ with coefficients in $\mathcal{M}$* is

$$H^n(\mathcal{A},\mathcal{M}) = Z^n(\mathcal{A},\mathcal{M})/B^n(\mathcal{A},\mathcal{M}) \quad .$$

When $\mathcal{A}$ is a dual space, one naturally considers dual normal bimodules $\mathcal{M}$ and restricts attention to bounded multilinear maps which are also weak*-weak* continuous. In this way, one obtains the cohomology spaces $H^n_w(\mathcal{A},\mathcal{M})$. An algebra $\mathcal{A}$ is called *amenable* if $H^n(\mathcal{A},\mathcal{M}) = 0$ for every dual normal bimodule $\mathcal{M}$.

Consider $n = 1$. If ϕ belongs to $Z^1(\mathcal{A},\mathcal{M})$, then ϕ is a linear map of $\mathcal{A}$ into $\mathcal{M}$ such that

$$0 = \partial\phi(A,B) = A\phi(B)-\phi(AB)+\phi(A)B \quad .$$

This is precisely the condition that ϕ be a derivation. The coboundaries are precisely the inner derivations. So $H^1(\mathcal{A},\mathcal{M}) = 0$ is a reformulation of the statement that every derivation of $\mathcal{A}$ into $\mathcal{M}$ is inner.

We now state some of the important results for C^*-algebras and von Neumann algebras. The first three are taken from Johnson, Kadison and Ringrose [1].

Theorem. *Let $\mathcal{A}$ be a C^*-algebra, and let $\mathcal{M}$ be an $\mathcal{A}''$ dual normal bimodule. Then*

$$H^n(\mathcal{A},\mathcal{M}) = H^n_w(\mathcal{A},\mathcal{M}) = H^n(\mathcal{A}'',\mathcal{M}) = H^n_w(\mathcal{A}'',\mathcal{M}) \quad .$$

Theorem. *Every AF von Neumann algebra $\mathcal{A}$ is amenable.* (*i.e.* $H^n(\mathcal{A},\mathcal{M}) = 0$ *for all* $n \geq 1$ *and every dual normal bimodule* $\mathcal{M}$.)

Corollary. *Every type I von Neumann algebra is amenable.*

The celebrated derivation theorem of Kadison [1] and Sakai [2] can be reformulated as:

Theorem. *For every von Neumann algebra $\mathcal{A}$, $H^1(\mathcal{A},\mathcal{A}) = 0$. (i.e. Every derivation of $\mathcal{A}$ into itself is inner.)*

The main result for nest algebras is due to Lance [2].

Theorem. *Let $\mathcal{N}$ be a nest, and let $\mathcal{M}$ be a weak* closed unital subspace of $\mathcal{B}(\mathcal{H})$ which is a $\mathcal{T}(\mathcal{N})$ bimodule. Then $H^n(\mathcal{T}(\mathcal{N}),\mathcal{M}) = 0$ for all $n \geq 1$.*

This theorem was used to prove the perturbation result Corollary 18.5. The idea was to prove it first for finite subnests. Then if $\mathcal{N}$ and $\mathcal{M}$ are close nests, a limit procedure produces an invertible linear map of $\mathcal{T}(\mathcal{N})$ onto $\mathcal{T}(\mathcal{M})$ close to the identity map. The trivial cohomology

$$
\begin{aligned}
0 &= H^1(\mathcal{T}(\mathcal{N}),\mathcal{B}(\mathcal{H})) = H^2(\mathcal{T}(\mathcal{N}),\mathcal{B}(\mathcal{H})) \\
&= H^2(\mathcal{T}(\mathcal{N}),\mathcal{T}(\mathcal{N})) = H^3(\mathcal{T}(\mathcal{N}),\mathcal{T}(\mathcal{N}))
\end{aligned}
$$

is required to apply two perturbation results of Raeburn and Taylor [1] to modify this map to produce an invertible S near I so that $Ad\,S$ takes $\mathcal{T}(\mathcal{N})$ onto $\mathcal{T}(\mathcal{M})$.

Exercises.

19.1 Let α_t, $t \geq 0$ be a norm continuous semigroup of automorphisms of a Banach algebra $\mathcal{A}$. For $t > 0$, define elements of $\mathcal{B}(\mathcal{A})$ by $\delta_t = t^{-1}(\alpha_t - \alpha_0)$ and $\mu_t = t^{-1}\int_0^t \alpha_u\,du$.

(a) Prove that $\delta_t \mu_s = \mu_t \delta_s$ for $s,t > 0$. Hence deduce that $\delta = \lim_{t \to 0} \delta_t$ exists.

(b) Prove that δ is a derivation, and $\alpha_t = e^{t\delta}$ for $t \geq 0$.

19.2 (Christensen) A weak* closed algebra $\mathcal{A}$ has the *automorphism implementation* property $AIP(r,s)$ for $r,s > 0$ if every automorphism α in $Aut\,\mathcal{A}$ such that $\|\alpha - id\| \leq r$ is of the form $\alpha = Ad\,A$ for some A in $\mathcal{A}$ with $\|A - I\| \leq s\|\alpha - id\|$. Show that every derivation of such an algebra is inner.

19.3 An alternate proof of Phillip's lemma: Let P be the projection of ℓ_∞^* onto ℓ_1 given by $P\phi = \phi\,|c_0$. Show that if ϕ_n is a sequence in ℓ_∞^* converging weak* to 0, then $P\phi_n$ converges weakly to 0 in ℓ_1. Hence deduce that $\lim_{n \to \infty} \|P\phi_n\| = 0$.

19.4 Extend Theorem 19.20 to all type I von Neumann algebras.

19.5 Show that the only compact derivation on $\mathcal{B}(\mathcal{H})$ is zero.

19.6 (a) Let $\mathcal{A}_4$ and $\mathcal{A}_6$ be the algebras of Exercise 17.5. Show that these algebras have derivations which are not inner. Compute $H^1(\mathcal{A},\mathcal{A})$.

(b) Let $\mathcal{A}_\infty$ be the algebra of Exercise 17.5 (c). Show that δ_T defines a continuous derivation which is not inner.

19.7 Consider the von Neumann algebra $\mathcal{D} = \sum_{k\geq 1} \oplus \mathcal{M}_{n_k}$.

(a) Show that if E is an idempotent in $\mathcal{M}_m(\mathcal{D})$, then $E = \sum_{k\geq 1} \oplus E_k$ where E_k is an idempotent in $\mathcal{M}_m(\mathcal{M}_{n_k}) \simeq \mathcal{M}_{mn_k}$. Hence $n_k^{-1} rank\, E_k$ is bounded.

(b) Let F be another idempotent in $\mathcal{M}_m(\mathcal{D})$ such that $rank\, F_k = rank\, E_k$ for all k. Show that F is similar to E in $\mathcal{M}_m(\mathcal{D})$.

(c) Hence deduce that $\mathcal{S}(\mathcal{D})$ is isomorphic to $\{g : \mathbb{N} \to \mathbb{N}_0$ such that $\sup n_k^{-1} g(k) < \infty\}$.

(d) Compute $K_0(\mathcal{D})$. Hence prove Corollary 19.27.

20. Representation and Dilation Theory

The purpose of this chapter is to analyze the structure of contractive representations of nest algebras. This generalizes the classical dilation theory for contractions of Sz. Nagy-Foias and Ando. For motivation, we develop this briefly first. The key tool is the notion of completely positive maps and the dilation theory due to Arveson. This will be developed in the second part of this chapter. Then we develop the nest algebra theory.

20.A Dilation Theory for Single Operators.

Given an operator A on a Hilbert space $\mathcal{H}$, we say that an operator T on a Hilbert space $\mathcal{K}$ containing $\mathcal{H}$ is a *dilation* of A provided $A^n = P(\mathcal{H})T^n \,|\mathcal{H}$ for all $n \geq 0$. When A is a contraction, we consider, in particular, unitary dilations of A.

20.1 Theorem. *Let A be a contraction on a Hilbert space $\mathcal{H}$. Then A has a unitary dilation U on a Hilbert space $\mathcal{K}$. Moreover, if U is minimal in the sense that $\mathcal{K} = \bigvee_{n=-\infty}^{\infty} U^n \mathcal{H}$, then U is unique (up to unitary equivalence fixing $\mathcal{H}$).*

Proof. For any contraction T, let $D_T = (I-T^*T)^{1/2}$. Form $\mathcal{K} = \sum_{n=-\infty}^{\infty} \oplus \mathcal{H}_n$ where each $\mathcal{H}_n = \mathcal{H}$. Define a unitary from $\mathcal{H}_{-1} \oplus \mathcal{H}_0$ to $\mathcal{H}_0 \oplus \mathcal{H}_1$ by $\begin{bmatrix} D_{A^*} & A \\ -A^* & D_A \end{bmatrix}$. Extend this to a unitary U on $\mathcal{K}$ by setting U to be the shift of $\mathcal{H}_n$ onto $\mathcal{H}_{n+1}$ for $n \neq 0,-1$ given by

$$U \sum_{n \neq -1,0} \oplus h_n = \sum_{n \neq 0,1} \oplus h_{n-1} \ .$$

It is clear that this is unitary. Moreover, the matrix of U is lower triangular with respect to the decomposition of $\mathcal{K}$ with A as its (0,0) entry. Hence the (0,0) entry of U^n is A^n. That is,

$$P(\mathcal{H}_0)U^n \,|\mathcal{H}_0 = A^n \quad \text{for all } n \geq 0 \ .$$

This may not be minimal, but $\mathcal{K}_0 = \bigvee_{n=-\infty}^{\infty} U^n \mathcal{H}_0$ is the smallest reducing subspace for U containing $\mathcal{H}_0$. The restriction $U_0 = U\,|\mathcal{K}_0$ is a minimal unitary dilation. Suppose that U_1 is another minimal unitary dilation on

K_1. For $n \geq m$ and x, y in $\mathcal{H}$,

$$(U_0^n x, U_0^m y) = (U_0^{n-m} x, y) = (T^{n-m} x, y) = (U_1^{n-m} x, y) = (U_1^n x, U_1^m y) \ .$$

By taking conjugates, this follows for $n < m$ as well. Hence

$$W\Big(\sum_{n=-N}^{N} U_0^n x_n\Big) = \sum_{n=-N}^{N} U_1^n x_n$$

is isometric from a dense subspace of K_0 to a dense subspace of K_1, and thus extends to a unitary (also denoted by W). Clearly, $W \mid \mathcal{H} = I$ and $WU_0 = U_1 W$ which establishes the desired equivalence. ■

20.2 Corollary (von Neumann's Inequality). *Let A be a contraction on a Hilbert space, and let p be a polynomial. Then*

$$\|p(A)\| \leq \|p\|_\infty = \sup_{|z| \leq 1} |p(z)| \ .$$

Thus the contractive representations of the disc algebra are parametrized by $b_1(B(\mathcal{H}))$, the unit ball of $B(\mathcal{H})$.

Proof. Let U be a unitary dilation of A. Then

$$\|p(A)\| = \|P(\mathcal{H}) p(U) \mid \mathcal{H}\| \leq \|p(U)\| \leq \|p\|_\infty \ .$$

If ϕ is a contractive representation of the disc algebra, then $A = \phi(z)$ is a contraction. The inequality shows that every A in the unit ball of $B(\mathcal{H})$ leads to a contractive representation extending the map $\pi(p) = p(A)$ for polynomials. ■

The disc algebra $A(\mathbf{D})$ is a unital subalgebra of $C(\mathbf{T})$, the continuous functions on the unit circle. A $C(\mathbf{T})$ dilation of a representation ρ of $A(\mathbf{D})$ on $\mathcal{H}$ is a $*$ representation π of $C(\mathbf{T})$ on K and an isometry V imbedding $\mathcal{H}$ in K such that

$$\rho(f) = V^* \pi(f) V \quad \text{for all } f \text{ in } A(\mathbf{D}) \ .$$

20.3 Corollary. *Every contractive representation ρ of $A(\mathbf{D})$ has a $C(\mathbf{T})$ dilation.*

Proof. Let $A = \rho(z)$ and let U be a unitary dilation of A on K. Define $\pi(f) = f(U)$ for f in $C(\mathbf{T})$ by the functional calculus. Then the properties of the dilation imply that

$$\rho(p) = P(\mathcal{H}) p(U) \mid \mathcal{H} = V^* \pi(p) V \quad \text{for every polynomial } p \ ,$$

where V is the natural imbedding of $\mathcal{H}$ into K. By continuity, π is a $C(\mathbf{T})$ dilation of ρ. ■

Now, consider two commuting contractions, or equivalently, two commuting representations of the disc algebra.

20.4 Ando's Theorem. *Let A_1 and A_2 be commuting contractions. Then they have commuting unitary dilations on a Hilbert space $\mathcal{K}$ such that*

$$P(\mathcal{H})U_1^nU_2^m \,|\mathcal{H} = A_1^nA_2^m \quad \text{for all } n,m \geq 0 \ .$$

Proof. The main step is to dilate A_1 and A_2 to commuting isometries. Let $\mathcal{K} = \sum_{n\geq 0} \oplus \mathcal{H}_n$ where $\mathcal{H}_n = \mathcal{H}$ and $\mathcal{H}$ is identified with $\mathcal{H}_0$. Let $D_i = D_{A_i}$, and define isometries V_i on $\mathcal{K}$ by

$$V_i \sum_{n=0}^{\infty} \oplus h_n = A_ih_0 \oplus D_ih_0 \oplus 0 \oplus \sum_{n\geq 3} \oplus h_{n-2} \ .$$

Note that, although V_1 and V_2 do not commute,

$$P(\mathcal{H}_0)V_1^nV_2^m \,|\mathcal{H}_0 = A_1^nA_2^m \quad \text{for all } n,m \geq 0 \ .$$

Also

$$V_1V_2\sum_{n=0}^{\infty} \oplus h_n = A_1A_2h_0 \oplus D_1A_2h_0 \oplus 0 \oplus D_2h_0 \oplus 0 \oplus \sum_{n\geq 5} \oplus h_{n-4}$$

and

$$V_2V_1\sum_{n=0}^{\infty} \oplus h_n = A_1A_2h_0 \oplus D_2A_1h_0 \oplus 0 \oplus D_1h_0 \oplus 0 \oplus \sum_{n\geq 5} \oplus h_{n-4} \ .$$

However,

$$\begin{aligned} \|D_1A_2h\|^2+\|D_2h\|^2 &= \|A_2h\|^2-\|A_1A_2h\|^2+\|D_2h\|^2 \\ &= \|h\|^2-\|A_1A_2h\|^2 \\ &= \|A_1h\|^2-\|A_2A_1h\|^2+\|D_1h\|^2 \\ &= \|D_2A_1h\|^2+\|D_1h\|^2 \ . \end{aligned}$$

Hence there is a unitary W of $\mathcal{H}^{(4)}$ onto itself such that

$$W(D_1A_2h \oplus 0 \oplus D_2h \oplus 0) = D_2A_1h \oplus 0 \oplus D_1h \oplus 0$$

for all h in $\mathcal{H}$. Let $\tilde{W}$ act on $\mathcal{K}$ by

$$\tilde{W}\sum_{n\geq 0} \oplus h_n = h_0 \oplus \sum_{k\geq 0} W(h_{4k+1} \oplus h_{4k+2} \oplus h_{4k+3} \oplus h_{4k+4}) \ .$$

Then $\tilde{V}_1 = \tilde{W}V_1$ and $\tilde{V}_2 = V_2\tilde{W}^*$ are isometries. Now $\tilde{V}_2\tilde{V}_1 = V_2V_1$ and

$$\tilde{V}_1\tilde{V}_2\sum_{n=0}^{\infty} \oplus h_n = \tilde{W}V_1V_2\tilde{W}^*\sum_{n\geq 0} \oplus h_n$$

$$= \tilde{W}(A_1A_2h_0 \oplus (D_1A_2h_0 \oplus 0 \oplus D_2h_0 \oplus 0)$$

$$\oplus \sum_{k\geq 1} W^*(h_{4k-3} \oplus h_{4k-2} \oplus h_{4k-1} \oplus h_{4k}))$$

$$= A_1A_2h_0 \oplus W(D_1A_2h_0 \oplus 0 \oplus D_2h_0 \oplus 0) \oplus \sum_{n\geq 5} \oplus h_{n-4}$$

$$= \tilde{V}_2\tilde{V}_1\sum_{n\geq 0} \oplus h_n \quad .$$

So $\tilde{V}_1$ and $\tilde{V}_2$ are the desired isometric dilations of A_1 and A_2. In particular, they have the form $\tilde{V}_i = \begin{bmatrix} A_i & 0 \\ * & * \end{bmatrix}$ on $\mathcal{H} \oplus (\mathcal{K} \ominus \mathcal{H})$.

Define

$$T_1 = \begin{bmatrix} \tilde{V}_1^* & 0 \\ 1-\tilde{V}_1\tilde{V}_1^* & \tilde{V}_1 \end{bmatrix} = \begin{bmatrix} * & 0 & 0 \\ * & A_1 & 0 \\ * & * & * \end{bmatrix}$$

on $\mathcal{K} \oplus \mathcal{K} \cong \mathcal{K} \oplus \mathcal{H} \oplus (\mathcal{K} \ominus \mathcal{H})$. This is unitary and dilates A_1. Extend $\tilde{V}_2$ to an isometry T_2 as follows. Set

$$T_2\sum_{n=0}^{N} T_1^{-n}k_n = \sum_{n=0}^{N} T_1^{-n}(\tilde{V}_2k_n) \quad .$$

This is well defined since $\sum_{n=0}^{N} T_1^{-n}k_n = 0$ implies

$$\sum_{n=0}^{N} T_1^{-n}(\tilde{V}_2k_n) = T_1^{-N}\sum_{n=0}^{N} \tilde{V}_1^{N-n}\tilde{V}_2k_n = T_1^{-N}\tilde{V}_2T_1^{N}\sum_{n=0}^{N} T_1^{-n}k_n = 0 \quad .$$

Moreover, $\sum_{n=0}^{N} T_1^{-n}k_n = T_1^{-N}\sum_{n=0}^{N} \tilde{V}_1^{N-n}k_n$ has the same norm as $T_1^{-N}\tilde{V}_2\sum_{n=0}^{N} \tilde{V}_1^{N-n}k_n$, and thus T_2 is isometric on this set. Extend T_2 by continuity to an isometry on $\bigvee_{n\geq 0} T_1^{-n}\mathcal{K} = \mathcal{K}_0 \oplus \mathcal{K}$. Define $T_2 \,|\, \mathcal{K}_0^{\perp} \oplus 0$ to be the identity. Then T_2 is an isometry on $\mathcal{K} \oplus \mathcal{K}$. Since $\mathcal{K}_0^{\perp} \oplus 0$ is invariant for T_1 and T_2, it is clear that they commute.

Now dilate T_1 to $U_1 = T_1^{(\infty)}$ on $(\mathcal{K} \oplus \mathcal{K})^{(\infty)}$, and dilate T_2 to U_2 on $(\mathcal{K} \oplus \mathcal{K})^{(\infty)}$ as in Theorem 20.1:

$$U_2\sum_{n=-\infty}^{\infty} \oplus x_n = \sum_{n<0} \oplus x_{n-1} \oplus (D_{T_2^*}x_{-1}+T_2x_0)$$

$$\oplus (D_{T_2}x_0-T_2^*x_{-1}) \oplus \sum_{n>1} \oplus x_{n-1} \quad .$$

By Fugelde's Theorem, T_1 commutes with T_2^*; and hence U_1 commutes with U_2. By construction, U_1 and U_2 have the form $\begin{bmatrix} * & 0 & 0 \\ * & A_i & 0 \\ * & * & * \end{bmatrix}$ with respect to the decomposition

$$(\sum_{n<0} \oplus (\mathcal{K} \oplus \mathcal{K}) \oplus \mathcal{K}) \oplus \mathcal{H} \oplus ((\mathcal{K} \ominus \mathcal{H}) \oplus \sum_{n \geq 1} \oplus (\mathcal{K} \oplus \mathcal{K})) \ .$$

From this, the identity $P(\mathcal{H})U_1^n U_2^m \,|\mathcal{H} = A_1^n A_2^m$ for $n,m \geq 0$ is immediate.■

We remark that the lower triangular form of U_i with respect to some decomposition is a consequence of the theorem even if it is not in evidence in the proof. To see this, note that one obtains a representation π of the bidisc algebra $A(\mathbf{D}^2)$ by sending a polynomial $p(z_1,z_2)$ to $p(U_1,U_2)$. This is contractive because U_i are commuting unitaries. The compression of this representation to $\mathcal{H}$ is still a representation. Hence $\mathcal{H}$ is semi-invariant for π by Theorem 2.16.

This argument also yields the generalization of von Neumann's inequality:

$$\|p(A_1,A_2)\| \leq \|p\|_\infty$$

for commuting contractions A_1, A_2 and polynomials p in $A(\mathbf{D}^2)$. This inequality fails for commuting triples.

20.5 Corollary (Sz. Nagy-Foias Lifting Theorem). *Let A and X be commuting contractions on $\mathcal{H}$, and let U be a unitary dilation of A on $\mathcal{K}$. Then X has a dilation Y on $\mathcal{K}$ commuting with U such that $\|Y\| = \|X\|$ and*

$$P(\mathcal{H})Y^n U^m \,|\mathcal{H} = X^n A^m \quad \text{for all} \quad n,m \geq 0 \ .$$

Proof. Normalize X to have $\|X\| = 1$. Let

$$U_A = \begin{bmatrix} * & 0 & 0 \\ * & A & 0 \\ * & * & * \end{bmatrix} \quad \text{and} \quad U_X = \begin{bmatrix} * & 0 & 0 \\ * & X & 0 \\ * & * & * \end{bmatrix}$$

be a commuting unitary dilation. The minimal unitary dilation U_0 of A is equivalent to the restriction of U_A to $\mathcal{K}_0 = \mathcal{K}_- \oplus \mathcal{H} \oplus \mathcal{K}_+$ where $\mathcal{K}_- = (\bigvee_{n \leq 0} U_A^n \mathcal{H}) \ominus \mathcal{H}$ and $\mathcal{K}_+ = (\bigvee_{n \geq 0} U_A^n \mathcal{H}) \ominus \mathcal{H}$, and $U_A \cong U_0 \oplus U_2$. Since U_X commutes with U_A, the compression Y_0 of U_X to $\mathcal{K}_0$ commutes with U_0. Moreover, because of the structure of U_X and $\mathcal{K}_0$, it follows that Y_0 has the form $\begin{bmatrix} * & 0 & 0 \\ * & X & 0 \\ * & * & * \end{bmatrix}$. The identity

$$P(\mathcal{H})Y_0^n U_0^m \,|\mathcal{H} = X^n A^m \quad \text{for all } n,m \geq 0$$

is now immediate. In general, any unitary dilation U of A decomposes as $U \cong V_0 \oplus V$ where V_0 acts on $\bigvee_{n=-\infty}^{\infty} U^n \mathcal{H}$ and is unitarily equivalent to U_0 by Theorem 20.1. Thus the desired dilation can be taken to by setting $Y \cong Y_0 \oplus 0$. ■

20.B Completely Positive Maps.

Let $\mathcal{B}$ be a C^*-algebra. A closed subspace $\mathcal{S}$ of $\mathcal{B}$ is an *operator system* if it contains the unit and is adjoint closed. For any closed subspace $\mathcal{A}$ of $\mathcal{B}$, let $\mathcal{M}_n(\mathcal{A})$ denote the subspace of the C^*-algebra $\mathcal{M}_n(\mathcal{B})$ consisting of all $n \times n$ matrices with coefficients in $\mathcal{A}$. If $\mathcal{S}$ is an operator system, then $\mathcal{M}_n(\mathcal{S})$ also has an order structure from the order in $\mathcal{M}_n(\mathcal{B})$. Suppose that ϕ is a linear map of $\mathcal{A}$ into a C^*-algebra $\mathcal{C}$. Define $\phi^{(n)}$ from $\mathcal{M}_n(\mathcal{A})$ into $\mathcal{M}_n(\mathcal{C})$ by $\phi^{(n)}([A_{ij}]) = [\phi(A_{ij})]$. The map ϕ is *completely bounded* if

$$\|\phi\|_{cb} \equiv \sup_{n \geq 1} \|\phi^{(n)}\| < \infty \ .$$

In particular, ϕ is *completely contractive* if $\|\phi\|_{cb} \leq 1$. When $\mathcal{A} = \mathcal{S}$ is an operator system, say that ϕ is *completely positive* if $\phi^{(n)}$ is positive for all $n \geq 1$. The connection between these two notions is quite strong.

20.6 Lemma. *Let ϕ be a completely positive map on an operator system $\mathcal{S}$. Then*

$$\|\phi\|_{cb} = \|\phi\| = \|\phi(1)\| \ .$$

Proof. Clearly, $\|\phi(1)\| \leq \|\phi\| \leq \|\phi\|_{cb}$. Let $S = (S_{ij})$ belong to $\mathcal{M}_n(\mathcal{S})$ with $\|S\| \leq 1$, and let $X = \phi^{(n)}(S)$. Note that the $n \times n$ identity matrix I_n belongs to $\mathcal{M}_n(\mathcal{S})$, and $P = \phi^{(n)}(I_n) = \phi(1)^{(n)}$. Since $\begin{bmatrix} I_n & S \\ S^* & I_n \end{bmatrix}$ is positive,

$$O_{2n} \leq \phi^{(2n)}\left(\begin{bmatrix} I_n & S \\ S^* & I_n \end{bmatrix}\right) = \begin{bmatrix} P & X \\ X^* & P \end{bmatrix} ,$$

where O_{2n} is the zero matrix. Thus

$$O_n \leq \begin{bmatrix} X^* & -\|X\| I_n \end{bmatrix} \begin{bmatrix} P & X \\ X^* & P \end{bmatrix} \begin{bmatrix} X \\ -\|X\| I_n \end{bmatrix}$$

$$= X^*PX + \|X\|^2 P - 2\|X\| X^*X \ .$$

Hence $2\|X\|^3 \leq 2\|X\|^2 \|P\|$, or $\|X\| \leq \|P\| = \|\phi(1)\|$. Thus $\|\phi^{(n)}\| = \|\phi(1)\|$ for all $n \geq 1$. ■

20.7 Lemma. *Let $\mathcal{A}$ be a unital subspace of a C^*-algebra $\mathcal{B}$. Let ϕ be a unital contraction from $\mathcal{A}$ into a C^*-algebra $\mathcal{C}$. Let $\mathcal{S} = \mathcal{A}+\mathcal{A}^*$, and define $\tilde{\phi}$ on $\mathcal{S}$ by*

$$\tilde{\phi}(A+B^*) = \phi(A)+\phi(B)^* \text{ for all } A, B \text{ in } \mathcal{A} \ .$$

Then $\tilde{\phi}$ is a well defined positive map.

Remark. It is easy to see, then, that $\tilde{\phi}$ is the *unique* positive extension of ϕ to $\mathcal{A}+\mathcal{A}^*$.

Proof. To show that $\tilde{\phi}$ is well defined, it suffices to show that if A and A^* belong to $\mathcal{A}$, then $\phi(A^*) = \phi(A)^*$. By considering Re A and Im A, it suffices to consider $A = A^*$. Let $\phi(A) = X = U+iV$ where U and V are self-adjoint. Now,

$$\max\{\|V+tI\|,\|V-tI\|\} = \|V\|+t \ .$$

Since $\|A \pm itI\|^2 = \|A\|^2+t^2$,

$$(\|A\|^2+t^2)^{1/2} \geq \|\phi(A \pm itI)\| = \|U+i(V \pm tI)\| \geq \|V \pm tI\| \ .$$

Thus, $\|V\|+t \leq (\|A\|^2+t^2)^{1/2}$ for all $t \geq 0$. In the limit as t increases, one obtains $\|V\| = 0$. Hence $\phi(A)$ is self-adjoint, and $\tilde{\phi}$ is well defined.

In fact, ϕ is positive on $\mathcal{A} \cap \mathcal{A}^*$. For if $A = A^*$ is positive, let $U = \phi(A)$ and $\epsilon = \inf\{\lambda : \lambda \in \sigma(U)\}$. For $t \geq \|A\|$,

$$t \geq \|A-tI\| \geq \|\phi(A-tI)\| = \|U-tI\| = t-\epsilon \ .$$

Thus $\epsilon \geq 0$ and U is positive.

Now let S be a positive element of $\mathcal{S}$. To show that $\tilde{\phi}(S)$ is positive, it suffices to show that $f(\tilde{\phi}(S)) \geq 0$ for every positive state f on $\mathcal{C}$. Clearly, $F = f \circ \phi$ is a norm one linear functional on $\mathcal{A}$ with $F(1) = 1$. Let $\tilde{F}$ be any Hahn-Banach extension of F to all of $\mathcal{B}$. The previous paragraphs applied to $\tilde{F}$ instead of ϕ shows that $\tilde{F}$ is a state on $\mathcal{B}$. In particular, it is positive. So for all A and B in $\mathcal{A}$,

$$\begin{aligned}\tilde{F}(A+B^*) &= F(A)+\overline{F(B)} = f(\phi(A))+\overline{f(\phi(B))} \\ &= f(\phi(A)+\phi(B)^*) = f(\tilde{\phi}(A+B^*)) \ .\end{aligned}$$

Thus, $f(\tilde{\phi}(S)) = \tilde{F}(S) \geq 0$ as desired. ∎

20.8 Corollary. *Assume the hypotheses of Lemma 20.7. Then ϕ is completely contractive if and only if $\tilde{\phi}$ is completely positive.*

Proof. If $\tilde{\phi}$ is completely positive, then $\tilde{\phi}$ is completely contractive by Lemma 20.6. So ϕ is completely contractive. Conversely, if $\phi^{(n)}$ is contractive, then since $\mathcal{M}_n(\mathcal{S}) = \mathcal{M}_n(\mathcal{A})+\mathcal{M}_n(\mathcal{A})^*$, it is easy to see that $\tilde{\phi}^{(n)} = (\phi^{(n)})^\sim$. Thus $\tilde{\phi}^{(n)}$ is positive by Lemma 20.7. So $\tilde{\phi}$ is completely positive. ∎

Now we require a Hahn-Banach theorem for completely positive maps. First, we do the matrix case.

20.9 Lemma. *Let S be an operator system in a C^*-algebra B, and suppose that ϕ is a completely positive map of S into $\mathcal{M}_n$. Then there is a completely positive map ψ of B into $\mathcal{M}_n$ extending ϕ.*

Proof. Let $e_1,\ldots,e_n$ be the standard basis for $\mathbb{C}^n$. Let $\xi = e_1 \oplus \ldots \oplus e_n$ be considered as a vector in $\mathbb{C}^n \oplus \ldots \oplus \mathbb{C}^n$. Define a linear functional Φ on $\mathcal{M}_n(S)$ by

$$\Phi([S_{ij}]) = (\phi^{(n)}[S_{ij}]\xi,\xi) \ .$$

This is the composition of two positive maps, and thus is positive. Let Ψ be any Hahn-Banach extension of Φ to $\mathcal{M}_n(B)$.

It is a well known result of Krein that Ψ is positive. To see this, note first that $\|\Phi\| = |\Phi(I_n)|$. For if $S = S^*$ is self-adjoint, norm one, then

$$\Phi(-I_n) \leq \Phi(S) \leq \Phi(I_n) \ .$$

Thus $|\Phi(S)| \leq |\Phi(I_n)|$. If S is an arbitrary norm one element, multiply it by a scalar λ of modulus one so that $\Phi(\lambda S) \geq 0$. Then since $\Phi(S^*) = \overline{\Phi(S)}$, one has

$$|\Phi(S)| = |\Phi(Re\, \lambda S)| \leq |\Phi(I_n)| \ .$$

Thus $\|\Psi\| = |\Psi(I_n)|$ as well. By Lemma 20.7 applied to $\Psi(I_n)^{-1}\Psi$, one sees that Ψ is positive.

Now define ψ from B into $\mathcal{M}_n$ as follows. Let E_{ij} be the matrix units for $\mathcal{M}_n$ with respect to the standard basis. Then $B \otimes E_{ij}$ represents the $n \times n$ matrix with B in the (i,j) entry and 0's elsewhere. Set

$$\psi(B) = \sum_{i=1}^{n} \sum_{j=1}^{n} \Psi(B \otimes E_{ij})E_{ij} \ .$$

A routine computation shows that for S in S,

$$\begin{aligned}\psi(S) &= \sum_{i=1}^{n} \sum_{j=1}^{n} \Phi(S \otimes E_{ij})E_{ij} \\ &= \sum_{i=1}^{n} \sum_{j=1}^{n} (\phi(S)e_j,e_i)E_{ij} = \phi(S) \ .\end{aligned}$$

So ψ extends ϕ.

We now verify that ψ is completely positive. Let $[B_{k\ell}]$ be a positive element of $\mathcal{M}_m(B)$. Let $x = x_1 \oplus \ldots \oplus x_m$ belong to $C^{nm} = \mathbb{C}^n \oplus \ldots \oplus \mathbb{C}^n$, and write $x_k = \sum_{i=1}^{n} x_{ki} e_i$. Then

$$\begin{aligned}(\psi^{(m)}([B_{k\ell}])x,x) &= \sum_{k,\ell=1}^{m} (\psi(B_{k\ell})x_\ell,x_k) \\ &= \sum_{k,\ell=1}^{m}\sum_{i,j=1}^{n} \Psi(B_{k\ell}\otimes E_{ij})x_{\ell j}\overline{x_{ki}} = \Psi([C_{ij}])\end{aligned}$$

where $C_{ij} = \sum_{k,\ell=1}^{m} x_{\ell j}\overline{x_{ki}}B_{k\ell}$. To prove that this is positive, it suffices to show that $C = [C_{ij}]$ is positive. Let $y = y_1 \oplus ... \oplus y_n$ be a vector in $\mathbb{C}^{n^2} = \mathbb{C}^n \oplus ... \oplus \mathbb{C}^n$. Then

$$\begin{aligned}(Cy,y) &= \sum_{i,j=1}^{n} (C_{ij}y_j,y_i) = \sum_{i,j=1}^{n}\sum_{k,\ell=1}^{m} (B_{k\ell}x_{\ell j}y_j,x_{ki}y_i) \\ &= ([B_{k\ell}]z,z) \geq 0\end{aligned}$$

where $z = z_1 \oplus ... \oplus z_m$ belongs to $\mathbb{C}^{nm}$, and $z_k = \sum_{i=1}^{n} x_{ki}y_i$. So $\psi^{(m)}$ is positive for all $m \geq 1$, and the lemma follows. ■

20.10 Arveson's Extension Theorem. *Let S be an operator system in a C^*-algebra $\mathcal{B}$, and suppose that ϕ is a completely positive map of S into $\mathcal{B}(\mathcal{H})$. Then ϕ has a completely positive extension ψ of $\mathcal{B}$ into $\mathcal{B}(\mathcal{H})$.*

Proof. Fix an increasing sequence P_n of projections of rank n with $\sup P_n = I$. Define $\phi_n(B) = P_n\phi(B)|P_n\mathcal{H}$. This is a completely positive map of S into $\mathcal{B}(P_n\mathcal{H})$ which is $*$ isomorphic to $\mathcal{M}_n$. So by Lemma 20.9, ϕ_n has a completely positive extension ψ_n of $\mathcal{B}$ into $\mathcal{B}(P_n\mathcal{H})$. Since $\|\psi_n\|_{cb} = \|\psi_n(1)\| = \|P_n\phi(1)P_n\| \leq \|\phi(1)\|$, this is a bounded sequence of c.b. maps. It remains to take a limit point in some appropriate topology.

The topology we use is the pointwise weak* topology (PW^*) on linear maps of $\mathcal{B}$ into $\mathcal{B}(\mathcal{H})$ given by pointwise weak* convergence. That is, Φ_α converges PW^* to Φ if

$$w^*\text{–}\lim \Phi_\alpha(B) = \Phi(B) \quad \text{for all } B \text{ in } \mathcal{B} .$$

As in the proof of the Banach-Alaoglu theorem, the map taking the unit ball of $\mathcal{B}(\mathcal{B},\mathcal{B}(\mathcal{H}))$ into $\prod_{B\in\mathcal{B}} (b_{\|B\|}(\mathcal{B}(\mathcal{H})),w^*)$ is a homeomorphism from the PW^* topology to a closed subset of a compact Hausdorff space. Hence the unit ball is PW^* compact.

It will be shown that the set of completely positive maps is PW^* closed. To this end, suppose that Φ_α are completely positive and Φ is the PW^*-limit of Φ_α. Let $[B_{ij}]$ be a positive element of $\mathcal{M}_n(\mathcal{B})$. Then

$$\Phi^{(n)}([B_{ij}]) = w^*\text{–}\lim_\alpha \Phi_\alpha^{(n)}([B_{ij}]) \geq 0$$

since the positive operators are weak* closed. Hence the c.p. maps of norm at most $\|\phi(1)\|$ forms a compact set in the PW^* topology.

Let ψ be any PW^* cluster point of the sequence ψ_n. Then ψ is completely positive. Since ϕ is the PW^* limit of ϕ_n, it follows that ψ is an extension of ϕ as desired. ■

20.11 Corollary. *Let $\mathcal{A}$ be a unital subspace of a C^*-algebra $\mathcal{B}$, and let ϕ be a unital complete contraction of $\mathcal{A}$ into $\mathcal{B}(\mathcal{H})$. Then there exists a completely positive map ψ of $\mathcal{B}$ into $\mathcal{B}(\mathcal{H})$ extending ϕ.*

Proof. By Corollary 20.8, ϕ extends to a completely positive map $\tilde{\phi}$ of $\mathcal{A}+\mathcal{A}^*$ into $\mathcal{B}(\mathcal{H})$. Apply Theorem 20.10 to $\tilde{\phi}$. ■

The next step is a basic structure theorem for unital completely positive maps. This theorem shows that completely positive maps *dilate* to $*$ representations. Equivalently, it says that all c.p. maps are corners of representations.

20.12 Stinespring's Dilation Theorem. *Let $\mathcal{B}$ be a C^*-algebra, and let ϕ be a unital completely positive map of $\mathcal{B}$ into $\mathcal{B}(\mathcal{H})$. Then there exists a unital $*$ representation π of $\mathcal{B}$ on a Hilbert space $\mathcal{K}$ and an isometry V of $\mathcal{H}$ into $\mathcal{K}$ such that*

$$\phi(B) = V^*\pi(B)V \text{ for all } B \text{ in } \mathcal{B} \ .$$

Proof. Consider the algebraic tensor product $\mathcal{B} \otimes \mathcal{H}$, and define a sesquilinear form by

$$<A \otimes x, B \otimes y> = (\phi(B^*A)x, y)$$

and extend by linearity. This is positive semidefinite since

$$<\sum_{i=1}^{n} A_i \otimes x_i, \sum_{i=1}^{n} A_i \otimes x_i> = \sum_{i=1}^{n} \sum_{j=1}^{n} (\phi(A_i^*A_j)x_j, x_i)$$
$$= (\phi^{(n)}([A_i^*A_j])x, x) \geq 0$$

where $x = x_1 \oplus \ldots \oplus x_n$. Thus this satisfies the Cauchy-Schwartz inequality. So

$$\mathcal{N} = \{u \in \mathcal{B} \otimes \mathcal{H} : <u,u> = 0\}$$
$$= \{u \in \mathcal{B} \otimes \mathcal{H} : <u,v> = 0 \text{ for all } v\}$$

is a subspace. The induced form on $\mathcal{B} \otimes \mathcal{H}/\mathcal{N}$ is thus positive definite. Let $\mathcal{K}$ be the Hilbert space completion of this quotient.

For A in $\mathcal{B}$, define a linear map $\pi_0(A)$ on $\mathcal{B} \otimes \mathcal{H}$ by $\pi_0(A)B \otimes x = AB \otimes x$ and extend linearly. This is an algebra homomorphism such that

$$<u,\pi_0(A)v> = <\pi_0(A^*)u,v> \ .$$

Hence for each u in $B \otimes \mathcal{H}$, $\rho_u(A) = <\pi_0(A)u,u>$ is a positive linear functional. Hence

$$\begin{aligned} <\pi_0(A)u,\pi_0(A)u> &= \rho_u(A^*A) \le \|A^*A\|\rho_u(1) \\ &= \|A\|^2<u,u> \ . \end{aligned}$$

So $\pi_0(B)$ leaves $\mathcal{N}$ invariant, and the induced map

$$\pi(A)(u+\mathcal{N}) = \pi_0(A)u+\mathcal{N}$$

defines a contractive $*$ representation on the quotient. Hence it extends by continuity to a $*$ homomorphism into $B(\mathcal{K})$.

Next, define V from $\mathcal{H}$ into $\mathcal{K}$ by $Vx = I \otimes x+\mathcal{N}$. Then

$$\|Vx\|^2 = (\phi(I)x,x) = \|x\|^2 \ ,$$

so V is an isometry. Finally, for A in B and x, y in $\mathcal{H}$,

$$\begin{aligned} (V^*\pi(A)Vx,y) &= <\pi(A)I \otimes x,I \otimes y> \\ &= <A \otimes x,I \otimes y> = (\phi(A)x,y) \ . \end{aligned}$$

Thus, $\phi(A) = V^*\pi(A)V$ as required. ∎

Let $\mathcal{A}$ be a unital subalgebra of a C^*-algebra B, and let ϕ be a unital homomorphism of $\mathcal{A}$ into $B(\mathcal{H})$. Say that ϕ has a *B-dilation* if there is a $*$-homomorphism π of B on a Hilbert space $\mathcal{K}$ and an isometry V of $\mathcal{H}$ into $\mathcal{K}$ such that $\phi(A) = V^*\pi(A)V$ for all A in $\mathcal{A}$. Clearly, ϕ is contractive.

20.13 Corollary (Arveson's Dilation Theorem). *Let $\mathcal{A}$ be a unital subalgebra of a C^*-algebra B. Let ϕ be a unital contractive representation of $\mathcal{A}$ into $B(\mathcal{H})$, and let $\tilde{\phi}$ be the positive extension of ϕ to $\mathcal{A}+\mathcal{A}^*$. The following are equivalent.*

(i) *ϕ has a B-dilation.*

(ii) *ϕ is completely contractive.*

(iii) *$\tilde{\phi}$ is completely positive.*

Proof. (ii) and (iii) are equivalent by Corollary 20.8. By Theorem 20.10, $\tilde{\phi}$ has a completely positive extension to all of B. Hence by Stinespring's Theorem 20.12, ϕ has a B-dilation. The converse is immediate since $*$-representations are contractive, whence $\phi^{(n)} = V^{*(n)}\pi^{(n)}V^{(n)}$ is contractive for all $n \ge 1$. ∎

20.C Representations of Nest Algebras.

By a *finite dimensional nest algebra*, we shall mean a nest subalgebra of some $\mathcal{M}_n$. The contractive representations of these algebras are simple and explicit; and are the key to the general analysis.

20.14 Theorem. *Let $\mathcal{A}$ be a finite dimensional nest algebra in $\mathcal{M}_n$. Let E_{ij} be the matrix units of $\mathcal{M}_n$ relative to a basis that puts $\mathcal{A}$ in block upper triangular form. Let ρ be a representation of $\mathcal{A}$ on $\mathcal{H}$ such that $\|\rho(E_{ij})\| \leq 1$ for every matrix unit in $\mathcal{A}$. Then ρ has an $\mathcal{M}_n$ dilation. (i.e. There is a $*$ representation π on a space $\mathcal{K}$ containing $\mathcal{H}$ such that $\rho(A) = P(\mathcal{H})\pi(A)\,|\,\mathcal{H}$ for all A in $\mathcal{A}$.)*

Remark. This shows that if $\|\rho(E_{ij})\| \leq 1$ just for the matrix units alone, then ρ is completely contractive.

Proof. First, assume that $\mathcal{A} = \mathcal{T}_n$ is the upper triangular algebra in $\mathcal{M}_n$. The operators $P_i = \rho(E_{ii})$ are norm one idempotents which annihilate each other. Thus they are pairwise orthogonal projections. So $\rho(1) = \sum_{i=1}^{n} P_i$ is an orthogonal projection. Since $\rho(A) = \rho(1)\rho(A)\rho(1)$, we may as well assume that $\rho(1) = I$. The ranges of P_i decompose $\mathcal{H}$ as $\mathcal{H}_1 \oplus \mathcal{H}_2 \oplus ... \oplus \mathcal{H}_n$. Now

$$\rho(E_{ij}) = \rho(E_{ii}E_{ij}E_{jj}) = P_i\rho(E_{ij})P_j \quad .$$

Let $T_i = P_i\rho(E_{i,i+1})\,|\,\mathcal{H}_{i+1}$. Then

$$\begin{aligned} P_i\rho(E_{ij})\,|\,\mathcal{H}_j &= P_i\rho(E_{i,i+1}E_{i+1,i+2}...E_{j-1,j})\,|\,\mathcal{H}_j \\ &= T_iT_{i+1}...T_{j-1} \quad . \end{aligned}$$

Thus ρ is determined by $\mathcal{H} = \mathcal{H}_1 \oplus ... \oplus \mathcal{H}_n$ and $\{T_1,...,T_{n-1}\}$; and any set of contractions $T_1,...,T_{n-1}$ yields a representation ρ with $\|\rho(E_{ij})\| \leq 1$ for all $i \leq j$.

Let us identify $\mathcal{H}_1, \ldots, \mathcal{H}_n$ with a fixed Hilbert space $\mathcal{R}$. If any $\mathcal{H}_i$ are finite dimensional, we can dilate ρ to the representation $\rho \oplus \delta$ where $\delta(E_{ij}) = E_{ij}^{(\infty)}$ is infinitely many copies of the identity representation. Now each T_i is a contraction in $\mathcal{B}(\mathcal{R})$. By Theorem 20.1, each T_i dilates to a unitary U_i on $\mathcal{K}_0 = \sum_{n=-\infty}^{\infty} \oplus \mathcal{R}_n$ with $\mathcal{R}_n \cong \mathcal{R}$ so that each U_i is lower triangular and the (0,0) entry is T_i. Identify $\mathcal{R}_0$ with $\mathcal{R}$. Note that

$$T_{ij} = T_iT_{i+1}...T_{j-1} = P(\mathcal{R})U_iU_{i+1}...U_{j-1}\,|\,\mathcal{R} \quad .$$

Now define a unitary operator

$$W = I \oplus U_1 \oplus U_1U_2 \oplus ... \oplus U_1...U_{n-1}$$

on $\mathcal{K} = \mathcal{K}_0^{(n)}$. Let π be the $*$ representation of $\mathcal{M}_n$ on $\mathcal{K}$ given by infinite inflation conjugated by W:

$$\pi([a_{ij}]) = W^*[a_{ij}I]W = [a_{ij}U_{ij}]$$

where $U_{ij} = U_iU_{i+1}...U_{j-1}$ if $i < j$, $U_{ii} = I$, and $U_{ji} = U_{ij}^*$ for $i < j$. Now $\mathcal{H} = \mathcal{R}^{(n)}$, so from the formula of the previous paragraph,

$$\rho(A) = P(\mathcal{H})\pi(A)\,|\,\mathcal{H}$$

as desired.

Now consider a general finite dimensional nest algebra $\mathcal{A}$. It contains a copy of $\mathcal{T}_n$. Restrict ρ to $\mathcal{T}_n$ and dilate as above. Then $\tilde{\rho}(A) = P(\mathcal{H})\pi(A)\,|\,\mathcal{H}$ is a completely positive extension of $\rho\,|\,\mathcal{T}_n$ to all of $\mathcal{M}_n$. But since $\mathcal{T}_n + \mathcal{T}_n^* = \mathcal{M}_n$, it is clear that the positive extension of ρ to $\mathcal{M}_n$ is unique. Now consider any matrix unit E_{ij} in $\mathcal{A}$ with $i > j$. Then the identities $\rho(E_{ij})\rho(E_{ji}) = \rho(E_{ii})$ and $\rho(E_{ji})\rho(E_{ij}) = \rho(E_{jj})$ together with the fact that these operators are contractions yields that $\rho(E_{ij})$ and $\rho(E_{ji})$ are partial isometries with

$$\rho(E_{ij}) = \rho(E_{ji})^* = \tilde{\rho}(E_{ji})^* = \tilde{\rho}(E_{ij}) \ .$$

The matrix units span $\mathcal{A}$, whence $\tilde{\rho}\,|\,\mathcal{A} = \rho$; and π dilates ρ as desired. ■

20.15 Corollary. *Every homomorphism ρ of a finite dimensional nest algebra $\mathcal{A}$ onto $\mathcal{B}(\mathcal{H})$ is similar to a completely contractive representation, and hence is completely bounded.*

Proof. By Corollary 17.2 and Exercise 17.1, the restriction of ρ to $\mathcal{A} \cap \mathcal{A}^*$ is similar to a $*$ representation. So replace ρ by $\rho_0 = Ad\ S_0 \circ \rho$ which is contractive on $\mathcal{A} \cap \mathcal{A}^*$. Let $F_1,...,F_k$ be the projections onto the atoms of $\mathcal{A}$ with $F_i \ll F_{i+1}$. Let

$$S_t = \sum_{j=1}^{k} t^{j-1}\rho_0(F_j) + \rho_0(I)^\perp \ ,$$

and let $\rho_t = Ad\ S_t \circ \rho_0$ for $t > 0$. Then $\rho_t(E_{ii+1}) = t\rho_0(E_{ii+1})$ if $E_{ii+1} = F_jE_{ii+1}F_{j+1}$ and $\|\rho_t(E_{ii+1})\| = 1$ if $E_{ii+1} = F_jE_{ii+1}F_j$. So for sufficiently small t, $\|\rho_t(E_{ij})\| \le 1$ for all matrix units in $\mathcal{A}$. Thus ρ_t is a completely contractive representation by Theorem 20.14. It is routine to verify that $(Ad\ S)^{(n)} = Ad\ S^{(n)}$ is bounded by $\|S\|\ \|S^{-1}\|$. Hence $\rho = Ad\,(S_tS_0)^{-1} \circ \rho_t$ is completely bounded, and

$$\|\rho\|_{cb} \le \|S_tS_0\|\ \|(S_tS_0)^{-1}\| \ .$$

■

The next step is to show that general nest algebras can be approximated by subalgebras isomorphic to finite dimensional nest algebras in such a way that the identity operator *approximately factors* through these algebras. By analogy to the corresponding property for von Neumann algebras, this property will be called *semidiscreteness*. The precise definition is the conclusion of the following theorem.

20.16 Theorem. *Every nest algebra $\mathcal{T}(\mathcal{N})$ is semidiscrete: There exist finite dimensional nest algebras $\mathcal{A}_n$, weak* continuous completely contractive maps $\phi_n : \mathcal{T}(\mathcal{N}) \to \mathcal{A}_n$, and weak* continuous completely isometric homomorphisms $\psi_n : \mathcal{A}_n \to \mathcal{T}(\mathcal{N})$ such that $\psi_n \circ \phi_n(A)$ converges weak* to A for all A in $\mathcal{T}(\mathcal{N})$.*

Moreover, it can be arranged that $K = \lim_{n \to \infty} \psi_n(\phi_n(K))$ in norm for every compact operator in $\mathcal{T}(\mathcal{N})$.

Proof. Let $\mathcal{N}$ be order isomorphic to a subset ω of $[0,1]$, and let μ be a Borel measure on $[0,1]$ mutually absolutely continuous to the spectral measure $E_{\mathcal{N}}$ of $\mathcal{N}$. By the results of Chapter 7 (Theorem 7.14 and Corollary 7.16), $\mathcal{N}$ is unitarily equivalent to the canonical nest on

$$\mathcal{H} = L^2(\mu_1) \oplus L^2(\mu_2) \oplus \dots$$

where $\mu_k = \chi_{J_k}\mu$ and J_k is a decreasing sequence of Borel subsets of ω.

Fix an integer N. Construct a partition of $[0,1]$ into disjoint intervals $I_1,\dots,I_p$ (in that order) and integer k_i in $[1,N]$ for $1 \le i \le p$ such that

i) $\mu(I_i) < N^{-1}$ or I_i is a singleton,

ii) $\mu_{k_i}(I_i) > (1-N^{-1})\mu(I_i)$ for $1 \le i \le p$, and

iii) $\mu_k(\bigcup\{I_i : k_i \ge k\}) > (1-N^{-1})\|\mu_k\|$.

This is possible since each μ_k is regular, and hence one can find finitely many disjoint intervals such that i) holds, $\mu_N(I_i) > (1-N^{-1})\mu(I_i)$, and

$$\mu(\bigcup I_i \cap J_N) > (1-N^{-1})\mu(J_N) \ .$$

These intervals will have $k_i = N$. Thus ii) holds for these intervals and iii) holds for $k = N$. Now expand the set of intervals so that i), ii) and iii) hold for $N-1$. Proceed in this way until the entire partition is defined.

For $1 \le i \le p$ and $1 \le k \le k_i$, let e_i^k be the positive multiple of $\chi_{I_i \cap J_{k_i}}$ of norm one in $L^2(\mu_k)$. For convenience of notation, set $e_i^k = 0$ if $k_i < k \le N$. Let $K_N = span\{e_i^k : 1 \le i \le p, 1 \le k \le N\}$. Let $\mathcal{A}_N$ be the finite dimensional nest algebra acting on K_N given by atoms $A_i = span\{e_i^k : 1 \le k \le k_i\}$ with the order $A_1 \ll A_2 \ll \dots \ll A_p$. It is clear that the map $\phi_N(T) = P(K_N)T\,|K_N$ maps $\mathcal{T}(\mathcal{N})$ into $\mathcal{A}_N$. Since this is just a compression map, it is completely contractive. We may also assume that the $(N+1)$st partition is a refinement of the Nth partition. Unfortunately, this does not imply that the subspaces K_N are nested; but this is not crucial.

Now we construct ψ_N. Note that $\mathcal{A}_N$ is spanned by the matrix units $E_{ij}^{k\ell} = e_i^k \otimes e_j^{\ell *}$, $1 \le k,\ \ell \le N$, $1 \le i \le j \le p$. Define $\psi_N(E_{ij}^{k\ell}) = E_{ij}^{k\ell}$ for $i < j$. For $i = j$ and $1 \le k,\ \ell \le k_i$, let $U_{ii}^{k\ell}$ be the canonical partial isometry from $L^2(I_i \cap J_{k_i}) \subseteq L^2(\mu_\ell)$ to $L^2(I_i \cap J_{k_i}) \subseteq L^2(\mu_k)$. Set

$\psi_N(E_{ii}^{k\ell}) = U_{ii}^{k\ell}$. Extend ψ_N to all of $\mathcal{A}_N$ by linearity. Note that $U_{ii}^{k\ell} e_i^\ell = e_i^k$, and that $\{U_{ii}^{k\ell} : 1 \le k,\ell \le k_i\}$ form matrix units for an algebra $*$ isomorphic to $\mathcal{B}(A_i)$. Moreover, $E_{ij}^{k\ell} = U_{ii}^{ks} E_{ij}^{st} U_{jj}^{t\ell}$ for all $i < j$, $1 \le k,s,t,\ell \le N$. It is now apparent that ψ_N is a representation of $\mathcal{A}_N$ which takes the matrix units to contractions. By Theorem 20.17, ψ_N is completely contractive.

It is clear from the construction that $\phi_N \circ \psi_N$ is the identity on $\mathcal{A}_N$.

Let us examine $\psi_N \circ \phi_N$. Properties i), ii) and iii) guarantee that $dist\,(x_k, span\,\{e_i^k : 1 \le i \le p\})$ tends to 0 for all x_k in $L^2(\mu_k)$ as N tends to infinity. Thus $dist\,(x, \mathcal{K}_N)$ tends to 0 as N tends to infinity for all x in $\mathcal{H}$. Hence $P(\mathcal{K}_N)$ converges strongly to I. Thus for any T in $\mathcal{T}(\mathcal{N})$, the sequence $P(\mathcal{K}_N)TP(\mathcal{K}_N)$ converges weak* to T. The point is that

$$\psi_N \circ \phi_N(T) - P(\mathcal{K}_N)TP(\mathcal{K}_N) = P(\mathcal{K}_N)^\perp \psi_N \circ \phi_N(T) P(\mathcal{K}_N)^\perp$$

because the only error occurs because $E_{ii}^{k\ell} \ne U_{ii}^{k\ell}$, and

$$U_{ii}^{k\ell} - E_{ii}^{k\ell} = P(\mathcal{K}_N)^\perp (U_{ii}^{k\ell} - E_{ii}^{k\ell}) P(\mathcal{K}_N)^\perp \ .$$

Since this difference is uniformly bounded by $\|T\|$, it tends to 0 in the weak* topology. Thus,

$$w^*\text{-}\lim \psi_N(\phi_N(T)) = T$$

for every T in $\mathcal{T}(\mathcal{N})$.

Now consider finite rank operators. Let P_s be the orthogonal projection onto $L^2(\mu_1) \oplus \ldots \oplus L^2(\mu_s)$. Let $R = x \otimes y^*$ be a norm one, rank one operator in $\mathcal{T}(\mathcal{N})$ with $R = P_s R P_s$ and fix $\epsilon > 0$. By Lemma 3.7, there is an element N in $\mathcal{N}$ such that $x = P(N)x$ and $y = P(N_-)^\perp y$. If $A = N \ominus N_-$ is a proper atom, choose an integer $N_0 \ge s$ so that A corresponds to a singleton in the partition of $[0,1]$ for $\mathcal{A}_N$, $N \ge N_0$. Otherwise, choose $N_0 \ge s$ sufficiently large that the partition $I_1,\ldots,I_p$ is sufficiently fine that there is an r with $\|E_{\mathcal{N}}(\bigcup_{i=1}^{r} I_i)^\perp x\| < \epsilon$ and $\|E_{\mathcal{N}}(\bigcup_{i=1}^{r} I_i) y\| < \epsilon$. In either case, one uses the fact that step functions are dense in $L^2(\mu)$ to choose $N_1 \ge N_0$ so that $\|P(\mathcal{K}_N)^\perp x\| < \epsilon$ and $\|P(\mathcal{K}_N)^\perp y\| < \epsilon$ for $N \ge N_1$. Now for $N \ge N_1$, one can find an integer m, and constants a_i^k and b_i^k so that

$$\|x - \sum_{i=1}^{m} \sum_{k=1}^{s} a_i^k e_i^k\| < 2\epsilon \text{ and } \|y - \sum_{i=m'}^{p} \sum_{k=1}^{s} b_i^k e_i^k\| < 2\epsilon$$

where $m' = m+1$ unless I_m is a singleton in which case $m' = m$. Then let x_N and y_N be the above approximates in $\mathcal{K}_N$. Note that $x_N \otimes y_N^*$ belongs to $\mathcal{A}_N$ and $\psi_N(x_N \otimes y_N^*) = x_N \otimes y_N^*$. Indeed, this is immediate except for the terms $a_m^k b_m^\ell E_{mm}^{k\ell}$. But since I_m is a singleton, $\psi_N(E_{mm}^{k\ell}) = E_{mm}^{k\ell}$, and all is well. Moreover,

$$\|x \otimes y^* - x_N \otimes y_N^*\| \le \|x - x_N\| \, \|y\| + \|x_N\| \, \|y - y_N\| < 5\epsilon \ .$$

Consequently, R is the limit of (rank one) operators R_N in the range of ψ_N. Thus

$$\begin{aligned}\|R - \psi_n(\phi_n(R))\| &\le \|R - R_n\| + \|\psi_n \circ \phi_n(R_n - R)\| \\ &\le 2\|R - R_n\| \ .\end{aligned}$$

Hence $\psi_n \circ \phi_n(R)$ converges to R in norm.

Clearly, every rank one operator in $\mathcal{T}(\mathcal{N})$ can be approximated by those of the form $P_s R P_s$ for some s. So by Corollary 3.13, $\psi_n \circ \phi_n(K)$ converges in norm to K for every compact operator in $\mathcal{T}(\mathcal{N})$. ∎

We are now ready to apply Arveson's dilation theory.

20.17 Corollary. *Every weak* continuous, contractive representation of a nest algebra is completely contractive.*

Proof. Let ρ be a weak* continuous contractive representation of $\mathcal{T}(\mathcal{N})$, and let ϕ_k and ψ_k be given by Theorem 20.16. Then $\rho \circ \psi_k$ are contractive representations of $\mathcal{A}_k$, and thus are completely contractive by Theorem 20.14. Hence $\rho \circ \psi_k \circ \phi_k$ are completely contractive. But ρ is weak* continuous and $\psi_k \circ \phi_k$ converges to the identity in the pointwise weak* (PW^*) topology. So $\rho \circ \psi_k \circ \phi_k$ converge to ρ in the PW^* topology, and thus ρ is completely contractive. ∎

20.18 Theorem. *Let $\mathcal{N}$ be a nest on a Hilbert space $\mathcal{R}$. Let ρ be a weak* continuous contractive representation of $\mathcal{T}(\mathcal{N})$ on a Hilbert space $\mathcal{H}$. Then there is an isometry V of $\mathcal{H}$ into $\mathcal{R}^{(\infty)}$ such that*

$$\rho(A) = V^* A^{(\infty)} V \text{ for all } A \text{ in } \mathcal{T}(\mathcal{N}) \ .$$

The range $V\mathcal{H}$ is a semi-invariant subspace of $\mathcal{T}(\mathcal{N})^{(\infty)}$.

Proof. Let $\mathcal{K}_1$ denote the unital C^*-algebra $\mathcal{K} + \mathbb{C}I$, and let $\mathcal{A} = \mathcal{T}(\mathcal{N}) \cap \mathcal{K}_1$. By Corollary 20.17, ρ is completely contractive. By Corollary 20.13, $\rho \mid \mathcal{A}$ has a $\mathcal{K}_1$ dilation. Namely, there is a representation π of $\mathcal{K}_1$ on a Hilbert space $\mathcal{H}'$ and an isometry V mapping $\mathcal{H}$ into $\mathcal{H}'$ such that

$$\rho(K) = V^* \pi(K) V \text{ for all } K \text{ in } \mathcal{K}_1 \ .$$

The representation decomposes as $\pi = \pi_a \oplus \pi_s$ on $\mathcal{K}_a \oplus \mathcal{K}_s$ where π_a is faithful on $\mathcal{K}$ and π_s annihilates $\mathcal{K}$. The range $V\mathcal{H}$ is contained in $\mathcal{K}_a$. To see this, let E_k be a sequence of finite rank projections tend weak* to the identity. Since ρ is weak* continuous,

$$h = w-\lim_{k\to\infty} \rho(E_k)h = \lim_{k\to\infty} V^*\pi(E_k)Vh$$
$$= V^*P(K_a)Vh \quad \text{for all} \quad h \in \mathcal{H} \; .$$

Thus, π can be replaced by π_a. The only representations of $\mathcal{K}$ are the ampliations $\pi_n(K) = K^{(n)}$, and there is no loss of generality in assuming that $n = \infty$. Thus $\rho(K) = V^*K^{(\infty)}V$ for K in $\mathcal{A}$. Since both sides of this identity are weak* continuous, and $\mathcal{A}$ is weak* dense in $\mathcal{T}(\mathcal{N})$ by Theorem 3.11, the identity extends to all A in $\mathcal{T}(\mathcal{N})$.

Since ρ is a homomorphism, Theorem 2.16 shows that the range $V\mathcal{H}$ is semi-invariant for $\pi(\mathcal{T}(\mathcal{N}))$. ■

20.19 Proposition. *Let $\mathcal{N}$ be a nest on a Hilbert space $\mathcal{R}$, and let ρ be a weak* continuous contractive representation of $\mathcal{T}(\mathcal{N})$ on $\mathcal{H}$. Then ρ has a minimal $\mathcal{B}(\mathcal{R})$ dilation π, and π is unique up to a unitary equivalence which fixes $\mathcal{H}$.*

Proof. Consider the dilation of Theorem 20.18. Identify $\mathcal{H}$ with $V\mathcal{H}$ and let $\mathcal{K}$ be the smallest reducing subspace for $\mathcal{B}(\mathcal{R})$ obtained by restriction to $\mathcal{K}$. Clearly, this is a minimal dilation of ρ. If π_1 is another minimal dilation of ρ on a space $\mathcal{K}_1$, define a map W from $\mathcal{K}$ to $\mathcal{K}_1$ as follows. Let $\tilde{\rho}$ be the unique completely positive, weak* continuous extension of ρ to $\mathcal{B}(\mathcal{R})$ (given by $\rho(T_1+T_2^*) = \rho(T_1)+\rho(T_2)^*$ and extending by continuity). For operators $T_1,...,T_n$ in $\mathcal{B}(\mathcal{R})$ and vectors $h_1,...,h_n$ in $\mathcal{H}$,

$$\|\sum_{j=1}^{n} \pi(T_j)h_j\|^2 = \sum_{j=1}^{n}\sum_{k=1}^{n}(\pi(T_k^*T_j)h_j,h_k)$$
$$= \sum_{j=1}^{n}\sum_{k=1}^{n}(\tilde{\rho}(T_k^*T_j)h_j,h_k)$$
$$= \sum_{j=1}^{n}\sum_{k=1}^{n}(\pi_1(T_k^*T_j)h_j,h_k) = \|\sum_{j=1}^{n}\pi_1(T_j)h_j\|^2 \; .$$

Thus there is an isometry W from $\mathcal{K} = span\{\pi(T)h : T \in \mathcal{B}(\mathcal{R}), h \in \mathcal{H}\}^-$ onto $\mathcal{K}_1 = span\{\pi_1(T)h : T \in \mathcal{B}(\mathcal{R}), h \in \mathcal{H}\}^-$ such that

$$W\pi(T)h = \pi_1(T)h \quad \text{for all} \quad T \in \mathcal{B}(\mathcal{R}), \; h \in \mathcal{H} \; .$$

Thus W is unitary, $Wh = h$ for all h in $\mathcal{H}$, and

$$W\pi(T)\pi(S)h = \pi_1(TS)h = \pi_1(T)\pi_1(S)h$$
$$= \pi_1(T)W\pi(S)h \; .$$

Thus $\pi_1(T) = W\pi(T)W^*$. ■

20.D Lifting Theorems for Nest Algebras.

In this section, we develop analogues of Ando's Theorem and the Sz-Nagy-Foias Lifting Theorem for representations of nest algebras.

20.20 Theorem. *Let $\mathcal{A}$ be a finite dimensional nest algebra in $\mathcal{M}_n$. Suppose that ρ is a unital contractive representation of $\mathcal{A}$ on $\mathcal{H}$ commuting with a contraction X. Then there is an $\mathcal{M}_n$ dilation π of ρ, and a commuting unitary dilation U of X on a space $\mathcal{K}$ such that*

$$P(\mathcal{H})U^n\pi(A)\,|\mathcal{H} = X^nA \quad \text{for all } n \geq 0, A \in \mathcal{A} \ .$$

Proof. First take $\mathcal{A} = \mathcal{T}_n$, the upper triangular algebra. Consider the explicit dilation of ρ constructed in Theorem 20.14. First $P_i = \rho(E_{ii})$ are pairwise orthogonal projections. We can assume that these are all the same rank, for otherwise replace ρ by $\rho \oplus \delta$ and X by $X \oplus 0$ where δ is infinitely many copies of the identity representation. Any dilation of this new pair will yield a dilation of (ρ, X). So $\mathcal{H}$ decomposes as $\mathcal{H} = \mathcal{H}_1 \oplus ... \oplus \mathcal{H}_n$ where each $\mathcal{H}_i \cong \mathcal{R}$, a fixed space. Since X commutes with ρ, $X = X_1 \oplus ... \oplus X_n$. As before, let $T_i = P_i\rho(E_{i,i+1})\,|\mathcal{H}_{i+1}$. Then $X_iT_i = T_iX_{i+1}$ for $1 \leq i \leq n-1$. By Theorem 20.1, each T_i has a unitary dilation U_i on $\mathcal{K}_0 = \sum_{n=-\infty}^{\infty} \oplus \mathcal{R}_n$ where $\mathcal{R}_n \cong \mathcal{R}$, $\mathcal{H}_i$ is identified with $\mathcal{R}_0$, and U_i is lower triangular with (0,0) entry T_i. Now $\mathcal{K} = \mathcal{K}_0^{(n)}$, and $W = I \oplus U_1 \oplus U_1U_2 \oplus ... \oplus (U_1U_2...U_{n-1})$ is a unitary operator on $\mathcal{K}$. $\mathcal{H}$ is identified with $\mathcal{R}_0^{(n)}$ in $\mathcal{K}$. The representation π is given by

$$\pi([a_{ij}]) = W^*[a_{ij}I]W = [a_{ij}U_{ij}]$$

where $U_{ij} = U_i...U_{j-1}$ for $i < j$, $U_{ii} = I$, and $U_{ij} = U_{ji}^*$ for $i > j$. Because each U_i is a lower triangular,

$$\begin{aligned} P(\mathcal{R}_0)\pi(E_{ij})\,|\mathcal{R}_0 &= P(\mathcal{R}_0)U_iU_{i+1}...U_{j-1}\,|\mathcal{R}_0 \\ &= T_iT_{i+1}...T_{j+1} = P_i\rho(E_{ij})\,|\mathcal{H}_j \ . \end{aligned}$$

Now X commutes with $A = \rho(E_{12}+E_{23}+...+E_{n-1,n})$. This latter operator has a unitary dilation on $\mathcal{K}$ given by

$$U = \begin{bmatrix} 0 & U_1 & 0 & 0 & 0 \\ 0 & 0 & U_2 & 0 & 0 \\ 0 & 0 & 0 & 0 & 0 \\ & & & & \\ 0 & 0 & 0 & 0 & U_{n-1} \\ U_0 & 0 & 0 & 0 & 0 \end{bmatrix}$$

where U_0 is the shift mapping $\mathcal{R}_n$ onto $\mathcal{R}_{n+1}$ for all n. (Note that U_0 is strictly lower triangular.) Apply the Sz. Nagy-Foias Lifting Theorem (Corollary 20.5) to obtain a dilation Y of X on $\mathcal{K}$ with $\|Y\| \leq 1$, $YU = UY$, and

$$P(\mathcal{H})Y^n U^m \,|\mathcal{H} = X^n A^m \quad \text{for} \quad n,m \geq 0 \ .$$

Now Y has an matrix form (Y_{ij}), and the commutation relation becomes $Y_{ij}U_j = U_{i+1}Y_{i+1,j+1}$. In particular, the diagonal part $D = Y_{11} \oplus ... \oplus Y_{nn}$ of Y also commutes with U. Moreover, $\mathcal{H}$ is semi-invariant for both Y and U, and thus Y and D both have the form $\begin{bmatrix} * & 0 & 0 \\ * & X & 0 \\ * & * & * \end{bmatrix}$ and U has the form $\begin{bmatrix} * & 0 & 0 \\ * & A & 0 \\ * & * & * \end{bmatrix}$ for the appropriate decomposition of $\mathcal{K}$. This shows that D also satisfies the dilation identity. So we assume that $Y = Y_1 \oplus ... \oplus Y_n$ is already diagonal.

Next, notice that Y commutes with π. Indeed Y now commutes with each $\pi(E_{ii})$ as it is diagonal, and with each $\pi(E_{i,i+1})$ since $Y_i U_i = U_i Y_{i+1}$. This identity implies that $Y = W^* Y_1^{(n)} W$ which clearly commutes with π. The rest is easy. Dilate Y to a lower triangular unitary V on $\mathcal{K}^{(\infty)}$ as in Theorem 20.1. Dilate π to $\pi^{(\infty)}$. Since π is self-adjoint, it commutes with Y^* as well, and hence $\pi^{(\infty)}$ commutes with V. The original space $\mathcal{H}$ is semi-invariant for both V and $\pi^{(\infty)}$ simultaneously because of the lower triangular forms. Thus the dilation identity holds.

Finally, consider a general nest algebra $\mathcal{A}$ in $\mathcal{M}_n$. This contains $\mathcal{T}_n$. Restrict ρ to $\mathcal{T}_n$ and dilate $(\rho\,|\mathcal{T}_n, X)$ to (π, V) as above. By the argument in the proof of Theorem 20.14, π dilates ρ. If E_{ij} belongs to $\mathcal{A}$ for $i > j$, then $T_{ij} = \rho(E_{ij})$ and $T_{ji} = \rho(E_{ji})$ satisfy $T_{ij}T_{ji} = P_i$ and $T_{ji}T_{ij} = P_j$. Hence $T_{ij} = T_{ji}^*$ is a partial isometry. Now the compression of $\pi\,|\mathcal{A}$ to $\mathcal{H}$ is a homomorphism, and thus $\mathcal{H}$ is semi-invariant for $\pi(\mathcal{A})$. Thus, it is a reducing subspace for $\pi(E_{ij})$ and $\pi(E_{ij})^*$. It is also semi-invariant for V, thus they have the form

$$\pi(E_{ij}) = \begin{bmatrix} * & 0 & * \\ 0 & T_{ij} & 0 \\ * & 0 & * \end{bmatrix} \quad \text{and} \quad V = \begin{bmatrix} * & 0 & 0 \\ * & X & 0 \\ * & * & * \end{bmatrix}$$

from which it is evident that $P(\mathcal{H})V^n \pi(E_{ij})\,|\mathcal{H} = X^n T_{ij}$. Since the matrix units span $\mathcal{A}$, (π, V) is the desired dilation. ∎

20.21 Corollary. *Let $\mathcal{A}$ be a finite dimensional nest algebra. Suppose that $\rho : \mathcal{A} \to \mathcal{B}(\mathcal{H})$ and $\sigma : A(\mathbf{D}) \to \mathcal{B}(\mathcal{H})$ are commuting contractive representations. Then the representation $\rho \otimes \sigma$ of $\mathcal{A} \otimes A(\mathbf{D})$ given by*

$$\rho \otimes \sigma([f_{ij}]) = \sum_{i,j} \sigma(f_{ij})\rho(E_{ij})$$

is completely contractive.

Proof. This is just a reformulation of Theorem 20.20 since σ is determined by the contraction $X = \sigma(z)$. The dilation obtained is completely positive and unital, hence completely contractive by Lemma 20.6. Thus, the compression to $\mathcal{H}$, $\rho \otimes \sigma$, is completely contractive. ∎

Now we extend this result to arbitrary nest algebras.

20.22 Theorem. *Let $\mathcal{N}$ be a nest on a Hilbert space $\mathcal{R}$. Suppose that ρ is a weak* continuous contractive representation of $\mathcal{A}$ on $\mathcal{H}$ which commutes with a contraction X in $B(\mathcal{H})$. Then there is an isometry V of $\mathcal{H}$ into $\mathcal{R}^{(\infty)}$ and a unitary U in $B(\mathcal{R}^{(\infty)})$ commuting with $\pi(A) = A^{(\infty)}$ for all A in $B(\mathcal{R})$ such that*

$$V^*U^n\pi(A)V = X^n\rho(A) \text{ for all } n \geq 0, A \in \mathcal{T}(\mathcal{N}) .$$

Proof. Let K_1 be the C^*-algebra $K+\mathbb{C}I$, and let $\mathcal{A} = \mathcal{T}(\mathcal{N}) \cap K_1$. Let $\mathcal{A}_n$, ϕ_n and ψ_n be given by Theorem 20.16. The representations τ_n of $\mathcal{A}_n \otimes A(\mathbf{D})$ given by

$$\tau_n([f_{ij}]) = \sum_{i,j} f_{ij}(X)\rho(\psi_n(E_{ij}))$$

is completely contractive by Corollary 20.21. Define a representation of $\mathcal{A} \otimes A(\mathbf{D})$ by $\tau(\sum_{i=1}^{N} K_i \otimes f_i) = \sum_{i=1}^{N} f_i(X)\rho(K_i)$. (The norm on $\mathcal{A} \otimes A(\mathbf{D})$ is given by the unique C^* norm on $K_1 \otimes C(\mathbf{T})$.) Now $K = \lim\psi_n(\phi_n(K))$ for K compact by Theorem 20.16, and $\rho(\psi_n(\phi_n(I))) = \rho(I)$. Thus

$$\tau(\sum_{i=1}^{N} K_i \otimes f_i) = \lim_{n\to\infty} \tau_n \circ (\phi_n \otimes id)(\sum_{i=1}^{N} K_i \otimes f_i) .$$

Now $\phi_n \otimes id$ is just a compression to K_N and thus is completely contractive, and τ_n is completely contractive, hence τ is completely contractive. So τ extends to all of $\mathcal{A} \otimes A(\mathbf{D})$.

By Arveson's Dilation Theorem 20.13, τ has a $K_1 \otimes C(\mathbf{T})$ dilation $\tilde{\tau}$ given by a $*$ representation π of K_1, and a commuting unitary operator $U = \tilde{\tau}(z)$ such that

$$P(\mathcal{H})U^n\pi(K)\,|\mathcal{H} = X^n\rho(K) \quad \text{for all } n \geq 0, \text{ and } K \in \mathcal{A} .$$

The representation π_1 splits into a direct sum $\pi = \pi_a \oplus \pi_s$ on $\mathcal{H}_a \oplus \mathcal{H}_s$ where π_a is faithful on K and π_s annihilates K. Since U commutes with π, it decomposes as $U = U_a \oplus U_s$. The weak* continuity of ρ implies that $\mathcal{H}$ is contained in $\mathcal{H}_a$. For if E_n is an increasing sequence of projections in K tending weak* to the identity, then for h in $\mathcal{H}$,

$$\begin{aligned} h &= w\text{-}\lim\rho(E_n)h = w\text{-}\lim P(\mathcal{H})\pi(E_n)h \\ &= P(\mathcal{H})P(\mathcal{H}_a)h \ . \end{aligned}$$

Thus, by restricting to $\mathcal{H}_a$, we may assume that π is faithful on K and thus is an ampliation $\pi(K) = K^{(n)}$. If $n < \infty$, this can be trivially dilated to $n = \infty$ by setting $U = I$ on the extra summands. Thus $\mathcal{H}$ is identified via an isometry V with a subspace of $\mathcal{R}^{(\infty)}$ such that

$$P(\mathcal{H})U^n\pi(K)\,|\mathcal{H} = X^n\rho(K) \text{ for } n \geq 0, K \in \mathcal{A} .$$

Now both sides are weak* continuous and $\mathcal{A}$ is weak* dense in $\mathcal{T}(\mathcal{N})$ by Theorem 3.11, so this dilation identity extends to all T in $\mathcal{T}(\mathcal{N})$. ∎

This analogue of Ando's Theorem yields the corresponding version of the Sz-Nagy-Foias Lifting Theorem.

20.23 Corollary. *Let $\mathcal{N}$ be a nest of subspaces of $\mathcal{R}$. Suppose that ρ is a weak* continuous contractive representation of $\mathcal{T}(\mathcal{N})$ on $\mathcal{H}$, and let π be a weak* continuous dilation to $B(\mathcal{R})$ on $\mathcal{K}$. If X is an operator on $\mathcal{H}$ commuting with ρ, then there is an operator Y on $\mathcal{K}$ commuting with π such that $\|Y\| = \|X\|$ and*

$$P(\mathcal{H})Y^n\pi(A)\,|\,\mathcal{H} = X^n\rho(A) \quad \text{for } n \geq 0 \text{ and } A \text{ in } \mathcal{T}(\mathcal{N}) .$$

Proof. Normalize so that $\|X\| = 1$. Let (π_1, U_1) be the dilation of (ρ, X) provided by Theorem 20.22. Let $\mathcal{K}_0$ be the smallest reducing subspace for π_1 containing $\mathcal{H}$. Then $\pi_0 = \pi_1\,|\,\mathcal{K}_0$ is, by Proposition 20.19, the unique minimal dilation of ρ. Let Y_0 be the compression of U to $\mathcal{K}_0$. Clearly, Y_0 is a contraction commuting with π_0, and $P(\mathcal{H})Y_0\,|\,\mathcal{H} = X$.

Now, $\mathcal{H}$ is semi-invariant for $(\pi(\mathcal{T}(\mathcal{N})), U)$. Let $\mathcal{K}^+$ denote the smallest subspace of $\mathcal{K}$ containing $\mathcal{H}$ and invariant for $\pi(\mathcal{T}(\mathcal{N}))$ and U. Similarly, let $\mathcal{K}^-$ denote the smallest subspace containing $\mathcal{H}$ invariant for $\pi(\mathcal{T}(\mathcal{N}))^*$ and U^*. Let $\mathcal{H}^+$ and $\mathcal{H}^-$ be the smallest subspaces of $\mathcal{K}^+$ and $\mathcal{K}^-$ respectively which contain $\mathcal{H}$ and are invariant for $\pi(\mathcal{T}(\mathcal{N}))$. Then $\pi\,|\,\mathcal{T}(\mathcal{N})$ has the form $\begin{bmatrix} * & 0 & 0 \\ * & \rho & 0 \\ * & * & * \end{bmatrix}$ and U has the form $\begin{bmatrix} * & 0 & 0 \\ * & X & 0 \\ * & * & * \end{bmatrix}$ with respect to the decomposition $(\mathcal{K}^+)^\perp \oplus \mathcal{H} \oplus (\mathcal{K}^+ \ominus \mathcal{H})$. From these triangular forms, it is evident that $(\mathcal{K}^+)^\perp \oplus \mathcal{H}$ is invariant for $\pi(\mathcal{T}(\mathcal{N}))^*$ and U^* and hence contains $\mathcal{K}^-$.

We wish to show that $\mathcal{K}_0 = (\mathcal{H}^- \ominus \mathcal{H}) \oplus \mathcal{H} \oplus (\mathcal{H}^+ \ominus \mathcal{H})$. To this end, it suffices to show that $\mathcal{H}^- + \mathcal{H}^+$ is π-invariant. As π is weak* continuous, it suffices to show that $\mathcal{H}^- + \mathcal{H}^+$ is invariant for $\pi(F)$ where F is finite rank. Let $\pi(T)h$, for T in $\mathcal{T}(\mathcal{N})$ and h in $\mathcal{H}$, be a typical element in the spanning set of $\mathcal{H}^+$. Then FT is finite rank and thus splits as a sum $FT = K_1 + K_2^*$ where K_i are compact operators in $\mathcal{T}(\mathcal{N})$ (Theorem 4.10). So

$$\pi(F)\pi(T)h = \pi(K_1)h + \pi(K_2^*)h \in \mathcal{H}^+ + \mathcal{H}^- .$$

A similar computation holds for vectors in $\mathcal{H}^-$. So $\mathcal{K}_0$ decomposes as claimed.

This decomposition implies that we have the lower triangular forms

$$\pi_0\,|\,\mathcal{T}(\mathcal{N}) = \begin{bmatrix} * & 0 & 0 \\ * & \rho & 0 \\ * & * & * \end{bmatrix} \quad \text{and} \quad Y_0 = \begin{bmatrix} * & 0 & 0 \\ * & X & 0 \\ * & * & * \end{bmatrix} .$$

Hence the dilation identity holds:

$$P(\mathcal{H})Y^n\pi(A)\,|\,\mathcal{H} = X^n\rho(A) \quad \text{for } n \geq 0 \text{ and } A \text{ in } \mathcal{T}(\mathcal{N}) .$$

For the general case, decompose π as $\pi_0 \oplus \pi'$ where π_0 is the minimal dilation of ρ. Dilate X to $Y_0 \oplus 0$. This completes the proof. ∎

20.24 Remark. A simple but useful trick generalizes this theorem to the case of an operator X in $\mathcal{B}(\mathcal{H}_2,\mathcal{H}_1)$ intertwining two representations ρ_i on $\mathcal{H}_i$, $i=1,2$ (i.e. $\rho_1(A)X = X\rho_2(A)$ for all A in $\mathcal{T}(\mathcal{N})$.) Suppose that π_i is a dilation of ρ_i on $\mathcal{K}_i$, $i=1,2$. Then $\rho_1 \oplus \rho_2$ is a representation of $\mathcal{T}(\mathcal{N})$ in $\mathcal{H}_1 \oplus \mathcal{H}_2$ with dilation $\pi_1 \oplus \pi_2$ on $\mathcal{K}_1 \oplus \mathcal{K}_2$. The operator $\tilde{X} = \begin{bmatrix} 0 & X \\ 0 & 0 \end{bmatrix}$ commutes with $\rho_1 \oplus \rho_2$, and thus there is an operator $\tilde{Y}$ on $\mathcal{K}_1 \oplus \mathcal{K}_2$ which commutes with $\pi_1 \oplus \pi_2$ and dilates $\tilde{X}$. Clearly, there is no loss in compressing $\tilde{Y}$ to its (1,2) entry, so we may suppose $\tilde{Y} = \begin{bmatrix} 0 & Y \\ 0 & 0 \end{bmatrix}$. Thus, Y intertwines π_1 and π_2 in that $\pi_1(B)Y = Y\pi_2(B)$ for all B in $\mathcal{B}(\mathcal{R})$. The triangular form obtained in the proof above shows that each $\mathcal{K}_i$ decomposes as $\mathcal{K}_i^- \oplus \mathcal{H}_i \oplus \mathcal{K}_i^+$ so that $\pi_i \,|\, \mathcal{T}(\mathcal{N})$ is lower triangular, and Y has the form $\begin{bmatrix} * & 0 & 0 \\ * & X & 0 \\ * & * & * \end{bmatrix}$ in $\mathcal{B}(\mathcal{K}_2,\mathcal{K}_1)$. This decomposition is canonical if we insist that $\mathcal{K}_1^+ \oplus \mathcal{H}_i$ be the smallest $\pi_i \,|\, \mathcal{T}(\mathcal{N})$ invariant subspace containing $\mathcal{H}_i$.

Now we turn to commuting representations of nest algebras.

20.25 Theorem. *Let $\mathcal{A}_1$ and $\mathcal{A}_2$ be nest subalgebras of $\mathcal{B}(\mathcal{R}_1)$ and $\mathcal{B}(\mathcal{R}_2)$. Let ρ_1 and ρ_2 be weak* continuous contractive representations of $\mathcal{A}_1$ and $\mathcal{A}_2$ on a common space $\mathcal{H}$ such that*

$$\rho_1(A_1)\rho_2(A_2) = \rho_2(A_2)\rho_1(A_1) \quad \text{for all } A_1 \in \mathcal{A}_1,\ A_2 \in \mathcal{A}_2 \ .$$

Then there is a Hilbert space $\mathcal{K}$ containing $\mathcal{H}$ and weak continuous $\mathcal{B}(\mathcal{R}_i)$-dilations π_i of ρ_i on $\mathcal{H}$, $i=1,2$, such that*

$$P(\mathcal{H})\pi_1(A_1)\pi_2(A_2)\,|\,\mathcal{H} = \rho_1(A_1)\rho_2(A_2) \qquad \text{for } A_1 \in \mathcal{A}_1,\ A_2 \in \mathcal{A}_2$$

and

$$\pi_1(B_1)\pi_2(B_2) = \pi_2(B_2)\pi_1(B_1) \qquad \text{for } B_1 \in \mathcal{B}(\mathcal{R}_1),\ B_2 \in \mathcal{B}(\mathcal{R}_2) \ .$$

Proof. Assume first that $\mathcal{A}_2$ is the upper triangular algebra $\mathcal{T}_n$ contained in $\mathcal{M}_n$. Let the standard matrix units for $\mathcal{M}_n$ be E_{ij}. Now ρ_1 commutes with $P_j = \pi(E_{jj})$ for $1 \le j \le n$. This decomposes ρ_1 as $\sum_{j=1}^{n} \oplus \rho_{1,j}$ on $\mathcal{H} = \sum_{j=1}^{n} \oplus \mathcal{H}_j$. Let $\pi_1 = \sum_{j=1}^{n} \oplus \pi_{1,j}$ be a weak* continuous $\mathcal{B}(\mathcal{R}_1)$ dilation of ρ_1 on $\mathcal{K} = \sum_{j=1}^{n} \oplus \mathcal{K}_j$. Now $T_j = P_j\rho_2(E_{i,i+1})\,|\,\mathcal{H}_{j+1}$ intertwines $\rho_{1,j}$ and $\rho_{1,j+1}$. By Remark 20.24, there are contractions Y_j in $\mathcal{B}(\mathcal{K}_{j+1},\mathcal{K}_j)$ which intertwine $\pi_{1,j}$ and $\pi_{1,j+1}$ and dilate X_j. The triangular form of Y_j is crucial. Each $\mathcal{K}_j$ decomposes as $\mathcal{K}_j^- \oplus \mathcal{H}_j \oplus \mathcal{K}_j^+$ where $\mathcal{H}_j \oplus \mathcal{K}_j^+$ is the smallest subspace containing $\mathcal{H}_j$ invariant for $\pi_{1,j} \mid \mathcal{A}_1$. Considering Y_j as an operator from $\mathcal{K}_{j+1}$ to $\mathcal{K}_j$ and $\pi_{1,j} \mid \mathcal{A}_1$ as a representation on $\mathcal{K}_j$ with this decomposition, we obtain the triangular forms

$$Y_j = \begin{bmatrix} * & 0 & 0 \\ * & T_j & 0 \\ * & * & * \end{bmatrix} \quad \text{and} \quad \pi_{1,j} \mid \mathcal{A}_j = \begin{bmatrix} * & 0 & 0 \\ * & \rho_{1,j} & 0 \\ * & * & * \end{bmatrix} .$$

Now consider each Y_j as an operator on $\mathcal{K}$. The triangular form implies

$$P(\mathcal{H})Y_i Y_{i+1} \ldots Y_{j-1} \mid \mathcal{H} = \rho_2(E_{ij}) \text{ for } 1 \leq i < j \leq n_2 .$$

Thus, $Y_1, \ldots, Y_{n-1}$ determine a representation π_2 of $\mathcal{T}_n$ on $\mathcal{K}$ which dilates ρ_2 commutes with the representation π_1 of $\mathcal{B}(\mathcal{R}_1)$. As usual, it is now easy to dilate the rest of the way up. Let σ_2 be the $\mathcal{M}_n$ dilation of π_2 constructed on $\mathcal{K}^{(\infty)}$ as in Theorem 20.14. Since π_1 is a $*$ representation, it commutes with $\pi_2(\mathcal{T}_n)^*$ also, and hence $\sigma_1 = \pi_1^{(\infty)}$ is an $\mathcal{B}(\mathcal{R}_1)$ dilation of π_1 commuting with σ_2. It is clear that $\sigma_1 \otimes \sigma_2$ dilates $\rho_1 \otimes \rho_2$ and thus the dilation identities hold.

Next, suppose that $\mathcal{A}_2$ is a general nest subalgebra of $\mathcal{M}_n$ containing $\mathcal{T}_n$. Dilate $\rho_1 \otimes (\rho_2 \mid \mathcal{T}_n)$ to a $\mathcal{B}(\mathcal{R}_1) \otimes \mathcal{M}_n$ representation $\sigma_1 \otimes \sigma_2$. The compression of $\sigma_1 \otimes \sigma_2$ to $\mathcal{H}$ is the unique (completely) positive extension of $\rho_1 \otimes (\rho_2 \mid \mathcal{T}_n)$ and hence agrees with $\rho_2 \otimes \rho_2$ on $\mathcal{A}_1 \otimes \mathcal{A}_2$. The restriction of ρ_2 to $\mathcal{A}_2 \cap \mathcal{A}_2^*$ is a $*$ representation with dilation σ_2, and hence $\mathcal{H}$ is invariant for $\sigma_2(\mathcal{A}_2 \cap \mathcal{A}_2^*)$. So as in the proof of Theorem 20.20, we see that $\sigma_1 \mid \mathcal{A}_1$ and $\sigma_2 \mid \mathcal{A}_2 \cap \mathcal{A}_2^*$ have the forms

$$\sigma_1 \mid \mathcal{A}_1 = \begin{bmatrix} * & 0 & 0 \\ * & \rho_1 & 0 \\ * & * & * \end{bmatrix} \quad \text{and} \quad \sigma_2 \mid \mathcal{A}_2 \cap \mathcal{A}_2^* = \begin{bmatrix} * & 0 & * \\ 0 & \rho_2 & 0 \\ * & * & * \end{bmatrix} .$$

Hence the dilation identity follows for each matrix unit E_{ij} in $\mathcal{A}$ with $i > j$.

Finally, let $\mathcal{A}_2$ be a nest algebra $\mathcal{T}(\mathcal{N})$ and proceed as in Theorem 20.22. Let $\mathcal{K}_1 = \mathcal{K} + \mathbb{C}I$, $\mathcal{A} = \mathcal{T}(\mathcal{N}) \cap \mathcal{K}_1$; and let $\mathcal{A}_n$, ϕ_n and ψ_n, $n \geq 3$, be given by Theorem 20.16. The representations $\rho_1 \otimes (\rho_2 \circ \psi_n)$ of $\mathcal{A}_1 \otimes \mathcal{A}_n$ have dilations and thus are completely contractive. Every K in $\mathcal{A}$ has the property that $\rho_2(K) = \lim_{n \to \infty} \rho_2(\psi_n(\phi_n(K)))$ and all maps are completely contractive. Thus, it follows that $\rho_1 \otimes (\rho_2 \mid \mathcal{A})$ is completely contractive. By Theorem 20.13, this has a $\mathcal{B}(\mathcal{R}_1) \otimes \mathcal{B}(\mathcal{R}_2)$ dilation $\pi_1 \otimes \pi_2$. As before, the weak* continuous part is itself a dilation of $\rho_1 \otimes (\rho_2 \mid \mathcal{A})$. The weak* continuity shows that it is a dilation of $\rho_1 \otimes \rho_2$. ■

20.26 Remark. Proceeding as in the proof of Theorem 20.23, we can establish a Sz. Nagy-Foias Lifting Theorem for nests. Namely, if ρ_1 and ρ_2 are commuting, weak* continuous, contractive representations of $\mathcal{T}(\mathcal{N}_1)$ and $\mathcal{T}(\mathcal{N}_2)$ on $\mathcal{H}$, and if π is a $\mathcal{B}(\mathcal{R})$ dilation of ρ_1 on $\mathcal{K}$, then there is a weak* continuous contractive representation σ of $\mathcal{T}(\mathcal{N}_2)$ on $\mathcal{K}$ commuting with π such that

$$P(\mathcal{H})\pi(A_1)\sigma(A_2)\,|\,\mathcal{H} = \rho_1(A_1)\rho_2(A_2) \text{ for } A_1 \in \mathcal{T}(\mathcal{N}_1),\, A_2 \in \mathcal{T}(\mathcal{N}_2)\ .$$

20.27 Examples. We finish this chapter with examples of Parrott and Paulsen which demonstrate that the dilation theorems do not work for triples. Let

$$U = \begin{bmatrix} 1 & 0 \\ 0 & -1 \end{bmatrix} \quad \text{and} \quad V = \begin{bmatrix} 0 & 1 \\ 1 & 0 \end{bmatrix} ,$$

and note that $UV+VU = 0$. Consider the three commuting 4×4 contractions

$$T_1 = \begin{bmatrix} 0 & I_2 \\ 0 & 0 \end{bmatrix} , \ T_2 = \begin{bmatrix} 0 & U \\ 0 & 0 \end{bmatrix} , \text{ and } T_3 = \begin{bmatrix} 0 & V \\ 0 & 0 \end{bmatrix} .$$

If (T_1,T_2,T_3) had a commuting unitary dilation (U_1,U_2,U_3), then the associated representation $\rho_1 \otimes \rho_2 \otimes \rho_3$ of $A(\mathbf{D}^3)$ would be completely contractive. Consider the matrix function

$$F(z_1,z_2,z_3) = \begin{bmatrix} z_1 & z_2 & 0 \\ z_3 & 0 & z_2 \\ 0 & z_3 & -z_1 \end{bmatrix} .$$

Then

$$\|F(z_1,z_2,z_3)\| = \| \begin{bmatrix} z_1 & 0 & 0 \\ 0 & z_3 & 0 \\ 0 & 0 & z_3\overline{z_2}z_1 \end{bmatrix} \begin{bmatrix} 1 & 1 & 0 \\ 1 & 0 & 1 \\ 0 & 1 & -1 \end{bmatrix} \begin{bmatrix} 1 & 0 & 0 \\ 0 & \overline{z_1}z_2 & 0 \\ 0 & 0 & \overline{z_3}z_2 \end{bmatrix} \|$$

The first and third matrices are unitary on the Shilov boundary $\mathbf{T}^3$, and hence

$$\|F\| = \| \begin{bmatrix} 1 & 1 & 0 \\ 1 & 0 & 1 \\ 0 & 1 & 1 \end{bmatrix} \| = \sqrt{3}\ .$$

However, $F(T_1,T_2,T_3)$ can be decomposed as a 2×2 matrix with $F(I,U,V)$ as its $(1,2)$ entry. But

$$F(I,U,V) = \begin{bmatrix} I & U & 0 \\ V & 0 & U \\ 0 & V & -I \end{bmatrix} = \begin{bmatrix} I & 0 & 0 \\ 0 & V & 0 \\ 0 & 0 & VU^* \end{bmatrix} \begin{bmatrix} I & I & 0 \\ I & 0 & I \\ 0 & I & I \end{bmatrix} \begin{bmatrix} I & 0 & 0 \\ 0 & U & 0 \\ 0 & 0 & V^*U \end{bmatrix} .$$

Again, the first and third entries are unitary, so

$$\|F(T_1,T_2,T_3)\| = \|F(I,U,V)\| = \| \begin{bmatrix} 1 & 1 & 0 \\ 1 & 0 & 1 \\ 0 & 1 & 1 \end{bmatrix} \| = 2 > \sqrt{3} .$$

Hence $\rho_2 \otimes \rho_2 \otimes \rho_3$ is not completely contractive.

It is true that $\|\rho_1 \otimes \rho_2 \otimes \rho_3\| = 1$. To briefly outline this, note that all products of any two of T_1,T_2,T_3 is in 0. Thus, if $f(z) = a_0 + \sum_{i=1}^{3} a_i z_i + r(z)$, where $r(z)$ has no constant or linear terms, then

$$f(T_1,T_2,T_3) = \begin{bmatrix} a_0 I & a_1 I + a_2 U + a_3 V \\ 0 & a_0 I \end{bmatrix}.$$

Computing $(f(T_1,T_2,T_3)\begin{pmatrix} x_1 \\ x_2 \end{pmatrix}, \begin{pmatrix} y_1 \\ y_2 \end{pmatrix})$ and straightforward estimates yields

$$\|F(T_1,T_2,T_3)\| \leq \| \begin{bmatrix} |a_0| & \sum_{i=1}^{3} |a_i| \\ 0 & |a_0| \end{bmatrix} \| .$$

This is a compression of the Laurent operator $g(z) = |a_0| + (\sum_{i=1}^{3} |a_i|)z + h(z)z^2$ to a 2×2 block, and hence

$$\| f(T_1,T_2,T_3)\| \leq \inf_{h \in A(\mathbf{D})} \| \, |a_0| + \sum_{i=1}^{3} |a_i| z + h(z)z^2 \| \leq \| f \|_\infty .$$

Now we will use this example to construct three commuting, contractive, unital representations of the triangular algebra $\mathcal{T}_2$ which do not dilate to commuting $\mathcal{M}_2$ representations. These representations will be unitarily equivalent to

$$\sigma_i \left(\begin{bmatrix} a & b \\ 0 & d \end{bmatrix} \right) = \begin{bmatrix} aI_4 & bT_i \\ 0 & dI_4 \end{bmatrix}^{(4)} .$$

For economy, we mark the non-zero entries of a typical element A of $\mathcal{A} = \mathcal{T}_2 \otimes \mathcal{T}_2 \otimes \mathcal{T}_2$ and a representation σ of $\mathcal{A}$ given by Schur multiplication ($[a_{ij}] \circ [b_{ij}] = [a_{ij}b_{ij}]$) by an operator matrix T.

$$A = \begin{bmatrix} * & * & * & * & * & * & * & * \\ \cdot & * & \cdot & * & \cdot & * & \cdot & * \\ \cdot & \cdot & * & * & \cdot & \cdot & * & * \\ \cdot & \cdot & \cdot & * & \cdot & \cdot & \cdot & * \\ \cdot & \cdot & \cdot & \cdot & * & * & * & * \\ \cdot & \cdot & \cdot & \cdot & \cdot & * & \cdot & * \\ \cdot & \cdot & \cdot & \cdot & \cdot & \cdot & * & * \\ \cdot & \cdot & \cdot & \cdot & \cdot & \cdot & \cdot & * \end{bmatrix} \text{ and } T = \begin{bmatrix} I_4 & T_1 & T_2 & 0 & T_3 & 0 & 0 & 0 \\ \cdot & I_4 & \cdot & T_2 & \cdot & T_3 & \cdot & 0 \\ \cdot & \cdot & I_4 & T_1 & \cdot & \cdot & T_3 & 0 \\ \cdot & \cdot & \cdot & I_4 & \cdot & \cdot & \cdot & T_3 \\ \cdot & \cdot & \cdot & \cdot & I_4 & T_1 & T_2 & 0 \\ \cdot & \cdot & \cdot & \cdot & \cdot & I_4 & \cdot & T_2 \\ \cdot & \cdot & \cdot & \cdot & \cdot & \cdot & I_4 & T_1 \\ \cdot & \cdot & \cdot & \cdot & \cdot & \cdot & \cdot & I_4 \end{bmatrix}.$$

The fact that $T_iT_j = 0$ for all i,j leads to the fact that σ is multiplicative. The three copies of $\mathcal{T}_2$ occur in $\mathcal{A}$ as

$$\begin{bmatrix} a & b & \cdot & \cdot & \cdot & \cdot & \cdot & \cdot \\ \cdot & d & \cdot & \cdot & \cdot & \cdot & \cdot & \cdot \\ \cdot & \cdot & a & b & \cdot & \cdot & \cdot & \cdot \\ \cdot & \cdot & \cdot & d & \cdot & \cdot & \cdot & \cdot \\ \cdot & \cdot & \cdot & \cdot & a & b & \cdot & \cdot \\ \cdot & \cdot & \cdot & \cdot & \cdot & d & \cdot & \cdot \\ \cdot & \cdot & \cdot & \cdot & \cdot & \cdot & a & b \\ \cdot & \cdot & \cdot & \cdot & \cdot & \cdot & \cdot & d \end{bmatrix}, \begin{bmatrix} a & \cdot & b & \cdot & \cdot & \cdot & \cdot & \cdot \\ \cdot & a & \cdot & b & \cdot & \cdot & \cdot & \cdot \\ \cdot & \cdot & d & \cdot & \cdot & \cdot & \cdot & \cdot \\ \cdot & \cdot & \cdot & d & \cdot & \cdot & \cdot & \cdot \\ \cdot & \cdot & \cdot & \cdot & a & \cdot & b & \cdot \\ \cdot & \cdot & \cdot & \cdot & \cdot & a & \cdot & b \\ \cdot & \cdot & \cdot & \cdot & \cdot & \cdot & d & \cdot \\ \cdot & \cdot & \cdot & \cdot & \cdot & \cdot & \cdot & d \end{bmatrix}, \text{ and } \begin{bmatrix} a & \cdot & \cdot & \cdot & b & \cdot & \cdot & \cdot \\ \cdot & a & \cdot & \cdot & \cdot & b & \cdot & \cdot \\ \cdot & \cdot & a & \cdot & \cdot & \cdot & b & \cdot \\ \cdot & \cdot & \cdot & a & \cdot & \cdot & \cdot & b \\ \cdot & \cdot & \cdot & \cdot & d & \cdot & \cdot & \cdot \\ \cdot & \cdot & \cdot & \cdot & \cdot & d & \cdot & \cdot \\ \cdot & \cdot & \cdot & \cdot & \cdot & \cdot & d & \cdot \\ \cdot & \cdot & \cdot & \cdot & \cdot & \cdot & \cdot & d \end{bmatrix}$$

respectively. The representation σ_i are all contractive. To show that the σ_i do not jointly dilate, it suffices to show that σ is not contractive. To this end, we define the following element A_0 in $\mathcal{A}$.

$$A_0 = \begin{bmatrix} 0&0&0&0&0&0&0&0 \\ 0&0&0&1&0&1&0&0 \\ 0&0&0&1&0&0&1&0 \\ 0&0&0&0&0&0&0&0 \\ 0&0&0&0&0&1&-1&0 \\ 0&0&0&0&0&0&0&0 \\ 0&0&0&0&0&0&0&0 \\ 0&0&0&0&0&0&0&0 \end{bmatrix}, \text{ and } \sigma(A_0) = \begin{bmatrix} 0&0&0&0&0&0&0&0 \\ 0&0&0&T_2&0&T_3&0&0 \\ 0&0&0&T_1&0&0&T_3&0 \\ 0&0&0&0&0&0&0&0 \\ 0&0&0&0&0&T_1&-T_2&0 \\ 0&0&0&0&0&0&0&0 \\ 0&0&0&0&0&0&0&0 \\ 0&0&0&0&0&0&0&0 \end{bmatrix}.$$

Then as before, we obtain $\|A_0\| = \sqrt{3}$ and $\|\sigma(A_0)\| = 2$. ∎

Notes and Remarks.

Classical dilation theory is developed in detail in the book by Sz. Nagy and Foias [2], and in the briefer monograph by Sz. Nagy [2]. A nice treatment of completely positive maps and the algebraic approach to dilation theory is in Paulsen [1]. Theorem 20.1 is due to Sz. Nagy [1]. Theorem 20.4 is taken from Ando [1]. Corollary 20.5 is a strengthened form of Sz. Nagy and Foias [1] due to Parrott [2]. Completely positive maps are introduced in Stinespring [1], where Theorem 20.12 is proved.

This theory was developed extensively by Arveson [4,5]. In particular, he proves 20.7, 20.8, Theorems 20.10 and Corollary 20.13. The proof of Theorem 20.10 based on Lemma 20.9 is taken from Paulsen [1].

The material on representations of nest algebras is taken from Paulsen, Power and Ward [1]. However, Theorem 20.14 was known, and is a special case of results of McAsey and Muhly [1]. They investigate certain non self-adjoint subalgebras of crossed products of C^*-algebras and von Neumann algebras. (For more information in this direction, see papers in the references by McAsey, Muhly, Saito and Solel.) The lifting theorems for nests are due to Paulsen and Power [1]. The finite dimensional case of Corollary 20.23 is due to Ball and Gohberg [2] who exhibit all operators Y commuting with π such that $\|Y\| = \|X\|$ and $P(\mathcal{H})Y|\mathcal{H} = X$. Example 20.27 is due to Parrott [1]. The extension of this example to commuting representations of $\mathcal{T}_2$ is due to Paulsen.

Exercises.

20.1 Verify that the unitary dilation of A constructed in Theorem 20.1 also satisfies $P(\mathcal{H})U^{*n}|\mathcal{H} = A^{*n}$ for $n \geq 0$.

20.2 (a) There is a notion of isometric dilation of a contraction. Show that there is a unique minimal isometric dilation of A, given by restricting its minimal unitary dilation U to $\mathcal{K}^+ = \bigvee_{n \geq 0} U^n \mathcal{H}$.

(b) Prove the analogue of the Sz. Nagy-Foias lifting theorem for isometric dilations.

20.3 Suppose A_i are contractions on $\mathcal{H}_i$ with unitary dilations U_i on $\mathcal{K}_i$. If X is a map in $\mathcal{B}(\mathcal{H}_2,\mathcal{H}_1)$ such that $A_1X = XA_2$, show that there is a map Y in $\mathcal{B}(\mathcal{K}_2,\mathcal{K}_1)$ such that $U_1Y = YU_2$, $\|Y\| = \|X\|$, and $P(\mathcal{H}_1)Y|\mathcal{H}_2 = X$.

20.4 Prove that if ϕ is a completely positive map from a C^*-algebra $\mathfrak{A}$ into a C^*-algebra $\mathcal{B}$, then

$$\phi(A)^*\phi(A) \leq \phi(A^*A) \text{ for all } A \text{ in } \mathfrak{A} .$$

Hint: Dilate ϕ using Stinespring's Theorem.

20.5 Prove that there is a unique minimal Stinespring dilation of a completely positive map up to the natural unitary equivalence.

20.6 Let ϕ be a map from a C^*-algebra $\mathfrak{A}$ into $\mathcal{M}_n$ such that $\phi^{(n)}$ is positive. Show that ϕ is completely positive. Hint: Examine the proof of Lemma 20.9.

20.7 Prove the lifting theorem for commuting representations of nests outlined in Remark 20.26.

20.8* If ϕ is a bounded representation of a nest algebra, is $\|\phi\|_{cb} = \|\phi\|$?

21. Operators in Nest Algebras

In this chapter, we survey two aspects of single operator theory related to nests. The first concerns the structure of unicellular operators. The second concerns which operators are unitarily equivalent to (small) compact perturbations of a fixed nest algebra. This chapter does not contain a complete account of either topic, but should serve as a useful introduction to the ideas involved.

21.A Unicellular Operators. An operator T is *unicellular* if $Lat\, T$ is a nest. It is an open problem to discover which maximal nests are the invariant subspace lattice of a single operator. (Such lattices are called *attainable.*) Theorem 5.5 shows that the Volterra operator V is unicellular. Thus the Volterra nest is an attainable lattice. By the Similarity Theorem, every continuous nest is attainable as the lattice of an operator similar to V. The simplest case is an atomic nest of order type $\omega+1 = \mathbb{N} \cup \{\infty\}$ or $(\omega+1)^* = \{-\infty\} \cup -\mathbb{N}$.

21.1 Proposition. *Let S be a backward shift on a basis $\{e_n, n \geq 0\}$ given by $Se_0 = 0$ and $Se_n = w_n e_{n-1}$ for $n \geq 1$. Suppose that w_n are positive, monotone decreasing, and belong to ℓ^p for some $p < \infty$. Then S is unicellular, and*

$$Lat\, S = \{M_n = span\{e_k, k \leq n\} : n \geq 0\} \ .$$

Proof. It is clear that each M_n is invariant for S. Also, if $x = \sum_{i=0}^{n-1} a_i e_i + e_n$, then the cyclic invariant subspace of S generated by x is precisely M_n. As $\bigcup_{n \geq 0} M_n$ is dense in $\mathcal{H}$, it suffices to show that if $x = \sum_{n \geq 0} a_n e_n$ does not belong to any M_n, then $L(x) \equiv span\{S^n x, n \geq 0\}$ equals $\mathcal{H}$.

First we show that e_0 belongs to $L(x)$. Fix an integer $n_0 > p$; and define $w_k^{(n_0)} = \prod_{j=1}^{n_0} w_{k+j}$. This is summable, so let $C = \sum_{k \geq 0} w_k^{(n_0)}$. Given $\epsilon > 0$, pick $n > n_0$ so that $w_n < \epsilon C^{-1} w_0^{(n_0+1)}$ and $|a_n| \geq \sup_{k>n} |a_k|$. Then

$$(a_n \prod_{j=1}^{n} w_j)^{-1} S^n x - e_0 = \sum_{k \geq 1} \frac{a_{n+k}}{a_n} \prod_{j=1}^{n} \frac{w_{j+k}}{w_j} e_k \quad .$$

Now for all $k \geq 1$,

$$\prod_{j=1}^{n} \frac{w_{j+k}}{w_j} = \frac{w_k^{(n_0)}}{w_0^{(n_0)}} \cdot \prod_{j=n_0+1}^{n-1} \frac{w_{j+k}}{w_{j+1}} \cdot \frac{w_{n+k}}{w_{n_0+1}}$$

$$< w_k^{(n_0)} \cdot \frac{w_{n+1}}{w_0^{(n_0+1)}} < \frac{\epsilon}{C} w_k^{(n_0)} \quad .$$

Hence

$$\|(a_n w_0^{(n)})^{-1} S^n x - e_0\| \leq \frac{\epsilon}{C} \sum_{k \geq 1} w_k^{(n_0)} < \epsilon \quad .$$

Suppose M_n is contained in $L(x_0)$. Write $\mathcal{H} = M_n \oplus M_n^{\perp}$ and $L(x) = M_n \oplus L_n(x)$. Then $L_n(x)$ is invariant for the compression S_n of S to $M_n^{\perp}$. Since S_n satisfies the same hypotheses as S, $L_n(x)$ contains e_{n+1}. By induction, $L(x)$ contains all $\{e_n, n \geq 0\}$ and hence equals $\mathcal{H}$. ■

To build more complicated unicellular operators, one requires a method for gluing them together. This is provided by the following theorem. By the *ordinal sum* $\mathcal{L} \dotplus \mathcal{M}$ of two subspace lattices $\mathcal{L}$ and $\mathcal{M}$, we mean $\{L \oplus 0 : L \in \mathcal{L}\} \cup \{\mathcal{H} \oplus M : M \in \mathcal{M}\}$.

21.2 Theorem. *Suppose S is an operator with cyclic vector g such that for some $s > 0$, $\|S^n\| = O((n!)^{-s})$. Let A be any forward unilateral weighted shift with weights w_n monotone decreasing and in ℓ^p for some $p < \infty$ such that*

$$\|S^n\| = O(\|A^{n+k}\|) \text{ for all } k \geq 0 \quad .$$

Then there is an operator T of the form $\begin{bmatrix} S & C \\ 0 & A \end{bmatrix}$ *such that* $\text{Lat}\, T = \text{Lat}\, S \dotplus (w+1)^*$*, and T has a cyclic vector.*

21.3 Remarks. This condition can be achieved by choosing $0 < r < s$ and taking $w_n = n^{-r}$, so that $\|A^n\| = (n!)^{-r}$. From Stirling's formula

$$((n!)^{-s}) = O(((n+k)!)^{-r}) \text{ for all } k \geq 0 \quad .$$

This norm condition implies a stronger version of itself, namely

$$\|S^n\| = O(n^{-1}\|A^{n+k}\|) \text{ for all } k \geq 0 \quad .$$

To see this, note that because w_n^p is a *monotone* decreasing summable sequence, $\lim_{n \to \infty} n w_n^p = 0$. Hence

$$\|A^{n+k+p}\| \le w_n^p \|A^{n+k}\| = O(n^{-1}\|A^{n+k}\|) \ .$$

Consequently,

$$\|S^n\| = O(\|A^{n+k+p}\|) = O(n^{-1}\|A^{n+k}\|) \ . \tag{21.2.1}$$

In fact, this shows that $\|S^n\|/\|A^{n+k}\| = O(w_n^p)$. Hence

$$\sum_{n \ge 1} \|S^n\|/\|A^{n+k}\| < \infty \quad \text{for all} \quad k \ge 0 \ .$$

To set the stage for the proof of Theorem 21.2, we describe the construction of C. For convenience, normalize S, A and g to have norm one. Choose a summable sequence c_n of positive real numbers such that

$$\lim_{n \to \infty} c_n^{-1}(\sum_{m > n} c_m) = 0 \ .$$

For example, one could take $c_n = (n!)^{-1}$. Now choose an increasing sequence of positive integers k_n such that

$$\lim_{n \to \infty} c_n^{-1} w_{k_n} = 0 \quad \text{and} \quad \lim_{n \to \infty} c_n^{-1} k_n \|S^{k_n - k_{n-1}}\| \, \|A^{k_n}\|^{-1} = 0 \ .$$

This choice is made inductively. The first condition follows if k_n is large enough. The second also follows because of condition (21.2.1). Now define C by

$$C = g \otimes \sum_{m \ge 1} c_m e_{k_m}^* \ .$$

21.4 Lemma. *Let* $T = \begin{bmatrix} S & C \\ 0 & A \end{bmatrix}$ *be defined as above. Let*

$$\lambda_n = (c_n \|A^{k_n}\|)^{-1} \ \textit{and} \ D = \sum_{n \ge 0} \|A^n\|^{-1} S^n g \otimes e_n^* \ .$$

Then $\lim_{n \to \infty} \lambda_n T^{k_n+1} = D$.

Proof. Write

$$\lambda_n T^{k_n+1} = \begin{bmatrix} \lambda_n S^{k_n+1} & X_n \\ 0 & \lambda_n A^{k_n+1} \end{bmatrix} \ .$$

Then

$$\lim_{n \to \infty} \|\lambda_n A^{k_n+1}\| = \lim_{n \to \infty} c_n^{-1} w_{k_n+1} = 0 \ .$$

Since $\|\lambda_n S^{k_n+1}\| = O(\lambda_n \|A^{k_n+1}\|)$, $\lim_{n \to \infty} \|\lambda_n S^{k_n+1}\| = 0$ also. A routine computation shows that $X_n = \lambda_n \sum_{j=0}^{k_n} S^j C A^{k_n - j}$. Set $C_m = g \otimes c_m e_{k_m}^*$ and

$X_{n,m} = \lambda_n \sum_{j=0}^{k_n} S^j C_m A^{k_n - j}$. Clearly, $C = \sum_{m \geq 1} C_m$ and $X_n = \sum_{m \geq 1} X_{n,m}$. Moreover,

$$S^j C_m A^{k_n - j} = c_m S^j g \otimes (A^{*k_n - j} e_{k_m})^* .$$

Thus if $m < n$, then $A^* e_{k_m} = 0$ for $j < k_n - k_m$. Thus

$$\begin{aligned} ||X_{n,m}|| &= \lambda_n c_m || \sum_{j=0}^{k_m} S^{k_n - j} g \otimes (A^{*j} e_{k_m})^* || \\ &\leq \lambda_n c_m (k_m + 1) ||S^{k_n - k_m}|| \, ||g|| \\ &< c_m (c_n^{-1} k_n ||S^{k_n - k_{n-1}}|| \, ||A^{k_n}||^{-1}) . \end{aligned}$$

Hence

$$\lim_{n \to \infty} \sum_{m=1}^{n-1} ||X_{n,m}|| \leq (\sum_{m=1}^{\infty} c_m) \lim_{n \to \infty} c_n^{-1} k_n ||S^{k_n - k_{n-1}}|| \, ||A^{k_n}||^{-1} = 0 .$$

Next, we have $||A^k|| = ||A^{*k} e_k||$. So $||A^{k_n}||^{-1} \, ||A^{*k_n - j} e_{k_n}|| = ||A^j||^{-1}$. Thus if $m > n$,

$$\begin{aligned} ||X_{n,m}|| &\leq \lambda_n c_m \sum_{j=0}^{k_n} ||S^j|| \, ||g|| \, ||A^{*k_n - j} e_{k_m}|| \\ &\leq c_n^{-1} c_m \sum_{j=0}^{k_n} ||S^j|| \, ||A^j||^{-1} . \end{aligned}$$

Hence

$$\lim_{n \to \infty} \sum_{m=n+1}^{\infty} ||X_{n,m}|| \leq \lim_{n \to \infty} c_n^{-1} (\sum_{m > n} c_m) \sum_{j=0}^{\infty} ||S^j|| \, ||A^j||^{-1} = 0 .$$

Finally, for $m = n$, $||A_{k_n}||^{-1} A^{*k_n - j} e_{k_n} = ||A^j||^{-1} e_j$. So

$$X_{n,n} = \lambda_n c_n \sum_{j=0}^{k_n} S^j g \otimes (A^{*k_n - j} e_{k_n})^* = \sum_{j=0}^{k_n} ||A^j||^{-1} S^j g \otimes e_j^* .$$

Since $\sum_{j \geq 0} ||A^j||^{-1} \, ||S^j|| < \infty$, $\lim_{n \to \infty} X_{n,n} = D$, and the lemma follows. ∎

Proof of Theorem 21.2. Let M be an invariant subspace for T. Let $M_0 = M \cap (\mathcal{H} \oplus 0)$. This is invariant for T, and hence (as a subspace of $\mathcal{H}$) is invariant for S. Thus if $M = M_0$, then M belongs to $Lat\, S + (\omega+1)^*$. If $M \neq M_0$, it must be shown that $M_0 = \mathcal{H} \oplus 0$. For then, $M = \mathcal{H} \oplus L$ where L belongs to $Lat\, A$ which is isomorphic to $(\omega+1)^*$ by Proposition 21.1. Whence $Lat\, T = Lat\, S + (\omega+1)^*$.

Let L be the closure of $P(0 \oplus \mathcal{H})M$. Then $M \vee (\mathcal{H} \oplus 0) = \mathcal{H} \oplus L$ is invariant for T, and thus L is a non-zero invariant subspace of A. By Proposition 21.2, there is an integer n_0 so that $L = span\{e_n, n \geq n_0\}$. By Lemma 21.5, the operator D belongs to the norm closed algebra generated by T, and therefore D leaves M invariant. For each $n \geq n_0$, choose x_k in M such that $\lim_{k\to\infty} P(0 \oplus \mathcal{H})x_k = e_n$. Then

$$\|A^n\|^{-1}S^n g = De_n = \lim_{k\to\infty} Dx_k \quad .$$

So M contains $span\{S^n g, n \geq n_0\}$ which has finite codimension in $\mathcal{H} \oplus 0$. As L has finite codimension in $0 \oplus \mathcal{H}$, it follows that M has finite codimension itself.

Let B be the restriction of T^* to $M^\perp$, which is invariant for T^*. Now T, and hence T^*, is quasinilpotent. Since $M^\perp$ is finite dimensional, B is nilpotent; say of order q. Then T^{*q} annihilates $M^\perp$, whence the range of T^q is contained in M. Let m be so large that $k_m \geq n_0$ and $k_m - k_{m-1} > q$. Then M contains

$$\begin{aligned} T^q e_{k_m-q+1} &= A^q e_{k_m-q+1} + \sum_{j=0}^{q-1} S^j C A^{q-1-j} e_{k_m-q+1} \\ &= \alpha e_{k_m+1} + \sum_{j=0}^{q-1} S^j C \alpha_j e_{k_m-j} = \alpha e_{k_m+1} + \alpha_0 g \quad , \end{aligned}$$

where $\alpha_0 = \omega_{k_m-q+1}^{(q-1)}$ and $\alpha = \alpha_0 \omega_{k_m}$. Since ω_{k_m} tends to 0, M contains g which is cyclic for S. So M contains $\mathcal{H} \oplus 0$ and equals $\mathcal{H} \oplus L$ as claimed∎

21.5 Corollary. *For any finite non-negative integers m and n, the lattice $\omega m+1+(\omega n)^*$ is attainable.*

Proof. Choose real numbers $p_1 > p_2 > \ldots > p_{n+m} > 0$. Let A_k be the unilateral weighted shift with weights n^{-p_k}, so that $\|A_k^n\| = (n!)^{-p_k}$. By induction, Theorem 21.2 provides operators C_k so that

$$S = \begin{bmatrix} A_1 & C_1 & 0 & & 0 & 0 \\ 0 & A_2 & C_2 & & 0 & 0 \\ 0 & 0 & A_3 & & & \\ & & & \ddots & & \\ 0 & 0 & 0 & & A_{m-1} & C_{m-1} \\ 0 & 0 & 0 & & 0 & A_m \end{bmatrix}$$

has $LatS \cong (\omega m+1)^*$, and S has a cyclic vector. The norm condition is verified in Exercise 21.1. Continue to use Theorem 21.2 to produce $C_m, \ldots, C_{n-1}$ so that

$$T = \begin{bmatrix} S^* & C_m & 0 & 0 & 0 \\ 0 & A_{m+1} & C_{m+1} & 0 & 0 \\ & & & & \\ 0 & 0 & 0 & A_{n-1} & C_{n-1} \\ 0 & 0 & 0 & 0 & A_n \end{bmatrix}$$

has $Lat\, T = Lat\, S + (n\omega+1)^* \cong \omega m + 1 + (\omega n)^*$. ■

A quite different approach is needed to produce an operator with lattice isomorphic to $\{-\infty\} \cup \mathbb{Z} \cup \{+\infty\} = (\omega+1)^* + (\omega+1)$. Like the last proof, it is computationally technical.

21.6 Theorem. *Let X be a bilateral shift with weights $w_n > 0$ given by $Se_n = w_n e_{n+1}$ for $n \in \mathbb{Z}$. Let $\pi_n = \prod_{k=0}^{n-1} w_n$ for $n \geq 0$ and $\pi_n = \prod_{k=-1}^{n} w_n$ for $n < 0$. Assume that $\{w_n, n \geq 0\}$ and $\{w_n, n \leq 0\}$ are monotone decreasing, and that*

$$\limsup_{n \to \infty} \frac{-\log \pi_n}{n^2} > \gamma > \log 3$$

and

$$\liminf_{|n| \to \infty} \frac{-\log \pi_n}{|n|^2} > \delta > 0 .$$

Then $Lat\, S = \{M_n = span\{e_k : k \geq n\} : n \in \mathbb{Z}\} \cup \{0, \mathcal{H}\}$.

Remark. One can take $w_n = 10^{-|n|}$ to satisfy the hypotheses of this theorem.

Proof. For convenience, normalize S to be norm one. Since $\mathcal{N} = \{M_n, n \in \mathbb{Z}; 0, \mathcal{H}\}$ is a complete lattice, it suffices to show that every cyclic invariant subspace for S is in $\mathcal{N}$. The conditions on the weights show that $\pi_n < e^{-n^2\delta}$ for $|n|$ large. Since w_n is monotone,

$$w_n \leq \pi_n^{1/|n|} < e^{-|n|\delta}$$

for $|n|$ large. Hence the weight sequence is summable. By Proposition 21.1, $Lat(S|M_n) = \{M_k : k \geq n\}$ and $Lat(S^*|M_n^\perp) = \{M_k^\perp : k \leq n\}$. Hence if $x \in M_n$, then the cyclic subspace $L(x)$ belongs to $\mathcal{N}$. Likewise, if $y \in M_n^\perp$, then the largest S invariant subspace orthogonal to y is in $\mathcal{N}$.

Now suppose that a unit vector $x = \sum a_n e_n$ generates a cyclic subspace which is not in $\mathcal{N}$. Let $y = \sum b_n e_n$ be a unit vector orthogonal to L. Then $a_n \neq 0$ for arbitrarily large negative n and $b_n \neq 0$ for arbitrarily large positive n. Define

$$\alpha_n = \begin{cases} \pi_n a_n, & n < 0 \\ \pi_n^{-1} a_n, & n \geq 0 \end{cases} \quad \text{and} \quad \beta_n = \begin{cases} \pi_n^{-1} \overline{b_n}, & n < 0 \\ \pi_n \overline{b_n}, & n \geq 0 \end{cases} .$$

Then for $k \geq 0$,

$$0 = (S^k x, y) = \sum_{n=-\infty}^{\infty} \Big(\prod_{i=0}^{k-1} w_{n+i}\Big) a_n \overline{b_{n+k}} = \sum_{n=-\infty}^{\infty} \alpha_n \beta_{n+k} .$$

Let $A_0 = 1$ and define $A_n > 0$ for $n < 0$ so that $\{\log A_n, n \leq 0\}$ is the least convex majorant of $\{0, \log|\alpha_n|, n < 0\}$. That is, $(n, \log A_n)$ lies on the boundary of $conv\{(0,0),(n,\log|\alpha_n|), n < 0\}$. Clearly, $|\alpha_n| \leq A_n$ for all $n \leq 0$. Since $(|\alpha_n|) = o(\pi_n)$ and $\log \pi_n < -n^2\delta$ for large n, it follows that $A_n = |\alpha_n|$ infinitely often. Similarly, let $B_0 = 1$ and define $B_n > 0$ for $n > 0$ so that $\{\log B_n, n \geq 0\}$ is the least convex majorant of $\{0, \log|\beta_n|, n \geq 1\}$. Again $E = \{n : B_n = |\beta_n|\}$ is infinite. The log convexity condition guarantees that $B_n^{-1}B_{n+1}$ and $A_{-n}^{-1}A_{-n-1}$ are monotone decreasing to 0 as n tends to infinity. Furthermore, the rapid growth condition implies that for infinitely many integers n,

$$B_n^{-1}B_{n+1} \leq e^{-2\gamma} B_{n-1}^{-1} B_n .$$

To see that, note that if $B_n^{-1}B_{n+1} \geq e^{-2\gamma}B_{n-1}^{-1}B_n$ for all $n \geq n_0$, then

$$B_{n+k}^{-1} B_{n+k+1} \geq e^{-2k\gamma} B_n^{-1} B_{n+1} .$$

Hence

$$B_{n+p} \geq B_n \prod_{k=0}^{p-1} e^{-2k\gamma} B_n^{-1} B_{n+1} = e^{-(p^2-p)\gamma} B_n^{1-p} B_{n+1}^p .$$

Taking logs yields

$$-\log \pi_{n+p} \leq -\log B_{n+p} \leq (p^2-p)\gamma + (p-1)\log B_n - p \log B_{n+1} .$$

Divide by $(n+p)^2$ and take the limit as $p \to \infty$, and

$$\limsup_{n \to \infty} \frac{-\log \pi_{n+p}}{(n+p)^2} \leq \gamma .$$

This contradicts the hypotheses, so our claim is valid.

Take a large integer n so that

$$L = B_n^{-1} B_{n+1} \leq e^{-2\gamma} B_{n-1}^{-1} B_n .$$

Then choose a negative integer m so that

$$M = A_m^{-1} A_{m-1} \leq e^{\gamma} L < A_{m+1}^{-1} A_m .$$

This implies that $\log A_m$ changes slope at m, hence $A_m = |\alpha_m|$. Likewise, $B_n = |\beta_n|$. Set $p = n - m$. Then

$$0 = \alpha_m^{-1}\beta_n^{-1}\sum_{k=-\infty}^{\infty}\alpha_k\beta_{k+p} = \alpha_m^{-1}\beta_n^{-1}\Big(\sum_{k<-p}+\sum_{k=-p}^{0}+\sum_{k>0}\Big)\alpha_k\beta_{k+p} \ .$$

These three terms, denoted S_1, S_2 and S_3, will be estimated separately.

For $k \geq 0$, $|\beta_k| \leq B_k$. When $k = n+s$

$$B_{n+s} = B_n\prod_{i=0}^{s-1}B_{n+i}^{-1}B_{n+i+1} \leq L^s B_n, \quad s \geq 0 \ .$$

Similarly, if $k = n-s$,

$$B_{n-s} = B_n\Big(\prod_{i=1}^{s}B_{n-i}^{-1}B_{n-i+1}\Big)^{-1} \leq e^{-2\gamma s}L^{-s}B_n, \quad 1 \leq s \leq n \ .$$

Likewise, for $k \leq 0$, $|\alpha_k| \leq A_k$; and

$$A_{m-s} \leq M^s A_m \leq e^{\gamma s}L^s A_m, \quad s \geq 0$$

and

$$A_{m+s} \leq M^{-s}A_m \leq e^{-\gamma s}L^{-s}A_m, \quad 1 \leq s \leq |m| \ .$$

Hence for $-n \leq s \leq |m|$,

$$A_m^{-1}B_n^{-1}A_{m+s}B_{n+s}\begin{cases}\leq e^{\gamma|s|}L^{|s|}e^{-2\gamma|s|}L^{-|s|} = e^{-\gamma|s|} & -n \leq s \leq -1\\ = 1 & s = 0\\ \leq e^{-\gamma s}L^{-s}L^s = e^{-\gamma s} & 1 \leq s \leq |m|\end{cases} \ .$$

Consequently,

$$|S_2| = \Big|\alpha_m^{-1}\beta_n^{-1}\sum_{k=-p}^{0}\alpha_k\beta_{k+p}\Big| \geq 1 - \sum_{\substack{k=-p\\ k\neq m}}^{0}A_m^{-1}B_n^{-1}A_kB_{k+p}$$

$$\geq 1-2\sum_{s=1}^{\infty}e^{-\gamma s} = \frac{1-3e^{-\gamma}}{1-e^{-\gamma}} > 0 \ .$$

This term is bounded below independent of the choice of n. So a contradiction will be obtained if the other two terms can be made arbitrarily small for n large.

The estimates above yield as special cases:

$$A_m^{-1}B_n^{-1}A_{-p}B_0 \leq e^{-\gamma n} \quad \text{and} \quad A_m^{-1}B_n^{-1}A_0B_p \leq e^{-\gamma|m|} \ .$$

Hence

$$A_m^{-1}B_n^{-1} \leq \min\{e^{-\gamma n}A_{-p}^{-1}, e^{-\gamma|m|}B_p^{-1}\} \ .$$

Since $B_n^{-1}B_{n+1}$ is monotone decreasing, $B_p^{-1}B_{p+k} \leq B^{-1}B_{k+1}$. So,

$$|S_3| = \Big|A_m^{-1}B_n^{-1}\sum_{k>0}\alpha_k\beta_{k+p}\Big| \leq e^{-\gamma|m|}B_p^{-1}\sum_{k>0}|\alpha_k|B_{k+p}$$

$$\leq e^{-\gamma|m|}B_1^{-1}\sum_{k>0}|\alpha_k|B_{k+1} \quad .$$

It remains to show that $\sum_{k>0}|\alpha_k|B_{k+1} < \infty$. But $\log\pi_k$ is a convex majorant of $\{\log|\beta_k|\}$ since w_k is monotone. Thus $B_{k+1} \leq \pi_{k+1}$. Therefore,

$$|\alpha_k|B_{k+1} \leq \pi_k^{-1}|a_k|\pi_{k+1} = w_k|a_k| < w_k \quad .$$

This is summable, so $|S_3|$ can be made arbitrarily small by a sufficiently large n (and hence $|m|$). Similarly, $|S_1|$ can be made small. This contradicts $S_1+S_2+S_3 = 0$, and hence $L(x)$ must belong to $\mathcal{N}$. ∎

21.B Unitary Orbits of Nest Algebras. Let $\mathcal{N}$ be a nest. Let $\mathcal{U}(\mathcal{N})$ denote the norm closure of $\{UTU^*: T \in \mathcal{T}(\mathcal{N}),\ U \text{ unitary}\}$. Similarly, let

$$\begin{aligned}\mathcal{U}_a(\mathcal{N}) &= \{UTU^*+K: T \in \mathcal{T}(\mathcal{N}),\ U \text{ unitary},\ K \text{ compact}\}\\ &= \{UTU^*: T \in \mathcal{QT}(\mathcal{N}),\ U \text{ unitary}\}\end{aligned}$$

and

$$\begin{aligned}\mathcal{U}_a^0(\mathcal{N}) = \{A \in \mathcal{B}(\mathcal{H})&: \text{for all } \epsilon > 0, \text{ there are } T \text{ in } \mathcal{T}(\mathcal{N}),\\ &U \text{ unitary, and } K \text{ compact such that } \|K\| < \epsilon\\ &\text{and } A = UTU^*+K\} \quad .\end{aligned}$$

Note the easy relations $\mathcal{U}_a^0(\mathcal{N}) \subseteq \mathcal{U}(\mathcal{N}) \subseteq \overline{\mathcal{U}_a^0(\mathcal{N})}$ and $\mathcal{U}_a^0(\mathcal{N}) \subseteq \mathcal{U}_a(\mathcal{N})$. The analysis of these sets requires detailed information about the structure of operators which cannot be dealt with in this book. A good reference for this material is Herrero [2].

Recall that an operator A is *quasitriangular* if there is an increasing sequence P_n of finite rank projections with $\bigvee_{n\geq 1} P_n = I$ such that $\lim_{n\to\infty}\|P_n^\perp AP_n\| = 0$. Let **QT** denote the set of quasitriangular operators. It is an easy exercise (Exercise 12.1) to show that every quasitriangular operator A can be expressed as $A = T_\epsilon+K_\epsilon$ where T_ϵ is triangular and K_ϵ is compact with $\|K_\epsilon\| < \epsilon$. Since **QT** is norm closed, it follows from this simple observation and its converse that if $\mathcal{N}$ is a nest of order type $\omega+1$ with only finite dimensional atoms, then

$$\mathcal{U}(\mathcal{N}) = \mathcal{U}_a^0(\mathcal{N}) = \mathcal{U}_a(\mathcal{N}) = \mathbf{QT} \quad .$$

Likewise, if $\mathcal{N}$ has order type $(\omega+1)^*$ with finite rank atoms, then

$$\mathcal{U}(\mathcal{N}) = \mathcal{U}_a^0(\mathcal{N}) = \mathcal{U}_a(\mathcal{N}) = \mathbf{QT}^* \quad .$$

Let $\sigma_0(T)$ denote the isolated eigenvalues of T of finite multiplicity. If λ belongs to $\sigma_0(T)$, let $E_T\{\lambda\}$ denote the Riesz projection corresponding to the eigenspace for λ. When X is a compact subset of the plane, let $X^\wedge$

denote the polynomially convex hull of X. The complement $\mathbb{C}\backslash X$ has a unique unbounded component O_∞, and $X^\wedge = \mathbb{C}\backslash O_\infty$. Thus the components of $X^\wedge$ are always simply connected. Define

$$B(\mathcal{H})_d = \{T \in B(\mathcal{H}): \sum_{\lambda \in \sigma_0(T)\backslash\sigma_e(T)^\wedge} rank\, E_T\{\lambda\} \leq d\} .$$

We can now state the main theorem.

21.7 Theorem. *Let $\mathcal{N}$ be a nest on a separable space.*

1) *If $\mathcal{N}$ is well ordered with finite dimensional atoms, then*

$$\mathcal{U}(\mathcal{N}) = \mathcal{U}_a^0(\mathcal{N}) = \mathcal{U}_a(\mathcal{N}) = \mathbf{QT} .$$

2) *If $\mathcal{N}^\perp$ is well ordered with finite dimensional atoms, then*

$$\mathcal{U}(\mathcal{N}) = \mathcal{U}_a^0(\mathcal{N}) = \mathcal{U}_a(\mathcal{N}) = \mathbf{QT}^* .$$

3) *If neither (1) nor (2) holds, let $d = \sum_{A \in \mathbf{A}} dim A$, where $\mathbf{A}$ denotes the set of atoms of $\mathcal{N}$. Then*

3∞) *when $d = \infty$, $\mathcal{U}(\mathcal{N}) = \mathcal{U}_a^0(\mathcal{N}) = \mathcal{U}_a(\mathcal{N}) = B(\mathcal{H})$;*

3d) *when $d < \infty$, $\mathcal{U}(\mathcal{N}) = \mathcal{U}_a^0(\mathcal{N}) = B(\mathcal{H})_d$ and $\mathcal{U}_a(\mathcal{N}) = B(\mathcal{H})$.*

We now state without proof some results that are required. A few of the easier results will be included in the Exercises.

21.8 Lemma. *Let T be a bounded operator on $\mathcal{H}^{(\infty)}$ with matrix (T_{ij}). Suppose T is block upper triangular ($T_{ij} = 0$ if $j < i$) and T_{ii} is quasitriangular for all $i \geq 1$. Then T is quasitriangular.*

21.9 Theorem (Apostol, Foias, Voiculescu). *An operator T in $B(\mathcal{H})$ is quasitriangular if and only if $\{\lambda : T-\lambda I$ is semi Fredholm, $ind(T-\lambda I) < 0\}$ is empty.*

21.10 Theorem (Weyl-von Neumann-Berg). *Let N and M be normal operators such that $\sigma_e(N) = \sigma_e(M)$ and $\sigma_0(N) = \sigma_0(M)$ including multiplicity. Then $N \underset{a}{\sim} M$.*

21.11 Theorem (Voiculescu). *Let T be any operator in $B(\mathcal{H})$. Suppose ρ is a representation of $C^*(T)+K/K$ and $A = \rho(\tilde{T})$. Then $T \underset{a}{\sim} T \oplus A^{(\infty)}$.*

The following lemma is an easy step in the proof of Theorem 21.9.

21.12 Lemma. *Let T be an operator in $\mathcal{B}(\mathcal{H})$. Given $\epsilon > 0$ and normal operators M_1 and M_2 such that $\sigma(M_1) \subseteq \sigma_{\ell e}(T)$ and $\sigma(M_2) \subseteq \sigma_{re}(T)$ (the left and right essential spectrum of T, respectively) , there is a compact operator K such that $\|K\| < \epsilon$ and*

$$T-K \simeq \begin{bmatrix} M_1 & * & * \\ 0 & * & * \\ 0 & 0 & M_2 \end{bmatrix} .$$

(Either M_1 or M_2 may be absent.)

21.13 Corollary. *If $\mathcal{N}$ has an infinite dimensional atom, then*

$$\mathcal{U}_a^0(\mathcal{N}) = \mathcal{U}(\mathcal{N}) = \mathcal{U}_a(\mathcal{N}) = \mathcal{B}(\mathcal{H}) .$$

Proof. Let the atom be $A = N_2 \ominus N_1$. Given T in $\mathcal{B}(\mathcal{H})$, pick a constant λ in $\sigma_e(T)$, and set $M_1 \simeq \lambda I \,|N_1$ and $M_2 \simeq \lambda I \,|N_2^\perp$. By the previous lemma, there is a compact operator K so that $\|K\| < \epsilon$, and

$$T-K \simeq \begin{bmatrix} \lambda I & * & * \\ 0 & * & * \\ 0 & 0 & \lambda I \end{bmatrix}$$

acting on $N_1 \oplus A \oplus N_2^\perp$. Clearly, this belongs to $\mathcal{T}(\mathcal{N})$. Hence $\mathcal{U}_a^0(\mathcal{N}) = \mathcal{B}(\mathcal{H})$, and the lemma follows. ∎

21.14 Lemma. *If $\mathcal{N}$ has an increasing sequence of atoms, then $\mathcal{U}_a^0(\mathcal{N})$ contains* **QT**.

Proof. It clearly suffices to show that **QT** is contained in $\mathcal{U}_a^0(\mathcal{N}_0)$ where $\mathcal{N}_0$ is a larger nest containing $\mathcal{N}$. So it may be assumed that there is an increasing sequence $A_n = N_n \ominus M_n$ of one dimensional atoms. Let T be a quasitriangular operator. Let ρ be any faithful representation of $C^*(\widetilde{T})$, and let $A = \rho(T)$. By Voiculescu's Theorem,

$$T \underset{a}{\sim} T \oplus A^{(\infty)} \simeq T \oplus (A^{(\infty)})^{(\infty)} .$$

Now if $A-\lambda I$ is left invertible, then $T-\lambda I$ is semi Fredholm and hence has $ind(T-\lambda I) \geq 0$. Thus $(T-\lambda I)^\sim$ is right invertible in the Calkin algebra, whence $A-\lambda I$ is invertible. Thus

$$\sigma(A^{(\infty)}) = \sigma_e(A^{(\infty)}) = \sigma_{\ell e}(A^{(\infty)}) .$$

In particular, $ind(A^{(\infty)}-\lambda I) = 0$ or $+\infty$. By Theorem 21.9, $A^{(\infty)}$ is quasitriangular.

Pick λ in $\sigma(A^{(\infty)})$, which is contained in $\sigma_e(T)$. Let $N_\omega = \bigvee_{n \geq 1} N_n$. Define normal operators $J_0 = \lambda I \,|N_\omega^\perp$ and $J_n = \lambda I \,|M_n \ominus N_{n-1}$ for $n \geq 1$. By Lemma 21.12, given $\epsilon > 0$, there are compact operators K_n,

$n \geq 0$ such that $\|K_n\| < 2^{-n-1}\epsilon$ such that

$$T - K_0 \simeq \begin{bmatrix} B_0 & * \\ 0 & J_0 \end{bmatrix} \quad \text{and} \quad A^{(\infty)} - K_n \simeq \begin{bmatrix} J_n & * \\ 0 & B_n \end{bmatrix} .$$

Moreover, it follows that B_n are quasitriangular. Hence we can assume that an additional compact perturbation has been included in K_n, $n \geq 0$ so that B_n are triangular with respect to bases $\{e_{n,k} : k \geq 0, n \geq 0\}$.

Let $K = \sum_{n \geq 0} \oplus K_n$. Since $\|K\| < \epsilon$, there is a compact operator K' with $\|K'\| < \epsilon$ such that

$$\begin{aligned} T - K' &\simeq T - K_0 \oplus \sum_{n \geq 1} \oplus A^{(\infty)} - K_n \\ &\simeq \begin{bmatrix} B_0 & * \\ 0 & J_0 \end{bmatrix} \oplus \sum_{n \geq 1} \oplus \begin{bmatrix} J_n & * \\ 0 & B_n \end{bmatrix} = T_0 . \end{aligned}$$

It remains to show that T_0 is unitarily equivalent to an operator in $\mathcal{T}(\mathcal{N})$. Let f_n be a unit vector in the atom A_n. Let U be a unitary such that

$$\begin{aligned} Ue_{n,k} &= f_{2^n + 2^{n+1}k} \qquad n \geq 0, k \geq 0, \\ U(dom\ J_0) &= N_\omega^\perp, \\ U(dom\ J_n) &= M_n \ominus N_{n-1},\ n \geq 1. \end{aligned}$$

Then UT_0U^* is scalar valued on the intervals $M_n \ominus N_{n-1}$ and $N_\omega^\perp$, and is upper triangular with respect to the nest

$$\{0 = N_0, M_1, N_1, \ldots, N_\omega, \mathcal{H}\} .$$

Thus UT_0U^* belongs to $\mathcal{T}(\mathcal{N})$. ∎

21.15 Corollary. *If $\mathcal{N}$ is well ordered with only finite dimensional atoms, then $\mathcal{U}(\mathcal{N}) = \mathcal{U}_a^0(\mathcal{N}) = \mathcal{U}_a(\mathcal{N}) =$* **QT**.

Proof. By Lemma 21.14, $\mathcal{U}_a^0(\mathcal{N})$ contains **QT**. Clearly, it suffices to show that $\mathcal{T}(\mathcal{N})$ is contained in **QT**. This will be shown by induction on the ordinal type of $\mathcal{N}$ (always a successor ordinal $\alpha+1$ since $\mathcal{H}$ is always the largest element of $\mathcal{N}$). This result has already been noted for the first case $\omega+1$. If $\alpha = \beta+1$ and it holds for nests isomorphic to $\beta+1$, then every T in $\mathcal{T}(\mathcal{N})$ has the form

$$T = \begin{bmatrix} T_1 & K_1 \\ 0 & K_2 \end{bmatrix}$$

where T_1 belongs to a nest order isomorphic to $\beta+1$, and K_2 acts on a finite dimensional space. Thus T_1 is quasitriangular by the induction hypothesis. Clearly, T is also quasitriangular. So the result follows for $\alpha+1$.

The other possibility is that α is a limit ordinal. Since α is countable, there are ordinals $\alpha_k < \alpha$ such that $\alpha = \sum_{k \geq 1} \alpha_k$. So every T in $\mathcal{T}(\mathcal{N})$ where $\mathcal{N}$ has order type $\alpha+1$ decomposes as

$$T \simeq \begin{bmatrix} T_1 & * & * & * \\ & T_2 & * & * \\ & & T_3 & * \\ & 0 & & \ddots \end{bmatrix}$$

where T_k belongs to $\mathcal{T}(\mathcal{N}_k)$ where $\mathcal{N}_k$ has order type α_k+1. By the induction hypothesis, each T_k is quasitriangular. Then Lemma 21.8 implies that T is also quasitriangular. Hence the result follows for α. The induction is complete and the corollary follows. ■

21.16 Corollary. *If $\mathcal{N}^\perp$ is well ordered with only finite dimensional atoms, then*

$$\mathcal{U}(\mathcal{N}) = \mathcal{U}_a^0(\mathcal{N}) = \mathcal{U}_a(\mathcal{N}) = \mathbf{QT}^* \ .$$

To progress further, we need more structural lemmas. These are proved using the techniques developed for Theorem 21.9. They will be stated here without proof. Caveat: Lemmas 21.17 and 21.18 are different.

21.17 Lemma. *Let T be an operator in $\mathcal{B}(\mathcal{H})$ and let $\epsilon > 0$. Then there is a compact operator K with $\|K\| < \epsilon$, and there are quasitriangular operators T_1 and T_2 in $\mathcal{B}(\mathcal{H})$ so that*

$$T-K \simeq \begin{bmatrix} T_1 & * \\ 0 & T_2^* \end{bmatrix} .$$

Furthermore, it may be arranged that either $\sigma_0(T_1)$ or $\sigma_0(T_2)$ is empty.

21.18 Lemma. *Let T be an operator in $\mathcal{B}(\mathcal{H})$, and let $\epsilon > 0$. Then there is a compact operator K with $\|K\| < \epsilon$, and there are quasitriangular operators T_1 and T_2 in $\mathcal{B}(\mathcal{H})$ such that*

$$T-K \simeq \begin{bmatrix} T_1^* & * \\ 0 & T_2 \end{bmatrix} .$$

Furthermore, it may be arranged that either $\sigma_0(T_1)$ or $\sigma_0(T_2)$ is empty.

21.19 Corollary. *If $\mathcal{N}$ has both an increasing sequence of atoms and a decreasing sequence of atoms, then $\mathcal{U}_a^0(\mathcal{N}) = \mathcal{U}(\mathcal{N}) = \mathcal{U}_a(\mathcal{N}) = \mathcal{B}(\mathcal{H})$.*

Proof. $\mathcal{N}$ must contain an element N_0 such that $\mathcal{N}_1 = \{N \in \mathcal{N}: N \leq N_0\}$ and $\mathcal{N}_2 = \{N \ominus N_0: N \in \mathcal{N}, N \geq N_0\}$ are infinite nests, one containing an increasing sequence of atoms and the other containing a decreasing sequence of atoms. By Lemma 21.14, one of $\mathcal{T}(\mathcal{N}_1)$ contains $\mathbf{QT}$ and the other contains $\mathbf{QT}^*$. So $\mathcal{U}_a^0(\mathcal{N})$ contains either

$$\begin{bmatrix} \mathbf{QT} & * \\ 0 & \mathbf{QT}^* \end{bmatrix} \text{ or } \begin{bmatrix} \mathbf{QT}^* & * \\ 0 & \mathbf{QT} \end{bmatrix}.$$

Hence Lemma 21.17 or Lemma 21.18 implies that $\mathcal{U}_a^0(\mathcal{N}) = \mathcal{B}(\mathcal{H})$. ∎

We state two more lemmas without proof.

21.20 Lemma. *Let T be an operator in $\mathcal{B}(\mathcal{H})$, and let $\epsilon > 0$. There is a compact operator K with $\|K\| < \epsilon$, an operator R such that*

$$\sigma_e(R) = \sigma_{\ell e}(R) = \sigma_{re}(R) = \sigma_e(T)^\wedge \text{ and } \sigma_0(R) = \sigma_0(T)\backslash\sigma_e(T)^\wedge ,$$

and bounded analytic functions h_n in H^∞ corresponding to Toeplitz operators T_{h_n} on H^2 such that

$$T - K \simeq \begin{bmatrix} \sum\limits_{n \geq 1} \oplus T_{h_n}^{(\infty)} & * & * \\ 0 & R & * \\ 0 & 0 & \sum\limits_{n \geq 1} \oplus T_{h_n}^{*(\infty)} \end{bmatrix}.$$

21.21 Lemma. *Suppose that T is a biquasitriangular operator ($T \in \mathbf{QT} \cap \mathbf{QT}^*$) such that $\sigma_0(T)$ is empty. Let $\epsilon > 0$ be given. Then there is a compact operator K with $\|K\| < \epsilon$ and there are normal operators N_j and block diagonal operators D_j, $1 \leq j \leq 4$, such that $\sigma(D_j) \subseteq \sigma(N_j) = \sigma_e(N_j) = \sigma_e(T)$ and*

$$T - K \simeq \begin{bmatrix} N_1 \oplus D_1 & * & * & * \\ 0 & N_2 \oplus D_2 & * & * \\ 0 & 0 & N_3 \oplus D_3 & * \\ 0 & 0 & 0 & N_4 \oplus D_4 \end{bmatrix}.$$

21.22 Lemma. *Let* $D = \sum_{n \geq 1} \oplus D_n$ *be a direct sum of finite rank operators, and let* N *be a normal operator such that* $\sigma(D) \subseteq \sigma_e(N) = \sigma(N)$. *Given* $\epsilon > 0$, *there is a compact operator* K *with* $\|K\| < \epsilon$ *such that*

$$N \oplus D - K \simeq \begin{bmatrix} \lambda_1 I & & & \\ & \lambda_2 I & & * \\ & & \lambda_3 I & \\ & 0 & & \ddots \end{bmatrix}$$

where the matrix acts on $\mathcal{H}^{(\infty)}$.

Proof. Put each D_n in triangular form with diagonal entries $\{\lambda_{n,j}: 1 \leq j \leq j_n\}$. Then $\sigma(D)$ contains $\bigcup_{n \geq 1} \sigma(D_n)$, and thus so does $\sigma(N)$. By the Weyl-von Neumann-Berg Theorem,

$$N \underset{a}{\sim} \sum_{n \geq 1} \oplus \lambda_{n,j} I \oplus \sum_{n \geq 1} \oplus \mu_k I$$

where $\{\mu_k, k \geq 1\}$ is a dense subset of $\sigma(N)$. Hence

$$N \oplus D \overset{a}{\simeq} \sum_{n \geq 1} \oplus \begin{bmatrix} \mu_n I & & & \\ & \lambda_{n,1} I & & 0 \\ & & \lambda_{n,2} I & \\ & 0 & & \lambda_{n,j_n} I \end{bmatrix} \oplus \begin{bmatrix} \mu_{n,1} & & \\ & \lambda_{n,2} & * \\ & 0 & \lambda_{n,j_n} \end{bmatrix}$$

$$\simeq \sum_{n \geq 1} \oplus \begin{bmatrix} \mu_n I & & & \\ & \lambda_{n,1} I & & * \\ & & \lambda_{n,2} I & \\ & 0 & & \lambda_{n,j_n} I \end{bmatrix}. \qquad \blacksquare$$

21.23 Corollary. *Let* $\mathcal{N}$ *be a continuous nest. Then* $\mathcal{U}_a^0(\mathcal{N})$ *contains* $\mathcal{B}(\mathcal{H})_0$. *Hence* $\mathcal{U}_a(\mathcal{N}) = \mathcal{B}(\mathcal{H})$.

Proof. Let T be an operator in $\mathcal{B}(\mathcal{H})$ such that $\sigma_0(T)$ is contained in $\sigma_e(T)^\wedge$. Given $\epsilon > 0$, apply Lemma 21.20 to produce a small perturbation

$$T\text{–}K \simeq \begin{bmatrix} \sum_{n\geq 1} \oplus T_{h_n}^{(\infty)} & * & * \\ 0 & R & * \\ 0 & 0 & \sum_{n\geq 1} \oplus T_{h_n}^{*(\infty)} \end{bmatrix},$$

and $\sigma(R) = \sigma_e(R) = \sigma_{re}(R) = \sigma_{\ell e}(R) = \sigma_e(T)^\wedge$. Theorem 21.9 implies that R is biquasitriangular. Thus an application of Lemmas 21.21 and 21.22 produces a small compact perturbation R_0 of R with an upper triangular form on $H^{(\infty)} \oplus H^{(\infty)} \oplus H^{(\infty)} \oplus H^{(\infty)}$ with scalar entries on the diagonal.

Split the continuous nest $\mathcal{N}$ into three intervals, each of which is a continuous nest $\mathcal{N}_i$, $1 \leq i \leq 3$. The nest $\mathcal{N}_2$ can be decomposed into intervals order equivalent to $\omega+\omega+\omega+\omega$. Let U_2 be a unitary carrying $H^{(\infty)} \oplus H^{(\infty)} \oplus H^{(\infty)} \oplus H^{(\infty)}$ onto the domain of $\mathcal{N}_2$ matching each copy of H to the corresponding interval. Then since the diagonal entries of R_0 are scalar, $U_2R_0U_2^*$ belongs to $\mathcal{T}(\mathcal{N}_2)$.

Let $S^{(\infty)}$ denote the unilateral shift of infinite multiplicity. Let $\mathcal{M}$ be any continuous nest, and partition $\mathcal{M}$ into intervals $\{E_k, k \geq 1\}$ with $E_{k+1} \ll E_k$. Then $S^{(\infty)}$ is unitarily equivalent to any strictly upper triangular isometry that carries each E_k onto E_{k+1}. So $S^{(\infty)}$ belongs to $\mathcal{T}(\mathcal{M}_0)$ where $\mathcal{M}_0$ is unitarily equivalent to $\mathcal{M}$. Hence so does every operator in the WOT closed algebra generated by $S^{(\infty)}$, which includes $T_h^{(\infty)}$ for every h in $H^{(\infty)}$. Now split $\mathcal{N}_1$ into countably many intervals F_n. By the above remarks, an operator unitarily equivalent to $T_{h_n}^{(\infty)}$ belongs to $\mathcal{T}(F_n\mathcal{N}_1)$. Hence $\sum_{n\geq 1} \oplus T_{h_n}^{(\infty)}$ is unitarily equivalent to an operator in $\mathcal{T}(\mathcal{N}_1)$.

Since $\mathcal{T}(\mathcal{N}_3)^* = \mathcal{T}(\mathcal{N}_3^\perp)$ is also a continuous nest algebra, $\sum_{n\geq 1} \oplus T_{h_n}^{*(\infty)}$ is unitarily equivalent to an operator in $\mathcal{T}(\mathcal{N}_3)$. Putting all the pieces together demonstrates that T belongs to $\mathcal{U}_a^0(\mathcal{N})$.

Every operator T has a compact perturbation with no isolated eigenvalues of finite multiplicity. Hence $\mathcal{U}_a(\mathcal{N}) = \mathcal{B}(\mathcal{H})$. ∎

21.24 Corollary. *Let $\mathcal{N}$ be a nest such that neither $\mathcal{N}$ nor $\mathcal{N}^\perp$ is well ordered. Then $\mathcal{U}_a^0(\mathcal{N})$ contains $\mathcal{B}(\mathcal{H})_0$ and hence $\mathcal{U}_a(\mathcal{N}) = \mathcal{B}(\mathcal{H})$. If the span of the atoms is infinite dimensional, then*

$$\mathcal{U}_a^0(\mathcal{N}) = \mathcal{U}(\mathcal{N}) = \mathcal{U}_a(\mathcal{N}) = \mathcal{B}(\mathcal{H}) .$$

Proof. First assume $d = \infty$. If $\mathcal{N}$ has an infinite dimensional atom, apply Corollary 21.13. If $\mathcal{N}$ has both an increasing sequence of atoms and a decreasing sequence of atoms, apply Corollary 21.19. Otherwise, the atoms of $\mathcal{N}$ are either well ordered or those of $\mathcal{N}^\perp$ are. Assume the former. Since $\mathcal{N}$ is not well ordered, it must have a continuous part. As the atomic part is well ordered, $\mathcal{N}$ contains an interval E_2 on which $\mathcal{N}$ is continuous. This splits $\mathcal{N}$ into intervals E_1, E_2, E_3. One of E_1 or E_3 contains an increasing sequence of atoms. For definiteness, say E_3 has this property.

By Lemma 21.18, for every operator T in $\mathcal{B}(\mathcal{H})$ and $\epsilon > 0$, there is a compact operator K_1 with $\|K_1\| < \epsilon$ such that

$$T-K_1 \simeq \begin{bmatrix} T_1^* & * \\ 0 & T_2 \end{bmatrix}$$

where T_2 is quasitriangular and $\sigma_0(T_1^*)$ is empty. Let λ be any point in $\sigma_e(T_1^*)$, and let $M_1 = \lambda I\,|E_1$. By Lemma 21.12, a second small compact perturbation yields

$$T-K_1-K_2 \simeq \begin{bmatrix} M_1 & * & * \\ 0 & B & * \\ 0 & 0 & T_2 \end{bmatrix}$$

where $\sigma_0(B) = \sigma_0(T_1^*)$ is empty. By Corollary 21.15, T_2 belongs to $\mathcal{U}_a^0(\mathcal{N}E_3)$. By Corollary 21.23, B belongs to $\mathcal{U}_a^0(\mathcal{N}E_2)$. Clearly, M_1 belongs to $\mathcal{T}(\mathcal{N}E_1)$. Hence T belongs to $\mathcal{U}_a^0(\mathcal{N})$; and $\mathcal{U}_a^0(\mathcal{N}) = \mathcal{B}(\mathcal{H})$.

If $d < \infty$ (i.e. $\mathcal{N}$ has finitely many, finite rank atoms), then $\mathcal{N}$ has an interval E_2 on which $\mathcal{N}$ is continuous. As in the proof of Corollary 21.13, every operator T has a small compact perturbation of the form

$$T-K \simeq \begin{bmatrix} \lambda I & * & * \\ 0 & T_0 & * \\ 0 & 0 & \lambda I \end{bmatrix} .$$

If $\sigma_0(T)\backslash\sigma_e(T)^\wedge$ is empty, the same follows for T_0. Then by Corollary 21.23, T_0 belongs to $\mathcal{U}_a^0(\mathcal{N}E_2)$ and hence T belongs to $\mathcal{U}_a^0(\mathcal{N})$. Hence $\mathcal{U}_a^0(\mathcal{N})$ contains $\mathcal{B}(\mathcal{H})_0$ and $\mathcal{U}_a(\mathcal{N}) = \mathcal{B}(\mathcal{H})$. ■

The full Theorem 21.7 has been proven except for the following:

21.25 Lemma. *Let $\mathcal{N}$ be a nest with finitely many finite atoms of total dimension d. Then $\mathcal{U}_a^0(\mathcal{N}) = \mathcal{U}(\mathcal{N}) = \mathcal{B}(\mathcal{H})_d$.*

Proof. Let T belong to $\mathcal{B}(\mathcal{H})$, and let

$$p = \sum_{\lambda \in \sigma_0(T)\backslash\sigma_e(T)^\wedge} \mathit{rank}\ E_T\{\lambda\} .$$

If $p > d$, there is a finite subset $S = \{\lambda_1,\dots,\lambda_n\}$ of $\sigma_0(T)\backslash\sigma_e(T)^\wedge$ with rank $E_T(S) > d$. The characteristic function of $\{\lambda_1,\dots,\lambda_n\}$ can be approximated uniformly in a neighbourhood of $\sigma(T)$ by polynomials (Runge's Theorem). Hence $E_T(S)$ belongs to the WOT closed algebra generated by T.

Consequently, if T belongs to $\mathcal{T}(\mathcal{N})$, so does $E_T(S)$. By Ringrose's Theorem 3.4, the atoms of $\mathcal{N}$ have total dimension at least rank $E_T(S)$, contrary to fact. Moreover, $\mathcal{B}(\mathcal{H})_d$ is closed because every operator sufficiently close to T will have isolated eigenvalues near S of the same dimension. Hence $\mathcal{U}_a^0(\mathcal{N})$ is contained in $\mathcal{B}(\mathcal{H})_d$, as is $\mathcal{U}(\mathcal{N})$ since it is contained in $\overline{\mathcal{U}_a^0(\mathcal{N})}$.

On the other hand, suppose $p \leq d$. Then there is an invertible operator S so that

$$STS^{-1} \simeq A \oplus T_0$$

where A acts on $\mathbb{C}^p$ and T_0 belongs to $\mathcal{B}(\mathcal{H})_0$. Clearly, T belongs to $\mathcal{U}_a^0(\mathcal{N})$ if and only if STS^{-1} belongs to $\mathcal{U}_a^0(S\mathcal{N})$. But $S\mathcal{N}$ has precisely the same atoms, so we may assume that $T = A \oplus T_0$. Since A can be put in triangular form, A is unitarily equivalent to an operator B in $\mathcal{T}(\mathcal{N})$ where $B = P(E)BP(E)$ and E is a projection in $\mathcal{N}'$ of rank p. Since $\mathcal{N}P(E)^{\perp}$ has a continuous part, Corollary 21.24 shows that T_0 belongs to $\mathcal{U}_a^0(\mathcal{N}P(E)^{\perp})$. Hence T belongs to $\mathcal{U}_a^0(\mathcal{N})$ as desired. ∎

Notes and Remarks.

Proposition 21.2 was proven for $\omega_n = 2^{-n}$ by Donoghue [1], and in this generality by Nikol'skii [1]. Nikol'skii also showed that the monotonicity is essential. Yakubovic [1] showed that w_n monotone decreasing to 0 is sufficient for unicellularity. Harrison and Longstaff [1] constructed a unicellular operator with lattice $\omega+\omega+1$. Corollary 21.6 is in Barria [1]. Theorem 21.3 is taken from Barria and Davidson [1]. They prove $\alpha+1+\beta^*$ is an attainable lattice for all countable ordinals α and β. Theorem 21.7 is due to Domar [1]. He also shows that bilateral weighted shifts are not unicellular if

$$\liminf_{|n|\to\infty} \frac{\log \pi_n}{|n|} + \log|n| > -\infty \ .$$

Other nests which are attainable are constructed in Barria [2] and Rosenthal [1].

Theorem 21.7 is due to Herrero [3] and [4] and answers a question raised by Arveson in 1981. The reader should consult these papers for proofs of Lemmas 21.17, 21.18 and 21.20 which are omitted in our text. The other omitted material can be found in Herrero's monograph [2], or in its original sources. Lemma 21.8 is due to Douglas and Pearcy [1]. (See Exercise 21.4). Theorem 21.9 appears in a long series of papers by Apostol, Foias and Voiculescu [1]. The Weyl-von Neumann-Berg Theorem was first proved by Berg [2]. (See Exercises 7.1 and 7.2). Theorem 21.11 is due to Voiculescu [2], and is related to the proof of Andersen's Theorem presented in Chapter 12. Lemma 21.21 is contained in Voiculescu [1].

Exercises.

21.1 Suppose that $\|A_i^n\| = O((n!)^{-p})$ for $i = 1,2$, and let

$$T = \begin{bmatrix} A_1 & C \\ 0 & A_2 \end{bmatrix} .$$

Use Stirling's formula to show that $\|T^n\| = O((n!)^{-(p-\epsilon)})$ for all $\epsilon > 0$.

21.2 Construct a unicellular operator with lattice isomorphic to $\omega+(\omega+1)^*+\omega+1+\omega^*$. Hint: Apply Theorems 21.3 and 21.7.

21.3* Is the canonical nest on $\ell^2(\mathbb{Q})$ attainable?

21.4 (a) Let

$$T = \begin{pmatrix} T_{11} & T_{12} \\ 0 & T_{22} \end{pmatrix}$$

act on $\mathcal{H} \oplus \mathcal{H}$, and suppose T_{11} and T_{22} are in $\mathbf{QT}$. Prove that T is quasitriangular. Hint: Suppose that T_{ii} are triangular with respect to $\{P_n\}$ and $\{Q_n\}$ respectively. Show that given $\epsilon > 0$, there are integers n_k so that $\sum_{k \geq 1} \|P_{n_k}^{\perp} T_{12} Q_k\| < \epsilon$.

(b) Prove Lemma 21.8.

21.5 (a) Suppose T is semi Fredholm of negative index which is bounded below by one on $(\mathit{ker}\, T)^{\perp}$. Let M be a finite dimensional subspace of $(\mathit{Ran}\, T)^{\perp}$ such that $\mathit{dim}\, M > \mathit{null}\,(T)$. Show that if N is any subspace containing $M \vee (\mathit{ker}\, T)$, then $\|P(N)^{\perp} T P(N)\| \geq 1$.

(b) Show that if T is quasitriangular, then whenever $T-\lambda I$ is semi Fredholm, $\mathit{ind}(T-\lambda I) \geq 0$.

IV. CSL ALGEBRAS

22. Commutative Subspace Lattices

In this chapter, we will examine the class of reflexive algebras with *commutative subspace lattice*, meaning a complete lattice of subspaces $\mathcal{L}$ such that $\{P(L):L \in \mathcal{L}\}$ is abelian. This includes all nest algebras, and some of the results of this chapter are not much easier in the nest case. It will be shown that $\mathcal{L}$ is reflexive (Theorem 22.9) and that there is a smallest weak* closed subalgebra of $Alg\,\mathcal{L}$ containing a maximal abelian von Neumann algebra, denoted $\mathcal{A}_{\min}$, with the property that $Lat\,\mathcal{A}_{\min} = \mathcal{L}$ (Theorem 22.16). In the case of a nest and in many other cases, one has $\mathcal{A}_{\min} = Alg\,\mathcal{L}$ (22.19-22.25). This generalizes some of the results of Chapter 15.

22.1 Example. Let (X,m) be a compact metric space with a finite regular Borel measure m. Let a pre-order $\leq$ given by a countable family f_n of continuous real valued functions by the rule $x \leq y$ if $f_n(x) \leq f_n(y)$ for all n. This may fail to be a partial order since $x \leq y$ and $y \leq x$ may not imply $x = y$. Such a pre-order will be called a *standard pre-order.*

A Borel subset E of X is *increasing* if $x \in E$ and $x \leq y$ implies $y \in E$. Let $\mathcal{L}(X, \leq, m)$ denote the set of subspaces of $L^2(X,m)$ of functions supported on increasing sets. This is in fact a complete lattice. For if A and B are increasing sets, so are $A \cap B$ and $A \cup B$. Let the subspace corresponding to A be denoted by L_A and $P(L_A) \equiv P_A$. Then $L_A \vee L_B = L_{A \cup B}$ and $L_A \wedge L_B = L_{A \cap B}$. If $\{A_\lambda : \lambda \in \Lambda\}$ is a collection of increasing sets, let s be the supremum of $m(\bigcup_{\lambda \in F} A_\lambda)$ as F ranges over finite subsets of Λ. One can choose a countable subset $\{A_n\}$ such that $A = \bigcup_{n=1}^{\infty} A_n$ has measure s. Clearly, A is increasing. It will be shown that $L_A = \bigvee_{\lambda \in \Lambda} L_{A_\lambda}$. Denote the closed span of $\{L_{A_\lambda} : \lambda \in \Lambda\}$ by M. Since M contains L_{A_n} for each A_n, M contains $L_A = \bigvee_{n \geq 1} L_{A_n}$. On the other hand, suppose that for some λ, L_{A_λ} is not contained in L_A. Then $m(A \cup A_\lambda) > s$ which contradicts the definition of s. So $M = L_A$. Similarly, $\mathcal{L}(X, \leq, m)$ is closed under arbitrary intersections.

22.2 Lemma. *$\mathcal{L}(X, \leq, m)$ is closed in the strong operator topology.*

Proof. The space $L^2(X,m)$ is separable, and hence the strong operator topology is metrizable. The set of projections is strongly closed, as is $L^\infty(X,m)$. So if A_n are increasing sets, and P_{A_n} converges strongly to P, then $P = P_A$ for some Borel set A. Let 1 denote the constant function. Then

$$0 = \lim_{n \to \infty} \| P_A - P_{A_n} 1 \|^2 = \lim_{n \to \infty} m(A \,\Delta\, A_n)$$

(where $A \,\Delta B$ denotes the symmetric difference $(A \backslash B) \cup (B \backslash A)$.) Drop to a subsequence so that $m(A \,\Delta\, A_n) \leq 2^{-n}$. Let $B = \bigcap_{n=1}^{\infty} \bigcup_{k \geq n} A_k$. This is an increasing set. Moreover,

$$m(A \backslash \bigcup_{k \geq n} A_k) \leq \lim_{k \to \infty} 2^{-k} = 0$$

for all n, so $P_B \geq P$. Also

$$m(\bigcup_{k \geq n} A_k \backslash A) \leq \sum_{k \geq n} 2^{-k} = 2^{1-n} \quad .$$

Thus $m(B \backslash A) = 0$ and $P_B \leq P$. So $P = P_B$ belongs to $\mathcal{L}(X, \leq, m)$. ∎

It is always possible to find a countable family $\{E_n, n \geq 1\}$ of closed increasing sets so that

$$x \leq y \text{ if and only if } \{n : x \in E_n\} \subseteq \{n : y \in E_n\} \quad .$$

For example, let r_k be an enumeration of the rationals and let

$$E_{n,k} = \{x : f_n(x) \geq r_k\} \quad .$$

On the other hand, if E_n is such a collection, then the complete lattice generated by $\{P_{E_n}, n \geq 1\}$ is all of $\mathcal{L}(X, \leq, m)$. At this point, we prove this only in a special case.

Let $X = 2^{\aleph_0}$ be the space of sequences (x_k) of zeros and ones with the product topology. The sets $F_n = \{x = (x_k) : x_n = 1\}$ are closed and open, and $\{F_n, F_n^c, n \geq 1\}$ is a sub-base for this topology. The "even pre-order" is given by

$$x \leq y \Leftrightarrow \{n \in \mathbb{N} : x \in F_{2n}\} \subseteq \{n \in N : y \in F_{2n}\}$$

$$\Leftrightarrow x_{2n} \leq y_{2n}, \quad n \geq 1 \quad .$$

22.3 Lemma. *Let $(X, \leq)$ be $2^{\aleph_0}$ with the even pre-order, and let μ be a regular Borel measure on X. Then the complete lattice generated by $\{P_{F_{2n}} : n \in \mathbb{N}\}$ is precisely $\mathcal{L}(X, \leq, \mu)$.*

Proof. Let C^+ be the smallest increasing set containing a set C. It will be shown that C^+ is closed if C is closed. The graph of the pre-order,

$$G = \{(x,y) \in X^2 : y \leq x\}$$

is closed because $X^2 \backslash G = \bigcup_{n \geq 1} F_{2n}^c \times F_{2n}$ is open. Let π_1 and π_2 be the two coordinate projections. Then

$$C^+ = \pi_1(\pi_2^{-1}(C) \cap G)$$

is the continuous image of a compact set, and hence is closed.

Now let P be a projection in $\mathcal{L}(X, \leq, \mu)$, and let A be an increasing Borel set such that $P = P_A$. By the regularity of μ, there is an increasing sequence of closed sets C_k contained in A such that $\mu(A \backslash \bigcup_{n \geq 1} C_n) = 0$. Hence C_n^+ is an increasing sequence of closed increasing sets contained in A such that $\bigvee_{n \geq 1} P_{C_n^+} = P_A$. So it suffices to prove that P_C belongs to the lattice generated by $\{P_{F_{2n}}, n \in N\}$ for every closed increasing set C.

Let $F_n^{(1)} = F_n$ and $F_n^{(-1)} = F_n^c$. For each finite sequence $\epsilon = \{\epsilon_n, 1 \leq n \leq 2N\}$ of ± 1's, let

$$F_\epsilon = \bigcap_{n=1}^{2N} F_n^{(\epsilon_n)} .$$

These sets form a base for the topology. Since C is closed, $X \backslash C$ is the union of $\mathcal{E} = \{F_\epsilon : F_\epsilon \subseteq X \backslash C\}$. Since C is increasing, it follows that if F_ϵ is disjoint from C, then the decreasing set

$$F_\epsilon^- = \{x \in X : x_{2n} \leq (\epsilon_{2n}+1)/2, 1 \leq n \leq N\}$$

is also disjoint from C. Now, $X \backslash F_\epsilon^- = \bigcup \{F_{2n} : \epsilon_{2n} = -1\}$. So

$$C = \bigcap_{\mathcal{E}} (X \backslash F_\epsilon^-)$$

belongs to the complete lattice generated by $\{F_{2n}, n \geq 1\}$. ■

22.4 Theorem. *Let $\mathcal{L}$ be a commutative subspace lattice on a separable Hilbert space. Then $\mathcal{L}$ is unitarily equivalent to some $\mathcal{L}(X, \leq, m)$ of Example 22.1 where X is a compact metric space, m is a regular Borel measure on X, and $\leq$ is a standard pre-order.*

Proof. Let $\mathcal{L}''$ denote the abelian von Neumann algebra generated by $\{P(L) : L \in \mathcal{L}\}$. Let $\mathcal{M}$ be any maximal abelian von Neumann algebra containing $\mathcal{L}''$. Since the unit ball of $\mathcal{B}(\mathcal{H})$ is separable and metrizable in the strong operator topology, one can choose countable families of projections $\{P_n, n \geq 1\}$ and $\{Q_n, n \geq 1\}$ which are strongly dense in $\{P(L): L \in \mathcal{L}\}$ and $Proj(\mathcal{M})$ respectively. Let

$$A = \sum_{n=1}^{\infty} 3.9^{-n} P_n + 9^{-n} Q_n \quad .$$

Since each P_n and Q_n can be recovered from the spectral projections of A, it follows that $C^*(A) = C^*\{P_n, Q_n, n \geq 1\}$ and $W^*(A) = \mathcal{M}$. Let X be the spectrum of A. By Theorem 7.8, there is a finite regular Borel measure m on X such that $\mathcal{M}$ is unitarily equivalent to $L^\infty(X,m)$ acting on $L^2(X,m)$ by multiplication, and A corresponds to M_x.

The projections P_n and Q_n correspond to multiplication by the characteristic functions of closed and open sets, which we denote E_{2n} and E_{2n-1} respectively. Since $\{P_n, Q_n\}$ generate $C^*(A)$, the functions $\{\chi_{E_n}, n \geq 1\}$ separate the points of X. Define a pre-order $\leq$ by

$$x \leq y \quad \text{if} \quad \{n : x \in E_{2n}\} \subseteq \{n : y \in E_{2n}\} \quad .$$

Then each set E_{2n} in an increasing set. The unitary equivalence carries $\mathcal{L}$ onto the strongly closed lattice generated by $\{P_{E_{2n}}, n \geq 1\}$. This will be shown to be all of $\mathcal{L}(X, \leq, m)$.

Define a map Φ from X onto $Y = 2^{\aleph_0}$ by $\Phi(x)(n) = 1$ if and only if $x \in E_n$. Clearly, $\Phi^{-1}(F_n) = E_n$ and $\Phi^{-1}(F_n^c) = E_n^c$. As these sets are a sub-base for the topology of Y, Φ is continuous. Since $\{E_n\}$ separate points, Φ is a homeomorphism onto its image $\Phi(X)$. The order on $\Phi(X)$ induced from X is precisely the restriction of the even pre-order to $\Phi(X)$. Let μ be the regular Borel measure on Y given by

$$\mu(A) = m(\Phi^{-1}(A)) \quad .$$

There is a natural unitary equivalence of $L^2(X,m)$ and $L^2(Y,\mu)$. Let C be an increasing set in X, and let $D = \Phi(X)$. Let D^+ be the smallest increasing set in Y containing D, namely $\{y \in Y : \exists x \in D, y \geq x\}$. Since D is increasing in the relative order on $\Phi(X)$, it follows that $D = D^+ \cap \Phi(X)$. On the other hand, if A is an increasing set in Y, then $\Phi^{-1}(A)$ is an increasing set in X. This establishes a unitary equivalence between $\{P_C : C \in \mathcal{L}(X, \leq, m)\}$ and $\{P_A : A \in \mathcal{L}(Y, \leq, \mu)\}$. This unitary equivalence takes $\{P_{E_{2n}} : n \geq 1\}$ onto $\{P_{F_{2n}} : n \geq 1\}$. By Lemma 22.3, $\{P_{F_{2n}}, n \geq 1\}$ generate $\mathcal{L}(Y, \leq, \mu)$ as a complete lattice. Hence $\{P_{E_{2n}}, n \geq 1\}$ generate $\mathcal{L}(X, \leq, m)$ as a complete lattice. ∎

Contained in this proof is the fact that if $\{E_{2n}, n \geq 1\}$ is a collection of increasing sets which determine the order on $(X, \leq)$, then $\{P_{E_{2n}}, n \geq 1\}$ generates $\mathcal{L}(X, \leq, m)$ as a complete lattice. Just insert $P_{E_{2n}}$ for P_n in the proof above. The proof establishes a unitary equivalence between the complete lattice generated by $\{P_{E_{2n}}, n \geq 1\}$ and $\mathcal{L}(Y, \leq, \mu)$, and hence with $\mathcal{L}(X, \leq, m)$.

Unlike the nest algebra case, there is no longer an abundance of rank one operators in $Alg\,\mathcal{L}$ to prove reflexivity. A substitute is needed. Let μ be a finite positive regular Borel measure on X^2. The marginal measures are the measures on X given by

$$\mu_1(A) = \mu(A \times X) \text{ and } \mu_2(A) = \mu(X \times A) .$$

Let $\mathbf{A}(X,m)$ denote the linear space of signed regular Borel measures μ on X^2 such that $|\mu|_1$ and $|\mu|_2$ are absolutely continuous with respect to m and set

$$||\mu|| = \max_{i=1,2} ||\frac{d|\mu|_i}{dm}||_\infty < \infty .$$

If $\leq$ is a standard pre-order on X, let $G(\leq) = \{(x,y) \in X^2 : y \leq x\}$ be the *graph* of the pre-order. We construct certain "*pseudo integral*" operators in $Alg\,\mathcal{L}$.

22.5 Theorem. *Let μ belong to $\mathbf{A}(X,m)$. Then there is a unique operator T_μ acting on $L^2(X,m)$ such that*

$$(T_\mu f, g) = \int\int_{X^2} f(y)\overline{g(x)}d\mu$$

for all f, g in $L^2(X,m)$. Moreover, $||T_\mu|| \leq ||\mu||$. The operator T_μ belongs to $Alg\,\mathcal{L}(X, \leq, m)$ if and only if

$$\text{supp}\mu \subseteq G = \{(x,y) \in X^2 : y \leq x\} .$$

Proof. Define a bilinear form on $L^2(X,m)$ by

$$<f,g> = \int\int_{X^2} f(y)\overline{g(x)}d\mu .$$

Then by the Cauchy-Schwartz inequality,

$$\begin{aligned} |<f,g>| &\leq \int\int_{X^2} |f(y)|\,|g(x)|d|\mu| \\ &\leq \left(\int\int_{X^2} |f(y)|^2 d|\mu|\right)^{1/2}\left(\int\int_X |g(x)|d|\mu|\right)^{1/2} \\ &= \left(\int_X |f(y)|^2 d|\mu|_2\right)^{1/2}\left(\int_X |g(x)|d|\mu|_1\right)^{1/2} \\ &\leq ||f||\,||g||\,||\mu|| . \end{aligned}$$

Thus there is a bounded operator T_μ with $||T_\mu|| \leq ||\mu||$ such that

$$(T_\mu f, g) = <f,g> \text{ for all } f,g \text{ in } L^2(X,m) .$$

Suppose that supp μ is contained in G. Let E be an increasing set, and let f,g belong to $L^2(X,m)$ so that $f = \chi_E f$ and $g = \chi_{E^c} g$. Then $f(y)\overline{g(x)}$ is supported on $E^c \times E$ which is disjoint from G. Hence $E^c \times E \cap G$ is empty, so

$$(T_\mu f, g) = \int\int_{E^c \times E \cap G} f(y)\overline{g(x)}d\mu = 0 \ .$$

So T_μ belongs to $Alg\,\mathcal{L}(X, \leq, m)$.

Conversely, suppose that T_μ belongs to $Alg\,\mathcal{L}(X, \leq, m)$. For every increasing set E, $f = \chi_E f$, and $g = \chi_{E_c} g$ in L^2, one has

$$0 = (T_\mu f, g) = \int\int_{E^c \times E} f(y)\overline{g(x)}d\mu \ .$$

Hence $|\mu|(E^c \times E) = 0$. There are countably many increasing sets E_n, $n \geq 1$ with the property that $x \geq y$ if and only if $y \in E_n$ implies $x \in E_n$ for all $n \geq 1$. Reformulating this, one obtains

$$X^2 \backslash G = \bigcup_{n \geq 1} E_n^c \times E_n \ .$$

Hence $|\mu|(X^2 \backslash G) = 0$, whence $\operatorname{supp}\mu$ is contained in the closed set G. ■

It can be shown that there is a convolution type multiplication on $\mathbf{A}(X,m)$ so that $T_\mu T_\nu = T_{\mu*\nu}$. Thus $\{T_\mu : \mu \in \mathbf{A}(X,m)\}$ is an algebra. We will not require this fact at this stage. The proof will be delayed until Appendix 22.31.

In order to prove reflexivity, we will show that $\mathbf{A}(X,m)$ is rich. A Borel subset K of X^2 will be called *marginally null* if there are sets N and M with $m(N) = m(M) = 0$ such that K is contained in $N \times X \cup X \times M$. Such sets obviously cannot support a non-zero measure in $\mathbf{A}(X,m)$. We require the converse.

22.6 Lemma. *Let S be a subspace of $C_{\mathbb{R}}(X)$ containing the constants. Let ϕ be a linear functional on S such that $1 = \phi(1) = \|\phi\|$ and let M_ϕ denote the set of Hahn-Banach extensions of ϕ to $C_{\mathbb{R}}(X)$. Then for every u in $C_{\mathbb{R}}(X)$,*

$$\sup\{\sigma(u) : \sigma \in M_\phi\} = \inf\{\phi(f) : f \geq u, f \in S\} \ .$$

Proof. Let α denote the right hand side. For any σ in M_ϕ, and f in S with $f \geq u$,

$$\sigma(u) \leq \sigma(f) = \phi(f) \ .$$

Hence the left hand side is dominated by α. On the other hand, one can define ϕ_0 on $S + \mathbb{R}u$ by $\phi_0(f+ru) = \phi(f)+r\alpha$. Suppose $\|f+ru\| \leq 1$ and $r \geq 0$. Then $f+ru \leq 1$, whence $u \leq r^{-1}(1-f)$. So

$$\alpha \leq r^{-1}(1-\phi(f)) \quad \text{or} \quad \phi_0(f+r\alpha) = \phi(f)+r\alpha \leq 1 \ .$$

On the other hand, if σ is any element in M_ϕ

$$-1 \leq \sigma(f+ru) = \phi(f)+r\sigma(u) \leq \phi(f)+r\alpha = \phi_0(f+ru) \ .$$

So $\|\phi_0\| \leq 1$. By the Hahn-Banach Theorem, ϕ_0 has a norm one extension σ_0 on $C_{\mathbb{R}}(X)$ and $\sigma_0(u) = \alpha$. ■

Each σ in M_ϕ is given by a measure $\sigma(f) = \int f d\sigma$. Thus it extends to all bounded Borel functions on X.

22.7 Corollary. *Let S and ϕ be as in Lemma 22.6, and let K be a compact subset of X. Then*

$$\sup\{\int_K d\sigma : \sigma \in M_\phi\} = \inf\{\phi(f) : f \in S, f \geq \chi_K\} \ .$$

Proof. As before, the right hand side is the larger of the two terms. Let the left hand side be δ, and fix $\epsilon > \delta$. Then since $\phi(1) = 1$,

$$\int_{X\setminus K} d\sigma \geq 1-\delta > 1-\epsilon$$

for every σ in M_ϕ. By the regularity of σ, there is a compact subset C of $X\setminus K$ with $\sigma(C) > 1-\epsilon$. By Urysohn's lemma, there is a continuous real valued function u_σ with $\chi_C \leq u_\sigma \leq \chi_{X\setminus K}$. Thus $\sigma(u_\sigma) > 1-\epsilon$. Let $O_\sigma = \{\sigma' \in M_\phi : \sigma'(u_\sigma) > 1-\epsilon\}$. The collection $\{O_\sigma : \sigma \in M_\phi\}$ is an open cover of the weak* compact set M_ϕ. Let $O_{\sigma_1}, \ldots, O_{\sigma_n}$ be a finite subcover, and let

$$u = \max\{u_{\sigma_1}, \ldots, u_{\sigma_n}\} \ .$$

Then $u \leq \chi_{X\setminus K}$ and $\sigma(u) > 1-\epsilon$ for all σ in M_ϕ. By Lemma 22.7,

$$\begin{aligned} \inf\{\phi(f) : f \in S, f \geq \chi_K\} &\leq \inf\{\phi(f) : f \in S, f \geq 1-u\} \\ &= \sup\{\sigma(1-u) : \sigma \in M_\phi\} < \epsilon \ . \end{aligned}$$

As $\epsilon > \delta$ is arbitrary, the nontrivial half of the inequality follows. ■

22.8 Theorem. *Let K be a compact subset of X^2 which is not marginally null. Then there is a non-zero measure μ in* $\mathbf{A}(X,m)$ *with* $\operatorname{supp}(\mu)$ *contained in K.*

Proof. Let $S = \{h \in C_{\mathbb{R}}(X^2) : h(x,y) = f(x)+g(y), f,g \in C_{\mathbb{R}}(X)\}$. Let $\phi(h) = \int h dm^2$. The set M_ϕ is precisely the set of positive measures such that $|\mu|_1 = |\mu|_2 = m$. It suffices to show that

$$\sup\{\int_K d\sigma : \sigma \in M_\phi\} > 0 \ .$$

For then, one takes σ in M_ϕ with $\int_K d\sigma > 0$ and sets $\mu(A) = \sigma(A \cap K)$ to obtain the desired measure.

Otherwise, one applies Corollary 22.7 to obtain a sequence h_n in S with $h_n \geq \chi_K$ and $\phi(h_n) < 2^{-n}$. Decompose $h_n(x,y) = f_n(x)+g_n(y)$. Make this decomposition unique by stipulating that $\min f_n(x) = 0$. Choose x_0 with $f_n(x_0) = 0$, and note that $g_n(y) = h_n(x_0,y) \geq 0$ for all y. Then

$$2^{-n} > \phi(h_n) = \int |f_n| + |g_n| dm \ .$$

Let $A_n = \{x \in X : |f_n(x)| + |g_n(x)| \geq ½\}$ and $B_n = \bigcup_{k \geq n} A_k$. Then

$m(A_k) < 2^{1-k}$, whence $m(B_n) < 2^{2-n}$; and

$$K \subseteq (A_n \times X) \cup (X \times A_n) \subseteq (B_n \times X) \cup (X \times B_n) \ .$$

Hence if $B = \bigcap_{n \geq 1} B_n$, one has $m(B) = 0$ and

$$K \subseteq \bigcap_{n \geq 1} (B_n \times X) \cup (X \times B_n) = (B \times X) \cup (X \times B) \ .$$

Consequently, K is marginally null, contrary to hypothesis. ∎

The set $\{T_\mu : \mu \in \mathbf{A}(X,m),\ \mathrm{supp}\,\mu \subseteq G(\leq)\}$ contains the multiplication algebra $\mathcal{M} = \{M_h : h \in L^\infty(X,m)\}$ and is an $\mathcal{M}$ bimodule. To see this, define a measure μ_h supported on the diagonal $\Delta = \{(x,x) : x \in X\}$ by

$$\mu_h(A) = \int_{\pi_1(A \cap \Delta)} h\,dm \ .$$

The marginal measures are $|\mu_h|_1 = |\mu_h|_2 = |h|dm$. So μ_h belongs to $\mathbf{A}(X,m)$. For any f and g in $L^2(X,m)$, one has

$$\begin{aligned}(T_{\mu_h} f, g) &= \int\int f(y)\overline{g(x)}d\mu_h \\ &= \int f(x)\overline{g(x)}h(x)dm(x) = (M_h f, g) \ .\end{aligned}$$

Let μ be any measure in $\mathbf{A}(X,m)$. The measure $\nu = h_1(x)h_2(y)\mu$ has support contained in $\mathrm{supp}\,\mu$, and marginals $|\nu|_1 \leq h_1\|\|h_2\|\ |\mu|_1$ and $|\nu|_2 \leq \|h_1\|h_2|\mu|_2$. Again, for f and g in $L^2(X,m)$, compute

$$\begin{aligned}(T_\nu f, g) &= \int\int f(y)\overline{g(x)}h_1(x)h_2(y)d\mu \\ &= (T_\mu h_2 f, \overline{h_1} g) = (M_{h_1} T_\mu M_{h_2} f, g) \ .\end{aligned}$$

So $M_{h_1} T_\mu M_{h_2} = T_\nu$.

Let $\mathcal{A}_{\min}(X, \leq, m)$ be the weak* closed algebra generated by

$$\{T_\mu : \mu \in \mathbf{A}(X,m),\ \mathrm{supp}\,\mu \subseteq G(\leq)\} \ .$$

This is, in fact, the weak* closed linear span of these operators. (see Appendix 22.31). The previous remarks show that $\mathcal{A}_{\min}(X, \leq, m)$ contains the multiplication algebra $\mathcal{M}$. The notation min will be justified in Theorem 22.16.

22.9 Theorem. *Lat* $\mathcal{A}_{\min}(X, \leq, m) = \mathcal{L}(X, \leq, m)$.

Proof. Let P be a projection onto an invariant subspace of $\mathcal{A}_{\min}$. This is also invariant for $\mathcal{M}$, and thus P belongs to $\mathcal{M}$; so $P = P(E)$ is multiplication by the characteristic function of a Borel set E. Choose an increasing sequence K_n of compact subsets of E so that $m(E \setminus \bigcup K_n) = 0$. Likewise, choose an increasing sequence C_n of compact subsets of E^c with $m(E^c \setminus \bigcup C_m) = 0$. For every T_μ in $\mathcal{A}_{\min}$,

$$\begin{aligned} 0 &= P(C_m)(P(E)^{\perp}T_{\mu}P(E))P(K_n) \\ &= P(C_m)T_{\mu}P(K_n) = T_{\mu_{mn}} \end{aligned}$$

where $\mu_{mn} = \mu \,|C_m \times K_n$. Thus there is no non-zero measure μ in $\mathcal{A}(X,m)$ supported on $(C_m \times K_n) \cap G(\leq)$. By Theorem 22.8, this set is marginally null. Now, $E^c \times E \cap G(\leq)$ is contained in

$$(E^c \backslash \bigcup_{m \geq 1} C_m) \times X \cup X \times (E \backslash \bigcup_{n \geq 1} K_n) \cup \bigcup_{m \geq 1} \bigcup_{n \geq 1} (C_m \times K_n) \cap G(\leq)$$

which is a countable union of marginally null sets, and thus is marginally null.

Let N be a Borel set with $m(N) = 0$ so that

$$E^c \times E \cap G(\leq) \subseteq (N \times X) \cup (X \times N) \ .$$

Let π_i be the coordinate projections of X^2 onto X. Define

$$\begin{aligned} F = (E \backslash N)^+ &= \{x : \exists y \in E \backslash N, y \leq x\} \\ &= \pi_1\{\pi_2^{-1}(E \backslash N) \cap G(\leq)\} \\ &= (E \backslash N) \cup \pi_1\{(E^c \cup N) \times (E \backslash N) \cap G(\leq)\} \\ &\subseteq E \backslash N \cup N = E \cup N \ . \end{aligned}$$

Thus F is an increasing set such that $m(E \Delta F) = 0$, whence $P(E) = P(F)$ belongs to $\mathcal{L}(X, \leq, m)$.

The inclusion of $\mathcal{L}(X, \leq, m)$ in $Lat\, \mathcal{A}_{\min}$ is immediate from Theorem 22.5, so the theorem follows. ∎

22.10 Corollary. *$\mathcal{L}(X, \leq, m)$ is reflexive.*

Proof.

$$\begin{aligned} \mathcal{L}(X, \leq, m) &\subseteq Lat\, Alg\, \mathcal{L}(X, \leq, m) \\ &\subseteq Lat\, \mathcal{A}_{\min}(X, \leq, m) = \mathcal{L}(X, \leq, m) \ . \end{aligned}$$ ∎

22.11 Corollary. *Every commutative subspace lattice on a separable Hilbert space is reflexive.*

Proof. Apply Theorem 22.4 and Corollary 22.10. ∎

We now turn to the consideration of $\mathcal{M}$ bimodules, where $\mathcal{M}$ is a masa containing $\mathcal{L}'' = \{P(L) : L \in \mathcal{L}\}''$, which have $\mathcal{L}$ as its lattice. Two such modules are $Alg\, \mathcal{L}$ and $\mathcal{A}_{\min}(\mathcal{L})$. As in Chapter 15, we will use the ***infinite ampliation*** $S \otimes I$ of a weak* closed module S in order to work with reflexive subspaces (Lemma 15.4).

Given a commutative subspace lattice $\mathcal{L}$ of $\mathcal{H}_1$ consider two lattices of $\mathcal{H}_1 \otimes \mathcal{H}$. Let $\mathcal{L} \otimes Proj(\mathcal{H})$ denote the strongly closed (complete) lattice of projections generated by $\{P(L) \otimes Q : L \in \mathcal{L}, Q \in Proj(\mathcal{H})\}$. Let $\mathcal{L} \underset{\text{conv}}{\otimes} Proj(\mathcal{H})$ denote the lattice of projections P in $Proj(\mathcal{H}_1 \otimes \mathcal{H})$ such that $(I \otimes E)P(I \otimes E)$ belongs to the closed convex hull $\overline{\text{conv}}\{P(L) \otimes E : L \in \mathcal{L}\}$ for every rank one projection E in $Proj(\mathcal{H})$. It is not immediate that this latter set is a lattice, but it is strongly closed and contains $P(L) \otimes Q$ for all L in $\mathcal{L}$ and Q in $Proj(\mathcal{H})$. So once it is shown to be a lattice, it will follow that it contains $\mathcal{L} \otimes Proj(\mathcal{H})$.

By Theorem 22.4, we can identify $\mathcal{L}$ with $\mathcal{L}(X, \leq, m)$ acting on $L^2(X,m)$. By Lemma 15.3, $L^2(X,m) \otimes \mathcal{H}$ is naturally identified with $L^2(X,\mathcal{H},m)$, the space of square integrable functions on X with values in $\mathcal{H}$. This identification takes $L^\infty(X,m) \otimes \mathcal{B}(\mathcal{H})$ onto $L^\infty(X,\mathcal{B}(\mathcal{H}),m)$. For example, if E is an increasing set and Q is a projection in $\mathcal{B}(\mathcal{H})$, then $P(E) \otimes Q$ is identified with the function $f = Q\chi_E$ which is an *increasing function* into $\mathcal{B}(\mathcal{H})_+$ in the sense that $x \geq y$ implies $f(x) \geq f(y)$. A function f is *essentially increasing* if it agrees with an increasing function a.e.(m).

22.12 Lemma. *The closed convex hull of $\mathcal{L}(X, \leq, m)$ is weak* closed, and coincides with the set of operators M_f for (essentially) increasing f in $L^\infty(X,m)$ such that $0 \leq f \leq 1$.*

Proof. Each χ_E for $P(E) = M_{\chi_E}$ in $\mathcal{L}(X \leq, m)$ is an increasing function with $0 \leq \chi_E \leq 1$. Thus the same holds for the weak* closed convex hull. Conversely, if f is increasing, then $E_r = \{x : f(x) \geq r\}$ is an increasing set. The operators

$$M_{f_n} = \frac{1}{n}\sum_{k=1}^{n} P(E_{k/n})$$

are in $\text{conv}(\mathcal{L}(X, \leq, m))$, and $0 \leq f - f_n \leq 1/n$. So M_f is in the norm closed convex hull of $\mathcal{L}(X, \leq, m)$. ∎

22.13 Corollary. *$\mathcal{L}(X, \leq, m) \underset{\text{conv}}{\otimes} Proj(\mathcal{H})$ consists of all (essentially) increasing, projection valued functions in $L^\infty(X,\mathcal{B}(\mathcal{H}),m)$. Consequently, it forms a complete lattice.*

Proof. Let P be an increasing projection valued function in $L^\infty(X,\mathcal{B}(\mathcal{H}),m)$ and let E be the rank one projection onto $\mathbb{C}\xi$ in $\mathcal{H}$. Then

$$(I \otimes E)P(I \otimes E)(x) = EP(x)E = (P(x)\xi,\xi)E \quad .$$

As $P(x)$ is increasing, $f(x) = (P(x)\xi,\xi)$ is an increasing function with $0 \leq f \leq 1$. So $f(x)E = M_f \otimes E$ belongs to $\overline{\text{conv}}\{P(L) \otimes E : L \in \mathcal{L}\}$ by Lemma 22.12.

Conversely, if P is a projection in $L^\infty(X,\mathcal{B}(\mathcal{H}),m)$ such that $(I \otimes E)P(I \otimes E)$ belongs to $\overline{\text{conv}}\{P(L) \otimes E : L \in \mathcal{L}\}$ for all rank one projections E, one deduces as above that $f_\xi(x) = (P(x)\xi,\xi)$ is increasing for every ξ in $\mathcal{H}$. Thus $x \geq y$ implies $(P(x)\xi,\xi) \geq (P(y)\xi,\xi)$ for all ξ; whence $P(x) \geq P(y)$. So P is an increasing projection valued function.

It is not difficult to see that this forms a complete, strongly closed lattice where the lattice operations are performed pointwise. (If one wishes to pick an arbitrary representing function instead of an increasing one, then one must deal with "almost everywhere", which is a technical nuisance but not a serious difficulty.) ∎

22.14 Lemma. *Let μ belong to $\mathbf{A}(X,m)$, and let F, G belong to $L^2(X,\mathcal{H},m)$. Then*

$$((T_\mu \otimes I)F,G) = \int\int_{X^2} (F(y),G(x))\,d\mu(x,y) \ .$$

Proof. The right hand side is absolutely integrable, uniformly on the unit ball of L^2 because it is dominated by

$$\int\int \|F(y)\|\,\|G(x)\|d\mu \leq \left(\int\int \|F(y)\|^2 d\mu\right)^{1/2}\left(\int\int \|G(x)\|^2 d\mu\right)^{1/2}$$
$$\leq \|F\|\,\|G\|\,\|\mu\| \ .$$

Suppose $F = \sum_{i=1}^{n} f_i \otimes u_i$ and $G = \sum_{j=1}^{m} g_j \otimes v_j$ where f_i, g_j belong to $L^2(X,m)$ and u_i, v_j belong to $\mathcal{H}$. Then

$$\begin{aligned} ((T_\mu \otimes I)F,G) &= \sum_{i=1}^{n}\sum_{j=1}^{n} (T_\mu f_i, g_j)(u_i,v_j) \\ &= \sum_{i=1}^{n}\sum_{j=1}^{n} \int\int_{X^2} f_i(y)\overline{g_j(x)}\,d\mu(x,y)(u_i,v_j) \\ &= \int\int_{X^2} (F(y),G(x))\,d\mu(x,y) \ . \end{aligned}$$

Such vectors are dense in $L^2(X,\mathcal{H},m)$ and both sides converge uniformly. Thus the formula is valid in general. ∎

22.15 Lemma. *$Lat(\mathcal{A}_{\min}(\mathcal{L}) \otimes I)$ contains $\mathcal{L} \underset{\text{conv}}{\otimes} Proj(\mathcal{H})$.*

Proof. By Theorem 22.4, we may assume that $\mathcal{L} = \mathcal{L}(X, \leq, m)$. Let T_μ belong to $\mathcal{A}_{\min}(X, \leq, m)$ and fix P in $\mathcal{L} \underset{\text{conv}}{\otimes} Proj(\mathcal{H})$. Let $F = PF$ and $G = P^\perp G$ be vectors in the range of P and $P^\perp$ respectively. Then for all (x,y) in $G(\leq)$, $y \leq x$ whence $P(y) \leq P(x)$; so $P(x)^\perp P(y) = 0$. Thus

$$\begin{aligned}((T_\mu \otimes I)F,G) &= \int\int_{X^2}(F(y),G(x))d\mu(x,y)\\ &= \int\int_{G(\leq)}(P(y)F(y),P(x)^\perp G(x))d\mu(x,y)\\ &= \int\int_{G(\leq)}(P(x)^\perp P(y)F(y),G(x))d\mu(x,y) = 0 \ .\end{aligned}$$

Consequently, $P^\perp(T_\mu \otimes I)P = 0$ as desired. ∎

The stage is now set to prove the minimality of $\mathcal{A}_{\min}$.

22.16 Lemma. *Let $\mathcal{A}$ be a weak* closed algebra containing a masa $\mathcal{M}$ with Lat $\mathcal{A} = \mathcal{L}$. Then Lat $\mathcal{A} \otimes I$ is contained in $\mathcal{L} \underset{\text{conv}}{\otimes} Proj(\mathcal{H})$.*

Proof. Let P be a projection onto some element of *Lat* $\mathcal{A} \otimes I$. This is, a fortiori, invariant for $\mathcal{M} \otimes I$. So by Theorem 15.1, P belongs to $\mathcal{M} \otimes \mathcal{B}(\mathcal{H})$. Also note that $\mathcal{M}$ commutes with $\mathcal{L}$, and hence contains $\mathcal{L}$. Let E be the projection of $\mathcal{H}$ onto $\mathbb{C}\xi$ for a given unit vector ξ. Then $(I \otimes E)P(I \otimes E) = M \otimes E$, where M is a positive contraction in $\mathcal{M}$. Let A belong to the unit ball of $\mathcal{A}$, and let x be a vector.

$$\begin{aligned}(AMA^*x,x) &= ((A \otimes I)(I \otimes E)P(I \otimes E)(A^* \otimes I)x \otimes \xi, x \otimes \xi)\\ &= \|P(A^* \otimes I)x \otimes \xi\|^2\\ &= \|P(A^* \otimes I)P(x \otimes \xi)\|^2\\ &\leq \|P(x \otimes \xi)\|^2\\ &= ((I \otimes E)P(I \otimes E)x \otimes \xi, x \otimes \xi) = (Mx,x) \ .\end{aligned}$$

Thus $AMA^* \leq M$.

Let $E_r = E_M[r,1]$ denote the spectral projection of M for the interval $[r,1]$, and let P_r be the least projection in $\mathcal{L}$ dominating E_r. This is the projection onto $\overline{\mathcal{A}E_r\mathcal{H}}$. If F is any projection in $\mathcal{M}$ with $F \leq P_r$, there is an operator A in $\mathcal{A}$ so that $FAE_r \neq 0$. Normalize so that $B = FAE_r$ has norm one. Then

$$rBB^* = B(rE_r)B^* \leq BMB^* \leq M \ .$$

Hence $\|FM\| \geq r\|FBB^*F\| = r$ for every $F \leq P_r$, and thus $M \geq rP_r$. This implies that $E_r \geq P_r \geq E_r$; that is, E_r belongs to $\mathcal{L}$ for all $r \geq 0$. Thus $M = \int_0^1 E_r dr$ belongs to $\overline{\text{conv}}(\mathcal{L})$. Therefore, P belongs to $\mathcal{L} \underset{\text{conv}}{\otimes} Proj(\mathcal{H})$. ∎

22.17 Theorem. *Let $\mathcal{A}$ be a weak* closed algebra containing a masa $\mathcal{M}$ with Lat $\mathcal{A} = \mathcal{L}$. Then $\mathcal{A}_{\min}(\mathcal{L}) \subseteq \mathcal{A} \subseteq Alg\,\mathcal{L}$. Conversely, every weak* closed algebra in this interval has Lat $\mathcal{A} = \mathcal{L}$, and contains $\mathcal{L}'$.*

Proof. By Lemmas 22.15 and 22.16, $Lat\,\mathcal{A} \otimes I$ is contained in $Lat\,\mathcal{A}_{\min} \otimes I$. By Lemma 15.4, both inflated algebras are reflexive. Hence $\mathcal{A}_{\min}(\mathcal{L})$ is contained in $\mathcal{A}$, which is in turn contained in $Alg\,\mathcal{L}$. By Theorem 22.9, $Lat\,\mathcal{A}_{\min} = \mathcal{L}$. Lemma 22.15 and 22.16 show that $Lat\,\mathcal{A}_{\min} \otimes I$ equals $\mathcal{L} \underset{\text{conv}}{\otimes} Proj(\mathcal{H})$ which is contained in $\mathcal{L}'' \otimes \mathcal{B}(\mathcal{H})$. By Theorem 15.1, the commutant of $\mathcal{L}'' \otimes \mathcal{B}(\mathcal{H})$ is $\mathcal{L}' \otimes I$. Another application of Lemma 15.4 shows that $\mathcal{A}_{\min}$ contains $\mathcal{L}'$. So the converse follows. ∎

22.18 Corollary. *If $\mathcal{A}$ is a weak* closed algebra containing a masa such that Lat $\mathcal{A} = \mathcal{L}$, then the diagonal $\mathcal{A} \cap \mathcal{A}^* = \mathcal{L}'$.*

Proof. $\mathcal{L}' \subseteq \mathcal{A}_{\min}(\mathcal{L}) \cap \mathcal{A}_{\min}(\mathcal{L})^* \subseteq \mathcal{A} \cap \mathcal{A}^* \subseteq Alg\,\mathcal{L} \cap (Alg\,\mathcal{L})^* = \mathcal{L}'$. ∎

The methods used above are closely related to the results of Chapter 15, part B. However, in the case of nests, we avoided the measure theoretic difficulties. Also, we still have not recovered the results of that chapter for two reasons. One is that we have only considered algebras, not general $\mathcal{M}$ bimodules. But more importantly, we have not discussed when $\mathcal{A}_{\min}(\mathcal{L}) = Alg\,\mathcal{L}$, even for nests.

22.19 Theorem. *Let $\mathcal{M}$ be a masa, and let $\mathcal{L}$ be a commutative subspace lattice. Suppose that $\mathcal{S}$ is a weak* closed $\mathcal{M}$-bimodule such that $\overline{\mathcal{S}x} = \overline{(Alg\,\mathcal{L})x}$ for all x in $\mathcal{H}$. Then $\mathcal{S}$ contains $\mathcal{A}_{\min}(\mathcal{L})$.*

Proof. Form the *algebra* $\mathcal{B}(\mathcal{S})$ of all operators of the form $\begin{bmatrix} M_1 & S \\ 0 & M_2 \end{bmatrix}$, where M_1, M_2 belong to $\mathcal{M}$ and S belongs to $\mathcal{S}$. This is weak* closed and contains the masa $\mathcal{M} \oplus \mathcal{M}$. A simple computation shows

$$\begin{aligned} Lat\,\mathcal{B}(\mathcal{S}) &= \{L_1 \oplus L_2 : L_i \in Lat\,\mathcal{M}, L_1 \supseteq \overline{\mathcal{S}L_2}\} \\ &= \{L_1 \oplus L_2 : L_i \in Lat\,\mathcal{M}, L_1 \supseteq \overline{(Alg\,\mathcal{L})L_2}\} \equiv \mathcal{L}_2 \ . \end{aligned}$$

By Theorem 22.17, $\mathcal{B}(\mathcal{S})$ contains $\mathcal{A}_{\min}(\mathcal{L}_2)$. From Theorem 22.9, it follows that $\mathcal{A}_{\min}(\mathcal{L})$ satisfies the hypothesis of this theorem. So $\mathcal{B}(\mathcal{A}_{\min}(\mathcal{L}))$ also contains $\mathcal{A}_{\min}(\mathcal{L}_2)$.

By Theorem 22.4 and its proof, we may suppose that $\mathcal{L} = \mathcal{L}(X, \leq, m)$ and $\mathcal{M} = L^\infty(X,m)$. Let $X^{(2)} = X \dot\cup Y$ denote the disjoint union of two copies of $X = Y$. Denote the elements of X by $x, x_1, x_2, \ldots$ and the elements of Y by $y, y_1, y_2, \ldots$. Define a partial order $\ll$ on $X^{(2)}$ by

$$x_1 \ll x_2 \Leftrightarrow x_1 = x_2 \ ,$$
$$y_1 \ll y_2 \Leftrightarrow y_1 = y_2 \ ,$$
$$x \not\ll y \ ,$$
$$y \ll x \Leftrightarrow y \leq x \ \text{ in } X \ .$$

One readily verifies that $\mathcal{L}_2 = \mathcal{L}(X^{(2)}, \ll, m \oplus m)$ and $\mathcal{M} \oplus \mathcal{M} = L^\infty(X^{(2)}, m \oplus m)$. By Corollary 22.13, $\mathcal{L}_2 \underset{\text{conv}}{\otimes} \mathcal{P}roj(\mathcal{H})$ consists of all projection valued functions $P = P_1 \oplus P_2$ in $L^\infty(X^{(2)}, \mathcal{B}(\mathcal{H}))$ such that

$$P_1(x) \geq P_2(y) \ \text{ if } \ x \geq y \ .$$

Consider an operator $T = \begin{bmatrix} M_1 & A \\ 0 & M_2 \end{bmatrix}$ in $Alg(\mathcal{L}_2)$. Then

$$P^\perp(T \otimes I)P = \begin{bmatrix} 0 & P_1^\perp(A \otimes I)P_2 \\ 0 & 0 \end{bmatrix} \ .$$

If $A = T_\mu$ for μ in $\mathbf{A}(X, m, \leq)$, let $F = P_2F$ and $G = P_1^\perp G$ in $L^2(X, m)$. One obtains by Lemma 22.14

$$((T_\mu \otimes I)F, G) = \int\int (F(y), G(x)) d\mu$$
$$= \int\int (P_1^\perp(x) P_2(y) F(y), G(x)) d\mu = 0 \ .$$

Thus $\begin{bmatrix} M_1 & T_\mu \\ 0 & M_2 \end{bmatrix}$ belongs to $\mathcal{A}_{\min}(\mathcal{L}_2)$. By the Appendix 22.31, these operators are weak* dense in $\mathcal{B}(\mathcal{A}_{\min}(\mathcal{L}))$. Hence $\mathcal{A}_{\min}(\mathcal{L}_2) = \mathcal{B}(\mathcal{A}_{\min}(\mathcal{L}))$. Consequently, $\mathcal{S}$ contains $\mathcal{A}_{\min}(\mathcal{L})$. ■

A commutative subspace lattice is called *synthetic* if $\mathcal{A}_{\min}(\mathcal{L}) = Alg(\mathcal{L})$. By Theorem 15.11 (or Proposition 22.23 below), every nest is synthetic. A lattice $\mathcal{L}$ is *complemented* if $L \in \mathcal{L}$ implies $L^\perp \in \mathcal{L}$. The lattice is *atomic* if $\mathcal{L}''$ is atomic. A lattice $\mathcal{L}$ will be called *finite width* if it is generated as a lattice by finitely many commuting nests. It will be called *width n* if n is the least number of nests required to generate $\mathcal{L}$ as a lattice.

22.20 Proposition. *Every complemented CSL is synthetic.*

Proof. If $\mathcal{L}$ is complemented, then $Alg\,\mathcal{L} = \mathcal{L}'$ is a von Neumann algebra. The proposition is immediate from Corollary 22.18. ■

22.21 Lemma. *Every Hilbert-Schmidt operator in $Alg(\mathcal{L})$ belongs to $\mathcal{A}_{\min}(\mathcal{L})$.*

Proof. Represent $\mathcal{L}$ as $\mathcal{L}(X, \leq, m)$ by Theorem 22.4. Each Hilbert-Schmidt operator K on $L^2(m)$ has an L^2 kernel $\mu = k(x,y)dm \times m$. This operator belongs to $Alg\,\mathcal{L}$ if and only if μ is supported on the graph of the pre-order $G(\leq)$ (compare with Theorem 22.5). The marginal measures are bounded if $k(x,y)$ is bounded. So let $k_n = \chi_n \circ k$ where χ_n is the characteristic function of $\{z : |z| \leq n\}$. Clearly, k_n converges to k in L^2. Thus, $K_n = T_{k_n dm \times m}$ belong to $\mathcal{A}_{\min}(\mathcal{L})$ and converge in norm to K. So K belongs to $\mathcal{A}_{\min}(\mathcal{L})$. ■

22.22 Proposition. *Every atomic CSL is synthetic.*

Proof. If $\mathcal{L}$ is atomic, then $\mathcal{L}'$ contains a sequence of finite rank projections P_n converging strongly to the identity. For each T in $Alg(\mathcal{L})$, TP_n is a sequence of finite rank operators in $Alg(\mathcal{L})$ converging strongly to T. By Lemma 22.20, TP_n belong to $\mathcal{A}_{\min}(\mathcal{L})$. Thus T does also. So $\mathcal{A}_{\min}(\mathcal{L}) = Alg(\mathcal{L})$. ■

22.23 Proposition. *Every nest is synthetic.*

Proof. By the Erdos Density Theorem 3.11, the finite rank operators in $\mathcal{T}(\mathcal{N})$ are strongly dense. By Lemma 22.21, these finite rank operators belong to $\mathcal{A}_{\min}(\mathcal{N})$. Whence $\mathcal{A}_{\min}(\mathcal{N}) = \mathcal{T}(\mathcal{N})$. ■

22.24 Lemma. *Let $\mathcal{L}$ be a synthetic CSL and let $\mathcal{N}$ be a nest commuting with $\mathcal{L}$. For every N in $\mathcal{N}$, $P(N) Alg\,\mathcal{L}\, P(N)^{\perp}$ is contained in $\mathcal{A}_{\min}(\mathcal{L} \vee \mathcal{N})$.*

Proof. By Theorem 22.4, we may assume that $\mathcal{L} \vee \mathcal{N} = \mathcal{L}(X, \leq, m)$ acting on $L^2(X,m)$. Let T_μ be a pseudo-integral operator in $Alg\,\mathcal{L}$. Then since $P(N)$ belongs to $(\mathcal{L} \vee \mathcal{N})'' \subseteq L^\infty(X,m)$. Since the pseudo-integral operators form an $L^\infty(X,m)$ bimodule, $P(N)T_\mu P(N)^{\perp}$ is a pseudo-integral operator T_ν. Furthermore, T_ν belongs to $Alg(\mathcal{L}) \cap \mathcal{T}(\mathcal{N}) = Alg(\mathcal{L} \vee \mathcal{N})$. By Theorem 22.5, T_ν belongs to $\mathcal{A}_{\min}(\mathcal{L} \vee \mathcal{N})$. Since $\mathcal{L}$ is synthetic, each T in $Alg(\mathcal{L})$ is the weak* limit of a net T_{μ_α} in $Alg(\mathcal{L})$. Thus if $P(N)T_{\mu_\alpha}P(N)^{\perp} = T_{\nu_\alpha}$, then $P(N)TP(N)^{\perp} = w^*\text{–}\lim T_{\nu_\alpha}$ belongs to $\mathcal{A}_{\min}(\mathcal{L} \vee \mathcal{N})$. ■

22.25 Lemma. *Let $\mathcal{L}$ be a synthetic CSL, and let $\mathcal{N}$ be a nest commuting with $\mathcal{L}$. If $\mathcal{L} \vee \mathcal{N} \vee \mathcal{N}^{\perp}$ is synthetic, then $\mathcal{L} \vee \mathcal{N}$ is also synthetic.*

Proof. Let $\{N_k, k \geq 1\}$ be a strongly dense subset of $\mathcal{N}$. For each n, let $E_1^{(n)},\dots,E_k^{(n)}$ be the atoms of the finite nest $\mathcal{F}_n$ consisting of $\{0, \mathcal{H}, N_1, \dots, N_n\}$. Fix T in $Alg(\mathcal{L} \vee \mathcal{N})$, and define $R_n = \Delta_{\mathcal{F}_n}(T)$ and

$$S_n = T - R_n = \sum_{i \neq j} E_i^{(n)} T E_j^{(n)} \ .$$

Now $E_i^{(n)} T E_j^{(n)} \neq 0$ only if $E_i^{(n)} \ll E_j^{(n)}$, so $E_i^{(n)} T E_j^{(n)}$ belongs to $\mathcal{A}_{\min}(\mathcal{L} \vee \mathcal{N})$ by the previous lemma. Thus S_n belongs to $\mathcal{A}_{\min}(\mathcal{L} \vee \mathcal{N})$.

Since $\|R_n\| \leq \|T\|$, we may drop to a subsequence so that $R = w^*\text{–}\lim R_{n_k}$ exists. Clearly, R commutes with $\mathcal{N}$ and belongs to $Alg(\mathcal{L} \vee \mathcal{N})$, and hence belongs to

$$Alg(\mathcal{L} \vee \mathcal{N} \vee \mathcal{N}^{\perp}) = \mathcal{A}_{\min}(\mathcal{L} \vee \mathcal{N} \vee \mathcal{N}^{\perp}) \ .$$

This is contained in $\mathcal{A}_{\min}(\mathcal{L} \vee \mathcal{N})$. Also, $S = T - R = w^*\text{–}\lim S_{n_k}$ belongs to $\mathcal{A}_{\min}(\mathcal{L} \vee \mathcal{N})$, whence T does. But T is arbitrary, so $\mathcal{L} \vee \mathcal{N}$ is synthetic. ■

22.26 Theorem. *Every finite width CSL lattice is synthetic.*

Proof. We show that if $\mathcal{L}$ has finite width and $\mathcal{B}$ is a Boolean algebra commuting with $\mathcal{L}$, the $\mathcal{L} \vee \mathcal{B}$ is synthetic. The special case $\mathcal{B} = \{0,I\}$ yields the desired result. Proceed by induction on n, the width of $\mathcal{L}$. For $n = 1$, $\mathcal{L}$ is a chain. So $\mathcal{B}$ and $\mathcal{B} \vee \mathcal{L} \vee \mathcal{L}^{\perp}$ are Boolean algebras, and hence are synthetic by Proposition 22.20. Hence $\mathcal{B} \vee \mathcal{L}$ is synthetic by Lemma 22.25. Now assume the result for width n, and let $\mathcal{L}$ have width $n+1$. Write $\mathcal{L} = \mathcal{L}_0 \vee \mathcal{N}$ where $\mathcal{L}_0$ has width n and $\mathcal{N}$ is a nest. Then $\mathcal{B}$ and $\mathcal{B} \vee \mathcal{N} \vee \mathcal{N}^{\perp}$ are Boolean algebras. By the induction hypothesis, both $\mathcal{L}_0 \vee \mathcal{B}$ and $\mathcal{L}_0 \vee \mathcal{B} \vee \mathcal{N} \vee \mathcal{N}^{\perp}$ are synthetic. Hence by Lemma 22.25, $\mathcal{L}_0 \vee \mathcal{B} \vee \mathcal{N} = \mathcal{L} \vee \mathcal{B}$ is synthetic. ■

There are examples of lattices $\mathcal{L}$ which are not synthetic. The known examples are all related to sets of spectral synthesis in commutative harmonic analysis. A discussion of this is contained in the notes at the end of this chapter.

We now consider the question of generators, and extend the result for nests (Corollary 15.17) to arbitrary CSL's.

22.27 Theorem. *Let $\mathcal{L}$ be a commutative subspace lattice, and let $\mathcal{M}$ be a masa containing $\mathcal{L}''$. Then $\mathcal{A}_{\min}(\mathcal{L})$ is singly generated as an $\mathcal{M}$ bimodule.*

Proof. By Theorem 22.4, we may represent $\mathcal{L}$ as $\mathcal{L}(X, \leq, m)$ and $\mathcal{M}$ as $L^{\infty}(X,m)$. The unit ball of $\mathcal{B}(\mathcal{H})$ in the weak* topology is metrizable and hence separable. So one can choose pseudo-integral operators $\{T_{\mu_n}, n \geq 1\}$ in the unit ball of $\mathcal{A}_{\min}(\mathcal{L})$ which are weak* dense in the ball of $\mathcal{A}_{\min}(\mathcal{L})$ (because of Appendix 22.31). Let $\mu = \sum_{n \geq 1} 2^{-n} \frac{|\mu_n|}{\|\mu_n\|}$. This is a finite measure in $\mathbf{A}(X, \leq, m)$. It has the property that if E and F are measurable subsets of X such that $(E \times F) \cap G(\leq)$ is not marginally null, then

$\mu(E \times F) > 0$. To see this, note that $\mu(E \times F) = 0$ forces $|\mu_n|(E \times F) = 0$ for all n, whence $P(E)T_{\mu_n}P(F) = 0$ for all n. From their density, we deduce that $P(E)\mathcal{A}_{\min}(\mathcal{L})P(F) = 0$. From Theorem 22.8, we get that $(E \times F) \cap G(\leq)$ is marginally null.

For any vector x in $\mathcal{H}$, let $P(E)$ be the projection onto $\overline{\mathcal{M}x}$, and let $P(L)$ be the projection onto $\overline{\mathcal{M}T_\mu\mathcal{M}x}$. Since $P(L)^\perp T_\mu P(E) = 0$, we have $P(L)^\perp \mathcal{A}_{\min}(\mathcal{L})P(E) = 0$. Hence

$$P(L)\mathcal{H} \supseteq \overline{\mathcal{A}_{\min}(\mathcal{L})x} \supseteq \overline{\mathcal{M}T_\mu\mathcal{M}x} = P(L)\mathcal{H} .$$

By Theorem 22.19, the weak* closure of $\mathcal{M}T_\mu\mathcal{M}$ contains $\mathcal{A}_{\min}(\mathcal{L})$. The reverse inclusion is obvious. So T_μ generates $\mathcal{A}_{\min}(\mathcal{L})$ as an $\mathcal{M}$ bimodule. ■

22.28 Corollary. *Let $\mathcal{L}$ be a commutative subspace lattice. Then $\mathcal{A}_{\min}(\mathcal{L})$ is doubly generated as a weak* closed algebra.*

Proof. Take $\mathcal{M}$ to be any masa containing $\mathcal{L}''$. Take A to be a self-adjoint generator of $\mathcal{M}$, and take T_μ as above. Then $\{A, T_\mu\}$ generate $\mathcal{A}_{\min}(\mathcal{L})$ as an algebra. ■

We wish to define the notation of *support* of an operator acting on $L^2(X,m)$. Then we will relate this to certain $L^\infty(X,m)$ bimodules. The *support* of T is the (closed set) of points (x,y) in $X \times X$ such that for every neighbourhood $U \times V$ of (x,y), $P(U)TP(V) \neq 0$. Equivalently, the complement of the support is the union of all open sets $U \times V$ of $X \times X$ such that $P(U)TP(V) = 0$.

Given a closed subset F of $X \times X$, let $\mathcal{A}_{\min}(F)$ denote the weak* closure of $\{T_\mu : \text{supp}(\mu) \subseteq F\}$. Let

$$\mathcal{A}_{\max}(F) = \{T \in \mathcal{B}(L^2(m)) : \text{supp}(T) \subseteq F\} .$$

Recall (15.B) that a subspace $\mathcal{S}$ of $\mathcal{B}(\mathcal{H})$ is called *reflexive* if $Tx \in \overline{\mathcal{S}x}$ for all x in $\mathcal{H}$ implies that T belongs to $\mathcal{S}$. The *reflexive hull* of $\mathcal{S}$, denoted $ref(\mathcal{S})$, is the smallest reflexive subspace containing $\mathcal{S}$. This is precisely $\{T \in \mathcal{B}(\mathcal{H}) : Tx \in \overline{\mathcal{S}x} \text{ for all } x \in \mathcal{H}\}$.

22.29 Theorem. *Let F be a closed subset of $X \times X$. Then $ref(\mathcal{A}_{\min}(F)) = \mathcal{A}_{\max}(F)$. If $\mathcal{S}$ is a weak* closed $L^\infty(X,m)$ bimodule such that $\overline{\mathcal{S}x} = \overline{\mathcal{A}_{\min}(F)x}$ for all x in $\mathcal{H}$, then $\mathcal{A}_{\min}(F) \subseteq \mathcal{S} \subseteq \mathcal{A}_{\max}(F)$.*

Proof. Suppose Tx belongs to $\overline{\mathcal{A}_{\min}(F)x}$ for every x in $\mathcal{H}$. Let $U \times V$ be an open set disjoint from F. Then for μ supported on F, f, g in $L^2(m)$,

$$(P(U)T_\mu P(V)f, g) = \int\int_{U \times V \cap F} f(y)\overline{g(x)}\,d\mu = 0 .$$

Thus $P(U)T_\mu P(V) = 0$ for all such μ, hence $P(U)TP(V) = 0$. Thus $\text{supp}(T)$ is contained in F. Hence $ref(\mathcal{A}_{\min}(F))$ is contained in $\mathcal{A}_{\max}(F)$.

Now suppose that $\operatorname{supp}(T)$ is contained in F. For any closed set $A \times B$ disjoint from F, a standard compactness argument produces a finite open cover $\{U_i \times V_i : 1 \leq i \leq n\}$ of $A \times B$ disjoint from F. Thus it follows from $P(U_i)TP(V_i) = 0$, $1 \leq i \leq n$, that $P(A)TP(B) = 0$. Next, suppose that $A \times B$ is a Borel rectangle disjoint from F. By the regularity of m, one can choose increasing sequences of closed sets A_n, B_n such that $m(A \backslash \bigcup A_n) = 0$ in $(B \backslash \bigcup B_n)$. Thus

$$P(A)TP(B) = s-\lim_{n \to \infty} P(A_n)TP(B_n) = 0 \ .$$

Fix a vector x in $\mathcal{H}$. Let $\overline{L^\infty(X,m)x} = L^2(B)$, $\overline{\mathcal{A}_{\min}(F)x} = L^2(A_1)$, and $\overline{\mathcal{A}_{\max}(F)x} = L^2(A_2)$; and set $A = A_2 \backslash A_1$. Since $P(A)T_\mu P(B) = 0$ for every μ in $\mathcal{A}(X,m)$ supported on F, it follows that $(A \times B) \cap F$ does not support any non-zero measure in $\mathbf{A}(X,m)$. If A_n and B_n are compact sets as above, Theorem 22.8 shows that $(A_n \times B_n) \cap F$ is marginally null for all n. Hence $(A \times B) \cap F$ is marginally null. Let N and M be null sets so that

$$(A \times B) \cap F \subseteq (N \times X) \cup (X \times M) \ .$$

Then F is disjoint from $(A \backslash N) \times (B \backslash M)$. By the previous paragraph, $P(A)TP(B) = 0$ for all T in $\mathcal{A}_{\max}(F)$. This occurs only if $P(A) = 0$. So $\mathcal{A}_{\max}(F)$ is contained in $ref(\mathcal{A}_{\min}(F))$, and equality follows.

We will now make use of Theorem 22.19 and its proof. As before, let $\mathcal{B}(\mathcal{S})$ denote the algebra of all operators of the form

$$\begin{bmatrix} M_1 & S \\ 0 & M_2 \end{bmatrix}, \quad M_i \in L^\infty(X,m) \ , \quad S \in \mathcal{S} \ .$$

This is weak* closed and contains the masa $L^\infty(X,m) \oplus L^\infty(X,m)$. Let $X^{(2)} = X \dot{\cup} Y$ denote the disjoint union of two copies of $X = Y$. Denote the elements of $X(Y)$ by x, x_1, x_2 (y, y_1, y_2). Define a pre-order $\ll$ on $X^{(2)}$ by

$$x_1 \ll x_2 \Leftrightarrow x_1 = x_2 \ ,$$

$$y_1 \ll y_2 \Leftrightarrow y_1 = y_2, \ ,$$

$$x \not\ll y,$$

$$y \ll x \Leftrightarrow (x,y) \in F \ .$$

It is immediately apparent that the graph of this pre-order is

$$\{(x,x) : x \in X\} \cup \{(y,y) : y \in Y\} \cup \{(x,y) : x \in X, y \in Y, (x,y) \in F\} \ .$$

Consequently, $\mathcal{A}_{\min}(\mathcal{L}(X^{(2)}, \ll, m \oplus m)) = \mathcal{B}(\mathcal{A}_{\min}(F))$.

By the proof of Theorem 22.19, if $\mathcal{S}$ is a weak* closed $L^\infty(X,m)$ bimodule such that $\overline{\mathcal{S}x} = \overline{\mathcal{A}_{\min}(F)x}$ for all x in $\mathcal{H}$, then $Lat\, \mathcal{B}(\mathcal{S}) = \mathcal{L}(X^{(2)}, \leq, m \oplus m)$. Hence $\mathcal{S}$ contains $\mathcal{A}_{\min}(F)$ and is contained in $ref(\mathcal{A}_{\min}(F)) = \mathcal{A}_{\max}(F)$. ∎

22.30 Corollary. *$Alg(\mathcal{L}(X, \leq, m))$ consists of all operators supported on $G(\leq)$.*

22.31 Appendix.

This section is devoted to the proof that there is a multiplication $*$ on $\mathcal{A}(X,m)$ such that $T_\mu T_\nu = T_{\mu*\nu}$. Hence it will follow that $\mathcal{A}_{\min}(X, \leq, m)$ equals the weak* closed linear span S_0 of $\{T_\mu : \mu \in \mathbf{A}(X,m),\ \mathrm{supp}(\mu) \subseteq G(\leq)\}$. The proof is long and technical, and we feel that a simpler proof should be possible. Indeed, the proof of Theorem 22.19 shows that S_0 consists of all operators T such that $Q^\perp(T \otimes I)P = 0$ whenever, P, Q are projections in $L^\infty(X, \mathcal{B}(\mathcal{H}))$ such that $x \leq y$ implies $P(x) \leq Q(y)$. Whereas Lemmas 22.15 and 22.16 show that $\mathcal{A}_{\min}(X, \leq, m)$ consists of all operators T such that $R^\perp(T \otimes I)P = 0$ whenever, P, R are projections in $L^\infty(X, \mathcal{B}(\mathcal{H}))$ such that $x \leq y$ implies $P(x) \leq R(x) \leq R(y)$. Since $\mathcal{L} \underset{\text{conv}}{\otimes} \mathit{Proj}(\mathcal{H})$ is a complete lattice, there is a least projection R with this property, namely the projection onto $\overline{(\mathcal{A}_{\min} \otimes I)\mathit{Ran}(P)}$. If it were shown that there is a set N with $m(N) = 0$ so that for $x \notin N$,

$$R(x) = \sup_{y \leq x, y \notin N} P(y)$$

then it follows that every Q as above satisfies $Q \geq R$. Hence $\mathcal{A}_{\min}$ equals S_0.

To prove this directly, we require a formula for $T_\mu f$. The following theorem is taken from Dunford and Schwartz [1], Theorem VI.8.6.

22.32 Theorem (Dunford-Pettis). *Let B be a separable Banach space, and let (X,m) be a σ-finite measure space. Let T be a bounded linear operator from $L^1(m)$ into B^*. There is an essentially unique Borel function $B : X \to B^*$ such that ess.sup $\|B(x)\| = \|T\|$, and for each $f \in L^1(m)$ and $b \in B$,*

$$(Tf)(b) = \int \langle B(x), b\rangle f(x)\,dm(x) \ .$$

This is the key ingredient in the following "disintegration of measures" theorem. For greater detail and greater generality, consult Bourbaki [1] Book VI, Chapter. 6, §3, Theorem 1.

22.33 Theorem (Disintegration). *Let X and Y be compact metric spaces, and let μ be a finite regular Borel measure on $X \times Y$. Let μ_1 be the marginal measure on X given by $\mu_1(A) = \mu(A \times Y)$. There is a Borel map $x \to \mu^x$ of X into the regular probability measures on Y such that*

$$\mu(S) = \int_X \left(\int_Y \chi_S(x,y)\,d\mu^x(y)\right) d\mu_1(x) \ .$$

Sketch of proof. Define a linear map T from $L^1(\mu_1)$ into $C(X \times Y)^*$ by

$$(Tf)(g) = \int g \cdot (f \circ \pi_1)\,d\mu$$

where π_1 is the first coordinate projection. One has

$$\|Tf\| = \|f \circ \pi_1 d\mu\| = \int |f \circ \pi_1| d\mu = \int |f| d\mu_1 = \|f\| \ .$$

So by the Dunford-Pettis Theorem, there is a Borel map B of X into $C(X \times Y)^* = M(X \times Y)$ such that

$$(Tf)(g) = \int \langle B(x), g \rangle f(x) d\mu_1(x) \ .$$

Now $B(t)$ is integration with respect to a measure ν^t on $X \times Y$, so we obtain

$$\int g(x,y) f(x) d\mu = \int \int g(x,y) d\nu^t(x,y) f(t) d\mu_1(t) \ .$$

If f and g are positive, the left hand side is positive, hence $\int g d\nu^t \geq 0$ $(a.e. d\mu)$. Whence $\nu^t \geq 0$ $(a.e. d\mu_1)$. Similarly, one obtains $\|\nu^t\| \leq 1$ $(a.e. d\mu_1)$.

Let D be a disc of positive radius ϵ on X. If f has support in D and g vanishes on $\pi_1^{-1}(D)$, then

$$0 = \int (\int g(x,y) d\nu^t(x,y)) f(t) d\mu_1(t) \ .$$

Hence $\int g(x,y) d\nu^t(x,y) = 0$ for almost all t in D. Since $C(X \times Y)$ is separable, choose a dense set g_n in $\{g : g \,|\pi_1^{-1}(D) = 0\}$. Then one obtains a set N of measure zero so that $\mathrm{supp}(\nu^t) \subseteq \pi_1^{-1}(D)$ for $t \in D \backslash N$. Utilize a countable collection of discs D_n with the property that for any ϵ, $\{D_n : diam\,(D_n) < \epsilon\}$ covers X to deduce that $\mathrm{supp}(\nu^t) \subseteq \pi_1^{-1}(t)$ $(a.e. d\mu_1)$. Thus there is a measure μ^t on Y such that $\int g d\nu^t = \int g(t,y) d\mu^t(y)$. Hence, one has

$$\int g(x,y) f(x) d\mu = \int \int g(x,y) d\mu^x(y) f(x) d\mu_1(x)$$

for all g in $C(X \times Y)$. This extends to all bounded Borel functions by Egoroff's Theorem and the Lebesgue Dominated Convergence Theorem. Uniqueness is easy. ■

22.34 Corollary. *Let X be a compact metric space with a finite regular Borel measure m. Let μ belong to $\mathbf{A}(X,m)$. Then there is a Borel map $x \longrightarrow \mu^x$ from X into the regular Borel measures on X such that*

$$\mu(S) = \int (\int \chi_S(x,y) d\mu^x(y)) dm(x)$$

and $ess.sup \|\mu^x\| \leq \|\mu\|$.

Sketch of proof. Apply Theorem 22.33 to $|\mu|$. Then by the Radon-Nikodym Theorem, there are bounded Borel functions so that $d\mu = h d|\mu|$ and $d|\mu|_1 = k dm$. Putting these together yields the corollary. ■

22.35 Corollary. $(T_\mu f)(x) = \int f(y)d\mu^x(y)$ $(a.e.m)$.

Proof.

$$\int T_\mu f(x)\overline{g(x)}dm(x) = (T_\mu f,g) = \int\int f(y)\overline{g(x)}d\mu(x)$$
$$= \int(\int f(y)d\mu^x(y))\overline{g(x)}dm(x) \ .$$

This holds for all g in $L^2(m)$, whence the result. ■

Naturally, one also deduces the existence of a Borel map $y \to \mu_y$ such that

$$\mu(S) = \int(\int \chi_S(x,y)d\mu_y(x)dm(y)$$

and

$$(T_\mu^* g)(y) = \int g(x)d\overline{\mu_y}(x) \quad (a.e.m)$$

where $\overline{\mu}(A) = \overline{\mu(A)}$ is the conjugate of μ.

22.36 Theorem. *For μ, ν in $A(X,m)$, the measure $\mu*\nu$ given by*

$$\mu*\nu(S) = \int(\mu_x \times \nu^x)(S)dm(x)$$

belongs to $\mathbf{A}(X,m)$, and $T_\mu T_\nu = T_{\mu*\nu}$.

Proof. Since $\|\mu^x\| \le \|\mu\|$ and $\|\nu^x\| \le \|\nu\|$, $\mu*\nu$ is a finite Borel measure. Moreover

$$(\mu*\nu)_1(A) = (\mu*\nu)(A \times X) = \int \mu_x(A)\|\nu^x\|dm(x)$$
$$\le \|\nu\|\mu_1(A) \le \|\nu\| \ \|\mu\|m(A) \ .$$

Similarly, $(\mu*\nu)_2 \le \|\mu\| \ \|\nu\|m$. So $\|\mu*\nu\| \le \|\mu\| \ \|\nu\|$, and $\mu*\nu$ belongs to $\mathbf{A}(X,m)$. For f, g in $L^2(m)$,

$$(T_\mu T_\nu f,g) = (T_\nu f,T_\mu^* g) = \int T_\nu f(x)\overline{T_\mu^* g(x)}dm(x)$$
$$= \int(\int f(y)d\nu^t(y))(\int \overline{g(x)}d\mu_t(x))dm(t)$$
$$= \int\int\int f(y)\overline{g(x)}d(\nu^t \times \mu_t)(y,x)dm(t)$$
$$= \int f(y)\overline{g(x)}d\mu*\nu = (T_{\mu*\nu}f,g) \ .$$
■

22.37 Corollary. *$\mathcal{A}_{\min}(X,m,\le)$ is the weak* closure of*

$$\{T_\mu : \mu \in \mathbf{A}(X,m), \ \mathrm{supp}(\mu) \subseteq G(\le)\} \ .$$

Proof. It suffices to show that if μ, ν have support in G, so does $\mu*\nu$. For then, $\{T_\mu:\mathrm{supp}(\mu) \subseteq G\}$ is closed under multiplication, and thus so is its weak* closure. We claim that μ_y is supported in $\{x : x \ge y\}$ $(a.e.m)$ and ν^x is supported on $\{y : x \ge y\}$ $(a.e.m)$. For

$$0 = \int \chi_{G^c} f d\mu = \int (\int \chi_{\{x : x \not\geq y\}} f(x,y) d\mu_y(x)) dm(y)$$

for all f in $C(X \times Y)$. Hence $\chi_{\{x : x \not\geq y\}} d\mu_y = 0$ $(a.e.m)$ as claimed. The same holds for ν^x. Thus, $\mu_x \times \nu^x$ is supported on $\{(y,z) : y \geq x \geq z\}$ which is contained in G for almost all x. Consequently, $\mu * \nu$ is supported in G as desired. ■

Notes and Remarks.

The entire content of this chapter is contained in Arveson's seminal paper [6]. A shorter proof on reflexivity (Corollary 22.10) valid for non-separable spaces as well is contained in Davidson [1]. It avoids the use of pseudo-integral operators, but does not yield the full strength of Theorem 22.9. This argument is outlined in the Exercises.

Arveson constructs a CSL $\mathcal{L}$ which is not synthetic based on failure of spectral synthesis in $\mathbb{R}^3$. Froelich [1] makes the connection even more explicit. Let G be a locally compact abelian group. The maximal ideal space of $L^1(G)$ is naturally homeomorphic to the dual group $\hat{G}$, and the Gelfand map is the Fourier transform. Given a closed subset E of $\hat{G}$, form two ideals in $L^1(G)$. Let

$$k(E) = \{f \in L^1(G) : \hat{f} \mid E = 0\}$$

and

$$I_0(E) = \{f \in L^1(G) : E \subseteq int \hat{f}^{-1}(0)\}$$

(i.e. the functions f such that $\hat{f}$ vanishes on E, or in a neighbourhood of E, respectively). A set E satisfies *spectral synthesis* if $I_0(E)$ is dense in $k(E)$.

L. Schwartz showed that S^{n-1} is not a set of spectral synthesis in $\mathbb{R}^n$ for $n \geq 3$. It is this example that Arveson uses to construct his example. Malliavin constructed Cantor sets in any nondiscrete locally compact abelian group which fail to admit spectral synthesis. This is a difficult area of harmonic analysis, and a characterization of sets with spectral synthesis would seem to be well beyond present day capability.

Let E be a closed subset of a LCA group $\hat{G}$. Form the algebra $\mathcal{A}_E$ of all operators of the form $\begin{bmatrix} M_1 & T_\mu \\ 0 & M_2 \end{bmatrix}$ where M_i belong to $L^\infty(G)$, and μ is a measure in $\mathbf{A}(G,m)$ supported on $\{(x,y) : x - y \in E\}$. Let $\mathcal{L}_E = Lat\, \mathcal{A}_E$. Clearly, $\mathcal{L}_E$ is a CSL and $\mathcal{A}_E = \mathcal{A}_{\min}(\mathcal{L}_E)$.

Theorem (Froelich [1]). *$\mathcal{L}_E$ is synthetic if and only if E is a set of spectral synthesis in $\hat{G}$.*

Froelich also translates other peculiar phenomena in harmonic analysis to produce CSL algebras with interesting properties. For example, he constructs algebras which contain operators in C_p for $p > 2$, but not C_2 operators; and an algebra which contains compact operators but no operators in any C_p class. See the end of Chapter 23.

Arveson's paper also contains a section devoted to lattice theoretic invariants of commutative subspace lattices. He uses this in some special classes to distinguish algebras up to similarity by their lattice invariants.

Exercises.

22.1 (Davidson) Let $\mathcal{L}$ be a finite CSL with atoms A_i, $1 \leq i \leq n$. Let $\ll$ be the partial order on the atoms induced from $\mathcal{L}$. Show that $Lat(Alg(\mathcal{L}) \otimes \mathbb{C}I)$ consists of the ranges of

$$\{P = \sum_{i=1}^{n} P(A_i) \otimes P_i : A_i \ll A_j \text{ implies } P_i \geq P_j\} \ .$$

22.2 (Davidson) Let $\mathcal{L}$ be a finite CSL. Let $R = u \otimes v^*$ be a rank one operator with $\|u\| = \|v\| = 1$. Let $d = dist(R, Alg\, \mathcal{L})$ and $\beta = \sup_{L \in \mathcal{L}} \|P(L)^{\perp} R P(L)\|$.

(a) Show there exists a unit vector e in the range of a projection P of the form given in Exercise 22.1 such that $d = \|P^{\perp}(R \otimes I)Pe\|$.

(b) Let P_v be the projection onto $\mathbb{C}v$. Define $t \geq 0$ and a unit vector y by $(P_v \otimes I)Pe = t(v \otimes y)$. Prove that

$$d \leq \|P^{\perp}(u \otimes y)\| \, \|P(v \otimes y)\| \ .$$

(c) Let a_i, b_i, t_i belong to $[0,1]$, $1 \leq i \leq n$ such that $\sum_{i=1}^{n} t_i = 1$ and $a_i b_i \leq \gamma$. Show that $(\sum_{i=1}^{n} t_i a_i)(\sum_{j=1}^{n} t_j b_j) \leq (1+\gamma^2)/2$.

(d) Apply (a), (b) and (c) to obtain $d^2 \leq (1+\beta^2)/2$.

22.3 (Davidson) Let $\mathcal{L}$ be a CSL and let $\mathcal{M}_k$ be the $k \times k$ matrices.

(a) Show that $Alg(\mathcal{L} \otimes \mathbb{C}^k) = Alg(\mathcal{L}) \otimes \mathcal{M}_k$ and $Lat(Alg(\mathcal{L}) \otimes \mathcal{M}_k) = \mathcal{L} \otimes \mathbb{C}^k$.

(b) Since $\mathcal{L}$ is closed in the strong operator topology, show that if P is any projection in $\mathcal{L}'$ not in $\mathcal{L}$, then there are unit vectors x, y in $\mathcal{H} \otimes \mathbb{C}^k$ (for some k) such that $(P \otimes I_k)x = x$ and $(P^{\perp} \otimes I_k)y = y$ and $\sup_{L \in \mathcal{L} \otimes \mathbb{C}^k} \|P(L)^{\perp} y\| \, \|P(L)x\| < 1$.

(c) Suppose P belongs to $Lat(Alg(\mathcal{L}))$ but is not in $\mathcal{L}$. Use (b) to construct a CSL $\mathcal{L}_1$ and a projection P_1 in $Lat(Alg(\mathcal{L}_1))$ and a rank one operator R such that $dist(R, Alg(\mathcal{L}_1)) = 1$ but $\sup_{L \in \mathcal{L}_1} \|P(L)^{\perp} R P(L)\| < 1$.

22.4 (Davidson) Let $\mathcal{L}$ be a CSL and let $\mathcal{F}$ denote an arbitrary finite sublattice of $\mathcal{L}$.

(a) For T in $\mathcal{B}(\mathcal{H})$, $dist(T, Alg\,\mathcal{L}) = \sup_{\mathcal{F} \subseteq \mathcal{L}} dist(T, Alg\,\mathcal{F})$.

(b) Use Exercises 22.2 and 22.3 to show that every CSL is reflexive.

23. Complete Distributivity and Compact Operators

In this chapter, we explore the relationship of compact operators in $Alg(\mathcal{L})$ to the lattice structure of $\mathcal{L}$. The main theorem is the equivalence of complete distributivity of $\mathcal{L}$ and the weak* density of the finite rank operators in $Alg(\mathcal{L})$.

An abstract lattice $\mathcal{L}$ is *distributive* if

$$x \vee (y \wedge z) = (x \vee y) \wedge (x \vee z) \text{ for all } x,y,z \text{ in } \mathcal{L} \quad \text{(D)}$$

$$x \wedge (y \vee z) = (x \wedge y) \vee (x \wedge z) \text{ for all } x,y,z \text{ in } \mathcal{L} \quad \text{(D*)}$$

The two distributive laws (D) and (D*) are equivalent (Exercise 23.1) so only one need be checked. If $\mathcal{L}$ is a CSL, then the identities

$$P(L \wedge M) = P(L)P(M) \quad \text{and}$$

$$P(L \vee M) = P(L)+P(M)-P(L)P(M)$$

readily yield both (D) and (D*).

A complete lattice $\mathcal{L}$ is completely distributive if it has a distributive law for arbitrary sets. Let $\{\Phi_\lambda : \lambda \in \Lambda\}$ be a collection of subsets of $\mathcal{L}$, and let $\Pi\Phi_\lambda$ denote the collection of functions $F: \Lambda \to \bigcup \Phi_\lambda$ such that $F(\lambda) \in \Phi_\lambda$. The distributive laws becomes

$$\bigwedge_{\lambda \in \Lambda}(\vee\Phi_\lambda) = \bigvee_{F \in \Pi\Phi_\lambda}(\bigwedge_{\lambda \in \Lambda} F(\lambda)) \quad \text{for all } \{\Phi_\lambda \subseteq \mathcal{L} : \lambda \in \Lambda\} \quad \text{(CD)}$$

$$\bigvee_{\lambda \in \Lambda}(\wedge\Phi_\lambda) = \bigwedge_{F \in \Pi\Phi_\lambda}(\bigvee_{\lambda \in \Lambda} F(\lambda)) \quad \text{for all } \{\Phi_\lambda \subseteq \mathcal{L} : \lambda \in \Lambda\} \quad \text{(CD*)}$$

A complete lattice is *completely distributive* if both (CD) and (CD*) hold.

The reader can easily verify that every complete sublattice of the lattice of subsets of a set is completely distributive. On the other hand, the Boolean algebra of idempotents in $L^\infty(m)$ for a non-atomic measure m is not completely distributive. To see this, consider $L^\infty(0,1)$. Identify each x in $(0,1)$ by its binary expansion $x = .\epsilon_1\epsilon_2...$ (which is a uniquely determined *a.e.m.*). Set $L_n = \{x : \epsilon_n = 0\}$, and let Λ consist of all infinite subsets S of $\mathbb{N}$. It is easy to verify that $\bigwedge_{n \in S} L_n = 0$ and $\bigvee_{n \in S} L_n = 1$. Thus

$$\bigwedge_{S \in \Lambda}(\bigvee_{n \in S} L_n) = 1 \ .$$

But the range of any function F in $\prod_{S \in \Lambda} S$ is infinite, so

$$\bigvee_{F \in \Pi S} \left(\bigwedge_{S \in \Lambda} L_{F(S)} \right) = 0 \quad .$$

So the projection lattice of $L^\infty(0,1)$ is not completely distributive.

Let $\mathcal{L}$ be a complete lattice. A *semi-ideal* of $\mathcal{L}$ is a subset M of $\mathcal{L}$ which is decreasing (i.e. $x \in M$ and $y \leq x$ implies $y \in M$). For x in $\mathcal{L}$, let

$$\phi(x) = \{\text{semi-ideals } M \text{ of } \mathcal{L} : x \leq \bigvee M\} \quad .$$

Define $K(x) = \bigcap \phi(x)$. Also, define

$$x_- = \bigvee\{y \in \mathcal{L} : x \nleq y\} \quad \text{and} \quad x_\# = \bigvee\{y \in \mathcal{L} : x \nleq y_-\} \quad .$$

23.1 Lemma. *For any lattice* $\mathcal{L}$, $K(x) = \{0\} \cup \{y \in \mathcal{L} : x \nleq y_-\}$. *Hence*

$$x_\# = \bigvee K(x) \leq x \quad .$$

Proof. For each y in $\mathcal{L}$, $\{z : y \nleq z\}$ is a semi-ideal which does not contain y. If $y \neq 0$ belongs to $K(x)$, then y belongs to every semi-ideal M with $\bigvee M \geq x$. Hence $y_- = \bigvee\{z : y \nleq z\} \ngeq x$. Conversely, if $x \nleq y_-$, suppose that M is an order ideal such that $x \leq \bigvee M$. Then M is not contained in $\{z : y \nleq z\}$. So M contains some $z \geq y$ and hence contains y. But M was arbitrary, so y belongs to $K(x)$. Since $M = \{y \in \mathcal{L} : y \leq x\}$ is a semi-ideal with $\bigvee M = x$, $x_\# = \bigvee K(x) \leq \bigvee M = x$. ■

23.2 Theorem. *For a complete lattice* $\mathcal{L}$, *the following are equivalent:*

(i) $\mathcal{L}$ *is completely distributive.*

(ii) $\mathcal{L}$ *satisfies* (CD).

(ii*) $\mathcal{L}$ *satisfies* (CD*).

(iii) $\mathcal{L}$ *is a complete homomorphic image of a complete lattice of subsets.*

(iv) $\mathcal{L}$ *has the splitting property: If* $x, y \in \mathcal{L}$ *and* $x \nleq y$, *then there exist* $p, q \in \mathcal{L}$ *with* $p \nleq y$ *and* $x \nleq q$ *such that* $\{z : z \geq p\} \cup \{z : z \leq q\} = \mathcal{L}$.

(v) $x = x_\#$ *for every* $x \in \mathcal{L}$.

Proof. Since (CD) and (CD*) are preserved by a complete lattice homomorphism, it follows that (iii) implies (i). Clearly, (i) implies (ii) and (ii*). For any complete lattice $\mathcal{L}$, let $S(\mathcal{L})$ denote the collection of all semi-ideals of $\mathcal{L}$. Then $S(\mathcal{L})$ forms a complete lattice of subsets of $\mathcal{L}$ under intersection and union. Define a map θ from $S(\mathcal{L})$ into $\mathcal{L}$ by $\theta(M) = \bigvee M$. The map θ is onto since $\theta\{y \in \mathcal{L} : y \leq x\} = x$. This map preserves arbitrary joins because

$$\theta(\bigcup M_\lambda) = \bigvee(\bigcup_{\lambda\in\Lambda} M_\lambda) = \bigvee_{\lambda\in\Lambda}(\bigvee M_\lambda) = \bigvee_{\lambda\in\Lambda}\theta(M_\lambda) \ .$$

Now consider $\bigcap_{\lambda\in\Lambda} M_\lambda$. This will be shown to equal $\{\bigwedge_{\lambda\in\Lambda} F(\lambda) : F \in \Pi M_\lambda\}$. Clearly, $\bigwedge_{\lambda\in\Lambda} F(\lambda) \le F(\lambda) \in M_\lambda$ for each M_λ; and hence belongs to $\bigcap_{\lambda\in\Lambda} M_\lambda$. On the other hand, if x belongs to $\bigcap_{\lambda\in\Lambda} M_\lambda$, then $F(\lambda) = x$ for all λ has $\bigwedge F(\lambda) = x$. Now suppose that L has (CD). Then

$$\theta(\bigcap_{\lambda\in\Lambda} M_\lambda) = \bigvee_{F\in\Pi M_\lambda}(\bigwedge_{\lambda\in\Lambda} F(\lambda)) = \bigwedge_{\lambda\in\Lambda}(\bigvee M_\lambda) = \bigwedge_{\lambda\in\Lambda}\theta(M_\lambda) \ .$$

Hence θ is a complete lattice isomorphism of $S(L)$ onto L. So (ii) implies (iii). Consequently, (ii) implies (ii*). These two notions are dual, so interchanging the role of meet and join shows that (ii*) implies (ii) as well.

Now consider (iii) $\Rightarrow$ (iv). If L is a complete lattice of subsets, and A and B are sets in L with $A \not\subseteq B$, let a be any element of $A\backslash B$. Let P be the least element of L containing a, and let Q be the largest element of L not containing a. Clearly, $P \not\subseteq B$ and $A \not\subseteq Q$. If C belongs to L, either $a \in C$, whence $P \subseteq C$ or $a \notin C$, whence $C \subseteq Q$. So L has the splitting property.

Now suppose L is a complete homomorphic image $\phi(M)$ of a complete lattice M of sets. If x, y belong to L and $x \not\le y$, let $A = \bigcap\phi^{-1}(x)$ and $B = \bigcup\phi^{-1}(y)$. So $\phi(A) = x$ and $\phi(B) = y$, and $A(B)$ is minimal (maximal) with this property. Let P and Q be obtained in M as above. Let $p = \phi(P)$ and $q = \phi(Q)$. Since ϕ is surjective, L equals the union $\{z : z \ge p\} \cup \{z : z \le q\}$. Suppose $p \le y$. Then $\phi(P \vee B) = p \vee y = y$. By the maximality of B, one has $P \subseteq B$ contrary to fact. So $p \not\le y$. Similarly $q \not\ge x$. Thus (iii) implies (iv).

Assume (iv) holds. If $x \ne x_\#$ for some x in L, then $x_\# < x$ by Lemma 23.1. So $x \not\le x_\#$. By the splitting property, there exist p, q in L with $p \not\le x_\#$ and $q \not\ge x$ such that $\{y : y \ge p\} \cup \{y : y \le q\} = L$. Thus

$$p_- = \bigvee\{y : y \not\ge p\} \le \bigvee\{y : y \le q\} = q \not\ge x \ .$$

As $p_- \not\ge x$ it follows from the definitions that $p \le x_\#$. This contradiction establishes (v).

Now suppose (v) holds. Define the map θ as in the first paragraph. It suffices to show that θ preserves meets. Given a set of semi-ideals $\{M_\lambda : \lambda \in \Lambda\}$, let $x = \bigwedge_{\lambda\in\Lambda}\theta(M_\lambda)$. Then each M_λ belongs to $\phi(x)$. Hence

$$x = \bigwedge \theta(M_\lambda) \geq \bigwedge_{M \in \phi(x)} \theta(M)$$
$$\geq \theta(\bigcap_{M \in \phi(x)} M) = \theta(K(x)) = x_\# = x \quad .$$

Thus

$$x = \bigwedge_{\lambda \in \Lambda} \theta(M_\lambda) \geq \theta(\bigcap_{\lambda \in \Lambda} M_\lambda) \geq \theta(K(x)) = x \quad .$$

So θ preserves meets, establishing (iii). ■

Now, consider subspace lattices again. First, we obtain an easy generalization of Lemma 3.7.

23.3 Lemma. *Let $\mathcal{L}$ be a subspace lattice. A rank one operator $x \otimes y^*$ belongs to $Alg(\mathcal{L})$ if and only if there is an L in $\mathcal{L}$ such that $x \in L$ and $y \in (L_-)^\perp$.*

Proof. Suppose $x \otimes y^*$ belongs to $Alg(\mathcal{L})$, and let L be the least element of $\mathcal{L}$ containing x. If M belongs to $\mathcal{L}$ and is not orthogonal to y, then,

$$x \in x \otimes y^*(M) \subseteq M \quad ,$$

hence $M \geq L$. Consequently, $L_- = \bigvee \{M \in \mathcal{L} : M \not\geq L\}$ is orthogonal to y.

Conversely, if $x \in L$ and $y \in (L_-)^\perp$, then for each M in $\mathcal{L}$, either i) $M \geq L$ whence $x \otimes y^*(M) \subseteq L \subseteq M$, or ii) $M \not\geq L$, whence $M \leq L_-$ so $x \otimes y^*(M) = \{0\} \subseteq M$. Hence $x \otimes y^*$ belongs to $Alg(\mathcal{L})$. ■

Let $\mathcal{R}_1(\mathcal{L})$ denote the algebra generated by the rank one operators in $Alg(\mathcal{L})$. Note that $\mathcal{R}_1(\mathcal{L})$ is a two sided ideal in $Alg(\mathcal{L})$.

23.4 Lemma. *For each vector u in $\mathcal{H}$, let $L(u)$ denote the least element of $\mathcal{L}$ containing u. Then $\overline{\mathcal{R}_1(\mathcal{L})u} = L(u)_\#$.*

Proof. Let M belong to $\mathcal{L}$. If $u \notin M_-$, then there is a vector y in $(M_-)^\perp$ such that $(u,y) \neq 0$. By Lemma 23.3, it then follows that $\mathcal{R}_1(\mathcal{L})u$ contains M. Moreover, that lemma shows that these are the only vectors which are the image of u under a rank one operator in $Alg(\mathcal{L})$. Thus

$$\overline{\mathcal{R}_1(\mathcal{L})u} = \bigvee\{M : u \notin M_-\}$$
$$= \bigvee\{M : L(u) \not\leq M_-\} = L(u)_\# \quad .$$

■

23.5 Corollary. *Lat* $\mathcal{R}_1(\mathcal{L})$ *consists of* $\{K : L_\# \le K \le L$ *for some* $L \in \mathcal{L}\}$. *In particular, every completely distributive lattice is reflexive.*

Proof. Let K be invariant for $\mathcal{R}_1(\mathcal{L})$, and let L be the least element of $\mathcal{L}$ containing K. Then since $L = \bigvee_{u \in K} L(u)$,

$$\begin{aligned} K \supseteq \overline{\mathcal{R}_1(\mathcal{L})K} &= \bigvee_{u \in K} \overline{\mathcal{R}_1(\mathcal{L})u} = \bigvee_{u \in K} L(u)_\# \\ &= \bigvee_{u \in K} \bigvee \{M \in \mathcal{L} : L(u) \not\le M_-\} \\ &= \bigvee \{M \in \mathcal{L} : L \not\le M_-\} = L_\# \ . \end{aligned}$$

Conversely, if $L_\# \le K \le L$, then $\mathcal{R}_1(\mathcal{L})K \subseteq \mathcal{R}_1(\mathcal{L})L = L_\# \subseteq K$. If $\mathcal{L}$ is completely distributive, then $L = L_\#$ so $Lat\, \mathcal{R}_1(\mathcal{L}) = \mathcal{L}$. ∎

23.6 Corollary. *If* $\mathcal{R}_1(\mathcal{L})$ *is weak* dense in* $Alg(\mathcal{L})$, *then* $\mathcal{L}$ *is completely distributive.*

Proof. By Lemma 23.4, $L_\# = \overline{\mathcal{R}_1(\mathcal{L})L} = \overline{(Alg\, \mathcal{L})L} = L$. Thus by Theorem 23.2, $\mathcal{L}$ is completely distributive. ∎

Let $C_2(\mathcal{L})$ denote $C_2 \cap Alg(\mathcal{L})$, the Hilbert-Schmidt operators in $Alg(\mathcal{L})$.

23.7 Theorem. *Let* $\mathcal{L}$ *be a commutative subspace lattice. The the following are equivalent.*

i) $\mathcal{L}$ *is completely distributive.*

ii) $\mathcal{R}_1(\mathcal{L})$ *is weak* dense in* $Alg(\mathcal{L})$.

iii) $C_2(\mathcal{L})$ *is dense in* $Alg(\mathcal{L})$ *in the weak operator topology.*

iv) *For every non-zero projection* P *in* $\mathcal{L}'$, $PC_2(\mathcal{L})P \ne 0$.

v) *For every non-zero projection* P *in* $\mathcal{L}''$, $P\mathcal{R}_1(\mathcal{L})P \ne 0$.

iv′) *If* $\mathcal{L} = \mathcal{L}(X, \le, m)$, *then* iv) *can be replaced by: For every subset* A *of* X *with* $m(A) > 0$, $m^2(A \times A \cap G(\le)) > 0$.

Proof. i) $\Rightarrow$ ii). If $\mathcal{L}$ is completely distributive, Lemma 23.4 shows that $\overline{\mathcal{R}_1(\mathcal{L})u} = \overline{Alg(\mathcal{L})u}$ for every vector u in $\mathcal{H}$. Since the weak* closure of $\mathcal{R}_1(\mathcal{L})$, $\overline{\mathcal{R}_1(\mathcal{L})}^{w*}$, is an ideal of $Alg(\mathcal{L})$, it is an $Alg(\mathcal{L})$ bimodule. Thus by Theorem 22.19, $\overline{\mathcal{R}_1(\mathcal{L})}^{w*}$ contains $\mathcal{A}_{\min}(\mathcal{L})$. In particular, it contains the identity operator. So it equals all of $Alg(\mathcal{L})$.

Clearly, ii) implies iii) implies iv). By Theorem 22.4, $\mathcal{L}$ can be represented as $\mathcal{L}(X, \le, m)$. Since $L^\infty(X,m)$ is contained in $\mathcal{L}'$, one obtains $P(A)C_2(\mathcal{L})P(A) \ne 0$ whenever $P(A) \ne 0$ (i.e. $m(A) > 0$). So $A \times A \cap G(\le)$ supports a Hilbert-Schmidt operator, and hence has positive m^2 measure. This establishes iv′).

Suppose iv′) holds, and fix a set A with $m(A) > 0$. Then by Fubini's Theorem,

$$0 < m^2(A \times A \cap G) = \int_A m\{x \in A : x \geq y\} dm(y) \ .$$

Let $E_y = \{x : x \geq y\}$. Then $B = \{y \in A : m(A \cap E_y) > 0\}$ has positive measure. Likewise

$$0 < m^2(B \times B \cap G) = \int_B m\{y \in B : y \leq x\} dm(x) \ .$$

Let $F_x = \{y : y \leq x\}$. Then $C = \{x \in B : m(F_x \cap B) > 0\}$ has positive measure. Choose any x_0 in C. Then $m(F_{x_0} \cap B) > 0$ and $m(A \cap E_{x_0}) > 0$. Hence

$$(E_{x_0} \cap A) \times (F_{x_0} \cap B) = \{(x,y) : x \in A, y \in B \text{ and } x \geq x_0 \geq y\} \ .$$

Thus this rectangle is contained in $G(\leq)$, and hence $P(E_{x_0} \cap A)\,\mathcal{B}(\mathcal{H})P(F_{x_0} \cap B)$ is contained in $Alg(\mathcal{L})$. Thus $P(A)\mathcal{R}_1(\mathcal{L})P(A) \neq 0$. This proves v).

Now assume v) holds. For any L in $\mathcal{L}$, one has by Lemma 23.4, $(L - L_{\#})\mathcal{R}_1(\mathcal{L})(L - L_{\#}) = 0$. Hence $L = L_{\#}$ for all L in $\mathcal{L}$. By Theorem 23.2, $\mathcal{L}$ is completely distributive. ∎

A simple reworking of the proof above yields:

23.8 Proposition. *Let $\mathcal{L}$ be a commutative subspace lattice. Then $\mathcal{K} \cap Alg(\mathcal{L})$ is weak* dense in $Alg(\mathcal{L})$ if and only if $P(\mathcal{K} \cap Alg(\mathcal{L}))P \neq 0$ for every non-zero projection P in $\mathcal{L}''$.*

Proof. The "only if" direction is obvious. Conversely, the weak* closure $\mathcal{A}$ of $\mathcal{K} \cap Alg(\mathcal{L})$ is an ideal in $Alg(\mathcal{L})$ and hence a bimodule. If x belongs to $\mathcal{H}$, and $L = \overline{Alg(\mathcal{L})x}$ and $M = \overline{\mathcal{A}x}$, one sets $P = P(L) - P(M)$. Then $P(\mathcal{K} \cap Alg\,\mathcal{L})P = 0$ and thus $P = 0$. By Theorem 22.19, $\mathcal{A}$ contains $\mathcal{A}_{\min}(\mathcal{L})$ including the identity. Since $\mathcal{A}$ is an ideal, $\mathcal{A} = Alg(\mathcal{L})$. ∎

23.9 Corollary. *Every completely distributive CSL is synthetic.*

Proof. Immediate from Theorem 23.7 and Lemma 22.20. ∎

There is a simple application of complete distributivity to tensor products of CSL algebras. As in Section 15.A, let $Alg\,\mathcal{L}_1 \otimes Alg\,\mathcal{L}_2$ denote the WOT closed span of the elementary tensors $A \otimes B$ acting on $\mathcal{H}_1 \otimes \mathcal{H}_2$. Similarly, let $\mathcal{L}_1 \otimes \mathcal{L}_2$ denote the complete lattice generated by $\{L_1 \otimes L_2 : L_1 \in \mathcal{L}_1, L_2 \in \mathcal{L}_2\}$.

23.10 Corollary. *Let $\mathcal{L}_1$ be a completely distributive CSL, and let $\mathcal{L}_2$ be any complete subspace latice. Then*

$$Alg\,\mathcal{L}_1 \otimes Alg\,\mathcal{L}_2 = Alg(\mathcal{L}_1 \otimes \mathcal{L}_2) \ .$$

If $\mathcal{L}_2$ is a CSL as well, then

$$Lat\,(Alg\,\mathcal{L}_1 \otimes Alg\,\mathcal{L}_2) = \mathcal{L}_1 \otimes \mathcal{L}_2 \ .$$

Proof. Since $A \otimes B$ leaves each $L_1 \otimes L_2$ invariant, one always has

$$Alg\,\mathcal{L}_1 \otimes Alg\,\mathcal{L}_2 \subseteq Alg(\mathcal{L}_1 \otimes \mathcal{L}_2) \ .$$

By Theorem 23.7, $\mathcal{R}_1(\mathcal{L}_1)$ is weak* dense in $Alg(\mathcal{L}_1)$. By Exercise 11.4, it follows that the unit ball of $\mathcal{R}_1(\mathcal{L}_1)$ is strong* dense in the ball of $Alg(\mathcal{L}_1)$. Let R_n be a sequence of contractions in $\mathcal{R}_1(\mathcal{L}_1)$ converging strong * to I. Let $S_{n,j}$ be rank one operators in $\mathcal{R}_1(\mathcal{L}_1)$ such that $R_n = \sum_{j=1}^{m_n} S_{n,j}$. Let us write $S_{n,j} = x_{n,j}y_{n,j}^*$ (suppressing the $\otimes$ which plays two roles here). Fix any operator T in $Alg(\mathcal{L}_1 \otimes \mathcal{L}_2)$. Then $(S_{n,i} \otimes I)T(S_{n,j} \otimes I)$ has the form $x_{n,i}y_{n,j}^* \otimes B_{ij}^{(n)}$ for some $B_{ij}^{(n)}$ in $\mathcal{B}\ (\mathcal{H}_2)$. Moreover, since this belongs to $Alg(\mathcal{L}_1 \otimes \mathcal{L}_2)$, $B_{ij}^{(n)}$ belongs to $Alg(\mathcal{L}_2)$. Further, if $B_{ij}^{(n)} \neq 0$, then $x_{n,i}y_{n,j}^*$ belongs to $Alg(\mathcal{L}_1)$. Thus

$$(R_n \otimes I)T(R_n \otimes I) = \sum_{i=1}^{m_n} \sum_{j=1}^{m_n} (S_{n,i} \otimes I)T(S_{n,j} \otimes I)$$

belongs to $Alg(\mathcal{L}_1) \otimes Alg(\mathcal{L}_2)$. This sequence converges to T in the weak* topology, and thus T belongs to $Alg(\mathcal{L}_1) \otimes Alg(\mathcal{L}_2)$. So equality holds.

If $\mathcal{L}_2$ is also a CSL, then by Theorem 22.11,

$$\mathcal{L}_1 \otimes \mathcal{L}_2 = Lat\,(Alg(\mathcal{L}_1 \otimes \mathcal{L}_2)) = Lat\,(Alg(\mathcal{L}_1) \otimes Alg(\mathcal{L}_2)) \ . \qquad \blacksquare$$

Finally, we introduce one more lattice theoretic property with connections to CSL lattices. By Exercise 23.4, every CSL lattice is *infinitely distributive*. Say that a complete, infinitely distributive lattice is Λ-*distributive* if for every subset $\{L_n, n \in \mathbb{N}\}$ of $\mathcal{L}$ and with $\Lambda = \{S \subseteq \mathbb{N} : |S| = \infty\}$, one has

$$\bigwedge_{S \in \Lambda} \big(\bigvee_{n \in S} L_n\big) \leq \bigvee_{S \in \Lambda} \big(\bigwedge_{n \in S} L_n\big) \ .$$

If $\mathcal{L}$ is completely distributive, then the left hand side equals $\bigwedge_{F \in \Pi S} (\bigwedge F(S))$. Since the range of each F must meet every infinite subset of $\mathbb{N}$, the range is cofinite. Thus this quantity is

$$\liminf L_n = \bigvee_{k \geq 1} \big(\bigwedge_{n \geq k} L_n\big) \ .$$

A similar argument shows that the inequality becomes

$$\liminf L_n \le \limsup L_n$$

in the completely distributive case.

On the other hand, if $\mathcal{L}$ fails to be Λ-distributive, then the left hand side, say F, is not smaller than the right hand side, E. Let $M_n = (L_n \vee E) \wedge F$. Then for every infinite subset S of $\mathbb{N}$

$$\bigvee_{n \in S} M_n = F \text{ and } \bigwedge_{n \in S} M_n = E \wedge F < F \ .$$

Conversely, the existence of such M_n clearly contradicts Λ-distributivity. As we saw at the beginning of this chapter, the projection lattice of $L^\infty(0,1)$ is not Λ-distributive.

23.11 Lemma. *Suppose E, F, P_n, $n \ge 1$ are commuting projections such that $E \le P_n \le F$ for all $n \ge 1$, and $w-\lim P_n = Q$ exists in the weak operator topology. Then the following are equivalent.*

1) $E = P(\ker(I-Q))$ *and* $F = P(\operatorname{Ran}(Q))$.

2) *For every infinite set S of positive integers,* $\bigvee_{n \in S} P_n = F$ *and* $\bigwedge_{n \in S} P_n = E$.

Proof. Assume 1) holds, and fix a set S. Then $\bigwedge_{n \in S} P_n \ge E$ is the projection onto $\{x : P_n x = x \text{ for } n \in S\}$. Since $\lim_{n \to \infty} (P_n x, x) = (Qx, x)$, this set is contained in $\{x : Qx = x\} = \ker(I-Q)$. Hence $\bigwedge_{n \in S} P_n \le E$ and equality follows. The identity $F = \bigvee_{n \in S} P_n$ is similar.

Conversely, one has $E \le Q \le F$ so

$$E \le P(\ker(I-Q)) \le P(\operatorname{Ran} Q) \le F \ .$$

Suppose, $P(\ker Q)^\perp = P(\operatorname{Ran} Q) < F$. Then there is a unit vector x such that $Fx = x$ and $Qx = 0$. Thus $\lim(P_n x, x) = 0$ and $(Fx, x) = 1$. Pick a subsequence $S = \{n_i, i \ge 1\}$ with $(P_{n_i} x, x) \le 3^{-i}$. Then

$$1 = (Fx, x) = (\bigvee P_{n_i} x, x) \le \sum_{i=1}^{\infty} (P_{n_i} x, x) \le \frac{1}{2} \ .$$

This is contradictory, so $F = P(\operatorname{Ran} Q)$. Similarly, $E = P(\ker(I-Q))$. ∎

23.12 Theorem. *Let $\mathcal{L}$ be a commutative subspace lattice. Then $\{P(L) : L \in \mathcal{L}\}$ is compact in the strong operator topology if and only if $\mathcal{L}$ is Λ-distributive. In particular, this holds if $\mathcal{K} \cap Alg(\mathcal{L})$ is weak* dense in $Alg(\mathcal{L})$.*

Proof. Suppose $\mathcal{L}$ is not compact in the strong operator topology. Then since the unit ball of $\mathcal{B}(\mathcal{H})$ is metrizable in the strong operator topology, there is a sequence $P_n = P(L_n)$ with no strongly convergent subsequence. But the unit ball is WOT compact and metrizable, so we may assume that $w\text{-}\lim P_n = Q$ exists. If Q were a projection, then this would converge strongly (Exercise 23.5). So Q is not a projection, whence $E = P(\ker(I-Q)) < F = P(\overline{Ran\, Q})$.

It will be shown that $\ker(I-Q)$ and $\overline{Ran\, Q}$ both belong to $\mathcal{L}$. For if $Qx = x$, then $\lim_{n \to \infty} P_n x = x$ (as in Exercise 23.5), so for every T in $Alg(\mathcal{L})$,

$$Tx = \lim_{n \to \infty} TP_n x = \lim_{n \to \infty} P_n TP_n x = w\text{-}\lim P_n Tx = QTx \ .$$

Hence Tx belongs to $\ker(I-Q)$. A similar computation shows that $T^*\ker(Q)$ is contained in $\ker Q$. As $\overline{Ran\,(Q)} = \ker(Q)^{\perp}$, it is also invariant for $Alg(\mathcal{L})$. So both subspaces belong to $Lat(Alg(\mathcal{L})) = \mathcal{L}$ by Corollary 22.10. Set

$$Q_n = (P_n \vee E) \wedge F = E + P_n(F-E) \ .$$

Clearly, Q_n belong to $\mathcal{L}$ and $w\text{-}\lim Q_n = Q$. By Lemma 23.10, $\bigvee_{n \in S} Q_n = F$ and $\bigwedge_{n \in S} Q_n = E$ for every infinite subset S of $\mathbb{N}$. By the remarks preceding the theorem, $\mathcal{L}$ is not Λ-distributive.

Conversely, suppose $\mathcal{L}$ is not Λ-distributive. Let Q_n and $E < F$ be elements of $\mathcal{L}$ such that $\bigvee_{n \in S} Q_n = F$ and $\bigwedge_{n \in S} Q_n = E$ for every infinite subset S of $\mathbb{N}$. This still holds if $\{Q_n\}$ is replaced by a subsequence with a WOT limit Q. By Lemma 23.11, $\ker(I-Q) = E\mathcal{H}$ and $Ran(Q) = F\mathcal{H}$. So Q is not a projection. Hence $\mathcal{L}$ is not WOT closed, and thus is not compact in the strong operator topology.

Now if $\mathcal{L}$ is not compact, let E, F, Q, and Q_n be as above. Let K be any compact operator in $Alg(\mathcal{L})$. For each x in $\mathcal{H}$, $w\text{-}\lim Q_n x = Qx$ whence $\lim KQ_n x = KQx$ in norm. Thus (as in the proof of the Aronszajn-Smith Theorem 3.1)

$$\begin{aligned} KQx &= \lim_{n \to \infty} KQ_n x = \lim_{n \to \infty} Q_n KQ_n x \\ &= \lim_{n \to \infty} Q_n KQx = w\text{-}\lim_{n \to \infty} Q_n KQx = QKQx \ . \end{aligned}$$

Consequently, K maps $\overline{Ran(Q)}$ into $\ker(I-Q)$. So

$$(F-E)(\mathcal{K} \cap Alg(\mathcal{L}))(F-E) = \{0\} \ .$$

By Proposition 23.8, $\mathcal{K} \cap Alg(\mathcal{L})$ is not weak* dense in $Alg(\mathcal{L})$. ∎

The proof above immediately yields the following corollary.

23.13 Corollary. *If $\mathcal{L}$ is a CSL containing elements L_n such that $\bigvee_{n\in S} L_n = \mathcal{H}$ and $\bigwedge_{n \in S} L_n = \{0\}$ for every infinite subset S of $\mathbb{N}$, then $Alg(\mathcal{L})$ contains no non-zero compact operators.*

23.14 Example. Let $X = 2^{\aleph_0}$. For $0 < p < 1$, let m_p be the measure on $\{0,1\}$ given by $m_p(\{0\}) = p$ and $m_p(\{1\}) = 1-p$. Let μ_p be the infinite product of m_p on each copy of $2 = \{0,1\}$. Let $\leq$ be the partial order $(x_n) \leq (y_n)$ if and only if $x_n \leq y_n$ for all n. Then $\mathcal{L} = \mathcal{L}(X, \leq, \mu_p)$ is the complete lattice generated by the subspaces L_k of functions supported on $E_k = \{x_n : x_k = 0\}$. For any finite subset S of $\mathbb{N}$,

$$\mu_p(\bigcap_{k\in S} E_k) = p^{|S|} \quad \text{and} \quad \mu_p(\bigcup_{k \in S} E_k) = 1-(1-p)^{|S|} .$$

Thus it follows that if S is any infinite subset of $\mathbb{N}$, then $\bigwedge_{n \in S} L_n = 0$ and $\bigvee_{n\in S} L_n = \mathcal{H}$. By Corollary 23.13, $Alg(\mathcal{L}) \cap \mathcal{K} = \{0\}$.

Now, we consider briefly the implications of having certain compact operators in $Alg(\mathcal{L})$. Exercise 23.7 shows that, unlike the nest case, even rank two operators need not be a sum of rank one operators in $Alg(\mathcal{L})$. However, the following theorem shows that finite rank operators in $Alg(\mathcal{L})$ imply the existence of rank one operators. Exercise 23.8 shows that there may be Hilbert-Schmidt operators in $Alg(\mathcal{L})$ but no finite rank operators.

23.15 Lemma. *Let $\mathcal{L}$ be a CSL lattice. Suppose F is a rank n operator in $Alg(\mathcal{L})$ such that PFQ has rank n or 0 for every pair of projections P, Q in $\mathcal{L}''$. Then F belongs to $\mathcal{R}_1(\mathcal{L})$.*

Proof. Write $F = \sum_{i=1}^{n} x_i \otimes y_i^*$, and let P, Q be projections in $\mathcal{L}''$. If $0 \neq PFQ = \sum_{i=1}^{n} Px_i \otimes Qy_i^*$, then $\{Px_i, 1 \leq i \leq n\}$ are linearly independent, as are $\{Qy_i, 1 \leq i \leq n\}$. Let L be the least element of $\mathcal{L}$ containing x_1, and hence all of $\{x_i, 1 \leq i \leq n\}$. Suppose M belongs to $\mathcal{L}$ and $M \not\geq L$. Then $P(M)^\perp x_1 \neq 0$ but $0 = P(M)^\perp F P(M)$. Since $\{P(M)^\perp x_i, 1 \leq i \leq n\}$ are independent, $\{P(M) y_i, 1 \leq i \leq n\}$ are not independent and thus are all zero. So y_i belongs to $M^\perp$, $1 \leq i \leq n$. So each y_i belongs to

$$\bigwedge\{M^\perp : M \not\geq L\} = (\bigvee\{M : M \not\geq L\})^\perp = (L_-)^\perp .$$

Hence $F = P(L) F P(L_-)^\perp$. By Lemma 23.3, each $x_i \otimes y_i^*$ belong to $Alg(\mathcal{L})$. Consequently, F is a combination of n rank one operators in $Alg(\mathcal{L})$. ∎

23.16 Theorem. *Let $\mathcal{L}$ be a commutative subspace lattice. The norm closure of $\mathcal{R}_1(\mathcal{L})$ contains all finite rank operators in $Alg(\mathcal{L})$.*

Proof. Proceed by induction on $n = rank(F)$. Clearly, if $n = 1$, then F belongs to $\mathcal{R}_1(\mathcal{L})$. Suppose that every F of rank at most $n-1$ belongs to $\overline{\mathcal{R}_1(\mathcal{L})}$. Fix F in $Alg(\mathcal{L})$ of rank n. Let P_0 be the supremum of all projections P in $\mathcal{L}''$ such that $rank(PF) < n$, and let Q_0 be the supremum of all projections Q in $\mathcal{L}''$ such that $rank(P_0^\perp FQ) < n$. Choose pairwise orthogonal projections P_k and Q_k in $\mathcal{L}''$ such that $rank(P_kF) < n$, $rank(P_0^\perp FQ_k) < n$ and $P_0 = \sum\limits_{k\geq 1} P_k$ and $Q_0 = \sum\limits_{k\geq 1} Q_k$. By the induction hypothesis, each P_kF and $P_0^\perp FQ_k$ belongs to $\overline{\mathcal{R}_1(\mathcal{L})}$. Since F is compact,

$$F - P_0^\perp F Q_0^\perp = \lim_{m\to\infty} \sum_{k=1}^{m} P_kF + P_0^\perp FQ_k$$

belongs to $\overline{\mathcal{R}_1(\mathcal{L})}$. Now $F_0 = P_0^\perp F Q_0^\perp$ has the property that for any projections P and Q in $\mathcal{L}''$, either $PF_0Q = 0$ or $rank(PF_0Q) = n$. By Lemma 23.15, F_0 belongs to $\mathcal{R}_1(\mathcal{L})$. Thus, F belongs to $\overline{\mathcal{R}_1(\mathcal{L})}$. ∎

We conclude this chapter with some interesting examples of CSL algebras. The properties of the algebras will be deduced from some results of commutative harmonic analysis which will be stated without proof. Let $\mathbf{T}$ denote the unit circle with Haar measure m. For any closed subset E of $\mathbf{T}$, let $G_E = \{(z,w) \in \mathbf{T}^2 : z\overline{w} \in E\}$. Consider the $L^\infty(\mathbf{T})$ bimodule $\mathcal{A}_E = \mathcal{A}_{\max}(G_E)$ defined in section 22.28, and the associated CSL algebra $\mathcal{B}_E$ consisting of all operators of the form

$$\begin{bmatrix} M_1 & A \\ 0 & M_2 \end{bmatrix}, \ M_1, M_2 \in L^\infty(\mathbf{T}),\ A \in \mathcal{A}_E \ .$$

Since $L^\infty(\mathbf{T})$ is non-atomic, the compact operators in $\mathcal{B}_E$ correspond to the compact operators in $\mathcal{A}_E$. They are never dense in $\mathcal{B}_E$.

Let μ be a finite regular (signed) Borel measure on $\mathbf{T}$ (i.e. an element of $M(\mathbf{T}) = C(\mathbf{T})^*$). Define a bounded operator in $L^2(\mathbf{T})$ by

$$C_\mu f(z) = f * \mu(z) = \int_{\mathbf{T}} f(z\overline{w})\,d\mu(w) \ .$$

23.17 Lemma. *Let μ be an element of $M(\mathbf{T})$ with closed support E. Then C_μ is a normal operator unitarily equivalent to $diag(\hat{\mu}(n))_{n\in\mathbb{Z}}$, and $supp(C_\mu) = G_E$.*

Proof. Let $e_n = z^n$, $n \in \mathbb{Z}$, be the standard basis for $L^2(\mathbf{T})$. Then

$$C_\mu e_n(z) = \int (z\overline{w})^n d\mu(w) = \int \overline{w}^n d\mu(w) z^n = \hat{\mu}(n) e_n(z) \ .$$

For arbitrary f, g in L^2,

$$(C_\mu f, g) = \int\int f(z\overline{w})\,d\mu(w)\overline{g(z)}\,dm(z) = \int\int_{\mathbf{T}^2} f(u)\overline{g(z)}\,d\mu(z\overline{u})\,dm(z) \ .$$

The measure $d\mu(z\bar{u})dm(z)$ has closed support G_E. Thus, if $f(u)\overline{g(z)}$ is supported on $U \times V$ disjoint from G_E, $(C_\mu f, g) = 0$ whence $supp(C_\mu)$ is contained in G_E. On the other hand, if $U \times V$ is open, and meets G_E, then $\chi_{U \times V} d\mu(z\bar{u})dm(z)$ is a non-zero measure, and thus does not annihilate all of $C(\mathbf{T}^2)$. Since $span\{f(u)\overline{g(z)}: f, g \in C(\mathbf{T})\}$ is dense in $C(\mathbf{T}^2)$ by the Stone-Weierstrass Theorem, it follows that $P(U)C_\mu P(V) \neq 0$. So $supp(C_\mu) = G_E$. ∎

23.18 Theorem. *Let E be a closed subset of $\mathbf{T}$. Let J be any ideal of compact operators on L^2. If $\mathcal{A}_E$ contains a compact operator in J, then there is a non-zero normal operator in $\mathcal{A}_E \cap J$.*

Proof. For w in $\mathbf{T}$, let D_w denote the unitary operator $D_w f(z) = f(z\bar{w})$. (This operator is C_{δ_w} where δ_w is the point mass at w.) We claim that if $supp(A) \subseteq G_E$, then $supp(D_w A D_{\bar{w}}) \subseteq G_E$. Suppose that $U \times V$ is an open set, disjoint from G_E. Thus $U \cap (VE)$ is empty. Let f be a function with support contained in V. Then $D_{\bar{w}} f$ has support contained in $V\bar{w}$. Thus $AD_{\bar{w}} f$ has support contained in $VE\bar{w}$, and $D_w A D_{\bar{w}} f$ has support in VE. This is disjoint from U, whence $U \times V$ is disjoint from $supp(D_w A D_{\bar{w}})$.

Let M_z denote the multiplication operator by the identity function. Define

$$\Delta_n(A) = \int_{\mathbf{T}} D_w M_z^n A D_{\bar{w}} dm(w) \ .$$

This converges in the norm of the ideal J, so is an element of $\mathcal{A}_E \cap J$. To see what this operator is explicitly, express A as a matrix (a_{nm}) with respect to the basis e_n. The operator D_w is diagonal with matrix $diag(w^n)$. Thus

$$(\Delta_0(A)e_n, e_m) = \int_{\mathbf{T}} w^{n-m} a_{nm} dm(w) = \delta_{nm} a_{nm} \ .$$

That is, $\Delta_0(A)$ is the diagonal of A with respect to the basis e_n. Since M_z^n is just a shift by n places in this basis, it follows that

$$\Delta_n(A) = diag(a_{k,k-n}) \ .$$

Since A is non-zero, at least one $\Delta_n(A) \neq 0$ as desired. ∎

This diagonal operator may not be the Fourier series of a measure, so this may not be some C_μ. In the terminology of harmonic analysis, this corresponds to a pseudo-measure. More precisely, the Fourier transform carries $\ell^1(\mathbb{Z})$ onto $A(\mathbf{T})$. The norm on $A(\mathbf{T})$ is precisely $\|f\|_A = \sum |\hat{f}(n)| = \|\mathcal{F}^{-1}(f)\|_1$. So the dual $PM(\mathbf{T}) = A(\mathbf{T})^*$ is isometrically isomorphic to $\ell^\infty(\mathbb{Z})$. In this disguise, the functionals are called *pseudo-measures*.

23.19 Theorem (Salem [1]). *There is a closed subset E of $\mathbf{T}$ of Lebesgue measure zero which supports a non-zero measure μ such that $\hat{\mu}$ belongs to ℓ^p for all $p > 2$.*

23.20 Corollary. *There is a CSL algebra which contains no Hilbert-Schmidt operators, yet contains an operator K belonging to C_p for all $p > 2$.*

Proof. Let E and μ be given by Salem's Theorem 23.19. Then by Lemma 23.17 $K = \begin{bmatrix} 0 & C_\mu \\ 0 & 0 \end{bmatrix}$ belongs to $\mathcal{B}_E$, and K belongs to C_p for all $p > 2$. On the other hand, $m^2(G_E) = m(E) = 0$. So $\mathcal{A}_E$, and thus $\mathcal{B}_E$, contains no Hilbert-Schmidt operators. ∎

23.21 Lemma. *Let A and B be operators in $\mathcal{A}_E$ and $\mathcal{A}_F$ respectively. Then AB belongs to $\mathcal{A}_{EF}$.*

Proof. Let V be an open subset of $\mathbf{T}$. Then $BP(V)$ has support contained in VE, and $AP(VE)$ has support contained in VEF. Thus,

$$ABP(V) = AP(VE)BP(V) = P(VEF)ABP(V) \ .$$

Consequently, if $U \times V$ is disjoint from G_{EF}, then $U \cap VEF$ is empty and $P(U)ABP(V) = 0$. ∎

The following result is taken from Graham and McGehee [1], Proposition 6.3.13.

23.22 Theorem. *There is a closed subset E of $\mathbf{T}$ such that $E^n = \{e_1 e_2 \dots e_n : e_i \in E\}$ has Lebesgue measure zero for all $n \geq 1$, and which supports a non-zero measure μ such that $\hat{\mu}(n)$ belongs to c_0.*

23.23 Corollary. *There is a CSL algebra which contains a non-zero compact operator, but contains no operators in C_p for any $p < \infty$.*

Proof. Let E and μ be given as in Theorem 23.22. By Lemma 23.17, $K = \begin{bmatrix} 0 & C_\mu \\ 0 & 0 \end{bmatrix}$ is a non-zero compact operator in $\mathcal{B}_E$. On the other hand, if $\mathcal{B}_E$ contains an operator in C_{2n}, it must have the form $\begin{bmatrix} 0 & A \\ 0 & 0 \end{bmatrix}$ where A belongs to $\mathcal{A}_E \cap C_{2n}$. By Theorem 23.18, $\mathcal{A}_E \cap C_{2n}$ contains a non-zero normal operator N. Thus N^n is a non-zero Hilbert-Schmidt operator. By Lemma 23.21, N^n is supported on G_{E^n} which has Lebesgue measure zero. This is absurd. Hence $\mathcal{B}_E \cap C_{2n} = \{0\}$ for all $n \geq 1$. ∎

Notes and Remarks.

The characterization Theorem 23.2 of complete distributivity is due to Raney [1, 2]. Lemmas 23.3-23.6 are due to Longstaff [2]. In Theorem 23.7, the equivalence of i), iii) and iv′) is due to Hopenwasser, Laurie and Moore [1] while the equivalence of i), ii) and v) is due to Laurie and Longstaff [1]. Corollary 23.10 is due to Kraus [3]. Theorem 23.12 is in Wagner [3], which acknowledges joint work with Froelich and Moore. Froelich [2] had earlier proved Corollary 23.13 directly and gave Example 23.14. The question remains whether weak* density of $\mathcal{K}$ in $Alg(\mathcal{L})$ or even the strong compactness of $\mathcal{L}$ is sufficient to imply complete distributivity. If not, what is the lattice condition corresponding to the weak* density of the compact operators in $Alg(\mathcal{L})$? Hopenwasser and Moore [1] show that the existence of finite rank operators in $Alg(\mathcal{L})$ implies $\mathcal{R}_1(\mathcal{L}) \neq 0$, and prove Theorem 23.16 in the finite width case without requiring the norm closure. The general result is new. The examples, 23.17-23.23, are due to Froelich [1]. He also uses the existence of a set E supporting pseudo-measures on ℓ^p for $p > q > 2$ but not for $p < q$ to product CSL algebras with operators in C^p for $p > q$ not $p < q$. A similar example gives a CSL with compact operators but no compact pseudo-integral operators.

Exercises.

23.1 Let $\mathcal{L}$ be a lattice satisfying the distributive law (D).

(i) Show that
$(x \wedge y) \vee (x \wedge z) \vee (y \wedge z) = (x \vee y) \wedge (x \vee z) \wedge (y \vee z)$.
Hint: substitute $(x \wedge y) \vee (x \wedge z)$ for x in (D).

(ii) Simplify (i) in the case $x \geq y$ to obtain the *modular law*

$$x \geq y \text{ implies } x \wedge (y \vee z) = y \vee (x \wedge z) \ . \qquad \text{(M)}$$

(iii) Prove (D*). Hint: Apply $(x \vee \ )$ to both sides of (i).

(iv) Show that (D), (D*), and (i) are equivalent.

23.2 (Longstaff) If $\mathcal{L}$ is a complete lattice, define

$$x_* = \bigwedge \{y_- : y \in \mathcal{L}, y \not\leq x\} \ .$$

(i) Prove that $x_\# \leq x_{*\#} \leq x \leq x_{\#*} \leq x_*$ for all x in $\mathcal{L}$.

(ii) Prove that $\mathcal{L}$ is completely distributive if and only if $x = x_*$ for all x in $\mathcal{L}$.

23.3 (Raney)

(i) An element x of a lattice $\mathcal{L}$ is *join-irreducible* if for every subset M of $\mathcal{L}$, $x \leq \bigvee M$ implies there exists $y \in M$, $x \leq y$. If $\mathcal{L}$ is a complete lattice of subsets of a set S, for each s in S, let $\ell(s) = \bigwedge \{x \in \mathcal{L} : s \in x\}$. Show that $\ell(s)$ is join-irreducible.

(ii) Hence show that if $\mathcal{L}$ is isomorphic to a complete lattice of subsets of S, then every element of $\mathcal{L}$ is the join of join-irreducible elements.

(iii) Prove the converse of (ii).

(iv) Show that $[0,1]$ with the usual order is completely distributive, but is not isomorphic to a complete lattice of subsets.

23.4 A complete distributive lattice $\mathcal{L}$ is called *meet continuous* if

$$x \wedge (\bigvee_{\lambda \in \Lambda} y_\lambda) = \bigvee_{\lambda \in \Lambda} (x \wedge y_\lambda) \text{ for all } x, y_\lambda \text{ in } \mathcal{L} \ .$$

Show that every complete sublattice of the Boolean algebra of idempotents in $L^\infty(m)$ is meet and join continuous for every measure m. Such lattices are called *infinitely distributive*.

23.5 Suppose that a sequence P_n of projections converges weakly to a projection Q. Prove that it converges strongly. Hint: Consider $\lim(P_n x, x)$ for $Qx = x$ and $Qx = 0$.

23.6 Show that if $K \cap Alg(\mathcal{L})$ is weak* dense in $Alg(\mathcal{L})$, then $Alg(\mathcal{L}) + K$ is closed. Hint: Corollary 11.7.

23.7 (Hopenwasser-Moore) Let $\mathcal{D}$ be the diagonal algebra with respect to an orthonormal basis $\{e_n, n \geq 1\}$. Let $\mathcal{A}$ be the algebra on $\mathcal{H} \oplus \mathcal{H}$ consisting of all operators of the form $\begin{bmatrix} D_1 & T \\ 0 & D_2 \end{bmatrix}$ where D_i belong to $\mathcal{D}$, T belongs to $\mathcal{B}(\mathcal{H})$, and $(Te_n, e_n) = 0$ for $n \geq 1$.

(i) Calculate $Lat\, \mathcal{A}$.

(ii) Let $x = \sum_{n \geq 1} a_n e_n$ belong to $\mathcal{H}$, and let $a_n \neq 0$ infinitely often. Let $y = \sum_{n \geq 1} n^{-1} a_n e_n$. Let $T = x \otimes y^* - y \otimes x^*$. Show that $\begin{bmatrix} 0 & T \\ 0 & 0 \end{bmatrix}$ is a rank two operator in $\mathcal{A}$ which is not a finite linear combination of rank one operators in $\mathcal{A}$.

(iii) Show that the span of the rank one operators of $\mathcal{A}$ is weak operator dense in $\mathcal{A}$, and in fact, the unit ball is the strong* closure of this span.

23.8 (Trent) Let K be a closed nowhere dense subset of $[0,1]$ with positive measure. Let

$$G = \{(x,x) : 0 \leq x \leq 1\} \cup \{(x,y) : 0 \leq x \leq ½ \leq y \leq 1, y - x \in K\} \ .$$

Show that G is the graph of a partial order $\leq$, and $m^2(G) > 0$. Hence deduce that $Alg(\mathcal{L}([0,1], \leq, m))$ contains $\mathcal{C}_2$ operators. Show that G does not contain any Borel rectangle. Hence deduce that the algebra contains no finite rank operators.

23.9 Prove that the operator C_μ of Lemma 23.15 belongs to $\mathcal{A}_{\min}(G_E)$.

24. Failure of the Distance Formula

In Chapter 9, distance formulae are established for nest algebras and for certain von Neumann algebras. These formulae are of critical importance in the proof of the Similarity Theorem and other results. So it is a significant obstruction that such a formula fails for certain CSL algebras. Indeed, there is no nice class of CSL algebras other than the two above which are known to have any distance formula. In this chapter, the counterexamples will be constructed.

A reflexive algebra $\mathcal{A} = Alg\,\mathcal{L}$ is *hyperreflexive* if there is a constant K such that for every T in $\mathcal{B}(\mathcal{H})$,

$$d(T,\mathcal{A}) \leq K \sup_{L \in \mathcal{L}} \|P(L)^{\perp}TP(L)\| \quad .$$

Note that for every A in $\mathcal{A}$ and L in $\mathcal{L} = Lat\,\mathcal{A}$, one has

$$\|P(L)^{\perp}TP(L)\| = \|P(L)^{\perp}(T-A)P(L)\| \leq \|T-A\| \quad .$$

Taking the infimum over A in $\mathcal{A}$ and the supremum over L in $\mathcal{L}$ yields

$$\sup_{L \in \mathcal{L}} \|P(L)^{\perp}TP(L)\| \leq d(T,\mathcal{A}) \quad .$$

Let the left hand quantity be denoted as $\beta_{\mathcal{A}}(T)$, or just $\beta(T)$ if $\mathcal{A}$ is understood.

Theorem 9.5 shows that equality holds for nest algebras. However, Example 9.7 shows that a constant greater than one is required even for 3×3 diagonal matrices. This example is the key to the construction of worse examples.

It is more convenient to work with bimodules over a masa rather than algebras. So we define hyperreflexivity for a subspace $\mathcal{S}$. Let

$$\beta_{\mathcal{S}}(T) = \sup_{\|x\|=1} d(Tx,\mathcal{S}x) \quad .$$

Recall from 15.B that $\mathcal{S}$ is reflexive if $\beta_{\mathcal{S}}(T) = 0$ implies T belongs to $\mathcal{S}$. Let

$$\kappa(\mathcal{S}) = \sup_{T \notin \mathcal{S}} \frac{d(T,\mathcal{S})}{\beta_{\mathcal{S}}(T)} \quad .$$

Say that $\mathcal{S}$ is *hyperreflexive* if $\kappa(\mathcal{S}) < \infty$. The reader should verify that if $\mathcal{S}$ is a unital algebra, then the two notions of hyperreflexivity coincide (Exercise 24.1). Moreover, one readily constructs an algebra containing a masa from $\mathcal{S}$ which is hyperreflexive precisely when $\mathcal{S}$ is (Exercise 24.2).

If S is a weak* closed subspace of $B(\mathcal{H})$, let $S_\perp$ denote its pre-annihilator in C_1. Then $S = (S_\perp)^\perp$ by the Hahn-Banach Theorem. Suppose S and $\mathcal{T}$ are two such subspaces of $B(\mathcal{H}_1)$ and $B(\mathcal{H}_2)$ respectively. Let $S*\mathcal{T}$ denote the subspace of $B(\mathcal{H}_1 \otimes \mathcal{H}_2)$ called the *dual product* of S and $\mathcal{T}$ given by

$$\begin{aligned} S*\mathcal{T} &= (S_\perp \otimes \mathcal{T}_\perp)^\perp \\ &= \{A \in B(\mathcal{H} \otimes \mathcal{H}) : tr(A(X \otimes Y)) = 0 \text{ when } X \in S_\perp, Y \in \mathcal{T}_\perp\} . \end{aligned}$$

These tensor products are spatial, and $S_\perp \otimes \mathcal{T}_\perp$ is endowed with the C_1 norm in $C_1(\mathcal{H}_1 \otimes \mathcal{H}_2)$. For the moment, we consider the special case $\mathcal{T} = \mathcal{D}_3$, the diagonal subalgebra of $\mathcal{M}_3$. Then

$$S*\mathcal{D}_3 = \left\{ \begin{bmatrix} X_1 & S_{12} & S_{13} \\ S_{21} & X_2 & S_{23} \\ S_{31} & S_{33} & X_3 \end{bmatrix} : X_i \in B(\mathcal{H}), S_{ij} \in S \right\} .$$

24.1 Lemma. *Let Y, X_1, X_2 and X_3 belong to $B(\mathcal{H})$. Then*

$$\left\| \begin{bmatrix} X_1 & Y & Y \\ Y & X_2 & Y \\ Y & Y & X_3 \end{bmatrix} \right\| \geq \frac{3}{2} \|Y\| .$$

Proof. First note that permuting the X_i's does not change the norm. Thus, the norm of the operator is at least as great as the average

$$\|T\| = \left\| \begin{bmatrix} X & Y & Y \\ Y & X & Y \\ Y & Y & X \end{bmatrix} \right\|$$

where $X = (X_1+X_2+X_3)/3$. Fix unit vectors x and y in $\mathcal{H}$, and let P_x and P_y be the projections onto $\mathbb{C}x \oplus \mathbb{C}x \oplus \mathbb{C}x$ and $\mathbb{C}y \oplus \mathbb{C}y \oplus \mathbb{C}y$ respectively. Then

$$\|T\| \geq \|P_y T P_x\| = \left\| \begin{bmatrix} \alpha & \beta & \beta \\ \beta & \alpha & \beta \\ \beta & \beta & \alpha \end{bmatrix} \right\|$$

where $\alpha = (Xx,y)$ and $\beta = (Yx,y)$. This latter operator is normal because it equals $3\beta Q+(\alpha-\beta)I$ where $Q = \frac{1}{3}\begin{bmatrix} 1 & 1 & 1 \\ 1 & 1 & 1 \\ 1 & 1 & 1 \end{bmatrix}$ is a rank one projection. This has spectrum $\{\alpha-\beta, \alpha+2\beta\}$. Thus its norm is

$$\max\{\,|\alpha-\beta\,|,|\alpha+2\beta\,|\} \geq (3/2)\beta \ .$$

Take the supremum over all unit vectors x,y to obtain $\|T\| \geq (3/2)\|Y\|$. ∎

24.2 Lemma. *Let S be a proper reflexive subspace of $B\,(\mathcal{H})$. Then*

$$\kappa(S{*}D_3) \geq (9/8)^{1/2}\kappa(S) \ .$$

Proof. Fix T in $B\,(\mathcal{H})$ and set $\tilde{T} = \begin{bmatrix} T & T & T \\ T & T & T \\ T & T & T \end{bmatrix}$. For each unit vector x in $\mathcal{H}$,

$$d\left(\tilde{T}\begin{pmatrix} x \\ 0 \\ 0 \end{pmatrix}, S{*}D_3\begin{pmatrix} x \\ 0 \\ 0 \end{pmatrix}\right) = d\left(\begin{pmatrix} Tx \\ Tx \\ Tx \end{pmatrix}, \mathcal{H} \oplus \overline{Sx} \oplus \overline{Sx}\right)$$

$$= \sqrt{2}\,d(Tx,\overline{Sx}) \leq \sqrt{2}\,\beta(T) \ .$$

Likewise, this obtains for unit vectors in $\mathcal{H}^{(3)}$ with one non-zero coordinate. Now, if $\|x_1\|^2+\|x_2\|^2 = 1$,

$$d\left(\tilde{T}\begin{pmatrix} x_1 \\ x_2 \\ 0 \end{pmatrix}, S{*}D_3\begin{pmatrix} x_1 \\ x_2 \\ 0 \end{pmatrix}\right) = d(T(x_1+x_2),\overline{Sx_1}+\overline{Sx_2})$$

$$\leq d(T(x_1+x_2),\overline{S(x_1+x_2)})$$

$$\leq \|x_1+x_2\|\beta(T) \leq \sqrt{2}\,\beta(T) \ .$$

Finally, if $x = \begin{pmatrix} x_1 \\ x_2 \\ x_3 \end{pmatrix}$ and all $x_i \neq 0$, then $\overline{S{*}D_3x} = \mathcal{H}^{(3)}$. So one obtains

$$\beta_{S*D_3}(\tilde{T}) \leq \sqrt{2}\,\beta(T) \ .$$

On the other hand, $S{*}D_3$ is weak* closed; so there is an operator $\tilde{S} = (S_{ij})$ in $S{*}D_3$ such that $\|\tilde{T}-\tilde{S}\| = d(\tilde{T},S{*}D_3)$. Let $\tilde{S}_\sigma = (S_{\sigma(i)\sigma(j)})$ for σ any permutation in S_3. By symmetry, $\tilde{S}_\sigma$ belongs to $S{*}D_3$. Since $\tilde{T} = \tilde{T}_\sigma$,

$$d(\tilde{T},S{*}D_3) = \|\tilde{T}-\tilde{S}\| = \|\tilde{T}-\tilde{S}_\sigma\| \geq \|\tilde{T}-\frac{1}{6}\sum_{\sigma\in S_3}\tilde{S}_\sigma\| \geq d(\tilde{T},S{*}D_3) \ .$$

Thus, $\tilde{S}$ can be replaced by the operator $1/6\sum\tilde{S}_\sigma$ which has the form $\begin{bmatrix} X & S & S \\ S & X & S \\ S & S & X \end{bmatrix}$ for some S in S and X in $B\,(\mathcal{H})$. Thus, by Lemma 24.1,

$$\|\tilde{T}-\tilde{S}\| = \left\| \begin{bmatrix} T-X & T-S & T-S \\ T-S & T-X & T-S \\ T-S & T-S & T-X \end{bmatrix} \right\| \geq \frac{3}{2}\|T-S\| \geq \frac{3}{2}d(T,S) \ .$$

Now it follows that

$$\kappa(S*D_3) \leq \sup_{T \in B(\mathcal{H})} \frac{d(\tilde{T},S*D_3)}{\beta_{S*D_3}(\tilde{T})}$$

$$\geq \sup_{T \in B(\mathcal{H})} \frac{3}{2\sqrt{2}} \frac{d(T,S)}{\beta(T)} = \left(\frac{9}{8}\right)^{1/2} \kappa(S) \ .$$

■

24.3 Corollary. *Let S_n denote the dual product of n copies of D_3. Then $\kappa(S_n) \geq (9/8)^{n/2}$.*

24.4 Corollary. *There is a CSL algebra which is not hyperreflexive.*

Proof. Let S_n be as above, and set $S = \sum_{n \geq 1} \oplus S_n$ to be the weak* closure of the direct sum. Let $D = \sum_{n \geq 1} \oplus D_{3^n}$ be the diagonal algebra, and note that S is a D bimodule. The algebra $\mathcal{A}$ consisting of all operators of the form $\begin{bmatrix} D_1 & S \\ 0 & D_2 \end{bmatrix}$ such that D_1 and D_2 belong to D and S belongs to S is a weak* closed algebra containing the atomic masa $D \oplus D$. Hence $\mathcal{L} = Lat\,\mathcal{A}$ is an atomic CSL. It is readily apparent that $\mathcal{A}$ contains the rank one operators of the form PKQ where P, Q are rank one projections in $D \oplus D$ and K belongs to $\mathcal{A}$. These are easily seen to span a weak* dense subspace of $Alg\,\mathcal{L}$. So $\mathcal{A}$ is reflexive, and hence is a CSL algebra.

By the easy parts of Exercises 24.2 and 24.3,

$$\kappa(\mathcal{A}) \geq \kappa(S) \geq \sup_{n \geq 1} \kappa(S_n) = \infty \ .$$

So $\mathcal{A}$ is not hyperreflexive. ■

To widen the class of counterexamples, we show how to embed S_n into certain CSL algebras. Let A_n be the $3^n \times 3^n$ matrix with $a_{ij} = 1$ if S_n has non-zero (ij) entries, and $a_{ij} = 0$ otherwise.

If $\mathcal{N}_1, \ldots, \mathcal{N}_k$ are nests on Hilbert spaces $\mathcal{H}_i$, let $\mathcal{N}_1 \otimes ... \otimes \mathcal{N}_k$ denote the CSL on $\mathcal{H}_1 \otimes ... \otimes \mathcal{H}_k$ generated by the subspaces $N_1 \otimes ... \otimes N_k$ for N_i in $\mathcal{N}_i$.

24.5 Lemma. *Suppose that $\mathcal{L} = \mathcal{N}_1 \otimes ... \otimes \mathcal{N}_k$ is the tensor product of k non-trivial nests. Let $A = (a_{ij})$ be a $k \times k$ matrix of 0's and 1's with distinct columns. Then $\mathcal{L}''$ contains two sets $\{E_1,...,E_k\}$ and $\{F_1,...,F_k\}$ of pairwise orthogonal, non-zero intervals of $\mathcal{L}$ such that*

$$E_i(Alg\,\mathcal{L})F_j = a_{ij}E_iB(\mathcal{H})F_j \quad \text{for} \quad 1 \leq i,j \leq k \quad .$$

Proof. Since $\mathcal{N}_i$ is a non-trivial nest, there are intervals $E_i^{(0)}$, $E_i^{(1)}$, $F_i^{(0)}$ and $F_i^{(1)}$ satisfying the conditions of the lemma for the matrix $\begin{pmatrix} 1 & 1 \\ 0 & 1 \end{pmatrix}$. To see this, note that if $\mathcal{N}_i$ can be split into four pairwise orthogonal intervals, one can choose $E_i^{(\epsilon)}$ and $F_i^{(\epsilon)}$ such that $E_i^{(0)} \ll F_i^{(0)} \ll E_i^{(1)} \ll F_i^{(1)}$. Otherwise, $\mathcal{N}_i$ consists of two or three atoms $A_i \ll A_2$ $(\ll A_3)$. So one can take $E_i^{(0)} = F_i^{(0)} = A_1$ and $E_i^{(1)} = F_i^{(1)} = A_2$.

For $1 \leq i \leq k$, define

$$E_i = E_i^{(0)} \otimes ... \otimes E_{i-1}^{(0)} \otimes E_i^{(1)} \otimes E_{i+1}^{(0)} \otimes ... \otimes E_k^{(0)}$$

and

$$F_j = F_1^{(a_{1j})} \otimes F_2^{(a_{2j})} \otimes ... \otimes F_k^{(a_{kj})} \quad .$$

The pairwise orthogonality of $\{F_1,...,F_k\}$ follows since the columns of A are distinct. Clearly, $E_i(Alg\,\mathcal{L})F_j = a_{ij}E_iB(\mathcal{H})F_j$ as desired. ■

24.6 Theorem. *Suppose that $\mathcal{N}_i$, $1 \leq i \leq 3^n$, are non-trivial nests, and $\mathcal{L}'$ is any CSL. Let $\mathcal{L} = \mathcal{N}_1 \otimes ... \otimes \mathcal{N}_{3^n} \otimes \mathcal{L}'$. Then*

$$\kappa(Alg\,\mathcal{L}) \geq (9/8)^{n/2} \quad .$$

Proof Let A_n be the $3^n \times 3^n$ matrix of 0's and 1's determining S_n. A simple inductive argument shows that the columns of A are distinct. By Lemma 24.5, there are intervals $E_1,...,E_{3^n}$ and $F_1,...,F_{3^n}$ of $\mathcal{L}_0 = \mathcal{N}_1 \otimes ... \otimes \mathcal{N}_{3^n}$ so that $E_i(Alg\,\mathcal{L}_0)F_i = a_{ij}E_iB(\mathcal{H}_0)F_j$, where $\mathcal{H}_0 = \mathcal{H}_1 \otimes ... \otimes \mathcal{H}_{3^n}$. Choose unit vectors x_i and y_j in the range of E_i and F_j, respectively. The rank one partial isometries $U_{ij} = x_i \otimes y_j^*$ belong to $Alg\,\mathcal{L}_0$ precisely when $a_{ij} = 1$. Let P and Q be the projections onto $span\,\{x_i, 1 \leq i \leq 3^n\}$ and $span\,\{y_j, 1 \leq j \leq 3^n\}$ respectively. Then $PB(\mathcal{H}_0)Q$ is linearly isometric to $\mathcal{M}_{3^n}$ in such a way that $P(Alg\,\mathcal{L}_0)Q$ corresponds to S_n. The inverse correspondence sends a matrix $T = (t_{ij})$ to the operator

$$\tilde{T} = \sum_{i,j=1}^{3^n} t_{ij}U_{ij} \quad .$$

Moreover,

$$d(\tilde{T},Alg(\mathcal{L}_0)) \geq d(P\tilde{T}Q, PAlg(\mathcal{L}_0)Q) = d(T,S_n) \ .$$

Now compute $\beta_{Alg\,\mathcal{L}_0}(\tilde{T})$. Fix a projection R in $\mathcal{L}_0$. Let $J = \{j : Ry_j \neq 0\}$ and $I = \{i : R^{\perp}x_i \neq 0\}$. Thus

$$I \times J = \{(i,j) : R^{\perp}U_{ij}R \neq 0\} \ .$$

Now if $a_{ij} = 1$, then U_{ij} belongs to $Alg\,\mathcal{L}_0$ whence $R^{\perp}U_{ij}R = 0$. So $\{a_{ij} : i \in I, j \in J\}$ consists of zeros. Let P_I be the projection onto $span\,\{x_i : i \in I\}$, and let Q_J be the projection onto $span\,\{y_j : j \in J\}$. Then

$$\|R^{\perp}\tilde{T}R\| = \|R^{\perp}(P_J\tilde{T}Q_J)R\| \leq \|P_J\tilde{T}Q_J\| \leq \beta_{S_n}(T) \ .$$

Consequently,

$$\kappa(Alg\,\mathcal{L}_0) \geq \sup_{T \in \mathcal{M}_{3^n}} \frac{d(\tilde{T},Alg\,\mathcal{L}_0)}{\beta_{Alg\,\mathcal{L}_0}(\tilde{T})}$$

$$\geq \sup_{T \in \mathcal{M}_{3^n}} \frac{d(T,S_n)}{\beta_{S_n}(T)} = \kappa(S_n) \geq \left(\frac{9}{8}\right)^{n/2} \ .$$

Finally, for $\mathcal{L} = \mathcal{L} \otimes \mathcal{L}'$, consider the operators $\tilde{T} \otimes I$. Now, $Alg\,\mathcal{L}$ is contained in $Alg\,\mathcal{L}_0 \otimes \mathcal{B}(\mathcal{H})$, so

$$d(\tilde{T} \otimes I, Alg\,\mathcal{L}) \geq d(\tilde{T} \otimes I, Alg\,\mathcal{L}_0 \otimes \mathcal{B}(\mathcal{H})) = d(\tilde{T},Alg\,\mathcal{L}_0) \ .$$

On the other hand, $\mathcal{L}$ is contained in $\mathcal{L}_0 \otimes \mathcal{E}$ where $\mathcal{E}$ is the σ-complete Boolean algebra generated by $\mathcal{L}'$. A strongly dense subset of $\mathcal{L}_0 \otimes \mathcal{E}$ is given by the projections of the form

$$R = \sum_{n \geq 1} R_n \otimes E_n$$

for R_n in $\mathcal{L}_0$ and a partition $\{E_n, n \geq 1\}$ of the identity in $\mathcal{E}$. As $\sum_{n \geq 1} E_n = I$,

$$R^{\perp} = \sum_{n \geq 1} I \otimes E_n - \sum_{n \geq 1} R_n \otimes E_n = \sum_{n \geq 1} R_n^{\perp} \otimes E_n \ .$$

Hence

$$\beta_{Alg\,\mathcal{L}}(\tilde{T} \otimes I) \leq \sup_{R \in \mathcal{L}_0 \otimes \mathcal{E}} \|R^{\perp}(\tilde{T} \otimes I)R\|$$

$$= \sup \| \sum_{n \geq 1} (R_n^{\perp} \otimes E_n)(\tilde{T} \otimes I) \sum_{n \geq 1} R_n \otimes E_n \|$$

$$= \sup \| \sum_{n \geq 1} (R_n^{\perp}\tilde{T}R_n) \otimes E_n \|$$

$$= \sup_{R \in \mathcal{L}_0} \|R^{\perp}\tilde{T}R\| = \beta_{Alg\,\mathcal{L}_0}(\tilde{T}) \ .$$

Consequently, $\kappa(Alg\,\mathcal{L}) \geq \kappa(Alg\,\mathcal{L}_0) \geq (9/8)^{n/2}$. ∎

24.7 Corollary. *The infinite tensor product of non-trivial nests fails to be hyperreflexive.*

We wish to establish a result more general than Lemma 24.2. To do this, a study of $S_\perp$ is required.

24.8 Proposition. *Let S be a subspace of $B(\mathcal{H})$, and let $\mathcal{R}_1$ denote the set of rank one operators in $S_\perp$. Then $ref(S) = \mathcal{R}_1^\perp$. Thus S is a reflexive subspace if and only if it is weak* closed and $\mathcal{R}_1$ spans $S_\perp$.*

Proof. Let $x \otimes y^*$ belong to $\mathcal{R}_1$. Then for all S in S,

$$0 = tr(Sx \otimes y^*) = (Sx, y) \ .$$

Thus y is orthogonal to $\overline{Sx}$. If T belongs to $ref(S)$, one has $Tx \in \overline{Sx}$. So $0 = (Tx, y) = tr(Tx \otimes y^*)$. Hence T belongs to $\mathcal{R}_1^\perp$. Conversely, $x \otimes y^*$ belongs to $\mathcal{R}_1$ for every y in $(Sx)^\perp$. Hence if T belongs to $\mathcal{R}_1^\perp$, $0 = (Tx, y)$ so Tx belongs to $\overline{Sx}$. Thus T is in $ref(S)$. The last statement of the proposition is a consequence of the Hahn-Banach Theorem. ■

24.9 Theorem. *Let S be a weak* closed subspace of $B(\mathcal{H})$. Let C denote the closed convex hull of the rank one elements in $b_1(S_\perp)$. Then S is hyperreflexive if and only if $r(S) > 0$, where*

$$r(S) = \sup\{r : b_r(S_\perp) \subseteq C\} \ .$$

In this case, $\kappa(S) = r(S)^{-1}$.

Proof. In the previous proof, it is established that $x \otimes y^*$ belongs to $\mathcal{R}_1$ if and only if $y \in (Sx)^\perp$. Thus for T in $B(\mathcal{H})$,

$$\begin{aligned}\sup_{C \in C} |tr(TC)| &= \sup_{x \otimes y^* \in b_1(\mathcal{R}_1)} |tr\, Tx \otimes y^*| \\ &= \sup_{\|x\|=1} \sup_{y \perp Sx, \|y\|=1} |(Tx, y)| \\ &= \sup_{\|x\|=1} \|P(Sx)^\perp Tx\| = \sup_{\|x\|=1} d(Tx, Sx) = \beta_S(T) \ .\end{aligned}$$

Also, by the Hahn-Banach Theorem,

$$d(T, S) = \sup_{C \in b_1(S_\perp)} |tr(TC)| \ .$$

So if C contains $b_r(S_\perp)$, one obtains

$$\beta_S(T) \geq \sup_{C \in b_r(S_\perp)} |tr(TC)| \geq r \ \ d(T, S) \ .$$

Thus, $\kappa(S) \leq r(S)^{-1}$ and S is hyperreflexive.

Conversely, suppose C does not contain $b_r(S_\perp)$ and in particular does not contain a trace class operator X in $b_r(S_\perp)$. By the Hahn-Banach separation theorem, there is an operator T such that

$$|tr(TX)| > \sup_{C \in C} \operatorname{Re} tr(TC) = \sup_{C \in C} |tr(TC)| = \beta_S(T) \ .$$

Hence

$$d(T,S) \geq r^{-1}|tr(TX)| > r^{-1}\beta_S(T) \ .$$

Thus, $\kappa(S) \geq r(S)^{-1}$ and equality is assured. ■

24.10 Lemma. *Let $S \subseteq B(\mathcal{H}_1)$ and $\mathcal{T} \subseteq B(\mathcal{H}_2)$ be reflexive subspaces. Then*

$$S * \mathcal{T} = ref(S \otimes B(\mathcal{H}_2) + B(\mathcal{H}_1) \otimes \mathcal{T}) \ .$$

Proof. By definition,

$$(S*\mathcal{T})_\perp = S_\perp \otimes \mathcal{T}_\perp = span\{A \otimes B : A \in S_\perp, T \in \mathcal{T}_\perp\} \ .$$

Since $S_\perp$ and $\mathcal{T}_\perp$ are spanned by rank one operators, it follows that $S_\perp \otimes \mathcal{T}_\perp$ is also spanned by rank one operators. Thus by Proposition 24.8, $S*\mathcal{T}$ is reflexive. If S belongs to S and X belongs to $B(\mathcal{H}_2)$, then

$$tr((S \otimes X)(A \otimes B)) = tr(SA)tr(XB) = 0$$

for every elementary tensor $A \otimes B$ in $S_\perp \otimes \mathcal{T}_\perp$. Hence $S \otimes X$ belongs to $S*\mathcal{T}$. Thus $S*\mathcal{T}$ contains $S \otimes B(\mathcal{H}_2)$, and also $B(\mathcal{H}_1) \otimes \mathcal{T}$ by symmetry.

To prove the lemma, it now suffices to show that if $x \otimes y^*$ annihilates $S \otimes B(\mathcal{H}_2) + B(\mathcal{H}_1) \otimes \mathcal{T}$, then it belongs to $S_\perp \otimes \mathcal{T}_\perp$. Now $\underline{S \otimes B(\mathcal{H}_2)}$ is a $\underline{I \otimes B(\mathcal{H}_2)}$ bimodule. Thus the projections onto $\overline{I \otimes B(\mathcal{H}_2)x}$ and $\overline{S \otimes B(\mathcal{H}_2)x}$ are invariant for $I \otimes B(\mathcal{H}_2)$, and hence belongs to $(I \otimes B(\mathcal{H}_2))' = B(\mathcal{H}_1) \otimes I$ by Theorem 15.1. Let them be denoted by $Q_1 \otimes I$ and $P_1 \otimes I$ respectively. Then P_1 is the projection onto $\overline{SQ_1\mathcal{H}_1}$. Similarly, there are projections Q_2 and P_2 so that $\overline{B(\mathcal{H}_1) \otimes Ix} = (I \otimes Q_2)\mathcal{H}$, $P_2\mathcal{H}_2 = \overline{\mathcal{T}Q_2\mathcal{H}_2}$ and $\overline{B(\mathcal{H}_1) \otimes \mathcal{T}x} = (I \otimes P_2)\mathcal{H}$. Consequently,

$$\begin{aligned}\overline{S \otimes B(\mathcal{H}_2) + B(\mathcal{H}_1) \otimes \mathcal{T}x} &= ((P_1 \otimes I) \vee (I \otimes P_2))\mathcal{H} \\ &= (P_1^\perp \otimes P_2^\perp)^\perp \mathcal{H} \ .\end{aligned}$$

Since $x \otimes y^*$ annihilates $S \otimes B(\mathcal{H}_2) + B(\mathcal{H}_1) \otimes \mathcal{T}$, it follows that $(P_1^\perp \otimes P_2^\perp)y = y$. On the other hand,

$$x = (Q_1 \otimes I)x = (I \otimes Q_2)x = (Q_1 \otimes Q_2)x \ .$$

As $SQ_1 = P_1SQ_1$, it follows that $Q_1C_1(\mathcal{H}_1)P_1^\perp$ is contained in $S_\perp$. Likewise, $Q_2C_1(\mathcal{H}_2)P_2^\perp$ is contained in $\mathcal{T}_\perp$. But

$$C_1(\mathcal{H}_1) \otimes C_1(C_2) = C_1(\mathcal{H}_1 \otimes \mathcal{H}_2) \ ,$$

whence it follows that $S_\perp \otimes \mathcal{T}_\perp$ contains

$$(Q_1 \otimes Q_2)C_1(\mathcal{H}_1 \otimes \mathcal{H}_2)(P_1^\perp \otimes P_2^\perp) \ .$$

In particular, $x \otimes y^*$ belongs to $\mathcal{S}_\perp \otimes \mathcal{T}_\perp$ as desired. This establishes the lemma. ■

24.11 Theorem. *Let $\mathcal{S}$ and $\mathcal{T}$ be proper reflexive subspaces of $\mathcal{B}(\mathcal{H}_1)$ and $\mathcal{B}(\mathcal{H}_2)$. Then $\kappa(\mathcal{S}*\mathcal{T}) \geq \kappa(\mathcal{S})\kappa(\mathcal{T})$.*

Proof. Fix A in $\mathcal{B}(\mathcal{H}_1)$ and B in $\mathcal{B}(\mathcal{H}_2)$. Compute

$$\begin{aligned} d(A \otimes B, \mathcal{S}*\mathcal{T}) &= \sup_{f \in b_1(\mathcal{S}_\perp \otimes \mathcal{T}_\perp)} |f(A \otimes B)| \\ &\geq \sup_{g \in b_1(\mathcal{S}_\perp), h \in b_1(\mathcal{T}_\perp)} |(g \otimes h)(A \otimes B)| \\ &= \sup_{g \in b_1(\mathcal{S}_\perp)} |g(A)| \sup_{h \in b_1(\mathcal{T}_\perp)} |h(B)| \\ &= d(A,\mathcal{S})d(B,\mathcal{T}) \ . \end{aligned}$$

On the other hand, if $S \in \mathcal{S}$ and $T \in \mathcal{T}$ such that $\|A-S\| = d(A,\mathcal{S})$ and $\|B-T\| = d(B,\mathcal{T})$, then $S \otimes B+(A-S) \otimes T$ belongs to $\mathcal{S}*\mathcal{T}$ and

$$\begin{aligned} d(A \otimes B, \mathcal{S}*\mathcal{T}) &\leq \|A \otimes B-(S \otimes B+(A-S) \otimes T)\| \\ &= \|A-S\| \, \|B-T\| = d(A,\mathcal{S})d(B,\mathcal{T}) \ . \end{aligned}$$

Let x be a unit vector in $\mathcal{H}_1 \otimes \mathcal{H}_2$, and define P_1, Q_1, P_2 and Q_2 as in the previous lemma. Then

$$\begin{aligned} d((A \otimes B)x, \mathcal{S}*\mathcal{T}x) &= d((A \otimes B)x, (P_1^\perp \otimes P_2^\perp)^\perp \mathcal{H}) \\ &= \|(P_1^\perp \otimes P_2^\perp)(A \otimes B)(Q_1 \otimes Q_2)x\| \\ &\leq \|P_1^\perp A Q_1\| \, \|P_2^\perp B Q_2\| \leq \beta_{\mathcal{S}}(A)\beta_{\mathcal{T}}(B) \ . \end{aligned}$$

Consequently, $\beta_{\mathcal{S}*\mathcal{T}}(A \otimes B) \leq \beta_{\mathcal{S}}(A)\beta_{\mathcal{T}}(B)$.

Hence

$$\begin{aligned} \kappa(\mathcal{S}*\mathcal{T}) &\geq \sup_{A \in \mathcal{B}(\mathcal{H}_1), B \in \mathcal{B}(\mathcal{H}_2)} \frac{d(A \otimes B, \mathcal{S}*\mathcal{T})}{\beta_{\mathcal{S}*\mathcal{T}}(A \otimes B)} \\ &\geq \sup_{A \in \mathcal{B}(\mathcal{H}_1)} \frac{d(A,\mathcal{S})}{\beta_{\mathcal{S}}(A)} \sup_{B \in \mathcal{B}(\mathcal{H}_2)} \frac{d(B,\mathcal{T})}{\beta_{\mathcal{T}}(B)} = \kappa(\mathcal{S})\kappa(\mathcal{T}) \ . \end{aligned}$$ ■

It is possible for the inequality in Theorem 24.11 to be strict (See Exercise 24.6). No upper bound for $\kappa(\mathcal{S}*\mathcal{T})$ is known (see Exercise 24.10).

Notes and Remarks.

The usefulness of distance formulae goes back to the results of Arveson and Christensen dealt with in Chapter 9. The term hyperreflexive was coined in the CBMS survey by Arveson [11]; but the notion has existed for some time. Proposition 24.8 and Theorem 24.9 are contained in Larson [3]. The main results of this section, 24.1-24.7, are due to Davidson and Power [1]. The notion of dual product and Theorem 24.11 are due to Larson [5].

Exercises.

24.1 Show that if a subspace S is hyperreflexive, then

$$\|P(SM)^{\perp}TP(M)\| \leq \kappa(S)d(T,S)$$

for every subspace M and every T in $\mathcal{B}(\mathcal{H})$. Hence deduce that the two notions of hyperreflexivity coincide for unital algebras.

24.2 Let S be a bimodule over a masa $\mathcal{M}$. Let $\mathcal{A}$ be the algebra of all operators of the form $\begin{bmatrix} M_1 & S \\ 0 & M_2 \end{bmatrix}$ such that $M_1, M_2 \in \mathcal{M}$ and $S \in S$. Show that

$$\kappa(S) \leq \kappa(\mathcal{A}) \leq \kappa(S)+2 \ .$$

24.3 Let S be the weak* closure of $\sum_{n \geq 1} \oplus S_n$ where S_n are weak* closed subspaces. Prove that

$$\sup_{n \geq 1} \kappa(S_n) \leq \kappa(S) \leq 3 \sup_{n \geq 1} \kappa(S_n) \ .$$

Hint: Note that S is contained in $\sum_{n \geq 1} \oplus \mathcal{B}(\mathcal{H})$, so Lemma 9.8 is relevant.

24.4 (Davidson-Power) Fix a basis $\{e_n, n \geq 1\}$ and let S consist of all operators with zero diagonal. This is a bimodule over the diagonal algebra $\mathcal{D}$. Construct an algebras $\mathcal{A}$ as in Exercise 24.2. Show that $\kappa(\mathcal{A}) \leq 3$. Show that $\mathcal{A}$ is not finite width.

24.5 (Larson) With the notation of Theorem 24.11, show that

$$\beta_{S*T}(A \otimes B) = \beta_S(A)\beta_T(B) \ .$$

24.6 (Larson) Let $S = \mathbb{C}\begin{bmatrix} 1 & 0 \\ 0 & 0 \end{bmatrix}$. Show that $\kappa(S) = 1$ but $\kappa(S*S) \geq (9/8)^{1/2}$.

24.7 (Larson) Let $\mathcal{A}_i$ be a proper reflexive subalgebra of $B(\mathcal{H}_i)$, $1 \leq i \leq n$. Let $\mathcal{A}$ be the algebra

$$\begin{bmatrix} \mathcal{A}_1 \otimes \mathcal{A}_2 \otimes ... \otimes \mathcal{A}_n & \mathcal{A}_1 * \mathcal{A}_2 * ... * \mathcal{A}_n \\ 0 & \mathcal{A}_1 \otimes \mathcal{A}_2 \otimes ... \otimes \mathcal{A}_n \end{bmatrix}.$$

Prove that $\mathcal{A}$ is a reflexive algebra with $\kappa(\mathcal{A}) \geq \kappa(\mathcal{A}_1)\kappa(\mathcal{A}_2)...\kappa(\mathcal{A}_n)$.

24.8 (Davidson-Power) Let $\mathcal{L}$ be a CSL such that $Alg\,\mathcal{L}$ is not hyperreflexive. Fix $0 < \epsilon < 1/6$, and choose T in $B(\mathcal{H})$ such that $d(T, Alg\,\mathcal{L}) = 1$ and $\beta(T) < \epsilon$. Set $V = I + \frac{1}{2}T$ and $\mathcal{M} = V\mathcal{L}$.

(i) Show that $Alg\,\mathcal{M}$ is similar to $\mathcal{L}$.

(ii) Let $\Theta(L) = VL$ be the natural lattice isomorphism of $\mathcal{L}$ onto $\mathcal{M}$. Show that $\|\Theta - id\| < 2\epsilon$.

(iii) Show that if S is any invertible operator such that $S\mathcal{L} = \mathcal{M}$, then $\|S - I\| \geq \frac{1}{4}$. Hint: Show that $S^{-1}V$ implements the identity automorphism of $\mathcal{L}$. Hence $S^{-1}V$ belongs to $Alg\,\mathcal{L}$.

24.9* Let $\mathcal{P} = \{P_n, n \geq 1\}$ where $P_n = span\{e_1, ..., e_n\}$ in $\ell^2 = span\{e_k, k \geq 1\}$. Let $\mathcal{P} \otimes \mathcal{P}$ be the CSL in $\ell^2 \otimes \ell^2$ generated by the two nests $\mathcal{P}_1 = \{P_n \otimes \ell^2, n \geq 1\}$ and $\mathcal{P}_2 = \{\ell^2 \otimes P_n, n \geq 1\}$. Does $Alg(\mathcal{P} \otimes \mathcal{P})$ have a distance formula?

24.10* If $\mathcal{S}$ and $\mathcal{T}$ are hyperreflexive subspaces, is $\mathcal{S} * \mathcal{T}$ hyperreflexive?

25. Open Problems

The purpose of this short chapter is to collect together some of the more important open problems in the area. Most of these deal with CSL algebras. In this subject, we are motivated in large part by the result for nest algebras. The wonted generalizations to arbitrary CSL algebras are usually difficult, and frequently not true in full generality. Thus exploring the fullest extent of these generalizations may lead to a better understanding of the structure of these interesting algebras. Many of these problems have been mentioned earlier in the text or exercises.

We begin with two questions about nest algebras. Let $\mathcal{P}$ denote the standard nest $P_n = span\{e_k : 0 \leq k \leq n\}$ on an orthonormal basis $\{e_k, k \geq 0\}$.

Problem 1. *Is $\mathcal{T}(\mathcal{P})^{-1}$ connected?*

It is surprising that this problem is open even for such a simple nest. It is conjectured by many that the invertibles in $\mathcal{T}(\mathcal{P})$ are not connected. The reason is based on an analogy with function theory. The algebra H^∞ of bounded analytic functions on the unit disk is a commutative Banach algebra. Hence the connected component of 1 in $(H^\infty)^{-1}$ is

$$(H^\infty)_0^{-1} = \{e^h : h \in H^\infty\} \ .$$

It is easy to construct invertible functions in H^∞ which have an unbounded logarithm, and hence are not connected to 1 in $(H^\infty)^{-1}$. For example, $h(z) = \dfrac{2i}{\pi}\log\dfrac{1+z}{1-z}$ is a conformal map of the disk onto $\{z : |\mathrm{Re}\, z\,| < 1\}$. Hence e^h is invertible in H^∞ but has unbounded logarithm h (or $h+2n\pi i$). Now the Toeplitz operator $T = T^*_{e^h}$ belongs to $\mathcal{T}(\mathcal{P})$ if one takes $\mathcal{H} = H^2$ and $e_k = e^{2\pi k i\theta}$ as the standard basis. The algebra $\{T^*_f : f \in H^\infty\}$ sits inside $\mathcal{T}(\mathcal{P})$. The discussion here shows that T is not connected to I through $\mathcal{T}(\mathcal{P})^{-1} \cap$ {Toeplitz operators}. It is a candidate for proving disconnectedness.

Another candidate is obtained using finite dimensional blocks. There exist nilpotent matrices V_k in $\mathcal{M}_k$ such that $\lim\limits_{k\to\infty} \|V_k\| = \infty$ and $\lim\limits_{k\to\infty} \|V_k - \mathrm{Re}\, V_k\| = 0$. For example, take a Toeplitz operator T_{h_k} where h_k is a conformal map of the disk onto the ellipse $\{(x,y) : (x/k)^2 + (ky)^2 < 1\}$ such that $h_k(0) = 0$. Then $\|T_{h_k}\| = k$ and $\|T_{h_k} - \mathrm{Re}\, T_{h_k}\| = k^{-1}$. The compression $V_{k,n}$ of T_{h_k} to P_n is always nilpotent (strictly upper triangular), $\|V_{k,n} - \mathrm{Re}\, V_{k,n}\| \leq k^{-1}$, and for sufficiently large n, $\|V_{n,k}\| > k - k^{-1}$.

Put these nilpotent matrices V_k in upper triangular form. Set $T_k = e^{iV_k}$ which is invertible in $\mathcal{T}_k$, the upper triangular matrices in $\mathcal{M}_k$. Then $T = \sum_{k\geq 1} \oplus T_k$ is an invertible upper triangular operator. (Indeed, it is "almost" unitary.) Suppose T_t is a rectifiable path connecting T to I (or one can even assume a piecewise linear path). Then the compression $T_{k,t}$ of T_t to the k-th block yields a path from T_k to I with uniform bounds on $\|T_{k,t}\|$, $\|T_{k,t}^{-1}\|$, and on the *length of the path*. The existence of such paths appears unlikely.

In finite dimensions, $\mathcal{T}_k^{-1}$ is always connected. If $T \in \mathcal{T}_k^{-1}$, then T is triangular with non-zero diagonal entries t_{ii}. As the diagonal invertible $\mathcal{D}_k^{-1}$ are connected (uniformly in k), we may assume that $t_{ii} = 1$ for all $1 \leq i \leq n$. Furthermore, the existence of a nice path as above yields a nice path with ones on the diagonal. There are two interesting paths from T to I. One is the straight line $T_t = (1-t)T+tI$. This is bounded by $\|T\|$ and short (length $\|T-I\|$). However, the norm $\|T_t^{-1}\|$ "blows up" as k increases in the example above. Another path, suggested by Vern Paulsen, is the path

$$\widetilde{T}_s = (t_{ij}s^{j-i}),\ 0 \leq s \leq 1 \ .$$

The map taking T to T_s is a homomorphism of $\mathcal{T}_k$ into itself, and thus preserves invertibility. It is also the Schur product of T with the positive matrix $S_s = (s^{|j-i|})$, and hence the map is contractive. Thus $\|\widetilde{T}_s\| \leq \|T\|$ and $\|\widetilde{T}_s^{-1}\| \leq \|T^{-1}\|$. However, the length of this path "blows up" as k increases.

A third approach involves K-theory. In the appendix to Chapter 19, we compute K_0 of a nest algebra. There is also the K_1 group of $\mathcal{T}(\mathcal{N})$ which deals precisely with connected components in $\bigcup_{n\geq 1} \mathcal{M}_n(\mathcal{T}(\mathcal{N}))$ where A in $\mathcal{M}_m(\mathcal{T}(\mathcal{N}))$ is identified with $A \oplus I_{n-m}$ in $\mathcal{M}_n(\mathcal{T}(\mathcal{N}))$. It is not difficult to show that if $K_1(\mathcal{T}(\mathcal{P})) \neq 0$, then $\mathcal{T}(\mathcal{P})^{-1}$ is not connected. However, this group is unknown. If $\mathcal{N}$ is a nest with only infinite dimensional atoms, then $K_1(\mathcal{T}(\mathcal{N})) = 0$. (For example, $\mathcal{N} = \mathcal{P}^{(\infty)}$ or $\mathcal{N}$ a continuous nest.) Even in these cases, it is not known if $\mathcal{T}(\mathcal{N})^{-1}$ is connected.

Problem 2. *Is every maximal nest equal to the invariant subspace lattice of a unicellular operator?*

This is the subject of the first half of Chapter 21. There are two basic approaches. One is to construct operators from simple ones, which is the approach taking in Chapter 21. The other is to use convolution type integral operators such as the Volterra operator. Let $\mathcal{Q}$ be the canonical nest on $\ell^2(\mathbb{Q})$. Since all infinite intervals of $\mathcal{Q}$ are essentially the same, it seems futile to try to build a unicellular operator for $\mathcal{Q}$ from smaller pieces. It is a natural test case for Problem 2. On the other hand, the constructive techniques may well work for countable nests.

The remaining problems concern CSL algebras. All lattices $\mathcal{L}$ in this chapter will be commutative subspace lattices.

Problem 3. *What is the Jacobson radical of Alg $\mathcal{L}$?*

Ringrose's characterization of the radical of a nest algebra, Theorem 6.7, is the natural place to start. It is not difficult to define analogous ideals I_ϕ in $Alg\,\mathcal{L}$ for every ϕ in $Hom(\mathcal{L},2)$, which is homeomorphic to the maximal ideal space of $C^*(\mathcal{L}) = C^*\{P(L):L \in \mathcal{L}\}$. So each condition 1)-4) of Theorem 6.7 can be reformulated for arbitrary CSL's. Hopenwasser [1] and Hopenwasser-Larson [1] prove that conditions 2), 3) and 4) are equivalent, and imply membership in the radical 1). So the problem becomes: Is $rad(Alg\,\mathcal{L}) = \bigcap\{I_\phi : \phi \in Hom(\mathcal{L},2)\}$?

Problem 4. *Is there a distance formula for $Alg(\mathcal{P} \otimes \mathcal{P})$?*

In Chapter 9, distance formulae for nests and nice von Neumann algebras are derived. These are crucial for the similarity theory, perturbations, and many other aspects of nest algebras. However, in the last chapter, we saw that the distance formula fails for arbitrary CSL algebras. Because these estimates are so powerful, it is important to decide if some larger class of CSL algebras has a distance formula. For example, is there a distance formula for finite width lattices? The natural test case for width two lattices is $\mathcal{P} \otimes \mathcal{P}$, the lattice on $\ell^2(\mathbb{N}^2)$ generated by $\{P_n \otimes \ell^2(\mathbb{N}), n \geq 0\}$ and $\{\ell^2(\mathbb{N}) \otimes P_n, n \geq 0\}$. This can be reformulated as a problem about a uniform estimate for the analogous finite dimensional algebras on $\mathbb{C}^{n^2}$.

Problem 5. *Are there examples of close CSL algebras which are not similar? or even not isomorphic?*

In Chapter 18, we saw that the distance formulae for nests and type I von Neumann algebras yield the good perturbation results of Lance and Christensen. Even for arbitrary CSL algebras, one has Theorem 18.2 which shows that close algebras have close lattices. However, Davidson and Power [1] use the failure of the distance formula to exhibit similar CSL algebras with close *lattices* but the similarity is far from the identity. Choi and Christensen [1] have constructed close, order isomorphic C^*-algebras which are not unitarily equivalent. Bad behaviour even occurs in finite dimensions for arbitrary matrix algebras (Choi-Davidson [1]). Information about perturbation results for CSL algebras should shed some light on their classification under similarity or approximate unitary equivalence.

Problem 6. *Classify commutative subspace lattices up to approximate unitary equivalence.*

The notion of similarity is not a good one for CSL's since similarity does not preserve orthogonality. If one wishes to study the similarity classes of CSL's, the appropriate class is the set of complete sublattices of σ-complete Boolean algebras of idempotents in $B(\mathcal{H})$. (See Davidson [7].) The notion of approximate unitary equivalence is better since it preserves the class of CSL's. The Similarity Theorem 13.20 shows that the classification of nests up to approximate unitary equivalence is given by dimension preserving order isomorphisms. On the other hand, Christensen's Theorem 18.12 shows that close abelian von Neumann algebras are unitarily equivalent. Now if CSL's $\mathcal{L}_1$ and $\mathcal{L}_2$ are complemented, approximate unitary equivalence yields approximate unitary equivalence of the abelian von Neumann algebras $span(\mathcal{L}_1)$ and $span(\mathcal{L}_2)$. By Christensen's Theorem, these lattices are unitarily equivalent. Even the compact perturbations of abelian von Neumann algebras are rigid because of the Johnson-Parrott Theorem 10.12. Arveson [10] develops a perturbation theory for lattices that is valid for certain "homogeneous" lattices which are compact in the strong operator topology (see Theorem 23.12). In particular, he gives a different proof of Andersen's Theorem 13.9.

What we are looking for here is a way of unifying the nest and Boolean algebra theory.

Problem 7. *If $\mathcal{K} \cap Alg\,\mathcal{L}$ is weak* dense in $Alg\,\mathcal{L}$, is $\mathcal{L}$ completely distributive?*

The main thrust of Chapter 23 was an analysis of the relationship between the lattice properties of $\mathcal{L}$ and the density of certain ideals of compact operators in $Alg\,\mathcal{L}$. What is missing is a lattice theoretic condition equivalent to the density of $\mathcal{K} \cap Alg\,\mathcal{L}$. It is possible that the notion of Λ-distributivity is equivalent to complete distributivity in this context, even though it is formally weaker. If not, any counterexample is likely to shed a useful light on the differences.

Problem 8. *If $\mathcal{L}_1$ and $\mathcal{L}_2$ are commuting synthetic lattices, is $\mathcal{L}_1 \vee \mathcal{L}_2$ synthetic?*

The real problem is to find an operator theoretic approach to synthesis. Problem 8 is one of several test questions. One can also ask if $\mathcal{L}_1 \cap \mathcal{L}_2$ and $\mathcal{L}_1 \otimes \mathcal{L}_2$ are synthetic. Arveson's original approach (Chapter 22) to this problem appears to be mired in measure theoretic technicalities. Froelich [1] has shown that there is a very tight connection with spectral synthesis in commutative harmonic analysis (see Notes 22.38). He has shown how certain results in harmonic analysis translate to exhibit interesting phenomena in CSL algebras. An operator theoretic result which implied something new about harmonic analysis would be very interesting.

Problem 9. *Does* $Alg\,\mathcal{L}_1 \otimes Alg\,\mathcal{L}_2 = Alg(\mathcal{L}_1 \otimes \mathcal{L}_2)$*?*

As in Chapter 15, the tensor products here are spatial and act on $\mathcal{H}_1 \otimes \mathcal{H}_2$. The $Alg\,\mathcal{L}_1 \otimes Alg\,\mathcal{L}_2$ is the WOT closed algebra generated by the elementary tensors $A_1 \otimes A_2$ where $A_j \in Alg\,\mathcal{L}_j, 1 \leq j \leq 2$. Similarly, $\mathcal{L}_1 \otimes \mathcal{L}_2$ is the complete lattice generated by the subspaces $L_1 \otimes L_2$ where $L_j \in \mathcal{L}_j, 1 \leq j \leq 2$. Clearly, $A_1 \otimes A_2$ is contained in $Alg(\mathcal{L}_1 \otimes \mathcal{L}_2)$; so the containment

$$Alg\,\mathcal{L}_1 \otimes Alg\,\mathcal{L}_2 \subseteq Alg(\mathcal{L}_1 \otimes \mathcal{L}_2)$$

is automatic.

If $\mathcal{L}_1$ and $\mathcal{L}_2$ are complemented, $\mathcal{A}_j = Alg\,\mathcal{L}_j$ are von Neumann algebras and $\mathcal{L}_j$ equals $Proj(\mathcal{A}'_j)$. So the answer to our question in this context becomes a celebrated theorem of Tomita [1]:

$$\mathcal{A}_1 \otimes \mathcal{A}_2 = (\mathcal{A}'_1 \otimes \mathcal{A}'_2)' \ .$$

If $\mathcal{L}_1$ is completely distributive, equality follows from Corollary 23.10. And if $\mathcal{L}_1$ is finite width, equality follows from Gilfeather-Hopenwasser-Larson [1]. These three cases suggest that if $\mathcal{L}_1$ is synthetic, then equality should hold. Kraus [3] has investigated this problem in even greater generality, and no counter example is known.

Problem 10. *Is every representation* ϕ *of a nest algebra completely bounded with* $\|\phi\|_{cb} = \|\phi\|$*?*

This is true for weak* continuous contractive representations of nest algebras (Theorem 20.18) and leads to an attractive dilation theory for representations. Paulsen, Power and Smith [1] have shown that this may fail for representations of completely distributive lattices even in finite dimensions. When ϕ is not weak* continuous, the problem is open even for contractive maps. The structure of representations of more general CSL algebras is relatively uncharted.

Note. The content of this chapter was the subject of a talk by the author at the GPOTS meeting in Lawrence, KA. A survey article on these same problems, Davidson [11], appears in the proceedings of that conference.

BIBLIOGRAPHY

Akemann, C.A. and Johnson, B.E.

[1] Derivations of non-separable C^*-algebras, J. Func. Anal. 33 (1979), 311-331.

Akemann, C.A. and Wright, S.

[1] Compact and weakly compact derivations of C^*-algebras, Pacific J. Math. 85 (1979), 253-259.

Alfsen, E.M. and Effros, E.G.

[1] Structure in real Banach spaces, Ann. Math. 96 (1972), 98-173.

Andersen, N.T.

[1] Compact perturbations of reflexive algebras, J. Func. Anal. 38 (1980), 366-400.

[2] Similarity of continuous nests, Bull. London Math. Soc. 15 (1983), 131-132.

Ando, T.

[1] On a pair of commuting contractions, Acta Sci. Math. (Szeged) 24 (1963), 88-90.

Apostol, C.

[1] The correction by compact perturbations of the singular behaviour of operators, Rev. Roum. Math. Pures et Apl. 21 (1976), 155-176.

Apostol, C. and Foias, C.

[1] On the distance to biquasitriangular operators, Rev. Roum. Math. Pures et Appl. 20 (1975), 261-265.

Apostol, C. and Davidson, K.R.

[1] Isomorphisms modulo the compact operators of nest algebras, II, Duke Math. J., to appear.

Apostol, C., Foias, C. and Voiculescu, D.

[1] Some results on non quasitriangular operators, II-VI, Rev. Roum. Pures et Appl. 18 (1973), 159-181, 309-324, 487-514, 1133-1140, 1473-1494.

Apostol, C. and Gilfeather, F.

[1] Isomorphism modulo the compact operators of nest algebras, Pacific J. Math. 122 (1986), 263-286.

Aronszajn, N. and Smith, K.T.

[1] Invariant subspaces of completely continuous operators, Ann. Math. 60 (1954), 345-350.

Arveson, W.B. (See also Fall, T.)

[1] A density theory for operator algebras, Duke Math. J. 34 (1967), 635-647.

[2] Analyticity in operator algebras, Amer. J. Math. 89 (1967), 578-642.

[3] Operator algebras and measure preserving automorphisms, Acta Math. 118 (1967), 95-109.

[4] Subalgebras of C^*-algebras, Acta Math. 123 (1969), 141-224.

[5] Subalgebras of C^*-algebras II, Acta Math. 128 (1972), 271-308.

[6] Operator algebras and invariant subspaces, Ann. Math. (2) 100 (1974), 433-532.

[7] Interpolation problems in nest algebras, J. Func. Anal. 20 (1975), 208-233.

[8] An invitation to C^*-algebras, Grad. Texts in Math. 39, Springer-Verlag, Berlin and New York, 1976.

[9] Notes on extensions of C^*-algebras, Duke Math. J. 44 (1977), 329-355.

[10] Perturbation theory for groups and lattices, J. Func. Anal. 53 (1983), 22-73.

[11] Ten lectures on operator algebras, CBMS series 55, Amer. Math. Soc., Providence, 1983.

Arveson, W.B. and Feldman, J.

[1] A note on invariant subspaces, Mich. Math. J. 15 (1968), 60-64.

Arveson, W.B. and Josephson, K.

[1] Operator algebras and measure preserving automorphisms II, J. Func. Anal. 4 (1964), 100-134.

Axler, S., Berg, I.D., Jewell, N. and Shields, A.

[1] Approximation by compact operators and the space $H^{\infty}+C$, Ann. Math. 109 (1979), 601-612.

Ball, J.A. and Gohberg, I.

[1] Shift invariant subspaces, factorization, and interpolation of matrices, Lin. Alg. Appl. 74 (1986), 87-150.

[2] A commutant lifting theorem for triangular matrices with diverse applications, Int. Operator Theory 8 (1985), 205-267.

Barría, J.

[1] On chains of invariant subspaces, Dissertation, Indiana University, 1974.

[2] The invariant subspaces of a Volterra operator, J. Operator Theory 6 (1981), 341-349.

Barría, J. and Davidson, K.R.

[1] Unicellular operators, Trans. Amer. Math. Soc. 284 (1984), 229-246.

Berg, I.D. (See also Axler, S.)

[1] An extension of the Weyl-von Neumann Theorem to normal operators, Trans. Amer. Math. Soc. 160 (1971), 365-371.

Bernstein, A.R. and Robinson, A.

[1] Solution of an invariant subspace problem of K.T. Smith and P.R. Halmos, Pacific J. Math. 16 (1966), 421-431.

Blackadar, B.

[1] K-theory for operator algebras, MSRI Pub. 5, Springer-Verlag, New York, 1986.

Bonsall, F.F. and Duncan, J.

[1] Complete normed algebras, Springer-Verlag, New York and Berlin, 1973.

Bourbaki, N.

[1] Eléments de mathematique, Part 1, Book VI, Intégration, Chapter 6, Intégration vectorielle, Hermann, Paris, 1959.

Brodskii, M.S.

[1] On a problem of I.M. Gelfand, Uspeki Mat. Nauk. 12 (1957), 129-132.

Calkin, J.W.

[1] Two sided ideals and congruences in the ring of bounded operators on Hilbert space, Ann. Math. (2) 42 (1941), 839-873.

Choi, M.D.

[1] Tricks or treats with the Hilbert matrix, Amer. Math. Monthly 90 (1983), 301-312.

Choi, M.D. and Christensen, E.

[1] Completely order isomorphic and close C^*-algebras need not be $*$-isomorphic, Bull. London Math. Soc. 15 (1983), 604-610.

Choi, M.D. and Davidson, K.R.

[1] Perturbations of matrix algebras, Mich. Math. J. 33 (1986), 273-287.

Choi, M.D. and Effros, E.

[1] Separable nuclear C^*-algebras and injectivity, Duke Math. J. 43 (1976), 309-322.

[2] Nuclear C^*-algebras and injectivity; the general case, Indiana J. Math. 26 (1977), 443-446.

Choi, M.D. and Feintuch, A.

[1] Operator factorization with respect to a nest algebra, preprint.

Christensen, E. (See also Choi, M.D.)

[1] Perturbations of type I von Neumann algebras, J. London Math. Soc. (2), 9 (1975), 395-405.

[2] Derivations of nest algebras, Math. Ann. 229 (1977), 155-161.

[3] Perturbations of operator algebras, Inv. Math. 43 (1977), 1-13.

[4] Perturbations of operator algebras II, Indiana Univ. Math. J. 26 (1977), 891-904.

[5] Extensions of derivations, J. Func. Anal. 27 (1978), 234-247.

[6] Near inclusions of C^*-algebras, Acta. Math. 144 (1980), 249-265.

[7] Extensions of derivations II, Math. Scan. 50 (1982), 111-122.

Christensen, E. and Peligrad, C.

[1] Commutants of nest algebras modulo the compact operators, Inv. Math. 56 (1980), 113-116.

Connes, A.

[1] Classification of injective factors, Ann. Math. 104 (1976), 73-116.

Conway, J.

[1] A course in functional analysis, Grad. Texts Math. no. 96, Springer-Verlag, New York and Berlin, 1985.

Davidson, K.R. (See also Apostol, C.; Barría, J.; Choi, M.D.)

[1] Commutative subspaces lattices, Indiana U. Math. J. 27 (1978), 479-490.

[2] Compact perturbations of reflexive algebras, Can. J. Math. 33 (1981), 685-700.

[3] Invariant operator ranges and reflexive algebras, J. Operator Theory 7 (1982), 101-107.

[4] Quasitriangular algebras are maximal, J. Operator Theory 10 (1983), 51-56.

[5] The essential commutant of CSL algebras, Indiana U. Math. J. 32 (1983), 761-771.

[6] Similarity and compact perturbations of nest algebras, J. reine angew. Math. 348 (1984), 286-294.

[7] Perturbations of reflexive operator algebras, J. Operator Theory 15 (1986), 289-305.

[8] Approximate unitary equivalences of continuous nests, Proc. Amer. Math. Soc. 97 (1986), 655-660.

[9] Continuous nests and the strange behaviour of Ext, Bull. London Math. Soc. 18 (1986), 485-492.

[10] The distance to the analytic Toeplitz operators, Illinois J. Math. 31 (1987), 265-273.

[11] Open problems in reflexive algebras, Rocky Mountain J. Math., to appear.

Davidson, K.R. and Fong, C.K.

[1] An operator algebra which is not closed in the Calkin algebra, Pac. J. Math. 72 (1977), 57-58.

Davidson, K.R. and Herrero, D.A.

[1] Quasisimilarity of nests, preprint.

Davidson, K.R. and Power, S.C.

[1] Failure of the distance formula, J. London Math. Soc. (2) 32 (1984), 157-165.

[2] Best approximation in C^*-algebras, J. reine angew. Math. 368 (1986), 43-62.

Davidson, K.R. and Wagner, B.

[1] Automorphisms of quasitriangular algebras, J. Func. Anal. 59 (1984), 612-627.

Davis, C., Kahan, W.M. and Weinberger, W.F.

[1] Norm preserving dilations and their applications to optimal error bounds, SIAM J. Numer. Anal. 19 (1982), 445-469.

Deddens, J.A.

[1] Another description of nest algebras, Hilbert space operators, Lect. Notes Math. 693, Springer-Verlag, N.Y., 1978, pp. 77-86.

Dixmier, J.

[1] Les opérateurs permutables á l'opérateur integral, Fas. 2, Portugal Math. 8 (1949), 73-84.

[2] Les moyennes invariant dans les semi-groupes et leur applications, Acta Sci. Math. (Szeged) 12A (1950), 213-227.

[3] Les fonctionnelles lin'earies sur l'ensemble des opérateurs bornés dans espace de Hilbert, Ann. Math. 51 (1950), 387-408.

[4] Les algèbres d'opérateurs dans l'espace Hilbertien, Gauthier-Villars, Paris, 1969.

[5] Les C^*-algèbres et leur représentations, Gauthier-Villars, Paris, 1969.

Domar, Y.

[1] Translation invariant subspaces and weighted ℓ^p and L^p spaces, Math. Scand. 49 (1981), 133-144.

Donoghue, W.F.

[1] The lattice of invariant subspaces of a completely continuous quasinilpotent transformation, Pacific J. Math. 7 (1957), 1031-1035.

Douglas, R.G.

[1] On extending commutative semigroups of isometries, Bull. London Math. Soc. 1 (1969), 157-159.

[2] Banach algebra techniques in operator theory, Academic Press, New York and London, 1972.

Douglas, R.G. and Pearcy, C.

[1] A note on quasitriangular operators, Duke Math. J. 37 (1970), 177-188.

Duncan, J. (See Bonsall, F.F.)

Dunford, N. and Schwartz, J.T.

[1] Linear operators I, II, III, Interscience Publishers, New York, 1957, 1963, and 1971.

Effros, E.G. (See also Alfsen, E.M.; Choi, M.D.)

Effros, E. and Lance, E.C.

[1] Tensor products of operator algebras, Adv. Math. 25 (1977), 1-34.

Elliot, G.A.

[1] Some C^*-algebras with outer derivations, Rocky Mountain J. Math. 3 (1973), 501-506.

[2] Some C^*-algebras with outer derivations II, Canad. J. Math. 26 (1974), 185-189.

[3] Some C^*-algebras with outer derivations III, Ann. Math. 106 (1977), 121-143.

Erdos, J.A.

[1] Some results on triangular operator algebras, Amer. J. Math. 89 (1967), 85-93.

[2] Unitary invariants for nests, Pacific J. Math. 23 (1967), 229-256.

[3] Operators of finite rank in nest algebras, J. London Math. Soc. 43 (1968), 391-397.

[4] An abstract characterization of nest algebras, Quart. J. Math. 22 (1971), 47-63.

[5] The triangular factorization of operators on Hilbert space, Indiana J. Math. 22 (1973), 939-950.

[6] On the trace of a trace class operator, Bull. London Math. Soc. 6 (1974), 47-50.

[7] Some questions concerning triangular operator algebras, Proc. Roy. Irish acad. 74A (1974), 223-232.

[8] Triangular integration on symmetrically normed ideals, Indiana Math. J. 27 (1978), 401-408.

[9] Non-selfadjoint operator algebras, Proc. Roy. Irish acad. 81A (1981), 127-145.

[10] On some ideals of nest algebras, Proc. London Math. Soc. (3) 44 (1982), 143-160.

[11] The commutant of the Volterra operator, Int. Eq. Operator Thy. 5 (1982), 127-130.

[12] Reflexivity for subspace maps and linear subspaces of operators, Proc. London Math. Soc. (3) 52 (1986), 582-600.

Erdos, J.A. and Hopenwasser, A.

[1] Essentially triangular algebras, Proc. Amer. Math. Soc. 96 (1986), 335-339.

Erdos, J.A. and Longstaff, W.E.

[1] The convergence of triangular integrals of operators on Hilbert space, Indiana J. Math. 22 (1973), 929-938.

Erdos, J.A. and Power, S.C.

[1] Weakly closed ideals of nest algebras, J. Operator Theory 7 (1982), 219-235.

Fall, T., Arveson, W. and Muhly, P.

[1] Perturbations of nest algebras, J. Operator Theory 1 (1979), 137-150.

Fan, K.

[1] Maximum properties and inequalities for the eigenvalues of completely continuous operators, Proc. Nat. Acad. sci. USA 37 (1951), 760-766.

Feeman, T.

[1] Best approximation and quasitriangular algebras, Trans. Amer. Math. Soc. 288 (1985), 179-187.

[2] M-ideals and quasitriangular algebras, Illinois J. Math. 31 (1987), 89-98.

Feintuch, A. (See Choi, M.D.)

Feintuch, A.

[1] Invertibility in nest algebras, Proc. Amer. Math. Soc. 91 (1984), 573-576.

Feldman, J. (See Arveson, W.B.)

Fillmore, P.A. and Williams, J.P.

[1] On operator ranges, Adv. Math. 7 (1971), 254-281.

Foias, C. (See also Apostol, C.; Sz. Nagy, B.)

[1] Invariant para-closed subspaces, Indiana U. Math. J. 21 (1972), 887-906.

Foias, C., Pearcy, C. and Voiculescu, D.

[1] The staircase representation of a biquasitriangular operator, Mich. Math. J. 22 (1975), 343-352.

Fong, C.K. (See Davidson, K.R.)

Froelich, J.

[1] Compact operators, invariant subspaces and spectral synthesis, Ph.D. Thesis, Univ. Iowa, 1984.

[2] Compact operators in the algebra of a partially ordered measure space, J. Operator Theory 10 (1983), 353-355.

[3] Compact operators, invariant subspaces and spectral synthesis, preprint.

[4] Invariant subspaces and thin-sets, preprint.

Gamelin, T., Marshall, D., Younis, R. and Zame, W.

[1] Function theory and M-ideals, Ark. Mat. 23 (1985), 261-279.

Gilfeather, F. (See also Apostol, C.)

[1] Derivations on certain CSL algebras, J. Operator Theory 11 (1984), 145-156.

[2] Derivations on certain CSL algebras II, Inter. Ser. Numer. Math. 65 (1984), 135-140, Birkhauser.

Gilfeather, F., Hopenwasser, A. and Larson, D.

[1] Reflexive algebras with finite width lattices: tensor products, cohomology, compact perturbations, J. Func. Anal. 55 (1984), 176-199.

Gilfeather, F. and Larson, D.R.

[1] Commutants modulo the compact operators of certain CSL algebras, in Topics in Mod. Operator Theory 2 (1981), Birkhauser.

[2] Nest subalgebras of von Neumann algebras, Adv. Math. 46 (1982), 171-199.

[3] Structure in reflexive subspace lattices, J. London Math. Soc. (2) 26 (1982), 117-131.

[4] Nest subalgebras of von Neumann algebras: commutants modulo compacts and distance estimates, J. Operator Theory 7 (1982), 279-302.

[5] Commutants modulo the compact operators of certain CSL algebras II, J. Int. Equations Operator Theory 6 (1983), 345-356.

Gilfeather, F. and Moore, R.L.

[1] Isomorphisms of certain CSL algebras, J. Func. Anal. 67 (1986), 264-291.

Gohberg, I. (See also Ball, J.)

Gohberg, I.C. and Krein, M.G.

[1] Introduction to the theory of linear nonselfadjoint operators, "Nauka", Moscow, 1965; English transl., Transl. Math. Monographs, 18, Amer. Math. Soc., Providence, RI, 1969.

[2] Theory and applications of Volterra operators in Hilbert space, "Nauka", Moscow, 1967; English transl., Transl. Math. Monographs, 24, Amer. Math. Soc., Providence, RI, 1970.

Graham, C.C. and McGehee, O.C.

[1] Essays in commutative harmonic analysis, Grund. Math. Wiss. 238, Springer-Verlag, New York, 1979.

Grothiendieck, A.

[1] Une résultat sur le dual d'une C^*-algebre, J. Math. Pures Appl. 36 (1957), 97-108.

Halmos, P.R.

[1] Invariant subspaces of polynomially compact operators, Pacific J. Math. 16 (1966), 433-437.

[2] A Hilbert space problem book, van Nostrand Reinhold, New York, 1967.

[3] Quasitriangular operators, Acta Sci. Math. (Szeged) 29 (1968), 283-293.

[4] Ten problems in Hilbert space, Bull. Amer. Math. Soc. 76 (1970), 887-933.

Harrison, K.J. and Longstaff, W.E.

[1] An invariant subspace lattice of order type $w+w+1$, Proc. Amer. Math. Soc. 79 (1980), 45-49.

[2] Automorphic images of commutative subspace lattices, Trans. Amer. Math. Soc. 296 (1986), 217-228.

Herrero, D.A. (See also Davidson, K.R.)

[1] Normal limits of nilpotent operators, Indiana Univ. Math. J. 23 (1974), 1097-1108.

[2] Approximation of Hilbert space operators, vol. 1, Res. Notes Math. #72, Pitman, 1982.

[3] Compact perturbations of continuous nest algebras, J. London Math. Soc. (2) 27 (1983), 339-344.

[4] Compact perturbations of nest algebras, Index obstructions, and a problem of Arveson, J. Func. Anal. 55 (1984), 78-109.

Herrero, D.A. and Larson, D.R.

[1] Ideals of nest algebras and models for operators, preprint.

Holmes, R., Scranton, B. and Ward, J.

[1] Approximation from the space of compact operators and other M-ideals, Duke Math. J. 42 (1975), 259-269.

Hopenwasser, A. (See also Erdos, J.A.; Gilfeather, F.)

[1] The radical of a reflexive operator algebra, Pacific J. Math. 65 (1976), 375-392.

[2] Compact operators in the radical of a reflexive algebra, J. Operator Theory 2 (1979), 127-129.

[3] The equation $Tx = y$ in a reflexive operator algebra, Indiana U. Math. J. 29 (1980), 124-126.

[4] Tensor products of reflexive subspace lattices, Mich. Math. J. 31 (1984), 359-370.

[5] Hypercausal ideals in nest algebras, J. London Math. Soc. (2) 34 (1986), 129-138.

[6] Hypercausal linear operators, SIAM J. Control and Opt. 22 (1984), 911-919.

Hopenwasser, A. and Kraus, J.

[1] Tensor products of reflexive algebras II, J. London Math. Soc. (2) 28 (1983), 359-362.

Hopenwasser, A. and Larson, D.R.

[1] The carrier space of a reflexive operator algebra, Pacific J. Math. 81 (1979), 417-434.

Hopenwasser, A., Laurie, C. and Moore, R.

[1] Reflexive algebras with completely distributive subspace lattices, J. Operator Theory 11 (1984), 91-108.

Hopenwasser, A. and Moore, R.

[1] Finite rank operators in reflexive operator algebras, J. London Math. Soc. (2) 27 (1983), 331-338.

Hopenwasser, A. and Plastiras, J.

[1] Isometries of quasitriangular operator algebras, Proc. Amer. Math. Soc. 65 (1977), 242-244.

Jewell, N. (See Axler, S.)

Johansen, T.J.

[1] Convergence of derivations on nest algebras, Math. Scand. 58 (1986), 46-54.

Johnson, B.E. (See also Akemann, C.A.)

[1] Continuity of homomorphism of algebras of operators II, J. London Math. Soc. (2) 1 (1969), 81-84.

[2] Cohomology in Banach algebras, Mem. Amer. Math. Soc. 127 (1972), Providence, RI.

[3] Perturbations of Banach algebras, Proc. London Math. Soc. (3) 34 (1977), 439-458.

[4] A counter example in the perturbation theory of C^*-algebras, Canad. Math. Bull. 25 (1982), 311-316.

Johnson, B.E., Kadison, R.V. and Ringrose, J.R.

[1] Cohomology of operator algebras III. Reduction to normal cohomology, Bull. Soc. Math. France 100 (1972), 73-96.

Johnson, B.E. and Parrott, S.K.

[1] Operators commuting with a von Neumann algebra modulo the set of compact operators, J. Func. Anal. 11 (1972), 39-61.

Josephson, K. (See Arveson, W.)

Kadison, R.V. (See also Johnson, B.E.)

[1] Derivations of operator algebras, Ann. Math. 83 (1966), 280-293.

Kadison, R.V. and Kastler, D.

[1] Perturbations of von Neumann algebras I, stability of type, Amer. J. Math. 94 (1972), 38-54.

Kadison, R.V., Lance, E.C. and Ringrose, J.R.

[1] Derivations and automorphisms of operator algebras II, J. Func. Anal. 1 (1967), 204-221.

Kadison, R.V. and Ringrose, J.R.

[1] Cohomology of operator algebras I, Type I von Neumann algebras, Acta Math. 126 (1971), 227-243.

[2] Algebraic automorphism of operator algebras, J. London Math. Soc. (2) 8 (1974), 329-334.

[3] Fundamentals of the theory of operator algebras, vol. 1, Academic Press, 1983; vol. 2, Academic Press, 1986.

Kadison, R.V. and Singer, I.M.

[1] Triangular operator algebras, Amer. J. Math. 82 (1960), 227-259.

Kahan, W.M. (See Davis, C.)

Karanasios, S.

[1] Full operators and approximation of inverses, J. London Math. Soc. (2) 30 (1984), 295-304.

[2] Triangular integration with respect to a nest algebra module, Indiana U. Math. J. 34 (1985), 299-317.

Kastler, D. (See Kadison, R.V.)

Knowles, G.J.

[1] C_p perturbations of nest algebras, Proc. Amer. Math. Soc. 92 (1984), 37-40.

[2] Quotients of nest algebras with trivial commutator, Pacific J. Math. 127 (1987), 121-126.

Kraus, J. (See also Hopenwasser, A.)

[1] W^*-dynamical systems and reflexive operator algebras, J. Operator Theory 8 (1982), 181-194.

[2] The slice map problem for σ-weakly closed subspaces of von Neumann algebras, Trans. Amer. Math. Soc. 279 (1983), 357-376.

[3] Tensor products of reflexive algebras, J. London Math. Soc. (2) 28 (1983), 350-358.

Kraus, J. and Larson, D.R.

[1] Some applications of a technique for constructing reflexive operator algebras, J. Operator Theory 13 (1985), 227-236.

[2] Reflexivity and distance formulae, J. London Math. Soc., (3) 53 (1986), 340-356.

Krein, M.G. (See Gohberg, I.C.)

Lambrou, M.S.

[1] Completely distributive lattices, Fund. Math. 119 (1983), 227-240.

Lance, E.C. (See also Effros, E.; Kadison, R.V.)

[1] Some properties of nest algebras, Proc. London Math. Soc. (3) 19 (1969), 45-68.

[2] Cohomology and perturbations of nest algebras, Proc. London Math. Soc. (3) 43 (1981), 334-356.

Larson, D.R. (See also Gilfeather, F.; Herrero, D.A.; Hopenwasser, A.; Kraus, J.)

[1] On the structure of certain reflexive operators, J. Func. Anal. 31 (1979), 275-292.

[2] A solution to a problem of J.R. Ringrose, Bull. Amer. Math. Soc. 7 (1982), 243-246.

[3] Annihilators of operator algebras, Topics in Modern Operator Theory 6, pp. 119-130, Birkhauser Verlag, Basel, 1982.

[4] Nest algebras and similarity transformations, Ann. Math. 121 (1985), 409-427

[5] Hyperreflexivity and a dual product construction, Trans. Amer. Math. Soc. 294 (1986), 79-88.

[6] Triangularity in operator algebras, Surveys of recent results in operator theory, J. Conway (ed.), Pitman Res. Notes in Math., Longman, to appear.

Larson, D.R. and Solel, B.

[1] Nests and inner flows, J. Operator Theory 16 (1986), 157-164.

Laurie, C. (See Hopenwasser, A.)

[1] Invariant subspace lattices and compact operators, Pacific J. Math. 89 (1980), 351-365.

[2] On density of compact operators in reflexive algebras, Indiana U. Math. J. 30 (1981), 1-16.

Laurie, C. and Longstaff, W.

[1] A note on rank one operators in reflexive algebras, Proc. Amer. Math. Soc. 89 (1983), 293-297.

Leggett, R.

[1] On the invariant subspace structure of compact, dissipative operators, Indiana U. Math. J. 22 (1973), 919-928.

Lidskii, V.B.

[1] Nonselfadjoint operators with trace, (Russian) Dokl. Akad. Nauk SSR 125 (1959), 485-487; English transl., Transl. Amer. Math. Soc. (2) 47 (1965), 43-46.

Livsic, M.S.

[1] On a certain class of operators in a Hilbert space, Mat. Sb. 19 (61) (1946), 239-262; English transl., Amer. Math. Soc. Transl. (2) 13 (1960), 61-83.

Loebl, R.I. and Muhly, P.S.

[1] Analyticity and flows in von Neumann algebras, J. Func. Anal. 29 (1978), 214-252.

Loginov, A.I. and Sulman, G.S.

[1] Hereditary and intermediate reflexivity of W^*-algebras, (Russian) Izv. Akad. Nauk. SSSR Ser. Mat. 39 (1975), 1260-1273; English transl., Math USSR-Izv. 9 (1975), 1189-1201.

Lomonosov, V.

[1] On invariant subspaces of families of operators commuting with a completely continuous operator, Funk. Anal. i Prilozen 7 (1973), 55-56. (Russian); English transl., Func. Anal. and Appl. 7 (1973), 213-214.

Longstaff, W.E. (See also Erdos, J.A.; Harrison, K.J.; Laurie, C.)

[1] Generators of reflexive algebras, J. Australian Math. Soc. Ser. A 20 (1975), 159-164.

[2] Strongly reflexive lattices, J. London Math. Soc. (2) 11 (1975), 491-498.

[3] Operators of rank one in reflexive algebras, Canad. J. Math. 28 (1976), 19-23.

Marshall, D. (See Gamelin, T.)

McAsey, M.

[1] Canonical models for invariant subspaces, Pacific J. Math. 91 (1980), 377-395.

[2] Invariant subspaces of nonselfadjoint crossed products, Pacific J. Math. 96 (1981), 457-473.

McAsey, M.J. and Muhly, P.S.

[1] Representations of nonselfadjoint crossed products, Proc. London Math. Soc. (3) 47 (1983), 128-144.

McAsey, M.J., Muhly, P.S. and Saito, K-S.

[1] Nonselfadjoint crossed products (Invariant subspaces and maximality), Trans. Amer. Math. Soc. 248 (1979), 381-409.

[2] Equivalence classes of invariant subspaces in nonselfadjoint crossed products, Publ. Res. Inst. Math. Sci., Kyoto U. 20 (1984), 1119-1138.

[3] Nonselfadjoint crossed products II, J. Math. Soc. Japan 33 (1981), 485-495.

[4] Nonselfadjoint crossed products III, J. Operator Theory 12 (1984), 3-22.

Moore, R.L. (See also Gilfeather, F.; Hopenwasser, A.)

Moore, R.L. and Trent, T.

[1] Extreme points of certain operator algebras, Indiana U. Math. J., to appear.

Muhly, P. (See also Fall, T.; Loebl, R.; McAsey, M.)

[1] Radicals crossed products and flows, Ann. Polon. Math. 43 (1983), 35-42.

Muhly, P.S. and Saito, K-S.

[1] Analytic crossed products and outer conjugacy, Math. Scand. 58 (1986), 55-68.

[2] Analytic subalgebras of von Neumann algebras, preprint.

Muhly, P.S., Saito, K-S. and Solel, B.

[1] Coordinates for triangular operator algebras, preprint.

Nikol'skii, N.K.

[1] Selected problems of weighted approximation and spectral analysis, Proc. Steklov. Inst. Math. 120 (1974), Amer. Math. Soc., Providence, 1976.

Nordgren, E., Radjabalipour, M., Radjavi, H. and Rosenthal, P.

[1] On invariant operator ranges, Trans. Amer. Math. Soc. 251 (1979), 389-398.

Nordgren, E., Radjavi, H. and Rosenthal, P.

[1] On Arveson's characterization of hyperreducible triangular operators, Indiana U. Math. J. 26 (1977), 179-182.

[2] Triangularizing semigroups of compact operators, Indiana U. Math. J. 33 (1984), 271-275.

Ong, S-C.

[1] Invariant operator ranges of nest algebras, J. Operator Theory 3 (1980), 195-201.

[2] Operator topologies and invariant operator ranges, Canad. Math. Bull. 24 (1981), 181-186.

[3] Converse to a theorem of Foias, and reflexive lattices of operator ranges, Indiana U. Math. J. 30 (1981), 57-63.

Parrott, S.K. (See also Johnson, B.)

[1] Unitary dilations for commuting contractions, Pacific J. Math. 34 (1970), 481-490.

[2] On a quotient norm and the Sz.-Nagy Foias lifting theorem, J. Func. Anal. 30 (1978), 311-328.

Paulsen, V.I.

[1] Completely bounded maps and dilations, Pitman Res. Notes Math. 146, Longman Sci. and Tech. 1986.

Paulsen, V.I. and Power, S.C.

[1] Lifting theorems for nest algebras, preprint.

Paulsen, V.I., Power, S.C. and Smith, R.R.

[1] Schur products and matrix completions, preprint.

Paulsen, V.I., Power, S.C. and Ward, J.

[1] Semidiscreteness and dilation theory for nest algebras, J. Func. Anal., to appear.

Pearcy, C. (See Douglas, R.G.; Foias, C.)

Peligrad, C. (See also Christensen, E.)

[1] Compact derivations of nest algebras, Proc. Amer. Math. Soc. 97 (1986), 668-672.

Phillips, J.

[1] Perturbations of type I von Neumann algebras, Pacific J. Math. 52 (1974), 505-511.

[2] Perturbations of C^*-algebras, Indiana U. Math. J. 23 (1974), 1167-1176.

Phillips, R.S.

[1] On linear transformations, Trans. Amer. Math. Soc. 48 (1940), 516-541.

Pitts, D.R.

[1] Factorization problems for nests, J. Func. Anal., to appear.

[2] On the K_0 groups of nest algebras, preprint.

Plastiras, J. (See also Hopenwasser, A.)

[1] Quasitriangular operator algebras, Pacific J. Math. 64 (1976), 543-549.

[2] Compact perturbations of certain von Neumann algebras, Trans. Amer. Math. Soc. 234 (1977), 561-577.

Popa, S.

[1] The commutant modulo the set of compact operators of a von Neumann algebra, J. Func. Anal. 71 (1987), 393-408.

Power, S.C. (See also Davidson, K.R.; Erdos, J.; Paulsen, V.I.)

[1] The distance to upper triangular operators, Math. Proc. Camb. Phil. Soc. 88 (1980), 327-329.

[2] Nuclear operators in nest algebras, J. Operator Theory 10 (1983), 337-352.

[3] Another proof of Liskii's theorem on the trace, Bull. London Math. Soc. 15 (1983), 146-148.

[4] On ideals of nest subalgebras of C^*-algebras, Proc. London Math. Soc. 50 (1985), 314-332.

[5] Commutators with the triangular projection and Hankel forms on nest algebras, J. London Math. Soc. 32 (1985), 272-282.

[6] Factorisation in analytic operator algebras, J. Func. Anal. 67 (1986), 413-432.

[7] The Cholesky factorisation in operator theory, to appear in Operator Theory: Advances and Applications, vol. 24, Birkhauser Verlag, 1987.

[8] Spectral characterisation of the Wold-Zasuhin decomposition and prediction-error operator, preprint, 1986.

[9] Analysis in nest algebras, Surveys of Recent Results in Operator Theory, J. Conway (ed.), Pitman Research Notes in Mathematics, Longman, to appear.

[10] Infinite tensor products of upper triangular matrix algebras, preprint, 1987.

Radjabalipour, M. (See Nordgren, E.)

Radjavi, H. (See also Nordgren, E.)

Radjavi, H. and Rosenthal, P.

[1] On invariant subspaces and reflexive algebras, Amer. J. Math. 91 (1969), 683-692.

[2] Invariant subspaces, Erg. Math. Grenz. 77, Springer-Verlag, New York, 1973.

Raeburn, I. and Taylor, J.L.

[1] Hochschild cohomology and perturbations of Banach algebras, J. Func. Anal. 25 (1977), 258-266.

Raney, G.N.

[1] Completely distributive lattices, Proc. Amer. Math. Soc. 3 (1952), 677-680.

[2] A subdirect-union representation for completely distributive lattices, Proc. Amer. Math. Soc. 4 (1983), 518-522.

[3] Tight Galois connections and complete distributivity, Trans. Amer. Math. Soc. 97 (1960), 418-426.

Rickart, C.E.

[1] General theory of Banach algebras, Van Nostrand Reinhold, 1960.

Ringrose, J.R. (See also Johnson, B.E.; Kadison, R.V.)

[1] Superdiagonal forms for compact linear operators, Proc. London Math. Soc. (3) 12 (1962), 367-384.

[2] On the triangular representation of integral operators, Proc. London Math. Soc. (3) 12 (1962), 385-399.

[3] On some algebras of operators, Proc. London Math. Soc. (3) 15 (1965), 61-83.

[4] On some algebras of operators II, Proc. London Math. Soc. (3) 16 (1966), 385-402.

[5] Compact nonselfadjoint operators, Math. Studies 35, Van Nostrand Reinhold, London, 1971.

[6] Automatic continuity of derivations of operator algebras, J. London Math. Soc. (2) 5 (1972), 432-438

[7] Cohomology of operator algebras, in Lectures on Operator Algebras, Lect. Notes in Math. 247, Springer-Verlag, 1972, pp. 355-434.

Robinson, A. (See Bernstein, A.R.)

Rosenoer, S.

[1] Distance estimates for von Neumann algebras, Proc. Amer. Math. Soc. 86 (1982), 248-252.

Rosenthal, P. (See also Nordgren, E.; Radjavi, H.)

[1] Examples of invariant subspace lattices, Duke Math. J. 37 (1970), 103-112.

Royden, H.L.

[1] Real Analysis, 2nd ed. MacMillan, New York, 1968.

Saito, K-S. (See McAsey, M.; Muhly, P.S.)

[1] The Hardy spaces associated with a periodic flow on a von Neumann algebra, Tohoku Math. J. 29 (1977), 69-75.

[2] Invariant subspaces and cocycles in nonselfadjoint crossed products, J. Func. Anal. 45 (1982), 177-193.

[3] Automorphisms and nonselfadjoint crossed products, Pacific J. Math. 102 (1982), 179-187.

Sakai, S.

[1] On a conjecture of Kaplansky, Tohoku Math. J. 12 (1960), 31-33.

[2] Derivations of W^*-algebras, Ann. Math. 83 (1966), 273-279.

Salem, R.

[1] On singular monotonic functions of the Cantor type, J. Math. and Phys. 21 (1942), 69-82.

Sarason, D.

[1] On spectral sets having connected complement, Acta Sci. Math. (Szeged) 26 (1965), 289-299.

[2] Invariant subspaces and unstarred operator algebras, Pacific J. Math. 17 (1966), 511-517.

[3] Generalized interpolation in H^∞, Trans. Amer. Math. Soc. 127 (1967), 179-203.

[4] Invariant subspaces, in Topics in Operator Theory, C. Pearcy (ed.), Math. Surveys 13, Amer. Math. Soc., Providence, 1974.

Schatten, R.

[1] A theory of cross spaces, Ann. Math. Studies 26, Princeton U. Press, 1950.

Schubert, C.F.

[1] The Corona Theorem as an operator theorem, Proc. Amer. Math. Soc. 69 (1978), 73-76.

Schwartz, J.

[1] Two finite, non-hyperfinite, non-isomorphic factors, Comm. Pure Appl. Math. 16 (1963), 19-26.

Scranton, B. (See Holmes, R.)

Shields, A. (See also Axler, S.)

[1] An analogue of a Hardy-Littlewood-Fejer inequality for upper triangular trace class operators, Math. Zeit. 182 (1983), 473-484.

Sherman, S.

[1] The second adjoint of a C^*-algebra, Proc. Int. Cong. Math., Cambridge 1 (1950), 470.

Singer, I.M. (See Kadison, R.V.)

Smith, R.R. (See also Paulsen, V.I.)

Smith, R.R. and Ward, J.D.

[1] M-ideal structure in Banach algebras, J. Func. Anal. 27 (1978), 337-349.

Solel, B. (See also Larson, D.R.; Muhly, P.S.)

[1] On derivations of certain algebras related to irreducible triangular algebras, Trans. Amer. Math. Soc. 277 (1983), 263-273.

[2] The invariant subspace structure of nonselfadjoint crossed products, Trans. Amer. Math. Soc. 279 (1983), 825-840.

[3] On some subalgebras of a von Neumann algebra crossed product, Trans. Amer. Math. Soc. 281 (1984), 297-308.

[4] Irreducible triangular algebras, Mem. Amer. Math. Soc. 290 (1984), Providence, RI.

[5] Analytic operator algebras (Factorization and an expectation), Trans. Amer. Math. Soc. 287 (1985), 799-817.

[6] Algebras with analytic operators associated with a epriodic flow on a von Neumann algebra, Canad. J. Math. 37 (1985), 405-429.

[7] Nonselfadjoint crossed products: invariant subspaces, cocycles and subalgebras, Indiana U. Math. J. 34 (1985), 277-298.

[8] Factorization in operator algebras, Proc. Amer. Math. Soc., to appear.

Spivak, M.

[1] Derivations and nest algebras on Banach space, Israel J. Math. 50 (1985), 193-200.

Stampfli, J.

[1] The norm of a derivation, Pacific J. Math. 33 (1970), 737-747.

Stinespring, W.

[1] Positive functions on C^*-algebras, Proc. Amer. Math. Soc. 6 (1955), 211-216.

Sulman, G.S. (See Loginov, A.I.)

[1] On reflexive operator algebras, (Russian) Mat. Sbornik 87 (1972), 179-187; English transl. Math. USSR Sb. 16 (1972), 181-190.

Sz. Nagy, B.

[1] Sur les contractions de l'espace de Hilbert, Acta Sci. Math. (Szeged) 15 (1953), 87-92.

[2] Unitary dilations of Hilbert space operators and related topics, CBMS 19, Amer. Math. Soc., Providence, 1974.

Sz. Nagy, B. and Foias, C.

[1] Dilatation des commutants d'opérateurs, C.R. Acad. Sci. Paris, Ser. A 266 (1968), 493-495.

[2] Harmonic analysis of operators on Hilbert space, American Elsevier, New York, 1970.

Takeda, Z.

[1] Conjugate spaces of operator algebras, Proc. Japan Acad. 28 (1954), 90-95.

Takesaki, M.

[1] A short proof of the commutation theorem $(\mathcal{M} \otimes \mathcal{N})' = \mathcal{M}' \otimes \mathcal{N}'$, Lectures on Operator Algebras, Springer Lect. Notes 247, Heidelberg, 1972, pp. 780-786.

[2] Theory of operator algebras I, Springer-Verlag, New York, 1979.

Taylor, J.L. (See also Raeburn, I.)

[1] Banach algebras and topology, in Algebras in Analysis, J.H. Williamson (ed.), Academic Press, London, 1975, pp. 118-186.

Tomita, M.

[1] Standard forms of von Neumann algebras, Vth Functional Analysis Symposium, Math. Soc. Japan, Sendai, 1967.

Tomiyama, J.

[1] On the projection of norm one in W^*-algebras, Proc. Japan Acad. 33 (1957), 608-612.

[2] On the projection of norm one in W^*-algebras II, Tohoku Math. J. 10 (1958), 204-209.

Voiculescu, D. (See also Apostol, C.; Foias, C.)

[1] Norm limits of algebraic operators, Rev. Roum. Math. Pures et Appl. 19 (1974), 371-378.

[2] A non commutative Weyl-von Neumann theorem, Rev. Roum. Math. Pures et Appl. 21 (1976), 97-113.

Wagner, B. (See also Davidson, K.R.)

[1] Automorphisms and derivations of certain operator algebras, Ph.D. dissertation, U. Calif., Berkeley, 1982.

[2] Derivations of quasitriangular algebras, Pacific J. Math. 114 (1984), 243-255.

[3] Weak limits of projections and compactness of subspace lattices, preprint.

Ward, J. (See Holmes, R.; Paulsen, V.I.; Smith, R.R.)

Weinberger, W.F. (See Davis, C.)

Wermer, J.

[1] Commuting spectral operators on Hilbert space, Pacific J. Math. 4 (1954), 355-361.

Williams, J.P. (See Fillmore, P.)

Wright, S. (See Akemann, C.A.)

Yakubovic, B.V.

[1] Conditions for unicellularity of weighted shift operators, Dokl. Akad. Nauk. SSSR 278 (1984), 821-824.

[2] Invariant subspaces of weighted shift operators, Zapiski Nauk. Sem. LOMI 141 (1985), 100-143.

Younis, R. (See Gamelin, T.)

Zame, W. (See Gamelin, T.)

INDEX

INDEX OF SYMBOLS

100 Optimal control of variational inequalities
V Barbu
101 Partial differential equations and dynamical systems
W E Fitzgibbon III
102 Approximation of Hilbert space operators. Volume II
C Apostol, L A Fialkow, D A Herrero and D Voiculescu
103 Nondiscrete induction and iterative processes
V Ptak and F-A Potra
104 Analytic functions – growth aspects
O P Juneja and G P Kapoor
105 Theory of Tikhonov regularization for Fredholm equations of the first kind
C W Groetsch
106 Nonlinear partial differential equations and free boundaries. Volume I
J I Díaz
107 Tight and taut immersions of manifolds
T E Cecil and P J Ryan
108 A layering method for viscous, incompressible L_p flows occupying R^n
A Douglis and E B Fabes
109 Nonlinear partial differential equations and their applications: Collège de France Seminar. Volume VI
H Brezis and J L Lions
110 Finite generalized quadrangles
S E Payne and J A Thas
111 Advances in nonlinear waves. Volume II
L Debnath
112 Topics in several complex variables
E Ramírez de Arellano and D Sundararaman
113 Differential equations, flow invariance and applications
N H Pavel
114 Geometrical combinatorics
F C Holroyd and R J Wilson
115 Generators of strongly continuous semigroups
J A van Casteren
116 Growth of algebras and Gelfand–Kirillov dimension
G R Krause and T H Lenagan
117 Theory of bases and cones
P K Kamthan and M Gupta
118 Linear groups and permutations
A R Camina and E A Whelan
119 General Wiener–Hopf factorization methods
F-O Speck
120 Free boundary problems: applications and theory, Volume III
A Bossavit, A Damlamian and M Fremond
121 Free boundary problems: applications and theory, Volume IV
A Bossavit, A Damlamian and M Fremond
122 Nonlinear partial differential equations and their applications: Collège de France Seminar. Volume VII
H Brezis and J L Lions
123 Geometric methods in operator algebras
H Araki and E G Effros
124 Infinite dimensional analysis–stochastic processes
S Albeverio
125 Ennio de Giorgi Colloquium
P Krée
126 Almost-periodic functions in abstract spaces
S Zaidman
127 Nonlinear variational problems
A Marino, L Modica, S Spagnolo and M Degiovanni
128 Second-order systems of partial differential equations in the plane
L K Hua, W Lin and C-Q Wu
129 Asymptotics of high-order ordinary differential equations
R B Paris and A D Wood
130 Stochastic differential equations
R Wu
131 Differential geometry
L A Cordero
132 Nonlinear differential equations
J K Hale and P Martinez-Amores
133 Approximation theory and applications
S P Singh
134 Near-rings and their links with groups
J D P Meldrum
135 Estimating eigenvalues with *a posteriori/a priori* inequalities
J R Kuttler and V G Sigillito
136 Regular semigroups as extensions
F J Pastijn and M Petrich
137 Representations of rank one Lie groups
D H Collingwood
138 Fractional calculus
G F Roach and A C McBride
139 Hamilton's principle in continuum mechanics
A Bedford
140 Numerical analysis
D F Griffiths and G A Watson
141 Semigroups, theory and applications. Volume I
H Brezis, M G Crandall and F Kappel
142 Distribution theorems of L-functions
D Joyner
143 Recent developments in structured continua
D De Kee and P Kaloni
144 Functional analysis and two-point differential operators
J Locker
145 Numerical methods for partial differential equations
S I Hariharan and T H Moulden
146 Completely bounded maps and dilations
V I Paulsen
147 Harmonic analysis on the Heisenberg nilpotent Lie group
W Schempp
148 Contributions to modern calculus of variations
L Cesari
149 Nonlinear parabolic equations: qualitative properties of solutions
L Boccardo and A Tesei
150 From local times to global geometry, control and physics
K D Elworthy

52 Wave propagation in viscoelastic media
F Mainardi
53 Nonlinear partial differential equations and their applications: Collège de France Seminar. Volume I
H Brezis and J L Lions
54 Geometry of Coxeter groups
H Hiller
55 Cusps of Gauss mappings
T Banchoff, T Gaffney and C McCrory
56 An approach to algebraic K-theory
A J Berrick
57 Convex analysis and optimization
J-P Aubin and R B Vintner
58 Convex analysis with applications in the differentiation of convex functions
J R Giles
59 Weak and variational methods for moving boundary problems
C M Elliott and J R Ockendon
60 Nonlinear partial differential equations and their applications: Collège de France Seminar. Volume II
H Brezis and J L Lions
61 Singular systems of differential equations II
S L Campbell
62 Rates of convergence in the central limit theorem
Peter Hall
63 Solution of differential equations by means of one-parameter groups
J M Hill
64 Hankel operators on Hilbert space
S C Power
65 Schrödinger-type operators with continuous spectra
M S P Eastham and H Kalf
66 Recent applications of generalized inverses
S L Campbell
67 Riesz and Fredholm theory in Banach algebra
B A Barnes, G J Murphy, M R F Smyth and T T West
68 Evolution equations and their applications
F Kappel and W Schappacher
69 Generalized solutions of Hamilton-Jacobi equations
P L Lions
70 Nonlinear partial differential equations and their applications: Collège de France Seminar. Volume III
H Brezis and J L Lions
71 Spectral theory and wave operators for the Schrödinger equation
A M Berthier
72 Approximation of Hilbert space operators I
D A Herrero
73 Vector valued Nevanlinna Theory
H J W Ziegler
74 Instability, nonexistence and weighted energy methods in fluid dynamics and related theories
B Straughan
75 Local bifurcation and symmetry
A Vanderbauwhede
76 Clifford analysis
F Brackx, R Delanghe and F Sommen
77 Nonlinear equivalence, reduction of PDEs to ODEs and fast convergent numerical methods
E E Rosinger
78 Free boundary problems, theory and applications. Volume I
A Fasano and M Primicerio
79 Free boundary problems, theory and applications. Volume II
A Fasano and M Primicerio
80 Symplectic geometry
A Crumeyrolle and J Grifone
81 An algorithmic analysis of a communication model with retransmission of flawed messages
D M Lucantoni
82 Geometric games and their applications
W H Ruckle
83 Additive groups of rings
S Feigelstock
84 Nonlinear partial differential equations and their applications: Collège de France Seminar. Volume IV
H Brezis and J L Lions
85 Multiplicative functionals on topological algebras
T Husain
86 Hamilton-Jacobi equations in Hilbert spaces
V Barbu and G Da Prato
87 Harmonic maps with symmetry, harmonic morphisms and deformations of metrics
P Baird
88 Similarity solutions of nonlinear partial differential equations
L Dresner
89 Contributions to nonlinear partial differential equations
C Bardos, A Damlamian, J I Díaz and J Hernández
90 Banach and Hilbert spaces of vector-valued functions
J Burbea and P Masani
91 Control and observation of neutral systems
D Salamon
92 Banach bundles, Banach modules and automorphisms of C*-algebras
M J Dupré and R M Gillette
93 Nonlinear partial differential equations and their applications: Collège de France Seminar. Volume V
H Brezis and J L Lions
94 Computer algebra in applied mathematics: an introduction to MACSYMA
R H Rand
95 Advances in nonlinear waves. Volume I
L Debnath
96 FC-groups
M J Tomkinson
97 Topics in relaxation and ellipsoidal methods
M Akgül
98 Analogue of the group algebra for topological semigroups
H Dzinotyiweyi
99 Stochastic functional differential equations
S E A Mohammed

151 A stochastic maximum principle for optimal control of diffusions
U G Haussmann

152 Semigroups, theory and applications. Volume II
H Brezis, M G Crandall and F Kappel

153 A general theory of integration in function spaces
P Muldowney

154 Oakland Conference on partial differential equations and applied mathematics
L R Bragg and J W Dettman

155 Contributions to nonlinear partial differential equations. Volume II
J I Díaz and P L Lions

156 Semigroups of linear operators: an introduction
A C McBride

157 Ordinary and partial differential equations
B D Sleeman and R J Jarvis

158 Hyperbolic equations
F Colombini and M K V Murthy

159 Linear topologies on a ring: an overview
J S Golan

160 Dynamical systems and bifurcation theory
M I Camacho, M J Pacifico and F Takens

161 Branched coverings and algebraic functions
M Namba

162 Perturbation bounds for matrix eigenvalues
R Bhatia

163 Defect minimization in operator equations: theory and applications
R Reemtsen

164 Multidimensional Brownian excursions and potential theory
K Burdzy

165 Viscosity solutions and optimal control
R J Elliott

166 Nonlinear partial differential equations and their applications. Collège de France Seminar. Volume VIII
H Brezis and J L Lions

167 Theory and applications of inverse problems
H Haario

168 Energy stability and convection
G P Galdi and B Straughan

169 Additive groups of rings. Volume II
S Feigelstock

170 Numerical analysis 1987
D F Griffiths and G A Watson

171 Surveys of some recent results in operator theory. Volume I
J B Conway and B B Morrel

172 Amenable Banach algebras
J-P Pier

173 Pseudo-orbits of contact forms
A Bahri

174 Poisson algebras and Poisson manifolds
K H Bhaskara and K Viswanath

175 Maximum principles and eigenvalue problems in partial differential equations
P W Schaefer

176 Mathematical analysis of nonlinear, dynamic processes
K U Grusa

177 Cordes' two-parameter spectral representation theory
D F McGhee and R H Picard

178 Equivariant K-theory for proper actions
N C Phillips

179 Elliptic operators, topology and asymptotic methods
J Roe

180 Nonlinear evolution equations
J K Engelbrecht, V E Fridman and E N Pelinovski

181 Nonlinear partial differential equations and their applications. Collège de France Seminar. Volume IX
H Brezis and J L Lions

182 Critical points at infinity in some variational problems
A Bahri

183 Recent developments in hyperbolic equations
L Cattabriga, F Colombini, M K V Murthy and S Spagnolo

184 Optimization and identification of systems governed by evolution equations on Banach space
N U Ahmed

185 Free boundary problems: theory and applications. Volume I
K H Hoffmann and J Sprekels

186 Free boundary problems: theory and applications. Volume II
K H Hoffmann and J Sprekels

187 An introduction to intersection homology theory
F Kirwan

188 Derivatives, nuclei and dimensions on the frame of torsion theories
J S Golan and H Simmons

189 Theory of reproducing kernels and its applications
S Saitoh

190 Volterra integrodifferential equations in Banach spaces and applications
G Da Prato and M Iannelli

191 Nest algebras
K R Davidson